NANOMEDICINE AND TISSUE ENGINEERING

State of the Art and Recent Trends

NANOMEDICINE AND TISSUE ENGINEERING

State of the Art and Recent Trends

Edited by

Nandakumar Kalarikkal, PhD, Robin Augustine, Oluwatobi Samuel Oluwafemi, PhD, Joshy K. S., and Sabu Thomas, PhD

AAP APPLE ACADEMIC PRESS

Apple Academic Press Inc.
3333 Mistwell Crescent
Oakville, ON L6L 0A2
Canada

Apple Academic Press Inc.
9 Spinnaker Way
Waretown, NJ 08758
USA

Exclusive worldwide distribution by CRC Press, a member of Taylor & Francis Group

International Standard Book Number-13: 978-1-77188-118-0 (Hardback)
International Standard Book Number-13: 978-1-4987-2642-9 (eBook)

Library and Archives Canada Cataloguing in Publication

Nanomedicine and tissue engineering : state of the art and recent trends / edited by Nandakumar Kalarikkal, PhD, Robin Augustine, Oluwatobi Samuel Oluwafemi, PhD, Joshy K. S., and Sabu Thomas, PhD.

Includes bibliographical references and index.
Issued in print and electronic formats.
ISBN 978-1-77188-118-0 (hardcover).--ISBN 978-1-4987-2642-9 (pdf)
1. Tissue engineering. 2. Nanomedicine. I. Augustine, Robin, author, editor
II. Kalarikkal, Nandakumar, author, editor III. Oluwafemi, Oluwatobi Samuel, editor IV. K. S., Joshy, editor V. Thomas, Sabu, author, editor
R857.T55N35 2016 612'.028 C2016-900366-3 C2016-900367-1

CIP data on file with US Library of Congress

Apple Academic Press also publishes its books in a variety of electronic formats. Some content that appears in print may not be available in electronic format. For information about Apple Academic Press products, visit our website at **www.appleacademicpress.com** and the CRC Press website at **www.crcpress.com**

ABOUT THE EDITORS

Nandakumar Kalarikkal, PhD

Dr. Nandakumar Kalarikkal obtained his master's degree in physics with a specialization in industrial physics and his PhD in semiconductor physics from Cochin University of Science and Technology, Kerala, India. He was a postdoctoral fellow at NIIST, Trivandrum, and later joined Mahatma Gandhi University. Currently, he is the Honorary Joint Director of the International and Inter University Centre for Nanoscience and Nanotechnology and Associate Professor in School of Pure and Applied Physics of Mahatma Gandhi University, Kottayam, Kerala, India. He was a visiting fellow in JNCASR, Bangalore. His research interests include nanostructured materials and their applications, nonlinear optics, laser plasma, and phase transition.

Robin Augustine

Dr. Robin Augustine is a postdoctoral researcher at Department of Materials Science and Engineering, Technion Israel Institute of Technology, Haifa, Israel. He has received his PhD in nanoscience and nanotechnology from International and Inter University Centre for Nanoscience and Nanotechnology, Mahatma Gandhi University, Kottayam, Kerala, India. He has received a bachelor's degree in botany from the University of Calicut, Kerala, India, and a master's degree in bioengineering from Madurai Kamaraj University, Tamilnadu, India. He is experienced in developing smart biomaterials such as wound dressings, tissue engineering scaffolds, drug delivery systems, and suture materials. Robin has many publications in international journals and conference proceedings to his credit.

Oluwatobi Samuel Oluwafemi, PhD

Oluwatobi Samuel Oluwafemi, PhD, is a Senior Lecturer at the Department of Chemistry and Chemical Technology, Walter Sisulu University, Mthatha Campus, Eastern Cape, South Africa. He has published many papers in internationally reviewed journals and has presented at several professional meetings. He is a fellow of many professional bodies, a reviewer for many international journals, and has received many awards for his excellent work in material research. His current research interests include application of nanoparticles in medicine, water treatment, polymer, LEDs, and sensors.

Joshy K. S.

K. S. Joshy is a researcher in the International and Inter University Centre for Nanoscience and Nanotechnology and C.M.S. College, Mahatma Gandhi University, Kottayam, Kerala, India. She has received a bachelor's degree in chemistry from Kerala University, Kerala, India, master's degree in polymer chemistry from School of Chemical Sciences Mahatma Gandhi University, Kottayam, Kerala, India and MPhil degree in polymer chemistry from Cochin University of Science and Technology, Kochi, Kerala, India. Joshy carried out her PhD work at Mahatma Gandhi University, Kottayam, Kerala, India. She is currently pursuing her research on the development of lipid and polymer nanoparticles for drug delivery applications. She has published in international journals and conference proceedings to her credit.

Sabu Thomas, PhD

Dr. Sabu Thomas is a Professor in School of Chemical Sciences and Honorary Director of the International and Inter university Centre for Nanoscience and Nanotechnology, Mahatma Gandhi University, Kottayam, Kerala, India. He joined Mahatma Gandhi University as a full-time faculty in 1987. He has been associated with several universities in Europe, China, Malaysia, and South Africa. Professor Thomas is a member of the Royal Society of Chemistry of London, a member of the New York Academy of Science, USA, and the recipient of awards from the Chemical Research Society of India and the Material Research Society of India (2013). Professor Thomas has supervised 65 PhD theses, and he has more than 530 publications, 43 books and four patents and 22775 citations to his credit. The h-index of Prof. Thomas is 73. Professor Thomas is listed as the 5th position in the list of Most Productive Researchers in India in 2008. His research focuses on polymer blends, recyclability, reuse of waste plastics and rubbers, fiber-filled polymer blends, nanocomposites, elastomers, pervaporation phenomena, and sorption and diffusion.

CONTENTS

LIST OF CONTRIBUTORS

Jesu Arockiaraj
Department of Biotechnology, Faculty of Science and Humanities, SRM University, SRM Nagar, Kattankulathur, Chennai-603203, Tamil Nadu, India

Robin Augustine
International and Inter University Centre for Nanoscience and Nanotechnology, Mahatma Gandhi University, Priyadarsini Hills, Kottayam-686560, Kerala, India

Varsha Banerjee
Department of Physics, Indian Institute of Technology, Hauzkhas-110016, New Delhi, India

Amitava Bhattacharyya
Advanced Textile and Polymer Research laboratory, PSG Institute of Advanced Studies, Coimbatore-641004, Tamil Nadu, India

Rupesh V. Chikhale
Department of Pharmaceutical Sciences, Rashtrasant Tukadoji Maharaj Nagpur University, Mahatama Jyotiba Fuley Shaikshanik Parisar, Amravati Road, Nagpur-440033, Maharashtra, India

Karthik Deekonda
Department of Biotechnology, Faculty of Science and Humanities, SRM University, SRM Nagar, Kattankulathur, Chennai-603203, Tamil Nadu, India

Anastasiya Derkachova
Institute of Physics, Polish Academy of Sciences, Al. Lotników 32/46, Warszawa, Poland

Dipali M. Dhoke
Department of Pharmaceutical Sciences, Rashtrasant Tukadoji Maharaj Nagpur University, Mahatama Jyotiba Fuley Shaikshanik Parisar, Amravati Road, Nagpur-440033, Maharashtra, India

J. Gopinathan
Advanced Textile and Polymer Research laboratory, PSG Institute of Advanced Studies, Coimbatore-641004, Tamil Nadu, India

Vinoy Jacob
Department of Microbiology, MCAS, Rasipuram, Nammakal-637408, Tamil Nadu, India

Daniel Jakubczyk
Institute of Physics, Polish Academy of Sciences, Al. Lotników 32/46, Warszawa, Poland

Ankit Jain
Pharmaceutics Research Projects Laboratory, Department of Pharmaceutical Sciences, Dr. Hari Singh Gour Central University, Sagar-470003, Madhya Pradesh, India

Sanjay K. Jain
Pharmaceutics Research Projects Laboratory, Department of Pharmaceutical Sciences, Dr. Hari Singh Gour Central University, Sagar-470003, Madhya Pradesh, India

Nandakumar Kalarikkal
International and Inter University Centre for Nanoscience and Nanotechnology, Mahatma Gandhi University, Priyadarsini Hills, Kottayam-686560, Kerala, India

Pramod B. Khedekar
Department of Pharmaceutical Sciences, Rashtrasant Tukadoji Maharaj Nagpur University, Mahatama Jyotiba Fuley Shaikshanik Parisar, Amravati Road, Nagpur-440033, Maharashtra, India

Krystyna Kolwas
Institute of Physics, Polish Academy of Sciences, Al. Lotników 32/46, Warszawa, Poland

Vimal Kumar
Department of Phytopharmaceutical & Natural Products
Institute of Pharmacy, Nirma University, Ahmedabad-382 481, Gujarat, India

Koshy M. Kymonil
Department of Pharmaceutics, Faculty of Pharmacy, Babu Banarasi Das National Institute of Technology & Management, Lucknow-227105, Uttar Pradesh, India

Ansuja P. Mathew
International and Inter University Centre for Nanoscience and Nanotechnology, Mahatma Gandhi University, Priyadarsini Hills, Kottayam-686560, Kerala, India

Sunil Menghani
Department of Pharmaceutical Sciences, Rashtrasant Tukadoji Maharaj Nagpur University, Mahatama Jyotiba Fuley Shaikshanik Parisar, Amravati Road, Nagpur-440033, Maharashtra, India

Deepa. P. Mohanan
International and Inter University Centre for Nanoscience and Nanotechnology, Mahatma Gandhi University, Priyadarsini Hills, Kottayam-686560, Kerala, India

P. Prakash
Department of Chemistry, Thiagarajar College, Madurai-625009, Tamil Nadu, India

Mamatha M. Pillai
Tissue Engineering Laboratory, PSG Institute of Advanced Studies, Coimbatore-641004, Tamil Nadu, India

Amit M. Pant
Department of Pharmaceutical Sciences, Rashtrasant Tukadoji Maharaj Nagpur University, Mahatama Jyotiba Fuley Shaikshanik Parisar, Amravati Road, Nagpur-440033, Maharashtra, India

Mala Rajendran
Department of Biotechnology, Mepco Schlenk Engineering College, Sivakasi-626005, Tamil Nadu, India

K. R. Rakhimol
International and Inter University Centre for Nanoscience and Nanotechnology, Mahatma Gandhi University, Priyadarsini Hills, Kottayam-686560, Kerala, India

Nilesh Rarokar
Department of Pharmaceutical Sciences, Rashtrasant Tukadoji Maharaj Nagpur University, Mahatama Jyotiba Fuley Shaikshanik Parisar, Amravati Road, Nagpur-440033, Maharashtra, India

K. Ravi Shankar
Department of Biotechnology, Faculty of Science and Humanities, SRM University, SRM Nagar, Kattankulathur, Chennai-603203, Tamil Nadu, India

Shubhini A. Saraf
Department of Pharmaceutical Sciences, School of Biosciences & Biotechnology, Babasaheb Bhimrao Ambedkar University, Lucknow- 226010, Uttar Pradesh, India

M. Saravanan
Institute of Biomedical Sciences, College of Health Science, Mekelle University, Mekelle-1871, Ethiopia

R. Selvakumar
Tissue Engineering Laboratory, PSG Institute of Advanced Studies, Coimbatore-641004, Tamil Nadu, India

Anantha S. Selvaraj
Department of Biotechnology, Mepco Schlenk Engineering College, Sivakasi-626005, Tamil Nadu, India

Neeraj K. Sharma
Department of Phytopharmaceutical & Natural Products
Institute of Pharmacy, Nirma University, Ahmedabad-382 481, Gujarat, India

Sandeep K Singh
Department of Pharmaceutics, Faculty of Pharmacy, Babu Banarasi Das National Institute of Technology & Management, Lucknow-227105, Uttar Pradesh, India

Vanchna Singh
Department of Physics, Indian Institute of Technology, Hauzkhas-110016, New Delhi, India and Department of Applied Science, Inderprastha Engineering College, Ghaziabad-201011, Uttar Pradesh, India

R. Sournaveni
Tissue Engineering Laboratory, PSG Institute of Advanced Studies, Coimbatore-641004, Tamil Nadu, India

Sabu Thomas
International and Inter University Centrefor Nanoscience and Nanotechnology, Mahatma Gandhi University, Priyadarsini Hills, Kottayam-686560, Kerala, India

Bhavana Venugopal
Department of Biotechnology, St. Joseph's College, Irinjalakuda, Thrissur-680 121, Kerala, India

Poonam Verma
Department of Pharmaceutics, Faculty of Pharmacy, Babu Banarasi Das National Institute of Technology & Management, Lucknow-227105, Uttar Pradesh, India

LIST OF ABBREVIATIONS

2D	two-dimensional
3D	three-dimensional
3DD	3D deposition
3DP	3D printing
5-FU	5-fluorouracil
AA	acetic acid
ABC	ATP-binding cassette
ABC	ATP-binding cassette
ACL	anterior cruciate ligament
ADM	adriamycin
aFGF	acidic fibroblast growth factor
AFM	atomic force microscope
AgCl NPs	silver chloride nanoparticles
AgNPs	silver nanoparticles
AIDS	acquired immunodeficiency syndrome
ALA	aminolevulinic acid
ALL	acute lymphoblastic leukemia
AMH	antimullerian hormone
AML	acute myelogenous leukemia
aq AA	aqueous acetic acid solution
ASGP-R	asialoglycoprotein receptor
ATP	adenosine triphosphate
AUC	area under the curve
BBB	blood–brain barrier
BC	bacterial cellulose
BCP	biphasic calcium phosphate
BCRP	breast cancer resistance protein
BCS	biopharmaceutical classification system
BDNF	brain-derived neurotrophic factor
bFGF	basic fibroblast growth factor
BGLAP	bone gamma-carboxyglutamic acid-containing protein

Bis-azo PC	1,2-Bis(4-n-butylphenylazo-4′- phenylbutyroyl)-l-α-phosphatidylcholine
Bis-sorbPC	1,2-Bis[10-(2′,4′-hexadienoyloxy)-decanoyl]-sn-phosphatidylcholine
BM-MSCs	bone marrow mesenchymal stem cells
BMPs	bone morphogenetic proteins
BNC	bacterial nanocellulose
BSA	bovine serum albumin
CA	cellulose acetate
CaCO3	calcium carbonate
CAD	computer-aided design
CaP	calcium phosphate
CDDP	cisplatin
cDNA	complimentary DNA
CEA	cultured epithelial autografts
CG	collagen
CG/PNIPAA/CS	collagen/poly(N-isopropyl acrylamide)/chitosan
CHEMS	cholesterylhemisuccinate
CHOL	cholesterol
CHX	chlorhexidine
CLEAs	cross-linked enzyme aggregates
CL-VTP	verteporfin-loaded cationic liposomes
Cmax	maximum serum concentration
CMCTS	carboxymethyl chitosan
CNTs	carbon nanotubes
COL-CS	collagen–chitosan
CPC	collagen-PRP-complex
CPPS	cell-penetrating peptides
CS	chitosan
CS	chondroitin sulfate
CSPS	calcium, sodium, and phosphosilicate
CT	computed tomography
CYP3A	cytochrome P450 3A
DBM	demineralized bone matrix
DC	decanoyl chloride
DC8,9PC	photopolymerizable diacetylene phospholipid
DCs	dendritic cells
Dex-HA	dextran-hyaluronate

Dex-MA	dextran methacrylate
DLS	dynamic light scattering
DMF	dimethyl formamide
DMSO	dimethyl sulfoxide
DNA	deoxy ribo nucleic acid
DOPC	1,2-dioleoyl-sn-glycero-3-phosphatidylcholine
DOPE	1,2-dioleoyl-sn-glycero-3-phosphoethanolamine
DOX	doxorubicin
DPHP	1-diphenyl-2-picrylhydrazyl
DPN	dip-pen lithography
DPPC	dipalmitoyl phosphatidylcholine
DPSCs	dental pulp stem cells
DRBRP	desktop robot-based RP
DS	diclofenac sodium
DSPC	distearoyl phosphatidylcholine
DSPE	distearoyl phosphatidylethanolamine
DTA	diphtheria toxin
EBL	electron beam lithography
EBs	embryoid bodies
EC	endothelial cell
ECM	extracellular matrix
EDAC	1-ethyl-3-(3-dimethylaminopropyl)-carbodiimide
EDC	1-ethyl-3-(3-dimethylaminopropyl) carbodiimide hydrochloride
EDTA	ethylenediaminetetraacetic acid
EGF	epidermal growth factor
ELISA	enzyme-linked immunosorbent assay
ELP4	elastin-like polypeptide-4
EM	electromagnetic
eNOS	endothelial nitric oxide synthase
EPI	epirubicin hydrochloride
EPO	erythropoietin
EPR	enhanced permeability and retention
FDA	Food and Drug Administration
FDM	fused deposition modeling
FETs	field-effect transistors
FGF	fibroblast Growth Factor
FR	folate receptor

FRET	fluorescence resonance energy transfer
FSH	follicle stimulating hormone
FTIR	fourier transform infrared spectroscopy
G	graphene
G1	first-generation
G2	second-generation
G3	third-generation
GA	glutaraldehyde
GAG	glycosaminoglycan
G-CSF	granulocyte colony stimulating factor
GDFs	growth and differentiation factors
GGF	glial growth factor
GI	gastrointestinal
GIT	gastrointestinal tract
GO	graphene oxide
GPA	glycerol-preserved allograft
GS	glucosamine sulfate
GSH	glutathione
HA	hyaluronan
HA	hydroxyapatite
HAp	hydroxyapatite
HBV	hepatitis B virus
HCAECs	human coronary artery endothelial cells
HDL	high-density lipoprotein
He-ALA	ALA-hexyl ester
HFIP	1,1,1,3,3,3-hexafluoro-2-propanol
HFP	1,1,1,3,3,3 hexafluoro-2-propanol
HGF	hepatocyte growth factor
HIFU	high-intensity focused ultrasound
HIV	human immunodeficiency virus
HIV-1	human immunodeficiency virus type 1
HLA	human leukocyte antigen
HmPVA-EPC	poly(vinyl alcohol)–epoxypropoxy coumarin conjugate
hMSCs	human bone marrow-derived mesenchymal stem cells
HPCs	hematopoietic progenitor cells
HSE	human skin equivalent
HSPC	hydrogenated soy phosphatidylcholine

HSPGs	heparan sulfate proteoglycans
HSV-1	herpes simplex virus type 1
HT	hyperthermia
hTf	human transferin
HUVEC	human umbilical vein endothelial cells
IAASF	interfacial activity assisted surface functionalization
ICAM	intercellular adhesion molecule
IDSA	Infectious Disease Society of America
IFN	interferon
iGDNF	immobilized glial cell-derived neurotrophic factor
IGF	insulin-like growth factor
IGF	insulin-like growth factor
IJP	ink jet printing
ILs	immunoliposomes
IM	intramuscular
IND	indomethacin
IPPSF	isolated perfused porcine skin flap
IV	intravenous
KB	human oral carcinoma cells
KTZ	ketoconazole
LaBP	laser-assisted bioprinting
LDHs	lactate dehydrogenases
LFA1	lymphocyte function-associated antigen1
LFUS	low frequency ultrasound
LH	luteinizing hormone
LHC	lidocaine hydrochloride
L-HSA	lactosaminated human albumin
LOM	laminated object manufacturing
LSPR	localized surface plasmon resonance
LSPs	localized surface plasmons
MAbs	monoclonal antibodies
MBC	minimum bactericidal concentration
MCL	medial collateral ligament
MDR	multidrug resistance
MDRAB	multidrug-resistant strains of *Acinetobacter baumannii*
MEMS	microelectromechanical systems
MFC	microfibrillated cellulose
MH	microcrystalline hydroxyapatite

MH	magnetic hyperthermia
MIC	minimum inhibitory concentration
MIT	Massachusetts Institute of Technology
MMPs	matrix metalloproteases
MNPs	magnetic nanoparticles
MPCs	mesenchymal progenitor cells
MPS	mononuclear phagocyte system
MRI	magnetic resonance imaging
MRP1	multidrug resistance-associated protein 1
MRSA	methicillin-resistant *Staphylococcus aureus*
MSC	mesenchymal stem cells
MTA	mineral trioxide aggregates
MWNTs	multi-walled carbon nanotubes
nanoHA	nanohydroxyapatite
NAP	naproxen
NBD	nucleotide-binding domain
NCC	nanocrystalline cellulose
NDDS	novel drug delivery system
NF PLLA	nanofibrous PLLA
NGF	nerve growth factor
nHA	nanohydroxyapatite
NHS	N-hydroxysuccinimide
NIH	National Institute of Health
NIL	nanoimprinting lithography
NIPAM	N-isopropylacrylamide
NLS	nuclear localization signals
NMPC	N-methylene phosphonic chitosan
NPCs	nuclear pore complexes
NPs	nanoparticles
NSL	nanosphere lithography
NT-3	neurotrophin-3
ORC	oxidized regenerated cellulose
OVA	ovalbumin
PAA	propylacrylic acid
PAMPS	pathogen-associated molecular patterns
PANI	polyaniline
PC	plastic compression
PC	phosphatidylcholine

PCEC	poly(εcaprolactone)
PCL	poly(ecaprolactone)
PCL	polycaprolactone
PCLEEP	poly (e-caprolactone-co-ethyl ethylene phosphate) polymer
PDGF	platelet-derived growth factor-bb
PDO	polydioxanone
PDT	photodynamic therapy
PE	phosphatidyl ethanolamine
PEG	polyethylene glycol
PEGMA	poly (ethylene glycol) methacrylate
PELCL	poly(ethylene glycol)-b-poly(L-lactide-co-caprolactone)
PELGA	poly(ethylene glycol)-b-poly(L-lactide-co-glycolide)
PEO	polyethylene oxide
PET	polyethylene terephthalate
PEVA	poly(ethylene-co-vinyl acetate)
PG	propylene glycol
PGA	poly(glycolic acid)
PGA	polyglycolide
P-gp	P-glycoprotein
PHA	polyhydroxyalkanoate
PHB	poly (β-hydroxybutyrate)
PHBV	poly-3-hydroxybutyrate-co-hydrovalerate
PHEMA	polyhydroxyethyl methacrylate
pHMGCL	poly-hydroxymethylglycolide-co-ε-caprolactone
PHO	polyhydroxyoctanoate
PHUE-O3	poly-3-hydroxy-10-undecenoate
PIBCA	poly (isobutylcyanoacrylate)
PLA	poly lactic acid
PLAGA	copolymers of PLA and PGA
PLCL	poly L-lactide-co-ε- caprolactone
PLD	polyethylene glycol-coated liposomes
PLGA	poly (D,L-lactide-co-glycolide)
PLLA	poly(a-Lalanine)
PLLA	poly (lactic acid-co-glycolide)
PLLA	poly-l-lactide
PMMA	polymethyl methacrylate

PMMA	poly (methyl methacrylate)
PMNs	polymorphonuclear neutrophils
PNaSS	poly (sodium styrene sulfonate)
PNIPAAm	poly (N-isopropylacrylamide)
PPF	poly (propylene fumarate)
PPy	polypyrrole
PRINT	particle replication in nonwetting templates
PRP	platelet-rich plasma
PRRs	pattern recognition receptors
PSLs	pH-sensitive liposomes
PSSLs	pH-sensitive stealth liposomes
PTA	phosphotungstic acid
PTFE	polyfluorethylene
PTMC	poly (trimethylene carbonate)
PTX	paclitaxel
PU	polyurethane
PVA	polyvinyl alcohol
PVP	poly(N-vinyl-2-pyrrolidone)
QD	quantum dots
QSAR	quantitative structure–activity relationships
RBCs	red blood cells
RES	reticuloendothelial system
RIF	rifampicin
RNAi	RNA interference
ROS	reactive oxygen species
RP	rapid prototyping
RPM	return point memory
rTE	recombinant human tropoelastin
SA	sodium alginate
SAP	self-assembling peptide
SDS	sodium dodecyl sulphate
SEM	scanning electrospun micrograph
SEM	scanning electron microscope
SERS	surface-enhanced Raman scattering
SF	silk fibroin
SiNW	silicon nanowire
siRNA	short-interfering RNA
siRNA	small interfering RNA

SLA	stereolithography
SLS	selective laser sintering
SMC	smooth muscle cell
SMS	scanning mass spectroscopy
SNEDDS	self-nanoemulsifying drug delivery system
SNEDDS	self nanoemulsifying drug delivery systems
SNPs	single nucleotide polymorphisms
SP	surface plasmon
SPC	soy phosphatidylcholine
SPION	superparamagnetic iron oxide nanoparticles
SPM	scanning probe microscope
SPM	superparamagnetic
SPP	surface plasmon polariton
SPR	surface plasmon resonance
SPU	segmented polyurethane
ssDNA	single strand DNA
SSF	solid freeform fabrication techniques
ST-gelatin	styrenated gelatin
SubQ	subcutaneous
SWNTs	single-walled nanotubes
SXR	steroid and xenobiotic receptor
TAT	transcriptional activator
TATp	TAT peptide
TCH	tetracycline hydrochloride
TCPs	tissue culture plates
TCRV	tacaribe virus
TE	tissue engineering
TE	transverse electric
TEM	transmission electron microscope
Tet	tetracycline
TFA	trifluoroacetic acid
TFE	trifluoroethanol
TGF	transforming growth factor
TGF-ß	transforming growth factor beta
TGF-β1	transforming growth factor beta 1
THF	tetrahydrofuran
TIM	transverse Ising model
TiO2	titanium dioxide

Tm	transition temperature
TM	transverse magnetic
TMD	transmembrane domains
TMHs	transmembrane helicles
TNF	tumor necrosis factor
TPE	thermoplastic elastomer
TSLs	thermosensitive liposomes
UC-MSCs	umbilical cord-derived mesenchymal stem cells
Und-ALA	ALA-undecanoyl ester
US FDA	United States Food and Drug Administration
UV	ultraviolet
VEGF	vascular endothelial growth factor
VPF	vascular permeability factor
VRE	vancomycin-resistant enterococcus
WD	wound dressing
WHO	World Health Organization
XPS	X-ray photoelectron spectroscopy
XRD	X-ray diffractometry
X-SCID	X-linked severe combined immunodeficiency
ZnO	zinc oxide
β-TCP	β-tricalcium phosphate

PREFACE

The book focuses on various aspects of nanomedicine and tissue engineering applications which employ molecular aspects of nanotechnology and tissue engineering to solve various issues, from the perspective of a future practitioner, including superparamagnetic nanoparticles for hyperthermia, silver nanoparticles in nanomedicine, optical diagnostics of molecules and cells using nanotechnology, nanoparticulate drug delivery system for antiviral drugs, liposomal drug delivery systems, nanoemulsifying drug delivery system (SNEDDS), functionalization of tissue engineering scaffolds, and many other recent advancements.

Nanomedicine is the medical application of nanotechnology. It can be described as the control, construction, monitoring, and repairing of human tissues at the basic molecular level by making use of the nanostructures and nanodevices. If the promise of nanomedicine holds true, we will be capable to reduce the side effects and have better response to therapy. Nanotechnology-based targeted drug delivery systems are much promising. We can look for drug targets on a cellular basis as opposed to a tissue or organ level, as we do now. Advances in biosensors and molecular probes will allow for more detailed examination of cellular processes. This will help in identifying molecular targets for drug development. Nanomedicine is well under way in oncology for the detection and treatment of cancer. The current and promising applications of nanomedicine include, but are not limited to, drug delivery, *in vitro* diagnostics, *in vivo* imaging, therapeutic applications, biomaterials, and tissue engineering.

Tissue engineering is a fascinating therapeutic strategy that combines cells, biomaterials, and the signals that stimulate the differentiation of cells to form tissues into surgically transplantable formats, aimed at the repair, replacement, maintenance, or enhancement of tissue function.

The very first two chapters of the book give a brief introduction to the nanomedicine and tissue engineering. These chapters outline the overall aspects covered in the book including the definitions, terminologies, and various technologies and approaches. Third chapter outlines the use of magnetic hyperthermia, where superparamagnetic (SPM) nanoparticles are used to generate heat, as a robust approach for the selective destruction of cancer cells without affecting normal cells. It provides a theoretical framework which yields precise relationships between the heat dissipated and external (laboratory) parameters (e.g., particle concentration, amplitude, frequency of the applied oscillating field, etc.) in the form of scaling laws.

Green synthesis of metal nanoparticles renders production of nanoparticles that are free from toxic contaminants. This can also avoid the liberation of hazardous waste materials into the environment. Microbial- and plant-mediated synthesis of nanoparticle from silver and silver derivatives and its biomedical application is the main perspective of fourth chapter of this book. From the point of view of optical diagnostics, noble metal nanoparticles are especially interesting because of their unusual optical properties which arise from their ability to resonate with the light field. Fifth chapter deals with the potential applications of noble metal nanoparticles for optical diagnostics of molecules and cells.

The surface functionalization of the nanoparticle using interfacial activity assisted surface functionalization (IAASF) technique might increase the efficacy, reduce drug- and formulation-related side effects, change the release kinetics of antivirals, increase the target specificity, increase their bioavailability, improve the pharmacokinetics, reduce treatment costs, and improve patient compliance. These features are particularly important in viral diseases where high drug doses are needed, the disease is chronic and fast remedy is required to overcome the illness, drugs are expensive and the success of a therapy is associated with a patient's adherence to the administration procedure. Sixth chapter describes the functionalization of polymeric nanoparticles using IAASF technique, for the targeted delivery of antiviral drugs.

Several nanocarriers are developed and explored so far to release drugs at desired sites. Among these, exhaustive research has established liposomes as an effective carrier to trigger the release of loaded cargo to site of action. Liposomes are one of the well-recognized and effective drug delivery vehicles of the present era that protects the drug from the body milieu as well as improves its stability, safety, and effectiveness. Seventh chapter of the book covers various aspects of triggering modalities such as temperature, pH, enzymes, and light to release drug from liposomes.

P-glycoprotein (P-gp) being the first identified ABC (ATP [adenosine triphosphate]-binding cassette) transporter utilizes the energy of ATP hydrolysis to transport a broad range of substrates, including anticancer agents, across biological membranes against concentration gradients. It has the ability to confer multidrug resistance (MDR) in various types of cancer cells. Its expression is often associated with adverse prognosis of patients and is supposed to be a major obstacle limiting the therapeutic efficacy of cancer chemotherapy. Eighth chapter provides a better understanding and advances in the development of P-gp inhibitors to reverse MDR using nanotechnology-based targeting strategies.

In recent years, much attention has been given to lipid-based formulations to improve the oral bioavailability of poorly water soluble drug compounds. Self-nanoemulsifying drug delivery systems (SNEDDS) are isotropic mixtures of oil, surfactants, and cosurfactants that form fine oil-in-water nanoemulsions upon mild agitation, followed by dilution with aqueous media. Ninth chapter deals with the development and characterization of SNEDDS to improve the oral bioavailability of water insoluble drug ketoconazole (KTZ).

Cells, scaffolds, and signaling molecules are the basic tools and pillars of tissue engineering. Culturing of cells in three-dimensional (3D) microenvironments simulates normal cellular compartment and enhances adhesion, proliferation, and differentiation of cells than in 2D. Different types of scaffold like porous, fibrous, extracellular matrix, hydrogel, microspheres, and native tissue scaffolds are available and they influence the characteristics of cellular processes. A variety of techniques for the fabrication of scaffolds were developed. Tenth chapter highlights the application of natural and synthetic biopolymers in 3D culture of cells, their fabrication techniques, and describes the biophysical, biochemical, and biomechanical properties of scaffold which influence cellular processes inside the scaffold.

Electrospinning is a unique, versatile, and facile approach for engineering nanofibrous scaffolds with characteristic micro-to-nanoscale topography and high porosity similar to the natural extracellular matrix (ECM). These nanofibrous matrices influence cellular activities both in vitro and in vivo. High surface-to-volume ratio and porous nature of the electrospun scaffolds help in the attachment of cells and incorporation of proteins, DNA, RNA, and drugs ranging from antibiotics to anticancer agents. Thus, they find promising applications in drug delivery, tissue engineering, and wound dressings. Eleventh chapter describes about the electrospinning process and its potential biomedical applications.

An effective cell adhesion, cell growth, and retention of differentiated cell's function in a tissue engineering scaffold depend on many factors such as biomimetic surface, oxygen tension, growth factor, immobilization or incorporation method of growth factor, controlled combinatorial activity of key signaling molecules or growth factors from scaffold or biomaterial, hydrophilicity of scaffold etc. In twelth chapter, an attempt has been made to bring out the various types and strategies of biomolecular functionalization of scaffolds involved in various tissue engineering applications like stem cell, vascular grafts, bone, skin, and nerve tissue engineering.

Wound dressings accelerate wound healing and restore the structural and functional integrity of skin. Wound colonization by multidrug resistant pathogens complicates and delays wound healing. Wound dressings incorporating different nanomaterials are available in the market with varying degrees of clinical outcome. Among the battery of antimicrobial nanomaterials available for wound dressings, silver finds a unique position due to its broad spectrum antimicrobial activity. Research is also focused on the use of zinc nanomaterial which functions as an antimicrobial and anti-inflammatory agent. Nanostructured cellulose scaffold with antiseptic property endowed by the incorporation of titanium dioxide nanomaterials are commercially available recently. One of the chapters of this book highlights the types of wounds, wound dressings, nanomaterials, their formulation, and their impact on wound healing. Thirteenth chapter gives an overview on skin, essential functions of the skin, and damages or injury which are common to the skin. Recent advancements in the wound care with a focus on skin substitutes and skin grafts are given in the last chapter. Advancement of polymeric substances as skin substitutes aiding the healing, acting as physical barriers, and giving other functionalities of the natural skin is also discussed.

In this book, topics on the molecular mechanism, diagnosis, prevention, and treatment options of various ailments using nanomedicine and tissue engineering approaches are discussed in a lucid manner, and these aspects are presented in a readily accessible form. As a result, each chapter provides an in-depth array of knowledge that will be satisfying to professional colleagues interested in nanomedicine and tissue engineering. We have given a great deal of attention on the recent trends in nanomedicine and tissue engineering in this book.

— Robin Augustine, Joshy K. S., Nandakumar Kalarikkal, and Sabu Thomas

CHAPTER 1

NANOMEDICINE: FROM CONCEPT TO REALITY

K. R. RAKHIMOL, ROBIN AUGUSTINE, SABU THOMAS, and NANDAKUMAR KALARIKKAL

ABSTRACT

Nanotechnology is an interdisciplinary field that covers many areas of physics, chemistry, and biology. The new field of nanotechnology: nanomedicine introduced a rapid change in medical field by opening new windows to improve human health and is still evolving as a new field of medicine by utilizing nanomaterials for monitoring, diagnosing, and treating diseases. The application of nanotechnology to the medical field helps the materials and devices to react with subcellular components with high specificity. Thus the diseases can be cured at genetic level. In this chapter, we discuss about the common medical applications of nanotechnology.

1.1 NANOMEDICINE

Increasing interest in the medical applications of nanotechnology has led to the emergence of a new field called nanomedicine. The combination of nanotechnology, molecular biology, and medicine will provide more efficient techniques for detection and diagnosis of diseases at the cellular level. Nanomedicine uses molecular tools and molecular knowledge of the human body to diagnose, treat, and prevent disease and traumatic injury, relieve pain, and preserve and improve human health. In short, nanomedicine is the application of nanotechnology to medicine (Freitas, 2005). The early genesis of the concept of nanomedicine sprang from the visionary idea that tiny nanorobots and related machines could be designed, constructed, and introduced into the human body to perform cellular repairs at the molecular level. It is very complex procedure to design, characterize, and apply the nanomaterials for the treatment of diseases (Nguyen, 2012). Figure. 1.1 depicts the various possible applications of nanomedicine.

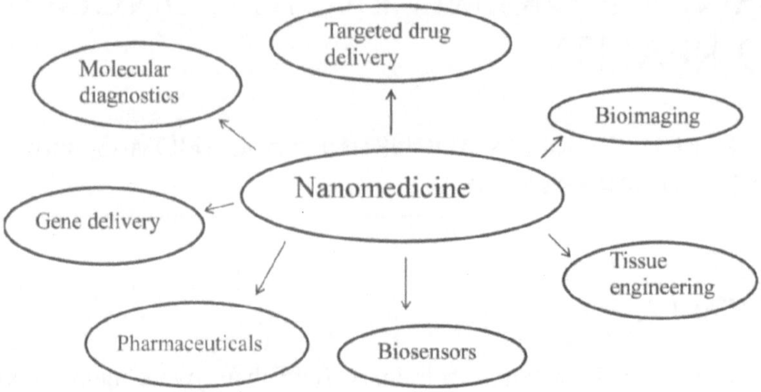

FIGURE 1.1 Schematic representation of various applications of nanomedicine.

Nanoparticle acts as the carriers for conventional drugs, antigens, proteins, enzymes, vaccines, peptides, etc. The nanocarriers are of two types: nanospheres and nanocapsules. In nanospheres, the drug will be dispersed throughout the carrier. But in nanocapsules, the drug will be covered by a capsule. The drug stays inside the cavity covered by polymeric membrane (Lamprecht, 2009).

Some antibiotic resistant bacteria produce antibiotic resistant proteins. Nanoparticle can be used in combination with the antibiotics to reduce the antibiotic resistance by interfering into the mechanism of protein action. For example, silver nanoparticle has been used in the treatment of wounds and burns.

1.1.1 CONCEPT OF NANOMEDICINE

Nanotechnology is the comprehending and control of matter within 1 to 100 nm to enable novel applications. Enclosing nanoscale science, technology, and engineering, nanotechnology includes manipulation of matter at this length scale. Nanotechnology is the common term to build and utilize functional structures with at least one dimension in nanometer scale (one billionth of a meter [10^{-9} m]). For the treatment of diseases, the drugs have to cross many biological barriers. Nanoparticles help the drugs to cross the barriers. Because of the very small size of nanoparticle it can also escape from endosomal digestion. This makes the nanomedicine to evolve as an important field in modern medicine.

1.1.2 ADVANTAGES OF NANOMEDICINE OVER OTHERS

Today, the biomedical science and engineering research at molecular level is using nanotechnological tools. One of the major advantages of nanomedicine over conventional methods is miniaturization: The small size of nanoparticles and nanocarriers can cross the biological membranes more easily than others and this helps the drugs to cross barriers like blood-brain barrier and treat the diseases. Another aspect is that only very small quantity of drugs is needed for therapeutical applications. Small size increases the surface area and reduces the amount of drugs needed.

2.1 NANOTECHNOLOGY FOR TARGETED DRUG DELIVERY

Most of the conventional drugs are organic molecules meant to degrade the desired protein formed by the pathogenic effect at the cellular level. A high concentration of drug is necessary for the degradation of the desired protein. At the same time this high concentration may cause harmful effects to the normal cells. To avoid this, the delivery of drugs only to the desired target is needed. A variety of physiological conditions of the diseased cells or tissue may be altered and it exploits the anatomical difference between normal and diseased tissues to achieve site specific and targeted delivery of drugs. A schematic representation of internalization of nanomedicine by the cell is given in Figure 1.2.

Nanomedicine carried by the nanocarrier is attached to a ligand which binds to the receptor on the cell membrane. The cell membrane forms an invagination at that site and internalizes the medicine into the cell by forming an endosome. The receptor is recycled to the cell membrane.

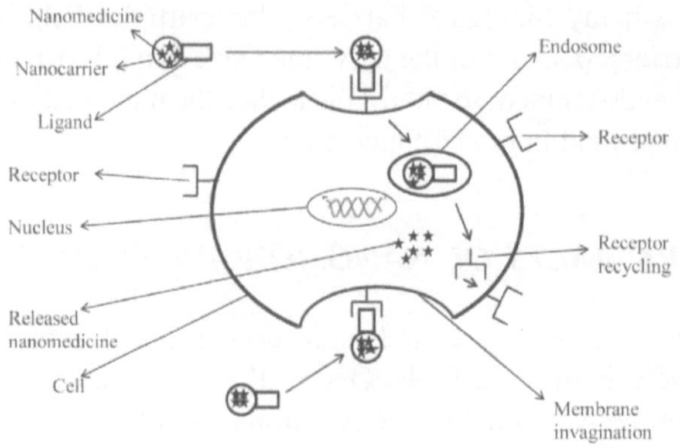

FIGURE 1.2 Schematic representation of internalization of nanomedicine by the cell.

The mechanism of the targeted drug delivery mainly involves different stages such as attachment of nanocarriers containing drug with a ligand, administration of this nanocarrier - ligand complex into the body, binding of this complex with the receptor present on target cell, and endocytosis of the complex into the cell. Nanoparticle can escape from endosomal digestion because of its small size and release the drug inside the cell. Drug delivery systems are particle carriers that can function as drug reservoirs and allow controlled release of drugs at the desired sites. They can also be used to increase local drug concentration by storing the drug within and controlling its release when they are bound to targets. An ideal targeting system should contain some basic characteristics which we now discuss.

a. Long circulating time

The drug should be present in circulation for a long time for reaching the target. After administration of nanoparticle into blood they are cleared by the macrophages, and this reduces the circulation time. The surface modification using polyethylene glycol (PEG) will increase the circulation time. Many other approaches have been developed to increase the circulation time of a nanoparticle. They are mechanical property modulation, engineering particle morphology, modification using CD47, and attaching on red blood cells (RBCs) (Yoo et al., 2010).

b. Appropriate concentration

An ideal drug delivery system can provide appropriate concentration of drugs to the target tissue. Below optimum level of drug concentration around the target will not provide a good result. In the case of tumor therapy, sufficient amount of drug is needed to kill the infected cell or tissue; otherwise, the complete destruction is not possible.

c. Activity

The drug delivery systems should not lose their activity during their circulation. It should reach the target with optimum activity. The drug delivery to highly specific targets needs the drug delivery systems which are much smaller than the target. This minute size can be achieved by the application of nanotechnology. The interaction of the nanoparticle with the cells depends on the source material, that is, the source material can be of two types, biological such as phospholipids or nonbiological such as cadmium. For the release of the drug in targeted site, the drug delivery system should be biodegradable. Nanoparticles are having unusual properties because of their size. Their minute size helps them to cross the biological membranes and barriers when compared with other micro-particles. Nanosystems can accumulate in the target at a higher concentration than normal drugs because of their small size. The increased vascular permeability allows the enhanced permeability and retention effect of nanosystems in tumors and other inflamed tissues (Sahoo et al., 2007).

2.1.1 DRUG DELIVERY VEHICLES

In traditional drug delivery systems such as oral administration, only small amount of drug will reach the target. To avoid wastage of drugs, targeted drug delivery systems can be used. The drug delivery systems can deliver a certain amount of drug to a targeted site for a prolonged time. There are different types of drug delivery systems such as liposomes, micelles, dendrimers, etc. An ideal drug delivery system must be biodegradable, non-toxic, biodegradable, and non immunogenic.

1.2.1.1 LIPOSOMES

Liposomes are complex structures made of phospholipids and small amount of other molecules. Liposomes vary in size from low molecular size range to tens of micrometer (Kale and Torchilin, 2007). Liposome is one of the most efficient drug delivery systems that protects the drug from the cellular environment and improves stability. They are biodegradable and nontoxic in nature. An advantage of increased circulation time of liposomes is the increased opportunity for selective localization to tumors and infected tissues. Both hydrophilic and hydrophobic drugs can be loaded in liposomes. Since liposomes can carry hydrophilic molecules within them, they have been used for the delivery of nucleic acid therapeutics. Liposomes can be modified by attaching antibodies or small molecules like folate for targeting tumors or for diagnostic applications (Webster, 2008).

Liposomes are considered as an efficient carrier and delivery system for pharmaceuticals. They have been used for the treatment of infectious diseases, autoimmune diseases, and cancer (Felnerova et al., 2004). Liposomes have also been used for the delivery of imaging agents to the targeted site (Weissig et al., 2000).

1.2.1.2 MICELLES

Amphiphilic polymers are polymeric carriers; they can carry the drug under a mild condition as a polymer micelle (Lee et al., 2007). They have two parts: hydrophilic and hydrophobic segments. The in vivo fate of the micelles depends on the properties of two parts and the ratio of the two parts (Maysinger et al., 2007). The nature of the hydrophobic segments in the micelles has a significant impact on the stability and biodistribution of the micelles. The longer length of the hydrophobic segment and the higher proportion of hydrophobic polymer award a greater thermodynamic stability (Bittner et al., 1998; Gaucher et al., 2005). Polymer micelles have been used extensively for delivery of therapeutics to cancer tissue.

1.2.1.3 DENDRIMERS

Dendrimers are also polymer based drug delivery vehicles. They consist of a central core, from which a number of highly branched tree-like arms originate in regular intervals to form a small, spherical, and very dense nanocarrier. The empty intermolecular cavity and the highly dense terminal groups can be used to entrap host molecules (Dufes et al., 2005; Pili et al., 2010). Earlier studies of dendrimers in drug delivery system focused on the encapsulation of drug molecules. But it was difficult to control drug release. One solution involves the attachment of pH-sensitive hydrophobic acetal groups on the dendrimer periphery. At mildly acidic pH acetal group will detach from the dendrimer and triggers the disruption of micelles and release of drugs. Another approach is by attaching the drugs to the periphery of the dendrimer by a degradable linkage between dendrimer and drug. New developments in dendrimer and polymer chemistry have introduced a new class of molecules called "dendronized polymers". They are linear polymers with dendrons at each repeat unit. They are different from linear polymers. They provide drug delivery advantages because of their longer circulation time and contain numerous sites for drug attachment (Gillies and Frechet., 2005).

1.2.1.4 SELF NANOEMULSIFYING DRUG DELIVERY SYSTEMS

Self nanoemulsifying drug delivery systems (SNEDDS) are anhydrous homogenous liquid mixtures of oil, surfactant, drug, and solubilizer. Upon dilution with water under gentle mixing, this forms oil-in-water nanoemulsion (~200 nm) spontaneously. The selection of SNEDDS components is determined by drug solubilization capacity, physico-chemical properties, and physiological fate. SNEDDS can also improve oral bioavailability of hydrophobic drugs. The conversion of liquid SNEDDS to solid SNEDDS minimizes the problems associated with liquid SNEDDS (Date et al., 2010). Usually the drug present in the SNEDDS is in the solution form during the gastrointestinal (GI) transit time. The drugs get converted into nanodroplets having increased surface area. So they help poorly soluble drugs for their absorption (Larsen et al., 2011).

1.2.1.5 NANOPARTICULATE DRUG DELIVERY SYSTEM FOR ANTIVIRAL DRUGS

Nanomedicines have given an innovative way to prevent the diseases caused by viruses by giving various therapeutic technologies that target the virus infected site in the body and has achieved success by curing the diseases. To achieve this target specificity, nanoparticles may be formulated using different biocompatible and biodegradable polymers. These polymers aim at targeting the antiviral drug to the virus infected cells or tissues by forming a ligand-receptor complex. However, it has been observed that some of the polymers like poly (D,L-lactide-co-glycolide) (PLGA) copolymer have limited types of functional groups available on the surface for conjugation to targeting ligands. In the current report, it is demonstrated that the interfacial activity assisted surface functionalization (IAASF) technique can be used to incorporate reactive functional groups such as maleimide onto the surface of PLGA nanoparticles which can be further conjugated with different peptides to achieve target specificity. The surface functionalization of the nanoparticle using IAASF technique might increase the efficacy, reduce drug and formulation related side effects, change the release kinetics of antivirals, increase the target specificity, increase their bioavailability, improve the pharmacokinetics, reduce treatment costs, and improve patient compliance. These features are particularly functional in viral diseases where high drug doses are needed, the disease is chronic and fast remedy is required to overcome the illness, drugs are expensive, and the success of a therapy is associated with a patient's adherence to the administration procedure.

1.3 NANOTECHNOLOGY FOR BIOIMAGING

Many nanoparticles have unique features suitable for biomedical imaging. Their unique chemical properties make them visible (Bulte and Modo, 2007). With the optical microscope scientists can observe only the large components of cells such as nucleus and mitochondria. The scanning electron microscope (SEM) and atomic force microscope (AFM) are used for the detection of nanometer-sized components. But all these are having limitations. They are used to observe the non-living samples and the image capture time takes up to several minutes. By the use of nanoscale imaging of living system we can observe the changes taking place inside the living system and it will help for the detection of diseases (Jain, 2012).

1.3.1 QUANTUM DOTS

Quantum dots (QD) are nanocrystals of semiconductor materials. They produce fluorescence when they are excited by a high-energy light of suitable wavelength. Because of their brightness and high photostability, QD can be used as molecular markers. When they are attached to a protein or any other target component, they will help to track their path. This helps to detect and diagnose the diseases. Their applications are now extended to cellular imaging, cell trafficking, tumor targeting, etc.

Nanoprobes based on semiconductor QD have been used for cancer detection. The technique involves the encapsulation of QD into the polymer and their attachment with the tumor-targeting ligands. Subcutaneous injection of this preparation leads to the binding of QD probes to the tumor specific cell surface biomarkers and accumulate in the tumor cells (Jain, 2012).

1.3.2 GOLD NANOPARTICLES

The fluorescent molecules are now used for many disease diagnostic tests such as enzyme-linked immunosorbent assay (ELISA). In these tests when a fluorescent labeled antibody reacts with the antigen, the compound produces fluorescence. Instead of the fluorescent molecule, the gold nanoparticle can be attached to the antibody, DNA, etc. Only one antibody can attach to one fluorescent molecule, but in the case of gold nanoparticle many copies of antibodies can be attached to the same nanoparticle. So the gold nanoparticle is more efficient and accurate than the fluorescent molecule (Jain, 2012).

Nanoparticle-based immunoassays have attained great interest in the past few years. Detection of immunoglobulins by gold nanoshells was achieved recently in saline, serum, and blood. When gold nanoparticles introduced into the samples containing appropriate antigen, the antigen-antibody interaction causes the aggregation of gold nanoshells, shifting the resonant wavelength further into infrared region (Hirsch et al., 2003). Antibodies labeled with magnetic nanoparticle produce magnetic signals when exposed to a magnetic field. Thus target-bound antibodies can be identified. At the same time unbound antibodies produce no net magnetic field and disperse in all directions.

1.4 NANOTECHNOLOGY IN TISSUE ENGINEERING

Tissue engineering is the interdisciplinary field for regeneration of tissues and organs. The physical and chemical properties of the materials used for artificial grafts depend on the applications. Commonly biodegradable and biocompatible materials are used to reduce the immune responses. The scaffolds can be made either by natural polymers or by synthetic polymers.

In tissue engineering, the common approach is to mimic the architecture of extra cellular matrix (ECM). The ECM plays an important role in regulating cellular behavior by influencing cells with biochemical signals and topographical cues. ECM has two main components: polysaccharides and fibrous proteins.

Nanotechnology helps the damaged tissues to reproduce their original structure and function. In tissue engineering, the artificially stimulated cell proliferation is taking place by the use of nanomaterial based scaffolds and growth factors. Tissue engineering can overcome the difficulties due to the organ transplants or artificial implants (Ikada, 2006). Nanofibre scaffolds have been used as potential tissue engineering platforms. A close imitation of ECM will provide more conductive environment for cellular functions such as adhesion, migration, etc. (Cao et al., 2009).

Nanofibers: Nanofibers are fibers with diameter of nanometer range. The development of nanofibers improves the scope of scaffolds which mimic the natural tissue in nanometer scale. They have high surface to volume ratio and microporous structure. This helps cell adhesion, migration, proliferation, and differentiation (Bhattarai et al., 2004; Ma et al., (2005). Hence current research focuses on nanofibrous systems for tissue engineering. Because of the similarity of three-dimensional (3D) structured nanofibers with that of natural ECM, they furnish micro/nano environment for cells to grow and execute their regular functions.

1.5 NANOBIOSENSORS

In sensors the use of biological components such as organisms, sub-cellular structures, and biomolecules are well known. The importance of biosensor is mainly because of the high specificity of biological reactions for detecting target analyte. Biosensor is a combination of a biological component specific to target analyte and a physical transducer that converts the biological recognition event to a measurable effect such as mechanical, optical, or an electrical signal. The signals from the transducer are passed to a detector where the signals are amplified and converted to concentration units and transferred to a display or storage device. Biosensors allow for the binding of specific analyte of interest to the sensor from other samples in a complex mixture. A schematic representation of a biosensor is shown in Figure.1.3.

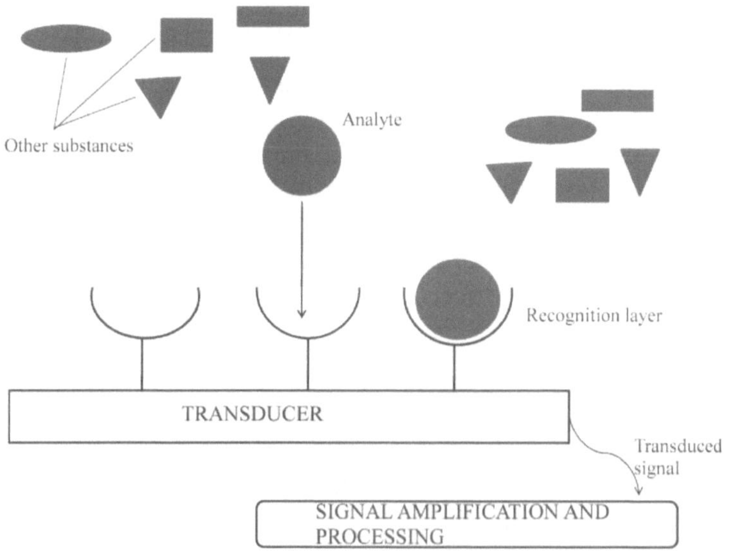

FIGURE 1.3 Schematic representation of a biosensor. Recognition layer of biosensor binds with the target analyte and give it to the transducer. Transduced signal is then sent for amplification and processing.

A recent advance in the field of biosensors is the development of optical nanobiosensors. A nanobiosensor is a biosensor that works on the nanometer size. The development of nanobiosensors by using optical fibers with submicron-sized dimensions opened a new efficient technique for cellular measurements. They are having very small sizes and suitable for monitoring intracellular physiological and biological parameters (Rai et al., 2012). Common biomedical applications of nanobiosensors are in various fields as discussed hereunder.

a. Diagnosis of diseases including cancer, diabetes, cardiovascular, and infectious diseases

The high surface to volume ratio of nanomaterials makes biosensors more sensitive by allowing single molecule detection which is very helpful in cancer diagnosis (early diagnosis will increase the survival rate). Conducting polymer nanowire and carbon nanotubes have been used recently for cancer biomarker diagnosis (Katz et al., (2004).

Nanotechnology-based solutions are used in the management of diabetes. Nanobiosensors have been developed to measure accurate level of glucose in blood. Nanotubes coupled with glucose oxidase have been used to serve as catalytic biomolecular sensors (Barone et al., 2005).

b. Immuno-assay (detection of Ag Ab reaction)

Nanoparticles can be used in conductivity-based sensors where they can change the signal when nanoparticle antibody conjugate bind with the specific antigen. Advances have been made in the sensors by using methods as diverse as light scattering, surface plasmon resonance (SPR), and electrical nanowires. The kinetics of antigen antibody interaction can be probed with SPR and fluorescent labeling (Riboh et al., 2003).

c. Detection of pathogenic bacteria

Researchers at the University of Pittsburg have synthesized a molecule from a hydrocarbon and an ammonium compound to find a unique nanotube structure having antimicrobial capability. The quaternary ammonium compound has the ability to disrupt cell membranes and causes cell death, whereas the hydrocarbon diacetylene changes color. The resulting molecule would have both the properties of biosensor and biocide (Lee *et al.*, 2004). Researchers at Purdue University have discovered nanocantilevers to design a new class of ultrasmall sensors for detecting viruses, bacteria, and other pathogens (Gupta et al., 2006).

1.5.1 TYPES OF NANOBIOSENSORS

Nanobiosensors have been using for several clinical applications. Depend on the complexity and type of particle to be detected different types of nanobiosensors is developed. Different types of nanobiosensors and their applications are depicted in Table 1.1.

TABLE 1.1 Different Nanobiosensors and their Applications

Types of Nanobiosensor	Applications
Optical biosensors	Inter and intracellular parameter analysis
Nanowire biosensors	Protein analysis
Nanotube-based biosensors	Enzyme substrate detection
Piezoelectric biosensors	Antigen antibody detection
Electrochemical biosensors	Enzyme substrate and antigen antibody detection
FRET-based DNA nanosensors	DNA detection
Quartz nanobalance biosensors	cDNA detection

1.5.1.1 OPTICAL NANOBIOSENSORS

Optical nanosensors make quantitative measurements within the intracellular environment by utilizing the sensitivity of fluorescence. The devices are small enough to insert into the living cells. The sensing component is protected from interfering components in the cells and the intracellular compartment is also protected from the toxic effects of sensing component. This makes optical nanosensors more advantageous than fluorescent dye (Aylott., 2003). The small sized optical fibers enable them to sense intracellular and intercellular parameters.

1.5.1.2 NANOWIRE BIOSENSORS

In disease diagnosis, proteomic studies, etc. protein analysis plays the central part. A silicon nanowire (SiNW) as field-effect transistors (FETs) has emerged as a tool for powerful protein detection without the need of labeling. They are highly sensitive. SiNWs have high surface to volume ratio. Field effect is the fundamental principle of nanowire electrical detection by which biomolecule binding on the surface of nanowire changes the electrical potential at the nanowire and that leads to the change of conducting charge carrier density inside the nanowire (Zheng and Lieber., 2011).

1.5.1.3 NANOTUBE-BASED BIOSENSORS

Carbon nanotubes (CNTs) possess excellent mechanical, electrochemical, and electrical properties which increases the use of carbon nanotubes in biosensors. Their high length to diameter ratio gives high surface to volume ratio. Through immobilization of enzymes, the selectivity and sensitivity can be improved. In such cases CNTs serve as transducers which communicate the signal more effectively between enzyme centers and substrate (Balasubramanian and Burghard., 2006).

1.5.1.4 PIEZO-ELECTRIC BIOSENSORS

Piezoelectric biosensors measure the frequency change that occurs when an antibody binds to the antigen. Piezoelectric immunosensor can detect antigen both in liquid and gas phases. They can also detect antigen in pictogram range. They are very sensitive to change in mass; because of this they are more dependable than the alternative biosensors like amplicor assay. They are more sensitive and faster than alternatives.

1.5.1.5 ELECTROCHEMICAL BIOSENSORS

Electrochemical biosensors are a subclass of chemical sensors, which combine the sensitivity of electrochemical transducers with high specificity of biological components. These biosensors contain biological components as recognition elements. They are of two types: biocatalytic sensors and affinity sensors. Biocatalytic biosensors contain enzymes as recognition element which reacts with the analyte and produces electroactive species. Affinity sensors contain antibodies or receptors which have selective affinity to the analyte.

1.5.1.6 FRET-BASED DNA NANOSENSORS

In the diagnosis of genetic diseases rapid and highly sensitive detection of DNA is necessary. Most of the DNA detection systems need the amplification of DNA or separation of unhybridized DNA strands from hybridized strands. Thus these are very complex procedures. An ultrasensitive nanosensor based on fluorescence resonance energy transfer (FRET) is capable of detecting small amount of DNA without any separation. The system contains QD linked DNA probes to bind with the DNA target and forming a FRET donor acceptor ensemble. The QD also amplifies the target signal by confining several targets in a nanoscale domain. Unbound nanosensors produce no background fluorescence but binding to a small amount of target DNA produces FRET signal by a separation free format (Zhang et al., 2005). Thus, FRET-based nanosensors are very useful to detect point mutation.

1.5.1.7 QUARTZ NANOBALANCE BIOSENSORS

This nanosensor contains quartz oscillators with thin films of single strand DNA (ssDNA) for sensing the presence of complimentary DNA (cDNA) sequences when hybridize with the immobilized ones. A quartz nanobalance is used to detect mass increment after the hybridization with cDNA probes (Nicolini et al., 1997).

1.6 NANOPARTICLES FOR GENE DELIVERY

Gene delivery is the process of introducing foreign DNA into cells. It is necessary for gene therapy and genetic modification of crops (Kamimura et al., 2011). For the treatment of genetic diseases, a suitable gene delivery system is needed. A schematic representation of polymeric gene delivery system is shown in Figure 1.4. Viral or non-viral vectors can be used for gene delivery. Non-viral vectors such as liposomes, cationic polymers, recombinant proteins, and inorganic nanoparticle have been widely using for the gene delivery. The efficiency of the gene delivery system depends on three factors.

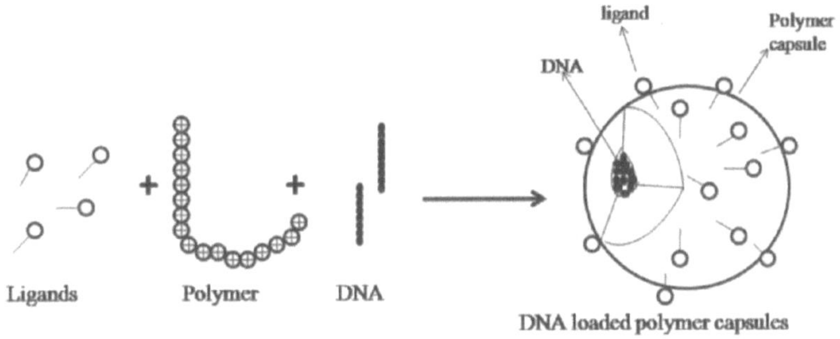

FIGURE 1.4 Schematic representation of polymeric gene delivery system.

Cellular uptake

Nanoparticles are efficiently internalized through a concentration and time dependent endocytic pathway. Specific intracellular delivery will considerably reduce drug toxicity and it will increase therapeutic effects. The positively or neutrally charged lactate dehydrogenases (LDHs) carrying anionic molecules are easily attached to the surface of a negatively charged plasma membrane. This facilitates the internalization into the cells (Choy et al., 2000, 2004).

a. Release of the gene from the endosome

Following the cellular uptake the nanoparticle will transport into primary endosome and then to sorting endosome. From sorting endosome a part is transported to the outside of the cell through the recycling endosome and rest of the contents entered into endolysosome. Following the intracellular uptake by endocytosis nanoparticles can escape rapidly to the cytoplasmic compartment from the endolysosomal compartment (Panyam et al., 2002). The escape is made possible by changing the surface charge of nanoparticle from anionic to cationic in the acidic pH of the endolysosomal compartment. Thus the nanoparticles interact with endolysosomal membrane and escape into cytoplasmic compartment. This rapid escape prevents the nanoparticle and encapsulated DNA from degeneration.

b. Nuclear targeting and transport

The nucleus is a very complex organelle containing double membrane which separates the nucleus from the cytoplasm. Exchange of materials is mainly possible through the nuclear pore complexes (NPCs) (Adam, 2001). They can transport DNA molecules of about 210–350 bp. The DNAs mainly used for gene delivery studies are larger than this size. Many transcription factors and polymerases can cross the nuclear membrane because of the presence of nuclear localization signals (NLS) which can react with transportin and form a complex and the protein is transferred into the nucleus. This nuclear localization signals could be attached to the drug delivery systems for the delivery of genes (Van Gaal et al., 2011).

Due to the controlled size and shape, the proper fusion of DNA with the nanoparticles and suitable release of DNA from these particles is possible (Xu et al., 2007).

Inorganic nanoparticles generally possess properties suitable for gene delivery such as

- Rich functionality
- Good biocompatibility
- Capability of target delivery
- Controlled release of gene

1.7 NANOMOLECULAR DIAGNOSIS

The use of molecular technologies to detect, diagnose, and monitor human diseases is known as molecular diagnosis. Nanomolecular diagnosis is the use of nanotechnology in molecular diagnosis and it is known as "nanodiagnostics". Because of the small sizes of nanoparticles, nanochips and nanoarrays have been using for molecular diagnosis. Biochips constructed with microelectromechanical systems (MEMS) are constructed by micromanipulation, whereas nanotechnology-based chips are by nanomanipulation. Even though microarrays/microchips detect the molecular interaction, they have some limitations. They need fluorescent or radioactive tag, which is time consuming and expensive procedure. But in the case of nanoarrays/nanochips the label-free method is used such as surface plasmon resonance (SPR) and quartz crystal microbalance which depends on mass detection (Jain, 2012).

Nanoparticles can be used for various molecular diagnostic applications such as

- Cancer diagnosis
- DNA sequencing and mapping
- Immune assays
- Pathogen detection
- Genotyping

1.8 NANOPORE SEQUENCING

This is one of the ultra rapid techniques for sequencing of DNA by nanopore engineering and assembly. In this method DNA can be passed through a 1–2 nm diameter pore in an α-hemolysin protein complex inserted into a lipid bilayer separating two conductive compartments. The current is recorded and it is translated into electronic signals to recognize each base. Through this technique it is possible to sequence more than 1,000 bases per second. This is very much useful for detecting single nucleotide polymorphisms (Moghimi et al., 2005).

1.9 NANOPHARMACEUTICALS

"Nanopharmaceuticals" include a wide range of area such as drug discovery, development, and delivery to the target site. The development of more accurate and cost-effective label-free technique is needed to detect diseases. It is possible with the use of materials whose physical properties change with the specific interactions. The nanomaterials like gold nanoparticles and QD can be used for drug discovery.

Researchers have created a chip of 1 cm^2 which contains thousands of tiny pores in it. The top of the pore is open and the bottom is closed and each having nanometer dimension (nanoporous). Each pore is capable of screening an individual drug, and thus emerged a new drug discovery method, "lab-on-a-chip".

The designing of a drug for a particular disease must need the knowledge about the interaction between the cells and complex biomolecules surrounding it and many of the recently used technique can only detect a small select group of molecules. Researchers at the Georgia Tech have created a nanoprobe, the scanning mass spectroscopy (SMS) probe which can detect the complex molecules and check the intracellular pathways in its natural environment (Kottke et al., 2009).

1.10 NANOTECHNOLOGY IN CANCER

The tumor expansion and metastasis is mainly caused by the tumor-associated angiogenesis. This realization and the development of cancer therapies and tumor imaging based on tumor-associated vasculature have opened the applications of nanotechnology into cancer therapeutics. The newly developing area of "nanohealth" ultimately allows detection of human health at the very early stages. The nanotechnology based injectable nanoscale delivery systems useful for cancer therapeutics have been defined as nanovectors. Nanovectors can reduce the side effects caused by the use of single chemotherapeutic agents and multimodality therapeutic regimens (Amiji, 2006).

One of the most important characteristics of the nanovector is their ability to cross the biological barriers. The biological barriers are very complex. The blood brain barrier prevents the entry of drugs into the brain for the tumor therapies. But the use of nanovector overcomes this biobarrier to treat the tumors. Another promising area of nanomedicine is in optics and imaging of tumors. Nanosized imaging agents that can be targeted to tumors in the body could provide a less invasive, more accurate, and faster way to diagnose tumors. A fundamental limitation of the molecular imaging is the signal to background ratio. The brightly glowing nanoparticle can overcome the background radiations caused by fluorescent proteins.

1.10.1 SILVER NANOPARTICLES FOR CANCER DETECTION

Early detection of cancer requires sensitive methods to monitor the intracellular changes. The probes for detecting the intracellular changes should be very small, bright, and stable for a long time in the intracellular environment without disturbing the normal functioning of the cell. Silver nanoparticle can be used to detect the intracellular changes. They are very small, bright (can be seen with optical microscope), and they do not photodecompose by extended illumination. The nanoparticle should be biocompatible by surface modification for their action (Kottke et al., 2009).

1.10.2 QDS FOR THE DIAGNOSIS OF CANCER

QDs coated with a polyacrylate cap linked to antibodies or streptavidin have been used for labeling breast cancer marker Her2 (Wu et al., 2003). QD labeling is highly specific, brighter, and stable than other fluorescent markers. The recent advances in QD bioconjugates raise possibilities to enable visualization of cancer cells in living animals (Gao and Nie, 2003).

1.10.3 CANCER DIAGNOSIS BY GOLD NANOPARTICLE

When exposed to infrared light of a particular wavelength, gold nanoparticles get heated. It causes a change in pressure near the particle. This will result in the generation of ultrasound by Plasmon resonance. Thus the light results in sound. By heating the gold nanoparticle attached to monoclonal antibodies (MAbs) specific to cancer cell, it is possible to detect the presence of cancer. The heating of gold nanoparticle increase the temperature up to 100°C. In photothermal therapy, the heated gold nanoparticle is used to destroy the tumor. The gold nanoparticle can also carry drugs to the cancerous cells, and when heated it will release the drugs to the target action (Jain, 2012).

1.11 NANOTECHNOLOGY IN SURGERY

In the ancient ages, surgery was a very complex procedure with very large instruments. Then the surgery is miniaturized as microsurgery such as key-hole surgery by reducing trauma to the body tissue during surgery. If the trauma during surgery is less, the recovery period is also less.

The main problem in the case of surgery is bleeding. During surgery a lot of blood is lost through bleeding. Surgery without bleeding is possible by nanotechnology. Scientists from the Massachusetts Institute of Technology (MIT) and Hong Kong University have discovered that some simple liquids composed of peptides which can be applied to the open wounds. The peptides will self-assemble to form a thin nanoscale protective layer, seal the wounds, and stop the bleeding. After healing of injury, the peptide will degrade and it will help to repair the tissue. Exact mechanism is unknown but the possibility is the interaction of peptide with extracellular matrix (ECM) around the cell.

Catheters with nanobiosensors can be used for the cardiovascular surgery to give the information about the environment around the surgical site. Femtosecond laser pulses opened a new area of surgery without causing any disturbance to the surroundings. It can remove an organelle from the cell such as mitochondria without causing any injury to the cell and other organelles in it. Nanolasers can also introduce drugs into the desired site through the skin.

"Nanorobots" or "nanobots" are also used for injury-free surgery. The nanobots can be introduced with the catheters. Nanobots are small robots which will move like an inchworm under the control of a surgeon. The device has a working channel, through which the various tools for treatment can be introduced (Kottke et al., 2009).

1.12 NANOTECHNOLOGY IN DENTISTRY

Nanotechnology provides more efficient methods to improve dental treatment, prevent oral diseases, and dental care. There have been numerous research works done in the application of nanotechnology in dental biomaterials, dental instruments, dental implants, and scaffolds for bone regeneration around dental implants and maxillofacial region, and nanodiagnostic tools to diagnose oral pathology. Nanotechnology offers tooth sealants and fillers (improve the strength and brightness of teeth and prevent the corrosion of teeth by using nanoparticle), restorative composite materials (Antimicrobial nanoparticles are used to prevent the tooth decay, e.g., silver nanoparticle). Laser- and digital-guided surgery combined with biotechnology will provide excellent dental care (Subramani and Ahmad., 2011).

1.12.1 NANOTECHNOLOGY ON DENTAL IMPLANTS

Recent studies show that on blood clot formation, adsorption of proteins and cell behaviors occurring upon implantation of dental implants, nanometer-controlled surfaces have great effect. Nanostructured surfaces can control the differentiation pathways into specific lineages and direct the nature of peri-implant tissues (Subramani and Ahmed., 2011).

1.12.2 NANOTECHNOLOGY IN ORAL SURGERY

Most of the dental procedures need local anesthesia. Because of the fear of injections, patients delay their treatment. Painless administration of local anesthesia is possible through nanotechnology. A colloidal suspension containing analgesic micron-size dental robots will be introduced on the patient's gum. The nanorobots would reach the pulp under the control of dentist and nanocomputer. Once installed, the nanorobots may shut down all sensitivity around the tooth which needs treatment. After the treatment the nanorobot may restore the sensation (Freitas., 2000; Love and Whitesides., 2001).

1.12.3 NANOTECHNOLOGY IN TOOTH DURABILITY AND APPEARANCE

Replacing of enamel with covalently bonded diamond and sapphire can improve the durability and appearance. They are biocompatible and have 20 times more hardness than the conventional ones. Accumulation of supra- and sub-gingival calculus can be prevented by delivering nanorobotic dentifrice through mouth wash or tooth paste. They metabolize the trapped organic matter into odorless and harmless vapors (Sivaramakrishnan and Neelakantan, 2014).

1.13 PROTEIN AND PEPTIDE DELIVERY

Proteins and peptides are considered as important potential drug of future. The diseases that can be cured by protein drugs include auto-immune diseases, cancer, hypertension, cardiovascular diseases, mental disorder, and metabolic diseases. Protein drug needs some requirements: they must be highly purified and concentrated, have short half-life, and should have a shelf life of at least two years. Using recombinant technology, it is possible to produce protein drugs for the treatment of life-threatening diseases such as hepatitis, diabetes, rheumatoid arthritis, etc. (Ratnaparkhi et al., 2011).

Protein and peptide drugs are commonly delivered through parenteral routs. The oral administration of peptides is not commonly used because the GI tract contains endo and exopeptidases, and they will act on the peptide and cleave it. Poor membrane permeability of hydrophilic peptides prevents them from entering into the cell. Structural modification of these compounds will overcome these difficulties. Various delivery systems like nanoparticulate and microparticulate delivery systems, prolease technology, microspheres, and mucoadhesive delivery have been developed for the delivery of proteins and peptides (Kumar et al., 2007).

Insulin is generally supplied by intravenous injection for diabetes treatment. It enters into the general circulation. So, only a fraction of injected dose reaches the liver. The use of nanoparticles overcomes this difficulty. Insulin can be encapsulated into poly (isobutylcyanoacrylate) (PIBCA) nanoparticles. The encapsulation protects insulin from proteolytic enzymes and increases absorption through intestinal mucosa. Recently insulin has been encapsulated in water-containing nanocapsules, which facilitates intestinal absorption when dispersed in a micro-emulsion (Pinto et al., 2006).

1.14 NANOPARTICLES FOR WOUND HEALING

Silver ions and compounds have been widely used for both healing and hygienic purposes for centuries, by using their strong bactericidal effects, and a broad spectrum antimicrobial activity (Klasen, 2000). By using the bactericidal property, silver compounds have been used for treating chronic wounds. The silver compounds were also used for treating ulcers, burns, etc. For treating wounds silver nanoparticles were either impregnated into the bandages or given as silver sulfadiazine containing cream (Hussain and Ferguson, 2006). Pure silver nanoparticles which are biostable can be produced through nanotechnology either by loading metallic particles into nanofibers or by photo-assisted reduction (Jain et al., 2009). Multilevel antibacterial effects (Fong and Wood, 2006) and low toxicity (Melaiye and Youngs, 2005) offers high antibacterial activity to silver. The chance of resistance against this antimicrobial effect is very low.

Because of the antimicrobial activity, silver nanoparticles have been used for the wound healing purpose for many years. Augustine et al. have used the silver nanoparticle in the sutures to prevent the microbial attack (Figure 1.5). On the bioresorbable sutures they have coated a layer of silver nanoparticles and they were immobilized by sodium alginate. They proved that this will prevent the growth of both gram-positive and gram-negative bacteria around the wound (Augustine and Rajarathinam, 2012).

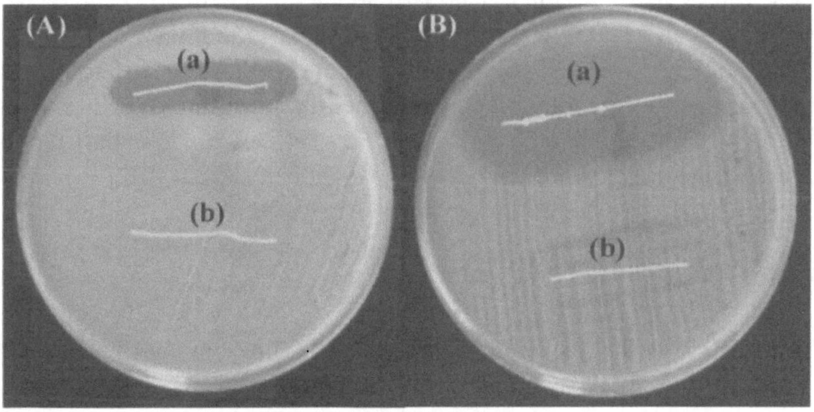

FIGURE 1.5 Plates showing in vitro antimicrobial activity of the suture. Plate (A): On *Staphylococcus aureus* culture, (a) Ag nano-alginate coated suture, (b) control without coating. Plate (B): On *E. coli* culture, (a) Ag nano-alginate coated suture, (b) control without coating.

1.14.1 NANOFIBROUS SCAFFOLDS FOR REGENERATION OF DAMAGED TISSUE

The scaffold for tissue regeneration should regenerate or replace the damaged tissue in combination with living cells. The scaffold should possess some properties such as enough porosity, controlled degradation, biocompatibility, controlled permeability, and it should act as the support for cell adhesion and proliferation. They closely resemble the ECM and so they promote attachment and proliferation of cells (Tran et al., 2009). The porosity enables the oxygen and water permeability and they prevent the entry of bacteria into the wounds. This makes electrospun nanofibers to act as a suitable material for wounds, burns, and diabetic ulcers. Electrospun membrane of polyurethane is very useful for wound dressing because of their ability to soak fluid from the wound and prevent the wound desiccation (Khil et al., 2003).

1.14.2 ZINC OXIDE NANOPARTICLES FOR WOUND HEALING

Augustine et al. studied the wound healing property of zinc oxide (ZnO) nanoparticle incorporated into polycaprolactone membrane scaffolds. They found that the ZnO nanoparticle incorporated polycaprolactone membrane can heal the wound without any scar formation. They suggested these membranes as a good skin substitute with fast wound healing by improved cell migration and cell proliferation (Augustine et al., 2014a). They also demonstrated the ability of ZnO to effectively vascularize polycaprolactone based skin substitutes (Augustine et al., 2014b).

1.15 TOXICITY

Nanocarriers help to overcome the toxicity of drugs and drug-induced side effects. But sometimes the nanocarriers themselves possess toxicity. A number of toxicology reports have shown that exposure to nanomaterials causes serious hazards to biological systems (Hoet et al., 2004). For example, when human keratinocytes are exposed to insoluble single walled carbon nanotube, oxidative stress and apoptosis occurs. The toxicity is more serious for intravenous injection of nanoparticles. Though cadmium selenide QDs are most commonly used for imaging, their metabolism and potential deleterious effects are little known (Moghimi et al., 2005). Cadmium selenide produces toxic cadmium ions under ultraviolet (UV) irradiation, which is lethal to cells. Some polymeric micelles can induce cell death by necrosis or apoptosis (Oberdörster et al., 2005). So it is important to check the mechanism of action of nanoparticles within the cellular environment before using them for biological applications.

1.16 FUTURE OF NANOMEDICINE

Nanotechnology is beginning to change the method and scale of drug delivery and vascular imaging systems. The National Institute of Health (NIH) Roadmap's "Nanomedicine initiatives" shows that within the next 10 years, nanomedicine will begin yielding more benefits. The National Cancer Institute has programs for producing nanoscale multifunctional entities that can diagnose, deliver therapeutic drugs, and monitor cancer treatment progress. The future of nanomedicine depends on design of nanomaterials and tools with thorough understanding of biological processes than finding the applications of commonly used nanomaterials (Moghimi et al., 2005).

1.17 CONCLUSION

Nanotechnology has taken the medical field to a new era. Many of the diseases now incurable will be curable with the help of nanotechnology. Over the next couple of years nanotechnology will reach all fields of medicine and improve human health. The toxicity of the nanoparticle is still a problem which affects the application of nanotechnology in medicine. It is expected that within few years, we will get more understanding about the coating and altering of nanoparticles for reducing their toxicity. All new technologies have to face ethical, legal, safety, environmental, and regulatory issues. According to nanotechnology, currently there is no Food and Drug Administration (FDA) regulation for nanobiotechnology products. This is because of the less number of nanobiotechnology products. If the number of products is increased, then the FDA regulations will take effect on nanobiotechnological products. At the same time there is an effort to standardize nanotechnology in general.

KEYWORDS

- **Biosensors**
- **Drug delivery systems**
- **Nanomedicine**
- **Nanoparticles**
- **Nanotechnology**

REFERENCES

Adam, Stephen A. "The nuclear pore complex." Genome Biology 2, no. 9 (2001): 1–6.

Amiji, Mansoor M., ed. Nanotechnology for cancer therapy. CRC Press, 2006.

Augustine, R., and K. Rajarathinam. "Synthesis and characterization of silver nanoparticles and its immobilization on alginate coated sutures for the prevention of surgical wound infections and the in vitro release studies." International Journal of Nano Dimension 2, no. 3 (2012): 205–212.

Augustine Robin, Edwin Anto Dominic, Indu Reju, Balarama Kaimal, Nandakumar Kalarikkal, and Sabu Thomas. "Electrospun polycaprolactone membranes incorporated with ZnO nanoparticles as skin substitutes with enhanced fibroblast proliferation and wound healing." RSC Advances 4 (2014a): 24777–24785.

Augustine Robin, Edwin Anto Dominic, Indu Reju, Balarama Kaimal, Nandakumar Kalarikkal, and Sabu Thomas. "Investigation on angiogenesis and its mechanism using zinc oxide nanoparticles-loaded electrospun tissue engineering scaffolds." RSC Advances (2014b). DOI:10.1039/C4RA07361D

Aylott, Jonathan W. "Optical nanosensors-an enabling technology for intracellular measurements." Analyst 128, no. 4 (2003): 309–312.

Balasubramanian, Kannan, and Marko Burghard. "Biosensors based on carbon nanotubes." Analytical and Bioanalytical Chemistry 385, no. 3 (2006): 452–468.

Barone, Paul W., Seunghyun Baik, Daniel A. Heller, and Michael S. Strano. "Near-infrared optical sensors based on single-walled carbon nanotubes." Nature Materials 4, no. 1 (2005): 86–92.

Bhattarai, Shanta Raj, Narayan Bhattarai, Ho Keun Yi, Pyong Han Hwang, Dong Il Cha, and Hak Yong Kim. "Novel biodegradable electrospun membrane: scaffold for tissue engineering." Biomaterials 25, no. 13 (2004): 2595–2602.

Bittner, Beate, Michael Morlock, Hans Koll, Gerhard Winter, and Thomas Kissel. "Recombinant human erythropoietin (rhEPO) loaded poly (lactide-co-glycolide) microspheres: influence of the encapsulation technique and polymer purity on microsphere characteristics." European Journal of Pharmaceutics and Biopharmaceutics 45, no. 3 (1998): 295–305.

Bulte, Jeff W. M., and Michel Modo. "Nanoparticles in biomedical imaging: emerging technologies and applications." Springer, 2007.

Cao, Haoqing, Ting Liu, and Sing Yian Chew. "The application of nanofibrous scaffolds in neural tissue engineering." Advanced Drug Delivery Reviews 61, no. 12 (2009): 1055–1064.

Choy, Jin-Ho, Seo-Young Kwak, Yong-Joo Jeong, and Jong-Sang Park. "Inorganic layered double hydroxides as nonviral vectors." Angewandte Chemie International Edition 39, no. 22 (2000): 4041–4045.

Choy, Jin-Ho, Ji-Sun Jung, Jae-Min Oh, Man Park, Jinyoung Jeong, Young-Koo Kang, and Ok-Jin Han. "Layered double hydroxide as an efficient drug reservoir for folate derivatives." Biomaterials 25, no. 15 (2004): 3059–3064.

Date, Abhijit A., Neha Desai, Rahul Dixit, and Mangal Nagarsenker. "Self-nanoemulsifying drug delivery systems: formulation insights, applications and advances." Nanomedicine 5, no. 10 (2010): 1595–1616.

Dufès, Christine, Ijeoma F. Uchegbu, and Andreas G. Schätzlein. "Dendrimers in gene delivery." Advanced Drug Delivery Reviews 57, no. 15 (2005): 2177–2202.

Felnerova, Diana, Jean-François Viret, Reinhard Glück, and Christian Moser. "Liposomes and virosomes as delivery systems for antigens, nucleic acids and drugs." Current Opinion in Biotechnology 15, no. 6 (2004): 518–529.

Fong, Joy, and Fiona Wood. "Nanocrystalline silver dressings in wound management: a review." International Journal of Nanomedicine 1, no. 4 (2006): 441.

Freitas, Robert A. "Nanodentistry." The Journal of the American Dental Association 131, no. 11 (2000): 1559–1565.

Freitas Jr, Robert A. "Nanotechnology, nanomedicine and nanosurgery." International Journal of Surgery 3, no. 4 (2005): 243–246.

Gao, Xiaohu, and Shuming Nie. "Molecular profiling of single cells and tissue specimens with quantum dots." Trends in Biotechnology 21, no. 9 (2003): 371–373.

Gaucher, Geneviève, Marie-Hélène Dufresne, Vinayak P. Sant, Ning Kang, Dusica Maysinger, and Jean-Christophe Leroux. "Block copolymer micelles: preparation, characterization and application in drug delivery." Journal of Controlled Release 109, no. 1 (2005): 169–188.

Gillies, Elizabeth R., and Jean M. J. Frechet. "Dendrimers and dendritic polymers in drug delivery." Drug Discovery Today 10, no. 1 (2005): 35–43.

Gupta, Amit K., Pradeep R. Nair, Demir Akin, Michael R. Ladisch, Steve Broyles, Muhammad A. Alam, and Rashid Bashir. "Anomalous resonance in a nanomechanical biosensor." Proceedings of the National Academy of Sciences 103, no. 36 (2006): 13362–13367.

Hirsch, L. R., J. B. Jackson, A. Lee, N. J. Halas, and J. L. West. "A whole blood immunoassay using gold nanoshells." Analytical Chemistry 75, no. 10 (2003): 2377–2381.

Hoet, Peter H.M., Irene Brüske-Hohlfeld, and Oleg V. Salata. "Nanoparticles–known and unknown health risks." Journal of Nanobiotechnology 2, no. 1 (2004): 12.

Hussain, Saiqa, and Craig Ferguson. "Silver sulphadiazine cream in burns." Emergency Medicine Journal 23, no. 12 (2006): 929–932.

Ikada, Yoshito. "Challenges in tissue engineering." Interface 3 (2006): 589–601.

Jain, Jaya, Sumit Arora, Jyutika M. Rajwade, Pratibha Omray, Sanjeev Khandelwal, and Kishore M. Paknikar. "Silver nanoparticles in therapeutics: development of an antimicrobial gel formulation for topical use." Molecular Pharmaceutics 6, no. 5 (2009): 1388–1401.

Jain, Kewal K. The handbook of nanomedicine. Springer, 2012.

Kale, Amit A., and Vladimir P. Torchilin. ""Smart" drug carriers: PEGylatedTATp-modified pH-sensitive liposomes." Journal of Liposome Research 17, no. 3–4 (2007): 197–203.

Kamimura, Kenya, Takeshi Suda, Guisheng Zhang, and Dexi Liu. "Advances in gene delivery systems." Pharmaceutical Medicine 25, no. 5 (2011): 293–306.

Katz, Eugenii, and Itamar Willner. "Biomolecule-functionalized carbon nanotubes: applications in nanobioelectronics." Chem Phys Chem 5, no. 8 (2004): 1084–1104.

Khil, Myung-Seob, Dong-Il Cha, Hak-Yong Kim, In-Shik Kim, and Narayan Bhattarai. "Electrospun nanofibrous polyurethane membrane as wound dressing." Journal of Biomedical Materials Research Part B: Applied Biomaterials 67, no. 2 (2003): 675–679.

Klasen, H. J. "A historical review of the use of silver in the treatment of burns. II. Renewed interest for silver." Burns 26, no. 2 (2000): 131–138.

Kottke, Peter A., Levent Degertekin, F. and Fedorov, Andrei G. "Scanning mass spectrometry probe: a scanning probe electrospray ion source for imaging mass spectrometry of submerged interfaces and transient events in solution." Analytical chemistry 82, no. 1 (2009): 19–22.

Kumar Malik, Dhirendra, Sanjula Baboota, Alka Ahuja, Sohail Hasan, and Javed Ali. "Recent advances in protein and peptide drug delivery systems." Current Drug Delivery 4, no. 2 (2007): 141–151.

Lamprecht, Alf, ed. Nanotherapeutics: drug delivery concepts in nanoscience. Pan Stanford Publishing, 2009.

Larsen, Anne T., Philip Sassene, and Anette Müllertz. "*In vitro* lipolysis models as a tool for the characterization of oral lipid and surfactant based drug delivery systems." International Journal of Pharmaceutics 417, no. 1 (2011): 245–255.

Lee, Sang Beom, Richard Koepsel, Donna B. Stolz, Heidi E. Warriner, and Alan J. Russell. "Self-assembly of biocidal nanotubes from a single-chain diacetylene amine salt." Journal of the American Chemical Society 126, no. 41 (2004): 13400–13405.

Lee, Sie Huey, Zhiping Zhang, and Si-Shen Feng. "Nanoparticles of poly (lactide)—tocopheryl polyethylene glycol succinate (PLA-TPGS) copolymers for protein drug delivery." Biomaterials 28, no. 11 (2007): 2041–2050.

Love, J. Christopher, and George M. Whitesides. "The art of building small." Scientific American 285, no. 3 (2001): 38.

Ma, Zuwei, Masaya Kotaki, Ryuji Inai, and Seeram Ramakrishna. "Potential of nanofiber matrix as tissue-engineering scaffolds." Tissue Engineering 11, no. 1–2 (2005): 101–109.

Maysinger, Dusica, Jasmina Lovrić, Adi Eisenberg, and Radoslav Savić. "Fate of micelles and quantum dots in cells." European Journal of Pharmaceutics and Biopharmaceutics 65, no. 3 (2007): 270–281.

Melaiye, Abdulkareem, and Wiley J. Youngs. "Silver and its application as an antimicrobial agent." Expert Opinion on Therapeutic Patents 15, no. 2 (2005): 125–130.

Moghimi, S. Moein, A. Christy Hunter, and J. Clifford Murray. "Nanomedicine: current status and future prospects." The FASEB Journal 19, no. 3 (2005): 311–330.

Nguyen, Ngoc Doan Trang, and Ly Thi Le. "Targeted proteins for diabetes drug design." Advances in Natural Sciences: Nanoscience and Nanotechnology 3, no. 1 (2012): 013001.

Nicolini, Claudio, Victor Erokhin, Paolo Facci, S. Guerzoni, Andrea Ross, and Pavel Paschkevitsch. "Quartz balance DNA sensor." Biosensors and Bioelectronics 12, no. 7 (1997): 613–618.

Oberdörster, Günter, Eva Oberdörster, and Jan Oberdörster. "Nanotoxicology: an emerging discipline evolving from studies of ultrafine particles." Environmental Health Perspectives 113, no. 7 (2005): 823.

Panyam, Jayanth, Wen-Zhong Zhou, Swayam Prabha, Sanjeeb K. Sahoo, and Vinod Labhasetwar. "Rapid endo-lysosomal escape of poly (DL-lactide-co-glycolide) nanoparticles: implications for drug and gene delivery." The FASEB Journal 16, no. 10 (2002): 1217–1226.

Pili, Roberto, Mark A. Rosenthal, Paul N. Mainwaring, Guy Van Hazel, Sandy Srinivas, Robert Dreicer, Sanjay Goel, Joseph Leach, Shirley Wong, and Peter Clingan. "Phase II study on the addition of ASA404 (vadimezan; 5, 6-dimethylxanthenone-4-acetic acid) to docetaxel in CRMPC." Clinical Cancer Research 16, no. 10 (2010): 2906–2914.

Pinto Reis, Catarina, Ronald J. Neufeld, António J. Ribeiro, and Francisco Veiga. "Nanoencapsulation II. Biomedical applications and current status of peptide and protein nanoparticulate delivery systems." Nanomedicine: Nanotechnology, Biology and Medicine 2, no. 2 (2006): 53–65.

Rai, Mahendra, Aniket Gade, Swapnil Gaikwad, Priscyla D. Marcato, and Nelson Durán. "Biomedical applications of nanobiosensors: the state-of-the-art." Journal of the Brazilian Chemical Society 23, no. 1 (2012): 14–24.

Ratnaparkhi, M. P., S. P. Chaudhari, and V. A. Pandya. "Peptides and proteins in pharmaceuticals."International Journal of Current Pharmaceuticals Research 3 (2011): 1–9.

Riboh, Jonathan C., Amanda J. Haes, Adam D. McFarland, Chanda Ranjit Yonzon, and Richard P. Van Duyne. "A nanoscale optical biosensor: real-time immunoassay in physiological buffer enabled by improved nanoparticle adhesion." The Journal of Physical Chemistry B 107, no. 8 (2003): 1772–1780.

Sahoo, S. K., S. Parveen, and J. J. Panda. "The present and future of nanotechnology in human health care." Nanomedicine: Nanotechnology, Biology and Medicine 3, no. 1 (2007): 20–31.

Sivaramakrishnan, S. M., and P. Neelakantan. "Nanotechnology in dentistry-what does the future hold in store." Dentistry 4, no. 198 (2014): 2161–1122.

Subramani, Karthikeyan, and Waqar Ahmed. Emerging nanotechnologies in dentistry: processes, materials and applications. William Andrew, 2011.

Tran, Phong A., Lijie Zhang, and Thomas J. Webster. "Carbon nanofibers and carbon nanotubes in regenerative medicine." Advanced Drug Delivery Reviews 61, no. 12 (2009): 1097–1114.

Van Gaal, Ethlinn V.B, Ronald S. Oosting, Roel van Eijk, Marta Bakowska, Dries Feyen, Robbert Jan Kok, Wim E. Hennink, Daan J.A. Crommelin, and Enrico Mastrobattista. "DNA nuclear targeting sequences for non-viral gene delivery." Pharmaceutical Research 28, no. 7 (2011): 1707–1722.

Webster, Thomas J. Safety of nanoparticles. Springer, 2008.

Weissig, Volkmar, John Babich, and Vladimir Torchilin. "Long-circulating gadolinium-loaded liposomes: potential use for magnetic resonance imaging of the blood pool." Colloids and Surfaces B: Biointerfaces 18, no. 3 (2000): 293–299.

Wu, Xingyong, Hongjian Liu, Jianquan Liu, Kari N. Haley, Joseph A. Treadway, J. Peter Larson, Nianfeng Ge, Frank Peale, and Marcel P. Bruchez. "Immunofluorescent labeling of cancer marker Her2 and other cellular targets with semiconductor quantum dots." Nature Biotechnology 21, no. 1 (2003): 41–46.

Xu, Zhi Ping, Tara L. Walker, Kerh-lin Liu, Helen M. Cooper, GQ Max Lu, and Perry F. Bartlett. "Layered double hydroxide nanoparticles as cellular delivery vectors of supercoiled plasmid DNA." International Journal of Nanomedicine 2, no. 2 (2007): 163.

Yoo, Jin-Wook, Elizabeth Chambers, and Samir Mitragotri. "Factors that control the circulation time of nanoparticles in blood: challenges, solutions and future prospects." Current Pharmaceutical Design 16, no. 21 (2010): 2298–2307.

Zhang, J., G. Yang, Yuan Cheng, Bo Gao, Qi Qiu, Y. Z. Lee, J. P. Lu, and Otto Zhou. "Stationary scanning X-ray source based on carbon nanotube field emitters." Applied Physics Letters 86, no. 18 (2005): 184104.

Zheng, Gengfeng, and Charles M. Lieber. "Nanowire biosensors for label-free, real-time, ultra-sensitive protein detection." In: Steven A. Toms andRobert J. Weil (Editors) Nanoproteomics, pp. 223–237. Humana Press, 2011.

CHAPTER 2

TISSUE ENGINEERING: PRINCIPLES, RECENT TRENDS AND THE FUTURE

ANSUJA P. MATHEW, ROBIN AUGUSTINE,
NANDAKUMAR KALARIKKAL, and SABU THOMAS

ABSTRACT

Since the time immemorial or for millennia, humans have been interested in manipulating things surrounding them to suit their needs, and with the advent of technology their inspiration to modify or alter life has increased. Tissue engineering is one such idea that has been practiced by humans for thousands of years and thus most of the achievements in this field that enthralls us today have its origin thousands of years back. Tissue engineering entitled as a multidisciplinary approach which involves the development and replacement of living tissues and organs for diseases and trauma. It applies the principles of both engineering and biology toward the development of viable substitutes which replace, restores, or improves the damaged, missing, or poorly functioning components of human tissues or organs. The success of skin tissue engineering and its commercialization made scientists to put on interest in expanding the research on tissue engineering and this will be one of the significant fields of science/health in the coming centuries. In this chapter, we discuss about the mechanisms involved in tissue engineering, the role of biomaterials, its fabrication methods, different approaches, current status, and various applications.

2.1 INTRODUCTION

Damages and degeneration of tissue happens in every organism now and then but rebuilding or treatment of these body parts becomes a major issue when we consider about the availability of the grafts needed for the particular replacement and various anatomical limitations. Rebuilding of body parts has a long tradition and was nurtured by the discovery and availability of new synthetics during World War II where, the introduction of various man-made materials for the reconstruction of the tissue damage happened. This reconstructive surgery which focused on the fabrication of living replacements in the laboratory is termed as tissue engineering (Lanza et al., 2011). The practice of tissue engineering seems relatively new but the idea of replacing of a tissue with another dates back to the16th century. It was only in 1988 at the_National Science Foundation workshop that the term "tissue engineering" was officially stated at first (O'Brien, 2011).

As discussed, tissue engineering is considered as a highly multidisciplinary approach as it draws experts from various fields like clinical medicine, mechanical engineering, material science, genetics, and life sciences.

2.2 PRINCIPLE OF TISSUE ENGINEERING

According to Langer and Vacanti, tissue engineering utilizes the basic principles of life sciences and engineering for the development of biological substitutes which then restore, maintain, or improve the tissue functions (Langer and Vacanti, 1993). The basic principle of tissue engineering depicted in Figure 2.1 can be demonstrated in this way: at first the cells have to be isolated from a source (allogenic, xenogenic, or autologous source), it is then expanded in a cell culture system or a bioreactor (expansion in vitro) and these expanded cells are then seeded onto a matrix/carrier which then provide structural support along with addition of proper medium (rich with nutrients and growth factors). Here the cells differentiate, proliferate and migrate to the carrier and replace the old tissues by forming new tissues. The tissue engineering construct formed as a result of this is then grafted back into the patient to function as the introduced replacement tissue (Vasita and Katti, 2006).

Tissue engineering can be performed by various approaches as listed hereunder.

1. In vitro tissue genesis for in vivo application
2. In vivo tissue genesis for in vivo application
3. In vitro tissue genesis for ex vivo application
4. In vitro tissue genesis for in vitro application (Irvin, 2003)

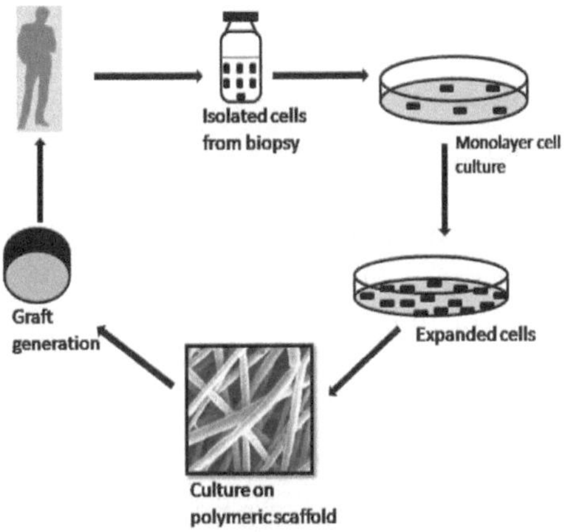

FIGURE 2.1 Basic principle of Tissue Engineering

As compared to the original biologic tissue which involves the co-ordination between the cells, extracellular matrix, and the signaling system, the tissue engineering system also needs a better understanding among the cells, scaffold, and the signal molecules and these three basic components together forms the "triad of the tissue engineering" and shown in Figure 2.2. The scaffold here serves as the mechanical platform onto which the cells get adhered, proliferate, and differentiate and thus it is better to develop a biocompatible and biodegradable scaffold (Rim et al., 2013).

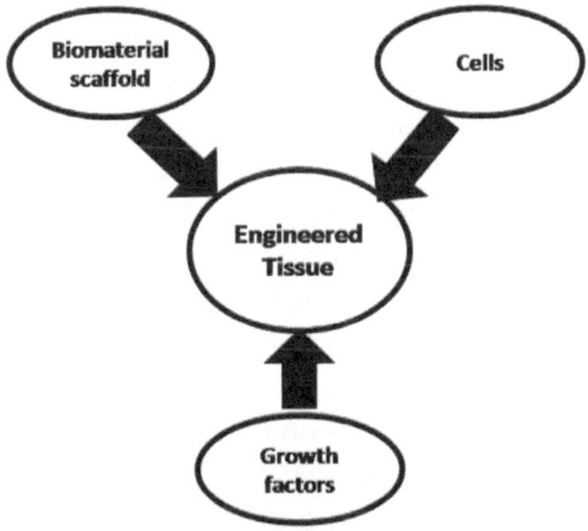

FIGURE 2.2 The Tissue engineering triad

2.3 INTERFACE BETWEEN NANOMEDICINE AND TISSUE ENGINEERING

Tissue engineering is another application of nanotechnology to medicine (nanomedicine). It helps to reproduce or replace damaged tissues by using suitable scaffolds made of nanomaterials and growth factors. Thus, tissue engineering is an area of nanotechnology where we focus on the construction of new tissues or organs for replacement. This effort may replace the existing conventional methods of treatment like organ transplantation or artificial implants.

The immune responses regarding these methods can be avoided by using autologous or isogenic cell sources in the scaffold constructs, by suppression of immune system of host, induction of tolerance in the host, or by immunomodulation of the tissue engineered construct (Fisher et al., 2007). The immune reactions also help in removing cellular debris caused by infections and injury that can lead to additional tissue damage. So to enhance regeneration, strategies have to be developed that exploit the beneficial aspects of the immune reactions while limiting the deleterious aspects. Appropriate application of technologies should be employed that have the potential to turn immune reactions to an asset for regeneration, differentiation, and more regenerative and less inflammatory.

2.4 SCAFFOLD AND TRANSPLANT

A transplant can be classified or defined in many ways; for example, it can be based on the relationship between the recipient and the donor, its location in a recipient, and so on. Different types of transplants include autograft (from one part of the body to other within one individual), isograft (within genetically identical species like identical twins), allograft (from different individuals of same species), and xenograft (from members of different species). The transplantation of organs or parts of organs has been considered as a conventional method for the curative treatment of end stage diseases of liver, kidney, heart, lung etc. Current methods of transplantation and reconstruction are time-consuming and involve very costly therapies like immunosupression therapy. The donor shortage is another problem apart from the serious side effects caused by the lifelong immunesupression therapies. Thus tissue engineering using scaffold biomaterials has taken an approach toward the replacement of lost tissue or organ function (Ratner, 2004).

An essential component in the tissue engineering approach is the carrier which is a highly porous artificial extracellular matrix (ECM), or scaffold which act as a template for tissue formation. This three dimensional (3D) scaffolds accommodate mammalian cells, regulate, and stimulate the cellular functions of adhesion, migration, growth, differentiation, and tissue organization and also guides their growth and regeneration in 3D. A scaffold in terms of both physical structure and chemical composition should mimic the structural and biological function of the native ECM as much as possible. It is due to the fact that the native ECM does not only just act as a physical support for cells but also provide a substrate with specific ligands and growth factors so that the cell can adhere and grow. Thus we can say that a tissue- engineered scaffold mimicking an ECM will surely play a similar role in promoting tissue regeneration in vitro as native ECM behaves in vivo (Ma et al., 2005).

The common source material for the natural scaffold is the ECM and there exist a practice of using these natural scaffolds as ECM for tissue engineering. These decellularized natural scaffolds derived from native tissues or complex organs by treating them with detergents to remove cellular material (decellularization) which can be then reseeded with healthy autologous or allogenic cells (Badylak et al., 2012). There are numerous examples of successful therapies based on natural ECM scaffolds but have some drawbacks also. These drawbacks can be challenged by using synthetic platforms with desired physical and biochemical properties of the natural ECM. The development and proper function of differentiated cells are contributed by a supportive 3D architecture of the scaffold.

Various biomaterials can be used as scaffolds to direct specific cell types to organize into 3D structures and to perform various differentiated functions of the targeted tissues. The most attractive scaffold materials are the synthetic bioresorbable polymers fully degradable into the body's natural metabolites under physiological conditions and which offer a possibility to create a completely natural organ or tissue equivalent. They overcome the issues like infection and fibrous tissue formation that are associated with permanent implants (Ratner, 2004). Thus a scaffold should be biocompatible with the cells and should guide the cell growth and hence the selection of a scaffold material is very important for a sovereign tissue engineered product. Most of the scaffolds are made from natural or synthetic polymers but in cases of certain hard tissues like bone and teeth, various ceramics like calcium phosphate compounds have been utilized (Park and Roderic, 2007). Natural polymeric materials used for the scaffold preparation includes collagen, chitosan, hyaluronic acid, silk fibroin, gelatin etc. and synthetic polymers include poly-lactic acid (PLA), polyethylene terephthalate (PET), poly-capro lactone (PCL), poly (lactic-co-glycolic acid (PLGA), poly (ethylene-co-vinylacetate (PEVA) etc.

2.5 BIOPOLYMER SCAFFOLDS FOR 3D CULTURE OF CELLS

Tissue engineering is a flourishing and promising biomedical engineering field which aims to develop viable substitutes to restore and maintain the function of damaged tissues. Culturing of cells in 3D micro-environments simulates normal cellular compartment and enhances adhesion, proliferation, and differentiation of cells than in 2D. Scaffolds can function as a delivery vehicle, a matrix for cell adhesion, and also serve as a mechanical barrier against infiltrating surrounding tissues which hampers tissue repair and regeneration. Scaffolds can be synthetic or natural in origin. Different types of scaffolds like porous, fibrous, customs, ECM, hydrogel, microspheres, and native tissue scaffolds are available and they influence the characteristics of cellular unit processes. A variety of techniques have been developed for fabrication of scaffolds. There are no scaffolds universally suitable for all cells and all applications. The following section of this chapter highlights the application of natural and synthetic biopolymers in 3D culture of cells, their fabrication techniques, and describes the biophysical, biochemical, and biomechanical properties of scaffold which influence cellular processes inside the scaffold.

2.6 BIOMATERIALS FOR TISSUE ENGINEERING

Biomaterials can be defined as materials that are natural or synthetic, and can be used therapeutically to repair, restore, or replace lost function. Such materials have been in use for decades, but the understanding of how the body interacts with implanted materials led to the progression of this field from the use of anything which was surgically available toward the use of materials which are deemed biocompatible. Biomaterials serve as 3D synthetic framework which is commonly referred to as scaffolds, matrices, or constructs and plays a very crucial role in tissue engineering.

Ceramics, synthetic polymers, and natural polymers are the three important groups of biomaterials that are used in fabrication of scaffolds with their own specific advantages and disadvantages. Thus the use of a composite scaffold comprising different phases is comparatively more (O'Brien, 2011). The biomaterials can be categorized into different sections as shown in Figure 2.3.

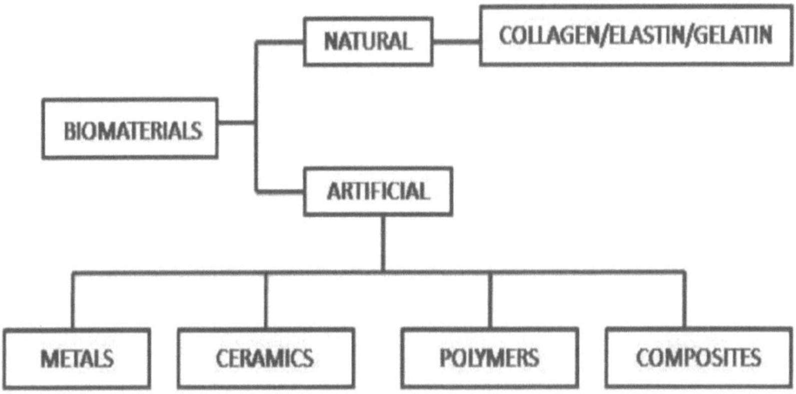

FIGURE 2.3 Classification of biomaterials

2.6.1 PROPERTIES OF A BIOMATERIAL

In order to function appropriately, the biomaterials have to satisfy certain properties like bulk properties and surface properties. The strength of the biomaterial implant can be determined by the bulk properties and the interaction with the biological system by the surface properties. To avoid failure of the implants, the properties of the biomaterial should be adjusted to its presumed function.

Bulk properties: The atomic organization as well as inter-atomic forces (ionic bonding, metallic bonding, and covalent bonding) that keep the atoms together determines the bulk properties of a material. Elastic modulus, yield stress, and ultimate stress are some properties that have to be considered from a mechanical point of view which determines the stiffness, strength, and deformability of the biomaterial. Fatigue that occurs through the cyclic stress is another important bulk property of a biomaterial.

Surface properties: Like intrinsic/bulk properties, surface properties are also important for the success of an implant. Energy, charge, release of ions, roughness, and composition are some of the factors which determine the surface characteristics of a biomaterial. The interaction between the material and biological constituents should not cause any deleterious effects to the surrounding cells, tissues, or organs and thus the material should not be antigenic, cytotoxic, carcinogenic, pyrogenic, or toxic to the living cells.

2.6.2 CHARACTERISTICS OF AN IDEAL SCAFFOLD

The selection of a scaffold is very important while considering the behavior of cells that has to be grown to form tissues or organs of specific dimensions. Some of the characteristics needed for an ideal scaffold are explained in the following sections.

2.6.2.1 POROSITY

The scaffold architecture should have highly interconnected pore structure and high porosity to ensure the cellular penetration and diffusion of nutrients to the cells within the scaffold and to the ECM formed by these cells. Porous nature also helps in the diffusion of waste products and the products of scaffold degradation to exit the body without interfering the surrounding tissues or organs (O'Brien, 2011). Porosity can be created intentionally by production process such as leaching of salt, sugar or starch crystals, sintering of beads, knitting, and weaving of fibers or it can also occur as a manufacturing artifact. Porous scaffolds are mainly used for artificial skin, blood vessels, drug delivery, bone and cartilage reconstruction, periodontal repair etc. each with fulfilling some specific requirements.

2.6.2.2 MECHANICAL STRENGTH

Mechanical strength is a key factor to consider in designing or determining the suitability of a scaffold in tissue engineering. The mechanical properties of the biomaterial used for making scaffold should match with that of the host tissue (Chan and Leong, 2008). For the creation of the scaffold, particularly in the construction of load-bearing hard tissues such as bone and cartilage that retains its structure even after implantation the mechanical strength is a very essential factor. The mechanical properties of the biomaterial should be adjusted to its proposed function to avoid failure. For example, for the fixation of a bone fracture it is necessary to have a required strength to avoid breakage. The intrinsic properties needed for a material from a mechanical point of view are elastic modulus, yield stress, and ultimate stress. These three properties determine the stiffness, deformability, and strength of a material. Another property is fatigue which is a process by which structures fail as a result of cyclic stress which is less than the ultimate tensile stress. Cyclic stresses are very common in human body in locations like pumping heart (artificial heart valves), connections of limbs (artificial hips) etc.

2.6.2.3 BIOCOMPATIBILITY

After the preparation of the required tissue-engineered construct, it should be successfully integrated into the living system and only after addressing the issues in the technologies for integration into the living system can achieve success in tissue engineering. This involves the issues of biocompatibility and immune acceptance (Lanza et al., 2011). Biocompatibility is the property of a material by which when it is introduced into a living body to perform a specific function, does not interfere or enhance the functioning of an organ and exerts neither local nor general toxicological actions in the body. After performing the functions it should get biodegraded and the products of the biodegradation should be fully eliminated and natural tissues should be regenerated in the place of the implant (Lipatova and Lipatova, 2000). In simple terms, biocompatibility can be defined as the ability for the performance of a medical device or material within an appropriate host response in a specific situation. The magnitude and duration of any adverse alterations in the homeotic mechanisms which determine the host response can be measured and is termed as biocompatibility assessment (Ratner, 2004). A schematic representation of the possible interactions between a biomaterial and biological components is shown in Figure 2.4.

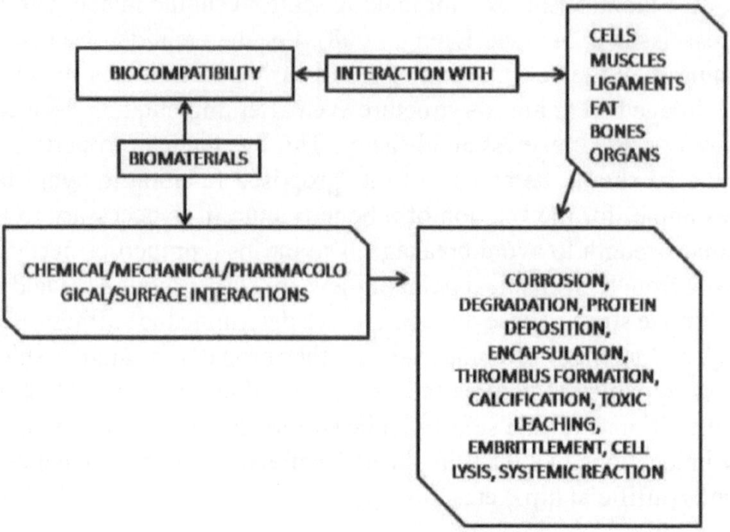

FIGURE 2.4 Interactions between a biomaterial and biological components

Potential biological hazards may include short-term effects, specific toxic effects, and long-term effects. Two perspectives for the in vivo assessment of the tissue compatibility are: first the tests for general biocompatibility of a biomaterial which is necessary for development of a biomaterial in further research and second is the biocompatibility tests for the final products, that is, a condition in which it is implanted. Various in vivo tests for determining the biocompatibility includes sensitization, irritation and intracutaneous reactivity, acute and sub-acute toxicity (systemic and sub chronic toxicity), genotoxicity, implantation, chronic toxicity, hemocompatibility, carcinogenicity, reproductive and developmental toxicity, biodegradation, and immune responses.

2.6.2.4 THE ROLE OF IMMUNE ACCEPTANCE

Tissue engineering of scaffold has become an important area in regenerative medicine and immune response is recognized as an important factor which influences regeneration. The immune reaction will start with an acute response to the injury followed by innate recognition of the foreign materials and a chronic immune response which involves specific recognition of the antigens like transplanted cells by adaptive immune response. All these together eventually leads to the rejection of the implants. The processes starting from transplantation of cells, implanting biomaterial scaffolds, or delivery of inductive factors may stimulate the immune reactions.

The biomaterial aims at creating a local environment for the growth of tissues but the injuries caused during the time of implantation and certain host inflammatory responses will have negative impacts in creating this environment. The repairing of these injuries can lead to fibrosis. Activation of complement proteins and cellular pattern recognition receptors (PRRs) initiate inflammatory cytokine and chemokine production which then leads to the recruitment of polymorphonuclear neutrophils (PMNs), monocytes, fibroblasts etc. to the injury sites. PMNs remove pathogen and cellular debris. They remove pathogen by phagocytosis, reactive oxygen species (ROS), and causes secondary damages to the surrounding tissues by various cytokines. Macrophages peaks for around one week and can persist at the injured site up to several months and can also secretes ROS and cytokines for the secondary damages. On the other hand, the presence of macrophages is necessary for the growth as they secrete growth factors and phagocyte cell debris (Boehler et al., 2011).

2.7 ENGINEERING BIOMATERIALS FOR TISSUE ENGINEERING

Different methods are available for engineering of biomaterials into a desirable form that is intended by the scaffold for tissue engineering and all of these techniques have advantages as well as drawbacks. Different types of techniques involved in the scaffold preparation are detailed as follows.

1. Solvent casting
   In this method a mold is prepared and this mold is then dipped into a polymer solution. Allowing sufficient time for the solvent evaporation and this will result in the formation of a layer of polymeric membrane. The main drawback of this technique is the toxicity caused by the solvent that is left after evaporation. Vacuum drying of the prepared scaffolds can overcome this to some extent but is very time-consuming.
2. Particulate leaching techniques

Similar to solvent casting method, in this technique also a mold is prepared but porogens like salt, wax, or sugars of desired size are added into the mold. The polymer solution is then added into this porogen-filled mold which is then evaporated and the salts are then leached out using water. The pore size can be controlled to some extent by controlling the size, shape, and amount of porogen taken for the experiment (Subia et al., 2010)

3. Gas foaming
   This technique uses gas as a porogen. A disc is prepared using desired polymer by compression molding and it is then placed in a chamber exposed to high pressure carbon dioxide for a few days during which pores are formed on the discs. Further, 3D porous structures are formed after the foaming process and the porosity can be further controlled by using salts or other porogens (Sachlos and Czernuszka, 2003).
4. Freeze drying
   It is based on the principle of sublimation and is used for preparation of porous scaffolds. Here a polymer is dissolved in an appropriate solvent and water is added into the polymer solution and mixed thoroughly to obtain an emulsion. Before separation of the phases, the emulsion is casted into a mold and frozen by immersing into liquid nitrogen and subsequently freeze dried to remove the solvent and water (Mandal and Kundu, 2009). This method does not use high temperature and no separate leaching step is needed but produces only low pore size and has long processing time.
5. Porogen leaching
   This method is performed by dispersing porogens like wax, salt, or sugar into either powdered materials or liquid particulates by the process of evaporation cross- linking etc. Scaffolds with up to 93% porosity can be produced using this method. It is very difficult to make scaffolds of accurate pore inter-connectivity and only wafers or membranes up to 3 mm thick can be produced by this method (Moore et al., 2004).
6. Fiber mesh (textile technologies)
   Individual fibers are either woven or interweaved into various 3D forms with variable pore sizes for scaffold fabrication (Martins et al., 2008). A polymer solution is poured over a nonwoven mesh of another polymer and is subsequently evaporated. Large surface

area and rapid diffusion of nutrient are some of the advantages while lack of structural stability is a drawback for this technique.

7. Phase separation
   It requires a temperature change that separates the polymer solution into two phases and also a solvent with low melting point and is easy to sublime. The temperature change will separate the polymer solution in the solvent into two separate phases. The solvent is later removed by extraction, evaporation, and sublimation resulting in a porous scaffold. An appropriate phase separation can also produce 3D fibrous structures with nanoscaled architecture.
8. Self-assembly
   When a disordered system forms an organized structure without any external direction, but by the local interaction between its own components, this can be termed as self-assembly. This is used for the production of 3D nanofiber structures. Hydrophobic and hydrophilic domains within the amphiphilic peptide sequence interact together due to the weak non-covalent bond and as the molecules come together they produces fast recovering hydrogel (Joshi et al., 2009; Zhang et al., 2005). Apart from peptides, synthetic polymer nanofibers with very thin diameter can also be formed by self-assembly methods. It is performed in aqueous salt or physiological solution and no solvents are used but the self-assembly process is a very elaborative process.
9. Electrospinning
   Electrospinning is a versatile technique producing continuous fibers of sub-micrometer-to-nanometer scale using electrostatic force. The polymer solution in a solvent is made to eject from a spinneret with high voltage to a collector with opposite or grounded charge forming a highly porous network after drying or solidification (Doshi and Reneker, 1993; Reneker and Chun, 1996). By varying the flow rate, distance between the needle and collector or the applied voltage, the scaffold architecture can be modified.
10. Rapid prototyping (RP)
   Controlling of porosity and pore size of a scaffold is a very difficult task and can be solved to a limit by using computer-assisted design and manufacturing techniques (CAD/CAM technologies). Using CAD software a 3D scaffold can be designed and using various algorithms in this software the porosity can be tailored (Melchels

et al., 2011). Fused deposition modeling (FDM), selective laser sintering (SLS), 3-D printing, or stereo lithography are some of the RP techniques used for the construction of 3D objects in a layer-by-layer layer method. RP technique can control the matrix architecture, mechanical property, biological effects etc. of scaffolds. Low resolution is a drawback of this technique.

11. Melt molding
    A teflon mold is filled with PLGA powder and gelatin microspheres (with specific diameter) and the mold is heated above the glass transition temperature of PLGA applying high pressure to the mixture making the PLGA particle to attach together (Thompson et al., 1995). The mold is then removed and the gelatin is then washed off and the scaffold is dried. This process can be modified to incorporate hydroxyapatite fibers.
12. Membrane lamination
    This method can be used for constructing 3D polymeric foam scaffolds that are biodegradable and have precise anatomical shapes. This uses a layer-by-layer fabrication process and generates porous 3D polymer foams with defined anatomical shapes using computer-assisted mechanisms. Disadvantage of this technique is lesser pore interconnectivity due to layering of porous sheets (Hutmacher, 2001). Only thin membranes are used and thus it is a time-consuming process.
13. Laser-assisted bioprinting (LaBP)
    Multicellular 3D patterns are made in natural matrix using laser-assisted bioprinting (LaBP). A fully cellularized substitute can be prepared using LaBP technique in an exact 3D spatial conformation by setting living suspensions of small cell volumes in patterns of high resolution. The generated tissue constructs may be studied in vivo (Lai et al., 2011). Matriderm® is an example of skin substitute successfully tried in mice (Michael et al., 2013).

2.8 COMMONLY USED BIOMATERIALS FOR SCAFFOLD FABRICATION

Different types of materials like natural, synthetic, biodegradable, or permanent materials have been investigated for the purpose of scaffolding in tissue engineering. Most of these materials have been already in use even before the advent of tissue engineering in various applications. Scaffolds should be absorbed by the surrounding tissues rather than surgical removal and thus degradability of scaffold is an essential factor, but the rate of degradation of scaffold should coincide with the rate of tissue formation. The problems related to long-term safety of permanently implanted structures can be circumvented by the use of degradable materials. Unlike non-degradable materials, degradable materials should fulfill requirements that are more stringent. Toxicity of the contaminants that leach out from the implants, subsequent metabolites, and degradation products must be taken into consideration and so only a few numbers of starting materials have been implied for the preparation of degradable biomaterials. As of 1999, the Food and Drug Administration (FDA) has approved only five synthetic degradable polymers for clinical application: PLA, PGA, polydioxanone, poly (caprolactone) (PCL), and poly (PCPP-SA anhydride). Some of the commonly used materials are explained in the following sections.

2.8.1 SILK

Silk is a fibrous protein derived from the silkworm, Bombyx mori. It is a natural biomaterial tough and strong, spun by insects and spiders. Different types of silk fibers are available and are composed of peptide molecules conferring distinct mechanical properties (Hinman et al., 2000). Silk fibroin is a potential natural biomaterial for preparing nanofibrous scaffolds. The in vitro biocompatibility of the silk nanofiber is proved by Min et al., 2004, and the fiber diameter, high porosity together with the cytocompatibility makes the silk nanofibers a suitable candidate material for scaffolding technology. Cell culture studies have shown a slow degradation of silk scaffolds in around four weeks as a result of proteolysis (Taddei et al., 2006). Silks can be either modified by using various chemical treatments or can be used with other materials for varying its mechanical properties and surface chemistry. The degradation behavior also can be modified according to the applications. Silk fibroin films and silk fibroin alginates are used in wound healing applications (Kearns et al., 2008).

More recently, spider silk has been explored in the field of both bone and cartilage tissue engineering. Silk, after the extraction of cericin is wound into strands and yarns and have been investigated for the ligament tissue engineering. Silk fibroin films that are heparinized and sulfonated show suitable mechanical properties for the tissue engineering of artificial blood vessels. Silkworm silk has also shown potential in liver tissue engineering and also supports Schwann cells and dorsal root ganglia and thus can be used as a potential scaffold for nerve tissue engineering (Cirillo et al., 2004; Yang et al., 2007). Thus, silk nanofibers can be a promising biomaterial for the tissue engineering applications.

2.8.2 COLLAGEN

Collagen is the most abundant protein present in the mammalian ECM and this natural polymer has been in use for a variety of tissue engineering applications. Collagen along with other proteins can be electrospun into nanofibers which resembles native state and can be used in tissue engineering applications (Jha et al., 2011). Collagen molecules have a triple helical structure and exist in a fibrillar form with elaborate 3D arrays in the ECM.

Collagen fibrils and their networks function as ECM (highly organized 3D architecture surrounding the cells) for most of the soft and hard connective tissues like bone, tendon, blood vessels, skin, cornea etc. Collagen encourages cellular growth and modifies the morphology, migration, adhesion, and differentiation of cells. Compared to other proteins, collagen is weakly antigenic and the immune reactions always depend on the species to which the tissue is implanted and the site of implantation. It is a biodegradable protein and the biodegradability can be either extended or completely suppressed by modifications. Peptide–amphiphile nanofibers produced by the process of self-assembly have been used for the preparation of hybrid bone implants. Nanofibrillar gels prepared were used to support neuronal cell attachment and differentiation of liver progenitor cells and also for brain repair. Skin substitutes based on cell seeding on 3D scaffolds of bovine collagen type I is a commercialized product. The interactions between cells and collagen have opened a great potential of using collagen with biocompatibility and controlled biodegradability in the field of tissue engineering (Chevallay and Herbage, 2000).

2.8.3 POLY LACTIC ACID (PLA)

PLA is a commonly used synthetic biomaterial. It is polyester and degrades within the human body to form lactic acid which can be easily removed from the body as it is a commonly occurring chemical in the body. According to Vert et al., (1991), the polymeric backbone of PLA is chemically degraded by simple hydrolysis and independent to any biological method and hence this cannot be described as biodegradation. PLA is a hydrophobic polymer and this hydrophobicity reduces the water uptake and thus reduces backbone hydrolysis also. There exist four distinct morphological polymers for PLA: D-PLA, L-PLA, racemic PLA, and meso-PLA due to the chiral nature of the lactic acid.

PLA is considered safe and non-toxic and its biocompatibility is proven by various regulatory agencies and has been used in a large number of clinically successful medical implants. Thus, devices prepared from PLA are more easily brought into market. A major drawback of PLA is that they are poor substrates for cell growth in vitro. Other factor is that the degraded product, lactic acid which is a relatively strong acid may get accumulated in the implant sites and delayed inflammatory response happens after six months to one year. Apart from all these, many porous scaffolds have been in use for tissue engineering application based on PLA and its copolymers (Netti, 2014).

Polyglycolide (PGA), which is a class of poly (α-esters), has been used to engineer tendons and cornea stroma in animal models. PGA fibers were also used to repair defects of cartilage tissues but a major drawback with PGA is its acidic degradation that results in unfavorable host response. Coral, calcium alginate, and demineralized bone matrix (DBM) are also successfully used as biomaterials for repairing bone tissues.

2.8.4 BIONANOMATERIALS IN TISSUE ENGINEERING

Bionanomaterials are any nanomaterial of biological origin. A myriad of biomedical applications have been associated with these bionanomaterials due to the advancement in their fields. Some of the bionanomaterials used in tissue engineering are described in the following sections.

2.8.4.1 NANOCHITOSAN/NANOCHITIN

Chitin (poly-β-(1-4)-N-acetyl-D-glucosamine), is a natural, renewable, and biodegradable polysaccharide and the rate of degradation is highly dependent on the molecular mass of the polysaccharide. Chitosan is an important derivative of chitin prepared by the partial deacetylation under alkaline condition. Both chitin and chitosan are non-toxic polymers and have found several biomedical applications in wound healing and tissue engineering.

Scaffolds prepared from chitin have been used in the regeneration of cartilage, bone, and tendon tissues (Wan and Taj, 2013). In order to use chitin in tissue engineering, it is needed to modulate its physical or biochemical properties for a better interaction in the biological environment. Nano-sized chitin/chitosan possesses a high performance because of its high surface area, quantum size effect, and small size compared to the traditional micro-sized chitin/chitosan materials (Qi and Xu, 2004). It can be prepared by many methods like coagulation, covalent cross-linking, precipitation, ionic cross-linking etc. (Berthold et al., 1996). Preparation of porous chitosan nanofibers ranging from nanometer to microns in diameter was performed by electrospinning technique and found many applications in tissue engineering. Blending of chitosan with other polymer makes an easier method for the formation of chitosan (nano) fibers. A wide range of experiments has been performed in case of chitin/chitosan as a biomaterial for tissue engineering but the use of nanosized chitin/chitosan needs further support for it to get established in the field of tissue engineering.

2.8.4.2 NANOCELLULOSE

Cellulose is a ubiquitous structural polymer which confers its mechanical properties to plant cells and is the most abundant organic polymer available on earth. It is an important structural component of cell wall of plants, some algae, bacteria etc. Nanocellulose or microfibrillated cellulose (MFC) is composed of nanosized cellulose fibrils having a high aspect ratio. Nanocellulose can be obtained by an acid hydrolysis of native fibers resulting in highly crystalline rigid nanoparticles (referred to as nanowhiskers) or nanocrystalline cellulose (NCC) (Peng et al., 2011). The purity, web-like nature, long fiber length, high degree of crystallinity, and the nanoscale fibril dimensions has made bacterial cellulose superior to plant-derived cellulose in tissue engineering. Based on the bacterial cellulose (BC) synthesized by Gluconacetobactor xylinus, a skin tissue repair material has been prepared. Nano-composites of BC and chitosan with a cohesive gel structure have showed excellent results promoting the healing of epithelial tissues and reduced inflammation (Fu et al., 2013).

Bacterial nanocellulose (BNC), being a novel non-degradable, biocompatible, and functionally competent material has been used recently in skin, bone, vascular, and cartilage tissue engineering. Nanocrystalline cellulose (NCC) formed from the acid-catalyzed degradation of cellulosic materials are able to create highly porous silica films and carbon films with chiral nematic organization and thus will have promising applications in the near future. Nanocellulose have been used to prepare 3D macroporous nanocellulose scaffolds by biofabrication technique using porogens and have shown ability to attract endothelial cells, chondrocytes, smooth muscle cells of various origin, urethral cells, as well as osteoprogenitor cells. By using bioprinting techniques, 3D porous nanocellulose scaffolds with large size, unique architecture, and with various surface modifications have been prepared to enhance the cell attachment and differentiation. A myriad of exciting new-cellulosic materials have been developed with nanoscale fibrillar structures having promising applications in the growing field of tissue engineering. On the other hand, the development of advanced biomedical applications for cellulose is still in its infancy (Dugan et al., 2013).

2.8.5 CARBON-BASED NANOMATERIALS

Carbon-based nanomaterials like graphene (G), carbon nanotube (CNT) etc. possess unique mechanical, electrical, and optical properties that present much interest in research fields like tissue engineering. They provide a similar microenvironment like biological ECM, both in terms of physical structure and chemical composition and thus a potential candidate for the development of artificial scaffolds. Typical graphene and CNT structures are shown in Figure 2.5.

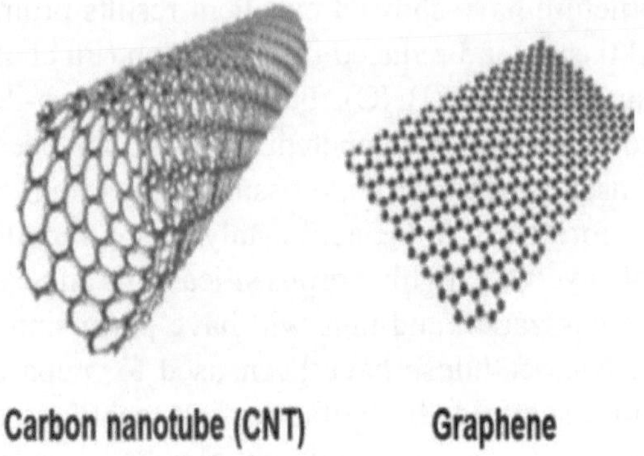

FIGURE 2.5 Structure of graphene and CNT

2.8.5.1 GRAPHENE

Graphene (G) is a 2D carbon crystal sheet of molecular thickness and composed of sp^2 hybrid bonded carbon. It poses a number of diverse and exceptional properties like optical, mechanical, electrical, and thermal qualities. With the rapid development of synthesis and functionalization approaches, graphene as well as its related derivatives have shown exceptional potential not only as physicochemical material but also are extended to biological uses. There is an intensive area of research focusing on the bio-applications of graphene and its related derivatives such as graphene oxide (GO) due to many intriguing properties (Guo and Dong, 2011).

Different types of osteoblast cells have been tested with graphene in the view of bone regeneration and bone treatment scaffolds. Graphene layers modified with artificial peroxidases and laminins have promoted cell adhesion and pseudopodial cell configuration, whereas coating graphene sheets with laminins alone showed eight times more proliferation than those with the pristine graphene (Guo et al., 2010). The high electrical conductivity of graphene could be used for modulating the behavior of neural cells or neural differentiation. The electrochemically active transduction by graphene helps in the bioelectrical signal transmissions in neurons which is essential for the neural activity. Graphene can also be used as electrode materials in neural prosthetic devices (Park et al., 2011).

Graphene polymer composites, Graphene hydrogel etc. were also used for the cell modulation along with structural scaffolding functions. The studies so far demonstrated the ability of graphene-based materials in supporting tissue engineering and other biomedical fields and this trend is believed to continue in the coming years.

2.8.5.2 CARBON NANOTUBE (CNT)

CNTs are graphite sheets that have been rolled into cylindrical tubes with length in nanometer or micrometer range and a diameter in nanoscale. CNT-based materials have high electrical conductivity, chemical stability, and physical strength with structural flexibility and thus they have been studied for neural cell adhesion and proliferation. The activities of the CNT surface can be altered by its chemical functionalization. The similarity of CNTs to that of ECM components makes it a potential candidate as a scaffold material for tissue engineering.

The CNT-polymer composites from the homogenous suspension of polymers and CNTs have been used to fabricate polymeric scaffolds for bone tissue engineering. Some examples include multi-walled carbon nanotubes (MWNTs)/polycaprolactone, ultra-short single-walled nanotubes (SWNTs)/poly (propylene fumarate) (PPF) etc. CNTs incorporated into polymeric nanofibers (both synthetic and natural) improved mechanical strength, thermal stability, and electroactivity. The CNT integration will increase the tensile strength and Young's modulus and decrease elongation at break. Several types of cells such as osteoblast, fibroblast, skeletal myoblast, and mesenchymal stem cells (MSCs) grew well on polymer/CNT nanofibrous scaffolds (Harrison and Atala, 2007)

CNT-inorganic composites and CNT-polymer-inorganic hybrid materials have been formulated and have shown positive results in tissue engineering. Even though the literature suggests that the surface functionalization of CNT can attenuate its toxicity in vivo, long term safety periods longer than six months have not been studied and thus it is very important to investigate its potential health risks. In spite of all these issues and challenges carbon-based materials are promising substrates for tissue engineering scaffolds because of their unique properties (Ku et al., 2013).

2.8.6 VARIOUS OTHER NANOPARTICLES USED IN TISSUE ENGINEERING

Other nanomaterials like ceramic, alumina or titania, and their composites have been used in the design of tissue engineering scaffolds, especially for bone and dental applications.

2.8.6.1 CALCIUM PHOSPHATE NANOPARTICLES

Calcium phosphate (CaP) is a major component of bone and is studied extensively as scaffolds for bone tissue engineering. Different types of CaPs exists but majority of the research is focused on hydroxyapatite (HA), β-tricalcium phosphate (β-TCP), or mixture of HA and β-TCP also known as biphasic calcium phosphate (BCP). CaPs are biocompatible, osteoconductive, and have the ability to bond directly to bone. Addition of dopants like SiO_2 and ZnO into CaP can control the dissolution rates, densification behavior, biocompatibility, mechanical strength, increased compressive strength, and cell viability (Fielding et al., 2012). Various types of calcium phosphate nanoparticles are involved in tissue engineering. However, HA is one of the most important and is explained in the following section.

2.8.6.1.1 HYDROXY APATITE NANOPARTICLES

Hydroxylapatite or hydroxyapatite (HA/HAp) is a mineral form of calcium apatite naturally occurring with the formula $Ca5(PO_4)_3$, usually written as $Ca_{10}(PO_4)_6(OH)_2$. They have the ability to integrate with the bone structures and support bone ingrowth and osseointegration without any local or systemic toxicity and inflammation thus proving its bioactive nature. Dense HA does not have the mechanical strength for long term load bearing applications. They also significantly increase the bioactivity and biocompatibility of the man-made materials. The bone tissue is a natural composite of HA nanocrystals embedded in collagen fibrils. Thus the incorporation of these HA nanofillers in a polymer matrix will mimic the structure of human bone. HAp can be used as a model component to study biomineralization in human body and is a choice for various biomedical applications like replacements for bone and periodontal defects, dental materials, bioactive coatings for osseous implants etc.

As compared with the conventional HAp, nanophase minerals showed improved cytophilicity and greater cell viability and proliferation and have stimulated great interest in tissue engineering application. A 3D scaffold made out of porous hydroxyapatite with interconnected pores have been developed by foam-gel technique and successfully showed osteoconduction with majority of pores filled with newly formed bone (Yoshikawa and Myoui, 2005). Hydroxyapatite macrochanneled porous scaffolds produced using polymer sponge templating method using reactive submicrometer powder, with optimized mechanical strength have shown good results for tissue engineering. Composites made with other materials like chitosan, collagen, and other polymers will reinforce the matrix along with osteoinduction.

2.8.6.2 ZINC OXIDE NANOPARTICLES

ZnO is a conventional semiconductor with a wide band gap and has been explored widely in multiple areas of science and is considered as a safe material by the FDA. The high stability, photo-luminescent properties, wide band gap semiconductor properties, absorption of ultraviolet UV radiation, and optical transparency of ZnO nanoparticles have gained applications in wide areas like photo-catalytic applications, biomedical applications etc. The applicability of ZnO nanoparticles in regenerative medicine and tissue engineering significantly increased the research in this area.

ZnO nanoflowers have shown induced proliferation and migration of endothelial cells leading to the formation of new blood vessels (Kumara Barui et., al., 2012). In another report ZnO nanoparticles with β-chitin hydrogel bandages showed antimicrobial and wound healing application with a good biocompatibility to human dermal fibroblast cells (Kumar et al., 2013). Scaffolds made up of electrospun polycaprolactone (PCL) with ZnO nanoparticles when used as a skin substitute have shown an enhanced rate of wound healing without any scar formation (Augustine et al., 2014b). They also demonstrated that ZnO nanoparticles can act as the key regulators of angiogenesis in the scaffolds in redox signaling mechanism (Augustine et al., 2014c). These reports have explained the significance of ZnO nanoparticles in tissue engineering and wound healing.

Nanofiber meshes prepared by sodium alginate/poly (vinyl alcohol) with different concentration of ZnO nanoparticles when cultured with mouse fibroblast showed good adhesion and spreading of fibroblast and also confirmed the antibacterial activity of the nanofibers improved by the increased concentration of ZnO nanoparticles. Electrospun membranes of polycaprolactone with varying concentration of ZnO nanoparticles have shown excellent fibroblast cell attachment and proliferation. This proved the efficiency of polycaprolactone/ZnO nanocomposites in tissue engineering applications like the regeneration of damaged skin where rigorous cell proliferation and antimicrobial properties are essential (Augustine et al., 2014a).

2.8.6.3 TITANIUM DIOXIDE NANOPARTICLES

Metals having a high metallic strength and exceptional fatigue resistance like titanium (Ti) have been widely used to produce porous metallic scaffolds. Titanium dioxide (TiO_2) has been studied extensively for bone replacement material due to its light weight and resistance toward corrosion. Bioactivity of the scaffolds can be increased by proper surface modification of these scaffolds (Das et al., 2008). The bio-inert surface of the TiO_2 makes the chemical bonding between the skeletal bones and the implant surface difficult.

The loose and powdery nature of TiO_2 nanoparticle makes it difficult to be used in scaffold; hence, modifications such as blending of TiO_2 nanoparticles with synthetic polymer are performed. TiO_2 NPs have been used as filler materials for biodegradable polymer matrices. Nanocrystals of TiO_2 with grain size<100 nm have showed the ability to stimulate nanometer surface topography and roughness in osseous tissues. Gerhardt et al., (2007) prepared poly (D,L lactic acid) composite films with different compositions of TiO_2 NPs. This showed an increased surface roughness in the films and an improved adhesion of osteoblast cells. Jayakumar et al., (2011) prepared a chitin-chitosan/TiO_2 NPs composite scaffolds for bone tissue engineering where the addition of TiO_2NPs decreases the pore size of the scaffold. Kim et al., (2014) prepared scaffolds from silk fibroin incorporated with TiO_2NPs resulting in a porous scaffold. The TiO_2 incorporation resulted in a decrease in pore size and swelling behavior and improvement in the mechanical property of the scaffolds. Thus TiO_2 can be used as an efficient filler material for the design of scaffolds for tissue engineering but a better result can be obtained with a 3D structure with inter-connective pores.

2.9 NANOFABRICATION TECHNOLOGIES IN TISSUE ENGINEERING

For the successful formation of a tissue-engineered construct, it is very important to consider about the scaffolds which serves as the mechanical platform for the adhesion, proliferation, differentiation etc. of the cells. There are various conventional methods for the fabrication of scaffolds, though there are some important methods too for the fabrication of scaffolds and are explained in the following sections.

2.9.1 ELECTROSPINNING

Electrospinning is an approach to prepare nanofibrous networks and is a cost- effective method for the fabrication of micro-to-nanometer scale diameter fibers, with very high specific surface area (Liang et al., 2007). Applications of electrospinning include tissue engineering scaffolds, catalytic nanofibers, filtration membranes, and fiber-based sensors.

A simple and inexpensive nature of the experimental set up is an attractive feature of the electrospinning. It consists of a syringe pump, a voltage source and a collector. During the process, a polymeric solution in the syringe is held at the tip of a needle due to surface tension. An electric field is applied using high voltage sources which provide a charge to the solution. With the increase in the electrical potential the solution overcomes the surface tension and forms a jet that is ejected out from the tip of the capillary tube or a syringe and gradually thins due to solvent evaporation and elongation and forms randomly oriented nanofibers which are collected on a stationary or rotating collector (Vasita and Katti, 2006). This process can be used successfully to spin synthetic or natural polymers into fibers of many kilometers in length. A schematic representation of electrospinning set up is shown in Figure 2.6.

The typical electrospinning set up can be modified to produce fibers with unique morphologies. Co-axial two capillary spinneret can be used to electrospin hollow nanofibers, and for aligned nanofibers a rotating drum collector can be used.

The process of electrospinning can be controlled or manipulated by many variables such as those as listed hereunder.

1. Solution properties: This includes viscosity, surface tension, polymer molecular weight, dipole moment, conductivity, dielectric

constant etc. Varying of any one of these parameters will affect the other and so the effect of these properties cannot be isolated.

2. Controlled variables: This includes flow rate, field strength of the electric field, design of the needle tip, distance between the tip and the collector, composition and geometry of the collector etc.
3. Ambient parameters: This involves velocity of the air, temperature, and humidity.

The electrospinning technique controls the thickness, composition, and also porosity of nanofibers with a simple experimental set up. The electrospun nanofiber with a high porosity and surface area allows favorable cell interactions and hence becomes potential candidates for tissue engineering application. Electrospinning provides a simple and cost-effective method to produce scaffolds with interconnecting pores and submicron range compared to other techniques like phase separation and self-assembly. It has been used for the preparation of 3D scaffolds using natural polymers, synthetic polymers, composite of both natural and synthetic polymers etc. Functionalized scaffolds for increasing the biocompatibility can also be prepared. However, in spite of the comprehensive experimental and theoretical studies explaining the ability to control fiber formation, fiber diameter uniformity is still a problem that needs to be addressed. Control of fiber morphology is necessary for the improved scaffold design that recreates functions of native ECM. For tissue engineering applications, designer scaffolds with dimensions that are clinically relevant and that support a homogeneous distribution of cells need to be addressed (Pham et al. 2006).

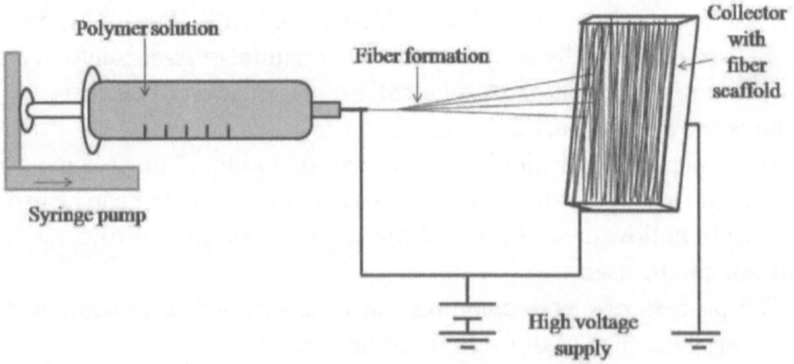

FIGURE 2.6 Schematic representation of electrospinning set up

2.9.2 NANOLITHOGRAPHY

The fabrication of nanometer-scale structures or patterns with at least one lateral dimension in nanometer scale (between the size of an atom and approximately 100 nm) is termed as nanolithography. Many techniques like X-ray, e-beam, imprint etching, and scanning probe lithographies have been used in patterning surfaces for tissue engineering (Laurencin and Nair, 2008). Various instruments used in nanolithography include scanning probe microscope (SPM), atomic force microscope (AFM) etc. for printing and etching in a single atom dimension on the surface. Nanolithography can be utilized in the semiconductor fabrication like integrated circuits, nanoelectromechanical systems, and other scientific fields in nanoresearch.

In electron beam lithography, highly focused beam of electrons are scanned over the surface of the substrate and with the help of a design editor and pattern generator these electron beams are guided. The surface contains an electron beam sensitive resist which generates a resist mask which is then used for the transfer of nanopattern. Dip-pen lithography (DPN) is another kind of lithography in which an AFM is used for the patterning even below 100 nm level. Here, the tip or cantilever of the AFM coated with a chemical compound or mixture (ink) acts as the pen and kept in a substrate contact. DPN emerges as a potential tool for manipulating cells at subcellular level resolutions. Activities like cell adhesion, patterning of subcellular ECM proteins, cell sorting etc. can be performed using DPN (Pulsipher and Yousaf, 2010).

Nanoimprinting lithography (NIL) is another kind of nanolithography used for fabrication of nanoscale patterns. The imprint resist used here is a formulation of monomer or polymer that can be cured by using UV light or heat. Many reports exist that explains the application of NIL in tissue engineering. Guillemette et al. have studied the effect of the surface topography and the interaction between cell-cell and cell-ECM interactions in cultured tissue by patterning polystyrene and a thermoplastic elastomer (TPE) using NIL and replication molding (replication molding is a variation of NIL) (Guillemette et al., 2009). Another work by Matsuzaka et al. (2003) fabricated polyester gratings using replication molding and studied the growth of osteoblast cells. Further, poly (methyl methacrylate) (PMMA) was patterned by NIL by another group for studying the reaction of neuronal process toward the grooves and ridges of the patterns (Johansson et al., 2006). Apart from the 1D gratings and groove structures,

2D pillar or hole structures were also studied and found that even similar features like circular holes or pillars arranged in different patterns like circular or hexagonal array have shown different cell responses toward it. Because of its capability for patterning substrates that are interesting for tissue engineering and also low cost and high throughput realization, NIL has a promising future in tissue engineering. The schematic representation of nanoimprinting lithography is shown in Figure 2.7.

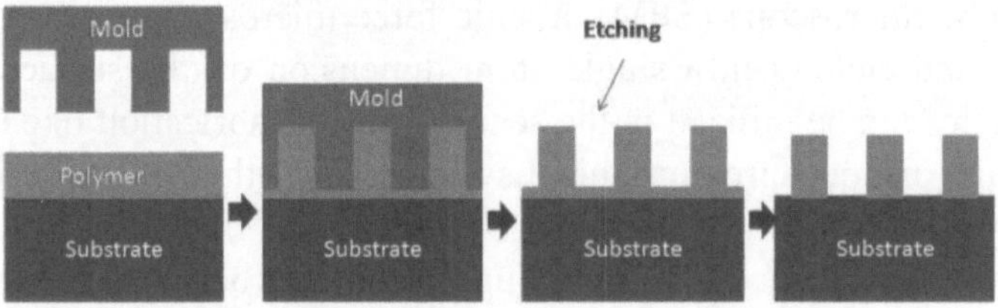

FIGURE 2.7 Schematic representation of nanoimprinting lithography

2.9.3 NANOPRINTING

Nanoprinting is another kind of nanofabrication technology for fabrication of 3D objects which is depicted in Figure 2.8. Further, 3D nanoprinting is a unique benefit of NIL that supports patterning of 3D structures. With this technique of 3D nanoprinting, the products and structures of our need can be constructed independent of the complexity of its shape. The concept of 3D printing at the nanoscale level will have various advantages like less wastage, economic viability, speed etc. (Li et. al., 2001).

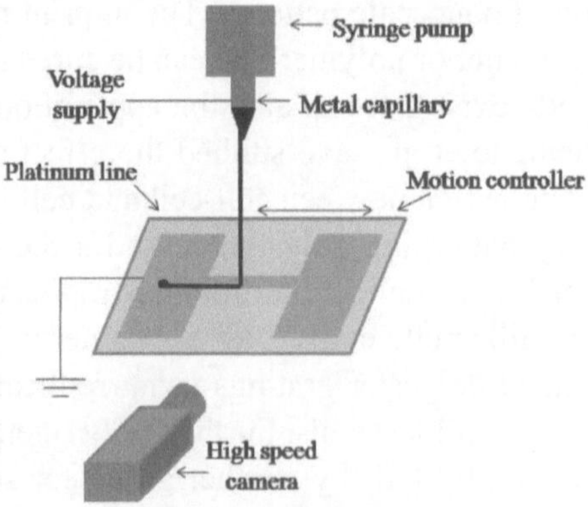

FIGURE 2.8 Schematic representation of nanoprinting

2.10 FUNCTIONALIZATION OF SCAFFOLDS WITH BIOMOLECULES FOR VARIOUS TYPES OF TISSUE ENGINEERING APPLICATIONS

Tissue engineering involves use of porous scaffold that can provide ambient conditions for the growth of target cells that are intended to grow inside or surface of the scaffold. Such growth inside the scaffold is mostly possible only when we follow a tissue engineering triad involving appropriate cells, relevant signaling molecules or biomolecules, and a proper porous scaffold. An effective cell adhesion as well as the growth and retention of differentiated cell's function in a scaffold depend on many factors such as bio-mimetic surface, oxygen tension, growth factor, immobilization or incorporation method of growth factor, controlled combinatorial activity of key signaling molecules or growth factors from scaffold or biomaterial, hydrophilicity of scaffold etc.

Scaffolds are having different surface properties like hydrophilicity or presence of functional groups on the surface which play a key role in cell adhesion, proliferation etc. For achieving an enhanced growth and regeneration in terms of cell attachment, cell proliferation, and matrix secretion, it is important to optimize the cell-biomaterial interactions and this can be achieved by the physical and chemical modification of the scaffolds. One such attempt is the modification with various cell surface receptors like integrin. Integrin receptors have arginine-glycine-aspartic acid (RGD) peptide sequence and mediate cell-matrix interactions and have been used in various studies for enhanced cell attachment (Orlando and Cheresh, 1991). PEG scaffolds with RGD domains also can direct cell regulation and proliferation. Another method is functionalizing the scaffolds with various functional groups like phosphates, amides, sulfonates etc. (Kuo et al., 2010). Addition of collagen-platelet composite (CPC) to a suture in a porcine model has shown an enhancement in suture repair via increased cellularity within the region of healing. Another example is the modification of silk fibroin performed by blending it with hyaluronan. Poly (sodium styrene sulfonate) (PNaSS) has been used as a functionalization agent for PET scaffolds because of the increased adherence shown by fibroblasts onto the surface compared with nonfunctionalized fibers (Ghasemi-Mobarakeh et al., 2010; Zhou et al., 2007). Figure 2.9 depicts the different types of scaffold modification.

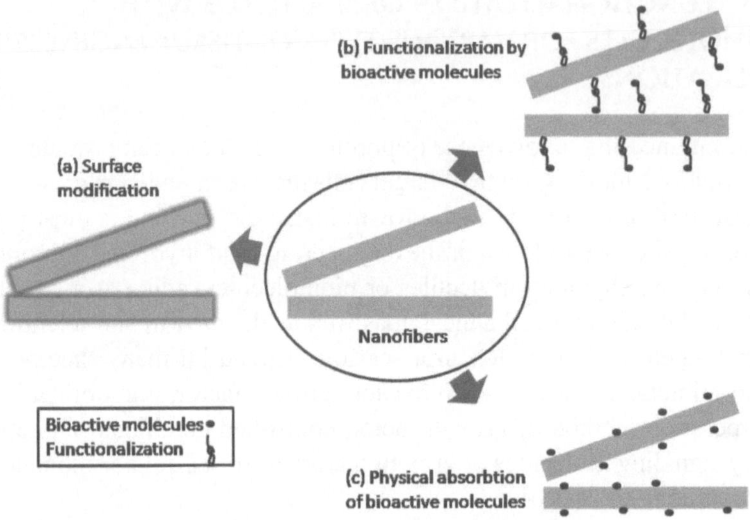

FIGURE 2.9 Different types of scaffold modification

2.11 3-D ARCHITECTURE AND CELL INCORPORATION

Tissue-engineered products traditionally involve seeding of cells on the scaffold, a structure that is capable of supporting 3D tissue organization and development. The scaffold which is either synthetic or biological or both in origin can be conjugated to bioactive materials like ECM or growth factors. But all type of materials available cannot be used for the preparation of 3D scaffolds because of their difference in chemical and physical properties and processability. The cell morphology is highly correlated with the cellular activities and the function and proliferation of cells are favored by strong cell adhesion and spreading. Maintenance of cell polarity is another important factor especially in case of epithelial cells and can be provided by a heterogeneous extracellular environment. In addition to cell morphology, functions of many organs are dependent on the 3D spatial relationship of cells with its ECM, for example, the relation between the shape of the skeletal tissues which is important for its function.

The regulation of gene expression is regulated differentially in 2D and 3D culture substrates. A suitable scaffold is critical that act as a template to direct the cell growth and ECM formation and development of 3D structure (Ratner, 2004). Obtaining uniform cell seeding at high densities and maintaining nutrient transport to cells in the scaffold are the major obstacles in the in vitro development of 3D cell-polymeric constructs. The hydrodynamic and biochemical factors in the cell environment should be controlled and is necessary to achieve a desired spatial and temporal distribution of cells. The scaffolds produced can be then placed in suitable culture systems like static culture, spinner flask culture, rotary vessel culture, perfusion culture etc. for seeding and growth.

As compared to the conventional approach of diffusion-based chemical modification of the cellular environment (i.e., by adding growth factors directly into the tissue culture media), the 3D physical microenvironment is a better option as a tool to control cell differentiation and significantly improved spatial control and reproducibility (Willerth and Sakiyama-Elbert, 2008). The growth factors needed for the growth and differentiation of the cells can also be incorporated into the 3D scaffolds so that it can be released in a controlled manner and there is no need for the addition of these factors to the culture media itself. This is an advantage when we consider the implantation of the scaffolds into the native environment where they have to support cell differentiation over a repair period in vivo. This acts in a way endogenous factors act in the ECM during development. The goal of creating 3D scaffolds for tissue engineering is to increase the ability to direct stem cell differentiation along specific paths and simultaneously improve the scalability of the cell production to large capacity needed for the clinical applications.

2.11.1 ROLE OF STEM CELLS IN TISSUE ENGINEERING

Tissue engineering involves the combination of living cells with a natural/synthetic support or scaffold to build a 3D living construct and the development of such a construct requires cautious selection of four key factors: scaffold, growth factors, ECM, and cells. The cells selected for tissue engineering should provide a long lasting repair of the damaged tissues. A sufficient number of cells must be produced to fill the defect and (1) they should differentiate into desired phenotypes, (2) must take up a 3D structural support/scaffold and produce ECM, (3) should be structurally and mechanically compliant with the native cells, (4) should be able to integrate with the native cells and must overcome immunological rejection, and (5) should be allied with minimal biological risks (Vats et al., 2002).

2.11.2 STEM CELLS

Stem cells are undifferentiated cells that can differentiate into specialized cells and divide (via mitosis) to produce more stem cells. They have a remarkable potential for developing into different cell types in the body during early life and growth and also serve as an internal repair system by dividing limitlessly and replenishing other cells till the person/animal is alive. When a stem cell divides, each new cell has the potential to remain as stem cell or some other type of cell with a more specialized function like a brain cell, red blood cell etc. Two important characteristics of stem cells are: (1) they are unspecialized cells that are capable of renewing themselves by frequent cell divisions even after a long period of inactivity and (2) they can be induced to form tissue-specific or organ-specific cells with specialized functions under special conditions.

Stem cells can be broadly divided into three types: (1) embryonic stem cells (grown in laboratory from cells of early embryo), (2) adult/tissue stem cells (found in our body for whole life), and (3) induced pluripotent stem cells/reprogrammed stem cells (similar to embryonic stem cells but formed from adult cells/specialized cells). Primary cells taken from the patient have been used in conjugation with scaffolds to produce tissues for re-implantation but the invasive nature of the cell collection and the chances of disease limit this strategy. Thus we focus on use of stem cells like embryonic stem cells (ESCs), bone marrow mesenchymal stem cells (BM-MSCs), umbilical cord-derived mesenchymal stem cells (UC-MSCs) etc. (Howard et al., 2008). The cells for tissue engineering can be from autologous, allogenic, or xenogenic sources. The cell sources can be further portrayed into adult stem cells or somatic stem cells, mature (non-stem) cells, ESCs, and totipotent stem cells or zygotes.

2.11.3 STEM CELLS IN TISSUE ENGINEERING

Because of the high proliferation capacity and the ability to acquire diverse cell fates depending upon the tissue in which they relay, stem cells have been used for tissue engineering and tissue regeneration studies. The stem cells can be (1) directly injected into the injured site or (2) grown in tissue culture flask and then conjugated with the scaffolds for the regeneration of the wound, or (3) can be used as a part for scaffold in therapeutic purposes.

The pluripotent cells utilized in tissue engineering are from hematopoietic cells derived from adult peripheral blood (bone marrow). These cells can be induced to differentiate into osteoblasts, adipocytes, myocytes etc. and are good candidates in tissue rebuilding (Stachowiak and Tzanakakis, 2011). Another type of pluripotent cell is mesenchymal stem cell which multiplies and gives rise to various other cell types and regenerates the damaged tissues. They also produce various compounds for the maintenance of the newly formed tissues and angiogenesis. All these will be mediated by various growth factors and proteins. Studies on a variety of systems highlights great prospects for the future of stem cell-based tissue engineering but only a few areas have shown its translation into clinical reality. A large number of tissue types ranging from epithelial surfaces like skin, mucosal membranes, and cornea to skeletal tissues can be engineered by using stem cells. The two main applications for the stem cells toward tissue engineering are: first the formation of 2D sheets like regeneration of skin and second is the reconstruction of a 3D structure like bone (Bianco and Robey, 2001).

Identification of novel stem cell technologies with materials that are able to deliver a combination of growth factors leading to reconstructive surgery or organ replacement are the important requirements for engineered tissue (Howard et al., 2008). In vitro bioreactors and the development and use of microfabrication technologies for creating vascularized tissues and organs are another important areas being investigated. Since the stem cell researches and its clinical applications are still debated, it is very necessary to consider the social, legal, and ethical issues regarding these experiments.

2.12 TISSUE ENGINEERING FOR REPLACING BODY PARTS

The applications of tissue engineering cover a broad range, but this term will be mostly associated with the repair or replacement of tissues or organs like bone, cartilage, skin etc. some of these applications are explained in the following sections.

2.12.1 ENGINEERING TISSUES FOR REPLACING SKIN

Skin is the largest organ of our body and is vital for the survival of the organism by acting as a barrier to the environment against adverse conditions.

It constantly undergoes regeneration and also possesses the capacity to repair wounds depending on the different types of stem cells in the skin. Engineered skin substitutes serve as an important medical application for the extensive burn wounds. The tissue constructs presently available lacks normal appendages of skin like sebaceous gland, sweat gland, hair follicles, and normal mechanical properties of the skin and thus cannot restore the normal skin anatomy (Wong and hang, 2009).

Skin damages mainly occur by chronic wounds (venous pressure and leg ulcers), burn injuries, skin excision, tumors, and other dermatological conditions. Tissue-engineered substitutes promote the regeneration of epidermis and dermis, prevents fluid loss, and provides protection from contaminations and can deliver ECM components, cytokines, growth factors, drugs etc. to the wound site that enhances the healing process and can be used with autografts. Bruke et al., in early 1980s successfully created artificial skin using fibroblast cells seeded on the collagen scaffolds for treatment of extensive burn injury and this is still being used. Skin substituents that are made from cell-seeded collagen have been commercialized extensively (Cen et al., 2008). Examples of commercially available skin substitutes are provided hereunder.

Epidermal substitutes- Epicel® and CellSpray® (cell based), Myskin™ and Laserskin® (scaffold-containing cells), ReCell® (autologous epidermal cell suspension)

Dermal substitutes- Integra®, Hyalomatrix PA® and AlloDerm® (cell free), Dermagraft®, TransCyte®, and Hyalograft 3D™ (scaffold-containing cells)

Dermoepidermal substitutes- OrCel®, Apligraf® (natural-based scaffold containing cells) and PolyActive® (synthetic scaffold-containing cells)

Bottom-up and top-down approaches are two important strategies for skin regeneration and repair where the top-down or scaffold approach uses a temporary scaffold as a substrate and bottom-up using cell aggregates to produce tissue-engineered construct without using scaffolds. Natural fibers like collagen, chitosan etc. and synthetic polymers like poly-L-lactide (PLLA), polycaprolactone (PCL), polyglycolic acid (PGA) etc. have been used either alone or in combination in skin tissue engineering. PCL membranes containing ZnO nanoparticles haveshown ability to promote wound healing without scar formation. Skin substitutes that are derived from skin stem cells hold promise for feasible gene therapy for skin. Skin tissue engineering is a maturing field that has benefited patients since the 1990s and it is hoped that new biomaterials will be produced to overcome many problems that exists in current approaches (MacNeil, 2008). Electrospun polycaprolactone membranes incorporated with ZnO nanoparticles perform as skin substitutes with enhanced fibroblast proliferation and wound healing (Augustine et al., 2014b).

2.12.2 ENGINEERING TISSUES FOR REPLACING LIVER

The liver has a complex structure and performs a myriad of functions in the body. It is a highly regenerative organ but the use of drugs, toxins, and viral infections leading to various diseases cause extensive damages to the hepatocytes that reduce its function and regeneration. Liver transplantation is the definitive treatment for the end stage liver failure but the shortage of organs limits the transplantation procedure (Palakkan et al., 2013). Organs like liver are found to be difficult to engineer, partly due to the lack of a well defined circulatory system. Liver cell (hepatocyte) transplantation offers a possible solution to overcome the organ shortage, one of the major limitations in organ transplantation. But the isolated liver cells suffer during the isolation and cryopreservation procedures which is one reason for the limited success of this transplantation procedure.

Tissue engineering approach created new liver tissue providing a potential solution to the obstacles that challenge liver cell transplantation. The primary cell sources for liver tissue engineering have been adult primary hepatocytes but the limited availability of quality human liver hepatocytes limited its use. Hepatocytes from different animals like rat, pig, mouse etc. were used for grafting liver construct. ECM has a very important role in maintaining the structure and function of the liver cells.

In vitro cultures in dishes which are coated with ECM like matrigel, fibronectin, collagen, as well as mixtures of collagen and fibronectin were able to preserve functions of hepatocyte for a short time (Castell and Gomez-lechon, 2009). Naturally derived polymers like collagen, alginate, chitosan etc. and synthetic biopolymers like the polymer family- poly(hydroxyl acid) which include PGA, PLA, PLGA copolymer, and their modified derivatives have been developed for liver tissue engineering. Scaffolds of porous sponge have been the most extensively used type of biodegradable scaffolds for the ex vivo culturing and transplantation of hepatocytes (Mikos et al., 1993, 1994). These sponges and bonded fiber structure of highly porous surface (83% porosity) are used to support growth of hepatocytes in vitro and in vivo. Generation of advanced bioreactors in the mid-1990's, where the liver cells can be cultured in 3D scaffolds, has revolutionized the liver tissue engineering (also known as bioreactors for tissue engineering) (Catapano et al. 2010). The advent in the field of tissue engineering is limited by the challenge of finding the right scaffold material and architecture to facilitate the function of hepatocyte and survival in vivo. The success in this field seems to provide the promise of creating engineered liver for the clinical transplantation in future (Uygun and Yarmush, 2013).

2.12.3 ENGINEERING TISSUES FOR REPLACING BONE

In populations where aging is coupled with poor physical activity and obesity, there is an expected increase in the incidence of bone disorders and other related conditions. Thus potential alternatives like engineered bone tissue has been viewed due to their infinite supply and lack of disease transmission. By the synergistic combination of biomaterials, cells and factor therapy, the bone tissue engineering aims at inducing new functional bone regeneration (Amini et al., 2012). In a wide range of clinical settings bone grafts are being used to augment bone repair and utilization. Though osseous tissue has a unique internal repair capacity of healing and remodeling without any scar formation, several conditions both acquired and congenital are necessary in bone replacement (Buckwalter et al., 1993).

The process of bone regeneration was uncovered by Prof. Marshall R. Urist, an orthopedic surgeon at the University of California, Los Angeles. Bone, being very alive, constantly rebuilds itself and the porous framework of bone is composed of collagen protein fibers which run through hydroxyapatite (hydroxyapatite is a mineral that makes up 70% of living bone). The problems like lack of bone availability for autografts, immune rejections from recipients, and transmission of diseases made it necessary to find substances more closely related to the real bone, like hydroxyapatite.

In 1992, the FDA approved a synthetic bone implant called "Pro Osteon" which is a calcium phosphate material that mimics hydroxyapatite.

However, this material lacks strength needed for weight-bearing bones but possess zero rejection. Another product "Megagraft 1000", a bioceramic processed from calcium metal, calcium hydroxide, and phosphoric acid encouraged faster bone growth. In 1993, another product "Collagraft", made out of hydroxyapatite/tricalcium phosphate and bovine collagen then mixed with the bone marrow of the patient have been approved by the FDA.

Functional bone tissue engineering is another technique which uses bone morphogenetic proteins (BMPs) which have great importance in increasing the bone regenerative potency and is done with the help of gene therapy. Synthetic polymers like PLA, PGA, PLGA, polyanhydrides such as poly (methacrylated 1, 6-bis (carboxyphenoxy) hexane), poly (methacrylated sebacic anhydride), a non-degradable polymer, poly (ethylene glycol) (PEG) etc. have shown good results in bone tissue engineering (Fisher and Reddi, 2003). A wide range of bioactive inorganic materials like tricalcium phosphate, bioactive glass, and HA and their combinations having a similar composition as that of bone are widely in use.

2.12.4 ENGINEERING TISSUES FOR REPLACING CARTILAGE

Articular cartilage exhibits very less capacity for intrinsic repair (poor regenerative properties), also it is an avascular tissue and so even very minor injuries may lead to progressive damage and osteoarthritic joint regeneration which results in pain and disability. Articular cartilage tissue engineering aims at repairing, regenerating, and/or improve diseased or injured articular cartilage functionality and holds great potential for improvement of the articular cartilage therapy (Zhang et al., 2009). Numerous attempts have been done for the development of grafts for repairing chondral and osteochondral defects but still remains with significant challenges in the clinical application of cartilage repair (Johnston et al., 2013).

The three types of cartilages in human body are hyaline cartilage (e.g., in diarthrodial joints), fibro cartilage (e.g., knee meniscus), and elastic cartilage (e.g., ear). Articular cartilage that covers bone surfaces is a soft and specialized hyaline cartilage which possesses superior lubrication, wear, and low friction properties and also reduces stresses in the joint. Various cell sources for the repair and regeneration of articular cartilage involves chondrocytes which is the sole source of cells, stem cells like mesenchymal stem cells, ESCs, dermis of the skin etc. Due to the biocompatibility of the natural biomaterials, scaffolds used in cartilage tissue engineering includes carbohydrate-based hyaluronic acid, alginate, chitosan, agarose and protein-based fibrin glue, collagen etc. are used and synthetic scaffolds like PLA, PGA, PLGA etc. are also used. Methods like electrospinning, particulate leaching, phase separation, and 3D printing techniques can be used for the preparation of nanofibrous 3D scaffolds (Zhang et al., 2009). Hydrogels based on hyaluron, PEGs etc. are also used for cartilage tissue engineering. Gene therapy is in an infant stage in cartilage tissue engineering, growth factors like insulin like growth factor-19 (IGF-1and transforming growth factor beta 1 (TGF-β1) have been successfully transfected within chondrocytes for the increase in the expression of collagen aggrecan. By the method of press-coating, an in vivo engineered cartilage construct was developed on PLA scaffold in a one-step method (Tuli et al., 2003). Self-assembling peptide hydrogel scaffolds are suitable candidates for cartilage tissue engineering. A better understanding for the development of clinically feasible designs in disease compromised animal models should be the future approach in the cartilage tissue engineering. Various evidences existing today represent a potentially sound approach to treat cartilage injury or trauma by the idea of tissue engineering.

2.12.5 ENGINEERING TISSUES FOR REPLACING TENDONS

Tendon is a connective tissue, physically binds muscles to skeletal structures, permits locomotion, and enhances joint stability. The structure has multiunit hierarchial collagen molecules, fiber bundles, fibrils, fascicles and tendon units, and resists tensile loads. Collagen I is the most abundant molecular component in tendon formed by self-assembly of collagen molecules.

Tendon injuries are difficult to treat and the classical surgical reconstructive methods have significant limitations especially when there is large tendon deficit. So transplantation is the other option which also has certain limitations and tissue engineering has become a newer option for an answer even though the role of tissue engineering in tendon healing is still unclear (Moshiri and Oryan, 2012). Tendon tissue engineering induces self-regeneration of tendon tissue in vivo, or produces functional tissue replacement in vitro which is then implanted in the body. The impact of tissue engineering in tendon healing can be increased by involving information regarding the structure, injury, healing, host immune response, biomaterial characteristics etc. in the tissue engineering approaches.

Natural scaffolds tested for tendon tissue engineering involves collagen and chitosan. Collagen as it is a major component of tendon has shown good results when used as collagen gels and also as composites with synthetic polymers which showed an improved mechanical property and cell migration. Chitosan-based hyaluronan composite fiber scaffolds also have shown improvements in in vivo models. Synthetic polymers like PLA, PGA, and PLGA were used for tendon repairs. Cell sources include bone marrow derived mesenchymal stem cells, tissue derived mesenchymal stem cells tenocytes, and tendon sheath fibroblasts. Growth factors directly injected into the wound sites showed enhancement in the tendon repair. Tendon injuries can be managed by gene therapy by the delivery of genes that help in healing to the site of injury. In vivo transfer of genes for the production of growth factors like BMP-12, BMP-IGF-1) and in vitro transfer of smad proteins (a group of intracellular proteins) co-expressed with BMP-2 has been practiced recently. Tendon tissue engineering is in an infant stage and translation of various investigations to clinics involves several important concerns like in vivo efficiency of the graft, host immune reactions etc. (Hampson et al., 2008).

2.12.6 ENGINEERING TISSUES FOR REPLACING LIGAMENTS

Like tendons, ligaments are also composed of collagen fibers but less densely packed and woven unlike the parallel arrangement of tendon. They connect two or more bones and are responsible for joint movement and stability. Rupture in ligament leads to abnormal joint kinematics and irreversible damage of the surrounding tissues which lead to degenerative diseases that do not heal naturally and cannot be repaired completely by conventional clinical methods. Various advantages of ligament tissue engineering are minimal patient morbidity, reliable fixation methods, infection or disease transmission, biodegradation along with adequate mechanical stability, rapid return to the preinjury functions etc.

Currently, anterior cruciate ligament (ACL) and medial collateral ligament (MCL) and glenohumeral ligaments are the most frequently practiced ligament tissues for tissue engineering, whereas all ligaments are in the pursuit of tissue engineering and studies are carried out to create functional replacements of tissues (Yilgor et al., 2011). Growth factors like IGF-1, TGF-β, vascular endothelial growth factor (VEGF), epidermal growth factor (EGF), platelet derived growth factor (PDGF), basic fibroblast growth factor (bFGF), platelet-rich plasma (PRP), and collagen-PRP-complex (CPC) are effective in ligament cell proliferation and matrix formation alone or in combination. The age and origin of the fibroblast also affects the proliferative response to PDGF and bFGF. The major cell source in tendon tissue engineering is MSC which can differentiate into various connective tissues. Both natural and synthetic materials are widely used in the form of gels, membranes, or 3D scaffolds. Collagen, hyaluronic acid, silk, poly (β-hydroxybutyrate) (PHB), poly-3-hydroxy-10-undecenoate (PHUE-O3), poly-3-hydroxybutyrate-co-hydrovalerate (PHBV) etc. are the examples for natural replacements and dacron polyester, PGA, PLLA, poly (lactic acid-co-glycolide) etc. are examples for synthetic materials used in ligament tissue engineering.

Ex vivo bioreactors can be used for the controlled biochemical and physical regulatory signals to guide the tissue development. In order to assemble ligament tissue structures, combinations like braiding, stratifying, knitting, or 3D braiding scaffolds and also merging of scaffolds with different material types and aligning its cellular content, functionalizing the surface, and adding mechanical stimulation have to be done (Yilgor et al., 2011).

2.13 GENE THERAPY AND TISSUE ENGINEERING

Gene therapy is a technique for correcting the defective genes that are responsible for the disease development by supplementing the defective gene with a functional one. Gene therapy can be either germ-line or somatic cell gene therapy. Somatic gene therapy is the most exclusive gene therapy and involves introduction of genes to somatic cells of an affected individual, whereas germ line gene therapy involves the permanent transmissible modification of the genome of a gamete, a zygote, or an early embryo. The prospect of human germ line gene therapy is not sanctioned currently.

It is been around 20 years since the clinical development of gene therapy started and at present over 320 ongoing clinical trials for gene therapy are going on regulated by the FDA. The products of gene therapy initially appear promising as they reflect strong scientific foundation and offers hope for treating rare and life-threatening disorders but despite all this the field of gene therapy has not yet produced any successful products that have gone through clinical trials and proved safe and effective for marketing approval. The results obtained from a trial for treating X-linked severe combined immunodeficiency (X-SCID) and several reports of encouraging results have created subsequent enthusiasm in the field of gene therapy (Takefman and Bryan, 2012).

By using intraperitoneal or intravenous injection, somatic cells are transfected, but after the implantation in the organism the survival of transfected cells is limited. By implanting vascular implants called organoids or neo-organs into the organism will serve as a support to these modified cells. This helps in the localization and the accessibility of the implant, to record the cell survival and progress of the implants and also improves the survival rate of the cells. If necessary, by removing the implants this treatment could be stopped. Angiogenisis was shown by introducing a sponge of type I collagen impregnated with acidic fibroblast growth factor (aFGF) in the abdominal cavity. In another experiment hepatocytes showed longer survival rates with the intraperitoneal implant of cultured microbeads of dextran (Thompson et al., 1989).

2.13.1 IN HEREDITARY DISEASES

Hereditary diseases like phenylketonuria, hypercholesterolaemia etc. where the organ concerned is not affected by the deficit circulatory protein and a partial correction will be enough to obtain an improvement in the disease conditions, gene therapy is the best suited solution. Neo-organs with polyfluorethylene (PTFE), type I collagen gel and recombinant human bFGF containing autologous cells were used for the long-term correction of genetic defect of β-glucuronidase gene (Moullier et al., 1993). Continuous secretion of erythropoietin and hemophilia factor VIII from a neo-organ in mice has also succeded. Matrices based on nylon and collagen along with cells transfected with genes have been implanted in athymic mice resulted in the production of human transferin (hTf). Another example is the formation of a collagen matrix with cells producing growth hormone implanted in hypophysectomized rats which showed a positive result.

2.13.2 IN CANCER

Neo-organs can be used for the treatment of cancer by increasing the immune response by stimulating immunogenic neo-antigens or immunity stimulating lymphokines like tumor necrosis factor (TNF), interlukin-4, interlukin-12, interferon (INF) α, β, γ etc. Cells transfected with G-CSF (granulocyte colony stimulating factor) gene and cultivated in collagen matrices successfully secreted G-CSF when implanted in mice (Chevallay and Herbage, 2000).

2.13.3 IN TISSUE REPLACEMENT

Tissue engineering constructs can be supplemented with various types of growth factors like BMPs, IGFs etc. by seeding of transfected cells (cells that are transfected with specific genes that codes for the production of various growth factors) into the scaffolds. This approach can in turn increase the rate of cell growth and adhesion and fast healing. But this method of gene transfection in tissue engineering is in a beginning stage and requires more investigations and clinical trials.

2.14 TISSUE ENGINEERING AS ALTERNATIVE TO DRUG SCREENING AND THERAPY

Production of a large number of healthy cells for repopulating a damaged site is the most important aspect of tissue engineering. It can be also considered as an alternative to drug therapy, gene therapy or whole organ transplantation. Tissue engineering can be utilized for the treatment of metabolic disorders. A metabolism can be said as a coordinated ensemble of chemical transformation controlled or regulated by various enzymes and a defective production of even a single enzyme can lead to various metabolic disorders. These missing links can be corrected or the effects of the disorder can be reduced by gene therapy or drug therapy.

The introduction of 3D cultures based on the combination of cells, scaffolds, and biomolecules have integrated microchip and microfluidic approaches to tissue engineering. The techniques such as replica molding, photolithography, and microcontact printing enable us to control cell position morphology and function by creating microscale-level structures.

The microfluidic approaches help in the manipulation of small amounts of fluids in hollow chambers, generates and tune the spatiotemporal gradients of nutrients and oxygen. This combination will lead to the organ-on-chip microdevices and represent a potential substitute for the animals in drug screening process. This in turn reduces the gap between 2D cultures and animal models (Huh et al., 2011).

As the results obtained with the animal models cannot be directly translated to humans because of the species specificity of drug action. 3D cultures are also being introduced in drug screening procedures to analyze the drug action, effectiveness, and to reduce the investment. Thus by transplanting human cells, significant efforts have been made for "humanizing" mice although it is expensive to adopt this in an assay format. Such experiments have been performed by culturing hepatocytes which then regains their morphology and protein expressions (Griffith and Swartz et al., 2006; Meli et al., 2012).

2.15 FUTURE

"Tissue engineering" is the term that represents a new concept focusing on regeneration of new tissues from cells with the support of biomaterials and growth factors. The term was coined around 30 years back for this interdisciplinary engineering method and has attracted much attention as a therapeutic means. This offers hope for a large number of patients with injuries, organ failures, and other clinical issues. These patients are treated currently also with transplantation of organs but as the number of patients increases day by day, there is a great need of donor organs. The new cases of organ failures increase each year and thus scientists in the field of regenerative medicine applies various principles of material sciences, bioengineering, and cell transplantation for constructing substitutes that restores and maintain normal function in diseased and injured tissues.

Various achievements acquired with the bladder, blood vessel, and tracheal replacements using tissue engineering have encouraged scientists to engineer other organs also in the laboratory that can cast light on various unsolved problems in tissue engineering. Some of the problems faced are the lack of innervations of tissues and organs which is a very important part for the full functionality of the neotissues or organs. The clinical application of these engineered constructs is still very limited, including skin, bone, cartilage, and capillary and periodontal tissues. Moreover tissue engineering is having a deep impact in the development of new therapies. For example, the developments of 3D cultures have reduced the gap between 2D cultures and animal models which facilitates a constant turnover of oxygen and nutrients under extended studies. But an unresolved issue is the translation of these 3D cultures to pharmaceutical level.

Tissue engineering has captured many advantages of normal cell culture over whole animal experiments. The relative transparency of the engineered tissue may allow the visualization of the structures and processes happening in the cells. It has a major role in physiological genomics, which link various pathways and products to phenotypes and physiological systems. The studies at the level of transcriptome, metabolome, proteome, and population levels have succeeded in identifying various genes responsible for the development as well as progression of various diseases. Later, various studies in recombinant, knock-out animals etc. assigned various phenotypes to the genome. Thus for various molecular level studies particularly those related to gene expression where the changes at the genomic level depicts changes in the phenotype, the 3D culture systems can be utilized. Several reports on the gene expression of the 3D cultured cells suggest that the expression levels of engineered constructs more closely parallels the in vivo situations. Many engineered skin substitutes utilized for toxicological testing and other clinical applications also showed similar level of expressions in case of both native and engineered tissues. All these reports points toward the use of engineered tissue as a model system for testing the gene expression and the effects of altered gene expression (Birgersdotter et al., 2005; Ghosh et al., 2005; Smiley et al., 2005).

Tissue engineering and human genomics together have a great potential in personalized medical care. There is a need of very intense analysis regarding the gene expression of healthy as well as diseased cells for creating molecular fingerprints of the disease stages. As the knowledge base regarding stem cell technology alarms, it may be also possible to produce engineered tissues by the differentiation of stem cells yielding an unlimited source of grafts for tissue replacement and repair. Since these grafts will be produced from the autologous cell source, the possibility of immune response and graft rejection can be minimized.

2.16 CONCLUSION

Taking into consideration the contributions given by tissue engineering toward science and human race and those that are going to happen in future, the 21st century can be considered as a revolutionary era that has marked its importance by this area of nanomedicine. In a wide sense, any manipulation that involves an alteration in structure and function of a tissue which may also include gene manipulation, surgical interference, hormonal therapy etc. can be considered as tissue engineering. A large number of obstacles and questions has to be resolved regarding the biocompatibility and establishment and functioning of the constructed tissues at present. The proper delivery of regulatory molecules remains another challenge to this, whereas the computer modeling for predicting the outcomes of the tissue product add up an advantage to the establishment of tissue engineering systems. The possible inevitable advantages of tissue engineering are the reduction in post-operative patient costs, improved patient care at less expense, and an enhancement in the quality of life with a reduction of cost. In conclusion, tissue engineering is an emerging field of science offering tremendous promise with the proper implementation of quality assurance.

KEYWORDS

- **Biomaterials**
- **Electrospinning**
- **Nanolithography**
- **Nanomedicine**
- **Stem cells**
- **Tissue engineering**

REFERENCES

Amini, Ami R., Cato T. Laurencin, and Syam P. Nukavarapu. "Bone tissue engineering: recent advances and challenges."*Critical Reviews™ in Biomedical Engineering* 40, no. 5 (2012): 363-408.

Augustine, Robin, Hruda Nanda Malik, Dinesh Kumar Singhal, Ayan Mukherjee, Dhruba Malakar, Nandakumar Kalarikkal, and Sabu Thomas. "Electrospun polycaprolactone/ZnO nanocomposite membranes as biomaterials with antibacterial and cell adhesion properties." *Journal of Polymer Research* 21, no. 3 (2014a): 1-17.

Augustine, Robin, Edwin Anto Dominic, Indu Reju, Balarama Kaimal, Nandakumar Kalarikkal, and Sabu Thomas. "Electrospun polycaprolactone membranes incorporated with ZnO nanoparticles as skin substitutes with enhanced fibroblast proliferation and wound healing."-*RSC Advances*-4 (2014b): 24777-24785.

Augustine, Robin, Edwin Anto Dominic, Indu Reju, Balarama Kaimal, Nandakumar Kalarikkal, and Sabu Thomas. "Investigation on angiogenesis and its mechanism using zinc oxide nanoparticles-loaded electrospun tissue engineering scaffolds."- *RSC Advances* (2014c). DOI:10.1039/C4RA07361D

Badylak, Stephen F., Daniel J. Weiss, Arthur Caplan, and Paolo Macchiarini. "Engineered whole organs and complex tissues."-*The Lancet*-379, no. 9819 (2012): 943-952.

Berthold, A., K. Cremer, and J. S. T. P. Kreuter. "Preparation and characterization of chitosan microspheres as drug carrier for prednisolone sodium phosphate as model for anti-inflammatory drugs."*Journal of Controlled Release*-39, no. 1 (1996): 17-25.

Bianco, Paolo, and Pamela Gehron Robey. "Stem cells in tissue engineering." *Nature*-414, no. 6859 (2001): 118-121.

Birgersdotter, Anna, Rickard Sandberg, and Ingemar Ernberg. "Gene expression perturbation in vitro—a growing case for three-dimensional (3D) culture systems." In-*Seminars in cancer biology*, vol. 15, no. 5, pp. 405-412. Academic Press, 2005.

Boehler, Ryan M., John G. Graham, and Lonnie D. Shea. "Tissue engineering tools for modulation of the immune response." *Biotechniques* -51, no. 4 (2011): 239.

Buckwalter, Joseph A., S. L. Woo, V. M. Goldberg, E. C. Hadley, F. Booth, T. R. Oegema, and D. R. Eyre. "Soft-tissue aging and musculoskeletal function." *The Journal of Bone & Joint Surgery*-75, no. 10 (1993): 1533-1548.

Castell, Jose V., and María José Gómez-Lechón. "Liver cell culture techniques." In-*Hepatocyte transplantation*, pp. 35-46. Humana Press, 2009.

Catapano, Gerardo, John F. Patzer II, and Jörg Christian Gerlach. "Transport advances in disposable bioreactors for liver tissue engineering." In-*Disposable bioreactors*, pp. 117-143. Springer, Berlin Heidelberg, 2010.

Cen, Lian, Wei Liu, Lei Cui, Wenjie Zhang, and Yilin Cao. "Collagen tissue engineering: development of novel biomaterials and applications."-*Pediatric research*-63, no. 5 (2008): 492-496.

Chan, B. P., and K. W. Leong. "Scaffolding in tissue engineering: general approaches and tissue-specific considerations."-*European Spine Journal*-17, no. 4 (2008): 467-479.

Chevallay, B., and D. Herbage. "Collagen-based biomaterials as 3D scaffold for cell cultures: applications for tissue engineering and gene therapy."-*Medical and Biological Engineering and Computing*-38, no. 2 (2000): 211-218.

Cirillo, B., M. Morra, and G. Catapano. "Adhesion and function of rat liver cells adherent to silk fibroin/collagen blend films."-*The International Journal of Artificial Organs*-27, no. 1 (2004): 60-68.

Das, Kakoli, Vamsi Krishna Balla, Amit Bandyopadhyay, and Susmita Bose. "Surface modification of laser-processed porous titanium for load-bearing implants."-*Scripta Materialia* -59, no. 8 (2008): 822-825.

Doshi, Jayesh, and Darrell H. Reneker. "Electrospinning process and applications of electrospun fibers."In-*Industry Applications Society Annual Meeting, 1993, Conference Record of the 1993 IEEE*, pp. 1698-1703. IEEE, 1993.

Dugan, James M., Julie E. Gough, and Stephen J. Eichhorn. "Bacterial cellulose scaffolds and cellulose nanowhiskers for tissue engineering." Nanomedicine -8, no. 2 (2013): 287-298.

Fielding, Gary A., Amit Bandyopadhyay, and Susmita Bose. "Effects of silica and zinc oxide doping on mechanical and biological properties of 3D printed tricalcium phosphate tissue engineering scaffolds." *Dental Materials*-28, no. 2 (2012): 113-122.

Fisher, J. P., and A. H. Reddi. "Functional tissue engineering of bone: signals and scaffolds."*Topics in Tissue Engineering*-1 (2003): 1-29.

Fisher,John P. Antonios G. Mikos, Joseph D. Bronzino. Tissue Engineering. CRC Press, (2007).

Fu, Lina, Jin Zhang, and Guang Yang. "Present status and applications of bacterial cellulose-based materials for skin tissue repair."-*Carbohydrate Polymers*-92, no. 2 (2013): 1432-1442.

Gerhardt, L-C., G. M. R. Jell, and A. R. Boccaccini. "Titanium dioxide (TiO_2) nanoparticles filled poly (D, L lactid acid) (PDLLA) matrix composites for bone tissue engineering."-*Journal of Materials Science: Materials in Medicine*-18, no. 7 (2007): 1287-1298.

Ghasemi-Mobarakeh, Laleh, Molamma P. Prabhakaran, Mohammad Morshed, Mohammad Hossein Nasr-Esfahani, and S. Ramakrishna. "Bio-functionalized PCL nanofibrous scaffolds for nerve tissue engineering."*Materials Science and Engineering: C*-30, no. 8 (2010): 1129-1136.

Ghosh, Sourabh, Giulio C. Spagnoli, Ivan Martin, Sabine Ploegert, Philippe Demougin, Michael Heberer, and Anca Reschner. "Three-dimensional culture of melanoma cells profoundly affects gene expression profile: a high density oligonucleotide array study."-*Journal of Cellular Physiology*-204, no. 2 (2005): 522-531.

Griffith, Linda G., and Melody A. Swartz. "Capturing complex 3D tissue physiology *in vitro*."-*Nature Reviews Molecular Cell Biology*-7, no. 3 (2006): 211-224.

Guillemette, Maxime D., Bo Cui, Emmanuel Roy, Robert Gauvin, Claude J. Giasson, Mandy B. Esch, Patrick Carrier et al. "Surface topography induces 3D self-orientation of cells and extracellular matrix resulting in improved tissue function."-*Integrative Biology*-1, no. 2 (2009): 196-204.

Guo, Chun Xian, Xin Ting Zheng, Zhi Song Lu, Xiong Wen Lou, and Chang Ming Li. "Biointerface by cell growth on layered graphene–artificial peroxidase–protein nanostructure for in situ quantitative molecular detection."-*Advanced Materials* 22, no. 45 (2010): 5164-5167.

Guo, Shaojun, and Shaojun Dong. "Graphene nanosheet: synthesis, molecular engineering, thin film, hybrids, and energy and analytical applications."*Chemical Society Reviews*-40, no. 5 (2011): 2644-2672.

Hampson, K., N. R. Forsyth, A. El Haj, and N. Maffulli. "Tendon tissue engineering.-*Topics in Tissue Engineering*-4 (2008): 1-21.

Harrison, Benjamin S., and Anthony Atala. "Carbon nanotube applications for tissue engineering." Biomaterials-28, no. 2 (2007): 344-353.

Hinman, Michael B., Justin A. Jones, and Randolph V. Lewis. "Synthetic spider silk: a modular fiber."-*Trends in Biotechnology*-18, no. 9 (2000): 374-379.

Howard, Daniel, Lee D. Buttery, Kevin M. Shakesheff, and Scott J. Roberts. "Tissue engineering: strategies, stem cells and scaffolds."-*Journal of Anatomy* 213, no. 1 (2008): 66-72.

Huh, Dongeun, Geraldine A. Hamilton, and Donald E. Ingber. "From 3D cell culture to organs-on-chips."-*Trends in Cell Biology*-21, no. 12 (2011): 745-754.

Hutmacher, Dietmar W. "Scaffold design and fabrication technologies for engineering tissues—state of the art and future perspectives."-*Journal of Biomaterials Science, Polymer Edition*-12, no. 1 (2001): 107-124.

Irvine, Darrell J. "BE. 462J Molecular principles of biomaterials, Spring 2003." (2003).

Jayakumar, R., Roshni Ramachandran, V. V. Divyarani, K. P. Chennazhi, H. Tamura, and S. V. Nair. "Fabrication of chitin–chitosan/nano TiO_2-composite scaffolds for tissue engineering applications."-*International Journal of Biological Macromolecules*-48, no. 2 (2011): 336-344.

Jha, Balendu Shekhar, Chantal E. Ayres, James R. Bowman, Todd A. Telemeco, Scott A. Sell, Gary L. Bowlin, and David G. Simpson. "Electrospun collagen: a tissue engineering scaffold with unique functional properties in a wide variety of applications."-*Journal of Nanomaterials*-2011 (2011): 7.

Johansson, Fredrik, Patrick Carlberg, Nils Danielsen, Lars Montelius, and Martin Kanje. "Axonal outgrowth on nano-imprinted patterns."-*Biomaterials*-27, no. 8 (2006): 1251-1258.

Johnstone, Brian, Mauro Alini, Magali Cucchiarini, George R. Dodge, David Eglin, Farshid Guilak, Henning Madry et al. "Tissue engineering for articular cartilage repair-the state of the art."-*European Cell & Materials*-25 (2013): 248-267.

Joshi, K. B., Prabhpreet Singh, and Sandeep Verma. "Fabrication of platinum nanopillars on peptide-based soft structures using a focused ion beam." *Biofabrication* -1, no. 2 (2009): 025002.

Kearns, V., A. C. MacIntosh, A. Crawford, and P. V. Hatton. "Silk-based biomaterials for tissue engineering." *Topics in Tissue Engineering* 4, no. 0 (2008): 5.

Kim, Jung-Ho, Faheem A. Sheikh, Hyung Woo Ju, Hyun Jung Park, Bo Mi Moon, Ok Joo Lee, and Chan Hum Park. "3D silk fibroin scaffold incorporating titanium dioxide (TiO_2) nanoparticle (NPs) for tissue engineering."-*International Journal of Biological Macromolecules*-68 (2014): 158-168.

Ku, Sook Hee, Minah Lee, and Chan Beum Park. "Carbon-based nanomaterials for tissue engineering."-*Advanced Healthcare Materials*-2, no. 2 (2013): 244-260.

Kumar, Sudheesh, Vinoth-Kumar Lakshmanan, Mincy Raj, Raja Biswas, Tamura Hiroshi, Shantikumar V. Nair, and Rangasamy Jayakumar. "Evaluation of wound healing potential of β-chitin hydrogel/nano zinc oxide composite bandage."-*Pharmaceutical Research*-30, no. 2 (2013): 523-537.

Kumará Barui, Ayan, Ajay Kumará Patel, and Chitta Ranjaná Patra. "Zinc oxide nanoflowers make new blood vessels."-*Nanoscale* -4, no. 24 (2012): 7861-7869.

Kuo, Catherine K., Joseph E. Marturano, and Rocky S. Tuan. "Novel strategies in tendon and ligament tissue engineering: advanced biomaterials and regeneration motifs."-*BMC Sports Science, Medicine and Rehabilitation*-2, no. 1 (2010): 20.

Lai, Yinzhi, Amish Asthana, and William S. Kisaalita. "-Biomarkers for simplifying HTS 3D cell culture platforms for drug discovery: the case for cytokines."-*Drug Discovery Today*-16, no. 7 (2011): 293-297.

Langer R, and J.P Vacanti. "Tissue engineering". *Science* 260, (1993):920-926.

Lanza, Robert, Robert Langer, and Joseph P. Vacanti, eds.-*Principles of tissue engineering*. Academic Press, 2011.

Laurencin, Cato T., and Lakshmi S. Nair, eds.-*Nanotechnology and tissue engineering: the scaffold*. CRC Press, 2008.

Li, Mingtao, Lei Chen, and Stephen Y. Chou. "Direct three-dimensional patterning using nanoimprint lithography."-*Applied Physics Letters*-78, no. 21 (2001): 3322-3324.

Liang, Dehai, Benjamin S. Hsiao, and Benjamin Chu. "Functional electrospun nanofibrous scaffolds for biomedical applications."*Advanced Drug Delivery Reviews*-59, no. 14 (2007): 1392-1412.

Lipatova, T. E., and Yu S. Lipatov. "Biocompatible polymers for medical application."In-*Macromolecular symposia*, vol. 152, no. 1, pp. 139-150. Wiley-VCH Verlag, 2000.

Ma, Zuwei, Masaya Kotaki, Ryuji Inai, and Seeram Ramakrishna. "Potential of nanofiber matrix as tissue-engineering scaffolds."-*Tissue Engineering*-11, no. 1-2 (2005): 101-109.

MacNeil, Sheila. "Biomaterials for tissue engineering of skin."-*Materials Today*11, no. 5 (2008): 26-35.

Mandal, Biman B., and Subhas C. Kundu. "Cell proliferation and migration in silk fibroin 3D scaffolds."-*Biomaterials*-30, no. 15 (2009): 2956-2965.

Martins, Ana M., Quynh P. Pham, Patrícia B. Malafaya, Rui A. Sousa, Manuela E. Gomes, Robert M. Raphael, F. Kurtis Kasper, Rui L. Reis, and Antonios G. Mikos. "The role of lipase and α-amylase in the degradation of starch/poly (ε-caprolactone) fiber meshes and the osteogenic differentiation of cultured marrow stromal cells."-*Tissue Engineering Part A*-15, no. 2 (2008): 295-305.

Matsuzaka, Kenichi, X. Frank Walboomers, Masao Yoshinari, Takashi Inoue, and John A. Jansen. "The attachment and growth behavior of osteoblast-like cells on microtextured surfaces."-*Biomaterials*-24, no. 16 (2003): 2711-2719.

McNeil, Scott E. "Nanotechnology for the biologist."-*Journal of Leukocyte Biology*-78, no. 3 (2005): 585-594.

Melchels, Ferry, Paul Severin Wiggenhauser, David Warne, Mark Barry, Fook Rhu Ong, Woon Shin Chong, Dietmar Werner Hutmacher, and Jan-Thorsten Schantz. "CAD/CAM-assisted breast reconstruction."-*Biofabrication* -3, no. 3 (2011): 034114.

Meli, Luciana, Eric T. Jordan, Douglas S. Clark, Robert J. Linhardt, and Jonathan S. Dordick. "Influence of a three-dimensional, microarray environment on human cell culture in drug screening systems."-*Biomaterials*-33, no. 35 (2012): 9087-9096.

Michael, Stefanie, Heiko Sorg, Claas-Tido Peck, Lothar Koch, Andrea Deiwick, Boris Chichkov, Peter M. Vogt, and Kerstin Reimers. "Tissue engineered skin substitutes created by laser-assisted bioprinting form skin-like structures in the dorsal skin fold chamber in mice."-*PloS one*-8, no. 3 (2013): e57741.

Mikos, Antonios G., Georgios Sarakinos, Susan M. Leite, Joseph P. Vacant, and Robert Langer. "Laminated three-dimensional biodegradable foams for use in tissue engineering."-*Biomaterials*-14, no. 5 (1993): 323-330.

Mikos, Antonios G., Amy J. Thorsen, Lisa A. Czerwonka, Yuan Bao, Robert Langer, Douglas N. Winslow, and Joseph P. Vacanti. "Preparation and characterization of poly (L-lactic acid) foams."-*Polymer*-35, no. 5 (1994): 1068-1077.

Min, Byung-Moo, Gene Lee, So Hyun Kim, Young Sik Nam, Taek Seung Lee, and Won Ho Park. "Electrospinning of silk fibroin nanofibers and its effect on the adhesion and spreading of normal human keratinocytes and fibroblasts *in vitro*."-*Biomaterials*-25, no. 7 (2004): 1289-1297.

Moore, Michael J., Esmaiel Jabbari, Erik L. Ritman, Lichun Lu, Bradford L. Currier, Anthony J. Windebank, and Michael J. Yaszemski. "Quantitative analysis of interconnectivity of porous biodegradable scaffolds with micro-computed tomography."-*Journal of Biomedical Materials Research Part A*-71, no. 2 (2004): 258-267.

Moshiri, A., and A. Oryan. "Role of tissue engineering in tendon reconstructive surgery and regenerative medicine: current concepts, approaches and concerns."-*Hard Tissue*-1, no. 2 (2012): 11.

Moullier, Philippe, Valérie Maréchal, Olivier Danos, and Jean Michel Heard. "Continuous systemic secretion of a lysosomal enzyme by genetically modified mouse skin fibroblasts." *Transplantation* 56, no. 2 (1993): 427-432.

Netti, Paulo, ed.-*Biomedical foams for tissue engineering applications*. Elsevier, 2014.

O'Brien, Fergal J. "Biomaterials & scaffolds for tissue engineering."-*Materials Today*-14, no. 3 (2011): 88-95.

Orlando, Robert A., and David A. Cheresh. "Arginine-glycine-aspartic acid binding leading to molecular stabilization between integrin alpha v beta 3 and its ligand."-*Journal of Biological Chemistry*-266, no. 29 (1991): 19543-19550.

Palakkan, Anwar A., David C. Hay, and James A. Ross. "Liver tissue engineering and cell sources: issues and challenges."-*Liver International*-33, no. 5 (2013): 666-676.

Park, Joon, and Roderic S. Lakes.-*Biomaterials: an introduction*. Springer, 2007.

Park, Sung Young, Jaesung Park, Sung Hyun Sim, Moon Gyu Sung, Kwang S. Kim, Byung Hee Hong, and Seunghun Hong. "Enhanced differentiation of human neural stem cells into neurons on graphene." *Advanced Materials*-23, no. 36 (2011): H263-H267.

Peng, B. L., N. Dhar, H. L. Liu, and K. C. Tam. "Chemistry and applications of nanocrystalline cellulose and its derivatives: a nanotechnology perspective."*The Canadian Journal of Chemical Engineering*-89, no. 5 (2011): 1191-1206.

Pham, Quynh P., Upma Sharma, and Antonios G. Mikos. "Electrospinning of polymeric nanofibers for tissue engineering applications: a review."-*Tissue Engineering*-12, no. 5 (2006): 1197-1211.

Pulsipher, Abigail, and Muhammad N. Yousaf. "Surface chemistry and cell biological tools for the analysis of cell adhesion and migration."*ChemBioChem* 11, no. 6 (2010): 745-753.

Qi, Lifeng, and Zirong Xu. "Lead sorption from aqueous solutions on chitosan nanoparticles."-*Colloids and Surfaces A: Physicochemical and Engineering Aspects*-251, no. 1 (2004): 183-190.

Ratner, Buddy D., ed.-*Biomaterials science: an introduction to materials in medicine*. Academic Press, 2004.

Reneker, Darrell H., and Iksoo Chun. "Nanometre diameter fibres of polymer, produced by electrospinning."-*Nanotechnology*-7, no. 3 (1996): 216.

Rim, Nae Gyune, Choongsoo S. Shin, and Heungsoo Shin. "Current approaches to electrospun nanofibers for tissue engineering."-*Biomedical Materials*-8, no. 1 (2013): 014102.

Sachlos, E., and J. T. Czernuszka. "Making tissue engineering scaffolds work. Review: the application of solid freeform fabrication technology to the production of tissue engineering scaffolds."-*European Cell & Materials* 5, no. 29 (2003): 39-40.

Smiley, Andrea K., Jennifer M. Klingenberg, Bruce J. Aronow, Steven T. Boyce, W. John Kitzmiller, and Dorothy M. Supp. "Microarray analysis of gene expression in cultured skin substitutes compared with native human skin."-*Journal of Investigative Dermatology*-125, no. 6 (2005): 1286-1301.

Stachowiak, Michal K., and Emmanuel S. Tzanakakis, eds.-*Stem cells: from mechanisms to technologies*. World Scientific, 2011.

Subia, B., J. Kundu, and S. C. Kundu. "Biomaterial scaffold fabrication techniques for potential tissue engineering applications."-In Daniel Eberli (editor), *Tissue Engineering* 524 (2010): 142-157, Intech (publisher)

Taddei, Paola, Takayuki Arai, Alessandra Boschi, Patrizia Monti, Masuhiro Tsukada, and Giuliano Freddi. "*In vitro* study of the proteolytic degradation of *Antheraea pernyi* silk fibroin."-*Biomacromolecules* -7, no. 1 (2006): 259-267.

Takefman, Daniel, and Wilson Bryan. "The state of gene therapies: the FDA perspective."-*Molecular Therapy*-20, no. 5 (2012): 877.

Thompson, Simon, Alan R. Clarke, Angela M. Pow, Martin L. Hooper, and David W. Melton. "Germ line transmission and expression of a corrected HPRT gene produced by gene targeting in embryonic stem cells."*Cell* 56, no. 2 (1989): 313-321.

Thomson, R. C., M. C. Wake, M. J. Yaszemski, and A. G. Mikos. "Biodegradable polymer scaffolds to regenerate organs."In-*Biopolymers Ii*, pp. 245-274. Springer, Berlin Heidelberg, 1995.

Tuli, Richard, Wan-Ju Li, and Rocky S. Tuan. "Current state of cartilage tissue engineering."-*Arthritis Research and Therapy*-5, no. 5 (2003): 235-238.

Uygun, Basak E., and Martin L. Yarmush. "Engineered liver for transplantation." *Current Opinion in Biotechnology*-24, no. 5 (2013): 893-899.

Vasita, Rajesh, and Dhirendra S. Katti. "Nanofibers and their applications in tissue engineering."-*International Journal of Nanomedicine* -1, no. 1 (2006): 15.

Vats, A., N. S. Tolley, J. M. Polak, and L. D. K. Buttery. "Stem cells: sources and applications."-*Clinical Otolaryngology & Allied Sciences*-27, no. 4 (2002): 227-232.

Vert, Michel, Suming Li, and Henri Garreau. "More about the degradation of LA/GA-derived matrices in aqueous media."-*Journal of Controlled Release*-16, no. 1 (1991): 15-26.

Wan, Andrew C. A., and Benjamin C. U. Tai. "CHITIN—a promising biomaterial for tissue engineering and stem cell technologies."-*Biotechnology Advances*-31, no. 8 (2013): 1776-1785.

Willerth, S. M., and Sakiyama-Elbert, S. E. "Combining stem cells and biomaterial scaffolds for constructing tissues and cell delivery." In: StemBook (editor), *The Stem Cell Research Community*, (July 09, 2008), StemBook, doi/10.3824/stembook.1.1.1, http://www.stembook.org.

Wong, D. J., and Chang, H. Y. "Skin tissue engineering." StemBook (editor), *The Stem Cell Research Community*, (March 31, 2009), StemBook, doi/10.3824/stembook.1.44.1, http://www.stembook.org.

Yang, Yumin, Xuemei Chen, Fei Ding, Peiyun Zhang, Jie Liu, and Xiaosong Gu. "Biocompatibility evaluation of silk fibroin with peripheral nerve tissues and cells in vitro."-*Biomaterials*-28, no. 9 (2007): 1643-1652.

Yilgor, Caglar, Pinar Yilgor Huri, and Gazi Huri. "Tissue engineering strategies in ligament regeneration."- *Stem Cells International* (2012), 1-9, doi:10.1155/2012/374676.

Yoshikawa, Hideki, and Akira Myoui. "Bone tissue engineering with porous hydroxyapatite ceramics."-*Journal of Artificial Organs*-8, no. 3 (2005): 131-136.

Zhang, Lijie, Jerry Hu, and Kyriacos A. Athanasiou. "The role of tissue engineering in articular cartilage repair and regeneration."-*Critical Reviews™ in Biomedical Engineering*-37, no. 1-2 (2009), 1-57.

Zhang, Shuguang, X. Zhao, and L. Spirio. "PuraMatrix: self-assembling peptide nanofiber scaffolds."-*Scaffolding in Tissue Engineering*-(2005): 217-238, www.3d-matrix.co.jp/dl_file/PuraMatrix_Introduction.pdf.

Zhou, Jie, M. Ciobanu, G. Pavon-Djavid, V. Gueguen, and V. Migonney. "Morphology and adhesion of human fibroblast cells cultured on bioactive polymer grafted ligament prosthesis." In-*Engineering in Medicine and Biology Society, 2007. EMBS 2007. 29th Annual International Conference of the IEEE*, pp. 5115-5118. IEEE, 2007.

CHAPTER 3

TAILORED HEATING BY SUPERPARAMAGNETIC NANOPARTICLES FOR HYPERTHERMIA APPLICATIONS

VANCHNA SINGH[1, 2] and VARSHA BANERJEE[1]

ABSTRACT

If present in sufficient number, superparamagnetic (SPM) nanoparticles interact via dipole – dipole coupling to produce a dipolar field and subsequently a permanent dipole moment. As a result, the assembly exhibits hysteresis on the application of an oscillating magnetic field yielding heat dissipation. When directed on a site of malignancy, these tiny heat generating machines hold the potential to destroy cells or introduce a modest rise in temperature to increase the efficacy of chemotherapy. This phenomenon is referred to as hyperthermia (MH) in the medical literature. Recently, there have been proposals to monitor hyperthermia using imaging techniques such as magnetic resonance imaging (MRI). This combined modality has immense utility in remedial procedures for cancer, but observations related to decreased heat dissipation have cast doubt on its applicability in a clinical laboratory. Through a first principle analysis, we provide practical procedures for tailored heat dissipation in the above procedures by an easy manipulation of laboratory parameters. Their usage has been largely empirical to date. Our calculations provide a firm grounding to these ad hoc methodologies.

[1]Department of Physics, Indian Institute of Technology, Hauz Khas, New Delhi, 110016, India.

[2]Department of Applied Sciences, Inderprastha Engineering College, Ghaziabad 201011, Uttar Pradesh, India.

3.1 INTRODUCTION

The tremendous potential of magnetic nanoparticles (MNPs) for diagnostic and therapeutic applications is primarily due to their wide applicability in diverse ways and means. These MNPs can be easily detected and manipulated by the application of external magnetic fields. They can be prepared in varying sizes ranging from nanometer to micrometers comparable to the dimensions of biological entities such as viruses, proteins, genes, and cells. As a result, MNPs provide exciting opportunities for site-specific drug delivery, improved quality of magnetic resonance imaging, manipulation of cell membranes, etc to name a few [1, 2]. Further, the presence of a magnetic moment opens up larger avenues in biomedicine. These particles, when subjected to an oscillating magnetic field, can be made to dissipate heat due to the phenomena of magnetic hysteresis. This phenomenon results in selective warming of the target site and is usually referred to as magnetic hyperthermia (MH) in the medical literature. Therefore, when directed on malignant tumors, magnetic nanoparticles hold the potential to destroy cells or introduce a modest rise in temperature to increase the efficacy of chemotherapy [3–6]. Consequently, last few years have evidenced intense investigations, both experimental as well as theoretical toward the understanding of heat dissipation in magnetic nanoparticles to combat cancers.

The clinical trials of magnetic hyperthermia have shown promising results but a serious short coming in conjuncture with the process is unwanted heating of collateral tissues due to particle diffusion. Another difficulty is in quantifying local concentration of the MNPs inside a tumor and monitoring its temperature rise when subjected to an oscillating magnetic field. These inputs are valuable to define and control treatment parameters and can be obtained from real-time imaging techniques such as magnetic resonance imaging (MRI) [7]. Using static magnetic fields, typically in the range 0.1–0.4 T, they can produce images and thermographs of soft tissues which yield details of the magnetic content. The possibility of performing MH in conjunction with MRI is therefore a topic of serious deliberations. However, reports of diminished heating in the combined therapy have been very disparaging [8–12]. To put it into the mainstream of cancer remedy, it is therefore important to understand the causes of quenching and alleviate them.

Single domain superparamagnetic (SPM) nanoparticles are most commonly used for remedial purposes due to a number of advantages associated with them. The SPM nanoparticles can be intravenously injected and then magnetically targeted to affected tissues or organs. Their nanometer size facilitates rapid clearance from the body by way of renal and biliary excretions [13]. This can be a menace though, if the process is extremely rapid. It often needs to be slowed down by conjugating nanoparticles with ligands, such as folic acid, avidin–biotin complex, peptides, and carbohydrates [14, 15]. Further, the particles get magnetized only on the application of a magnetic field. So the risks of aggregation and embolism are substantially lower for paramagnetic nanoparticles vis–à–vis their ferromagnetic counterparts. However, each nanomagnet dissipates only a femtowatt of heat when subjected to an oscillating magnetic field! An often-used thumb rule is that a heat deposition of 100 m Wcm^{-3} is sufficient to raise the tissue temperature to ~45 C desired for hyperthermia and chemotherapy applications [1]. Thus substantially large concentrations are required for desired effects.

The dangers of aggregation and safe levels of toxicity then are a matter of great concern. It is therefore a challenge to choose laboratory parameters which result in the required heat dissipation but at the same time maintain a nontoxic and nonaggregating system.

In this chapter, we provide a theoretical framework which yields precise relationships between the heat dissipated and external (laboratory) parameters (e.g., particle concentr-ation, amplitude, and frequency of the applied oscillating field, etc.) in the form of scaling laws. They make it possible to choose parameters in *a number of ways* to obtain the desired heat dissipation. We then extend this framework to mimic the environment of the combined modalities of MH and MRI. Once again with the help of area scaling laws, we are able to mitigate the quenching of heat by an easy manipulation of laboratory parameters. To illustrate the usage of our procedures, we provide prototypical calculations using magnetite (Fe_3O_4) which are most commonly used in therapeutic procedures.

The chapter is organized as follows. Section 3.2 provides the mean field theory for calculating hysteresis loops, area scaling laws, and their utility to tailor heat dissipation. In Section 3.3, we extend our framework to MRI-monitored MH and demonstrate how parameter selection can alleviate the quenching of heat. Finally in Section 3.4, we conclude the chapter with a summary and discussion.

3.2 TAILORED HEATING IN MAGNETIC HYPERTHERMIA

3.2.1 RELAXATION MECHANISMS IN SPM NANOPARTICLES

A ferromagnetic sample such as iron for instance, comprises of domains, each having spontaneous magnetization pointing in a different direction. If the size of the sample is reduced, there comes a point beyond which a single domain state becomes preferable. The direction of magnetization of the single domain particle does not remain fixed in time though, but undergoes fluctuations or "relaxations" as the magnetic moment rotates between the crystallographic anisotropy axes. As a result the time averaged magnetization is still zero and the particle is paramagnetic. It is called "super paramagnetic" because each particle has a giant magnetic moment arising due to a large number ($\sim 10^5$) of individual atomic moments.

In the case of uniaxial anisotropy (in the z-direction say), the magnetic energy of a single domain, SPM particle is given by [16, 17]

$$E = VK \sin^2 \Phi \tag{3.1}$$

where $V = 4\pi r_c^3 / 3$ is the magnetic volume of a particle with radius r_C, K is the effective magnetic anisotropy constant, and Φ is the angle between the z-axis and direction of the "super" magnetic moment of the single-domain particle. The minimum energy occurs at $\Phi = 0$ and π and these angles define the two equilibrium orientations of the magnetic moment. A schematic depiction of the particle and corresponding energy is provided in Figure 3.1. If $VK >> k_B T$, where k_B is the Boltzmann constant and T the absolute temperature the magnetic moment is mostly locked in two minimum energy orientations resulting in the so-called Ising limit. In this limit, $\Phi(t)$ may be viewed as a dichotomic Markov process in which it

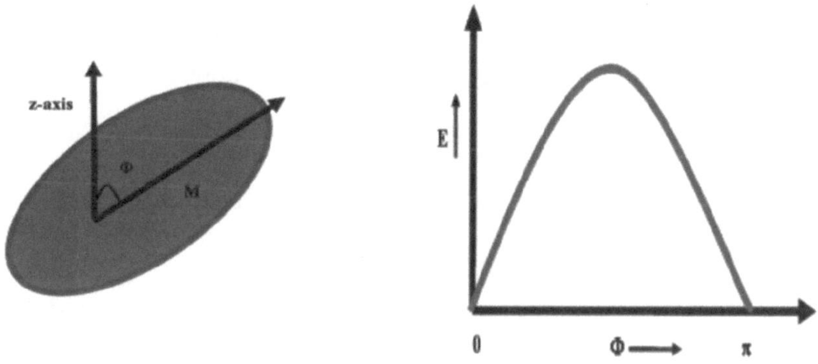

FIGURE 3.1 Schematic depiction of a single domain SPM nanoparticle and the free energy of the particle with the two energy minima positions.

jumps at random between the angles 0 and π. The jump rate is governed by the Arrhenius–Kramer formula

$$\lambda_{0\to\pi} = \lambda_{\pi\to 0} = \lambda_0 \exp\left(-\frac{VK}{k_B T}\right) \tag{3.2}$$

where λ_O is the "attempt" frequency. The reciprocal of the jump rate is the familiar Néel relaxation time:

$$\tau_N = \tau_0 \exp\left(VK / k_B T\right) \tag{3.3}$$

where τ_O is related to the inverse of the attempt frequency of magnetic reversal. As can be seen from Eq. (3.3), the Néel relaxation time can be manipulated by K and V. This is tantamount to changing laboratory parameters such as composition, shape, size, etc. of the nanoparticles. Size-selection has usually been the easiest route to tailor relaxation time [18]. Suspended particles can also relax via a physical rotation of the particle. This is referred to as Brownian relaxation and can be blocked by size selection [19].

3.2.2 HEAT DISSIPATION IN SPM ASSEMBLIES

When sufficiently large in number, the supermoments start interacting via dipole–dipole coupling. To incorporate the effects of dipolar interactions, it is essential to add the contribution due to the "dipolar Hamiltonian" to the magnetic energy of the particle defined by Eq. (3.1). In the limit of large anisotropy, the dipolar Hamiltonian is given by the following form [20–22]

$$H_d = -\sum_{i,j=1}^{N} J_{ij}^d \sigma_i \sigma_j = -\hbar^2 \mu^2 V^2 \sum_{i,j} \gamma_i \gamma_j \frac{\left(1-3\cos^2\theta_{ij}\right)}{\left|r_{ij}\right|^3} \cos\Phi_i \cos\Phi_j \qquad (3.4)$$

Here, the dipolar interaction $J_{ij}^d = \gamma_i \gamma_j \hbar^2 \left(1-3\cos^2\theta_{ij}\right) / \left|r_{ij}\right|^3$ where γ_i and γ_j are the

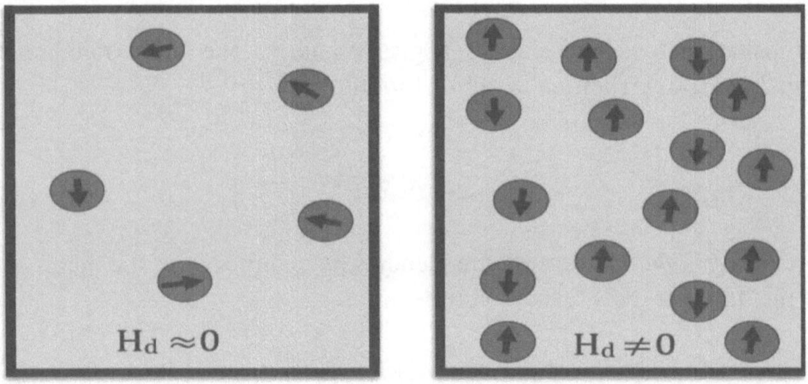

FIGURE 3.2 Schematic picture showing the effect of dipolar field H_d in a magnetic bead. At low concentration, H_d is extremely small (left); at appreciable concentration, H_d leads to ordering in the bead (right).

gyromagnetic ratio of the jth and th particle respectively, $\vec{r}_{ij}$ is the distance between the two particles, θ_{ij} is the angle between $\vec{r}_{ij}$ and the anisotropy axis and Φ is as defined in Eq. (3.1).

A mean field approach is invoked to treat the complicated interaction in Eq. (3.4). Each particle is visualized to be experiencing an effective local magnetic field H_d due to the surrounding (magnetic) medium. Further, the Ising limit allows for the replacement of cos Φ by a two-state variable σ. These simplifications yield the mean-field dipolar Hamiltonian

$$H_d^{MF} = -\mu V \sigma H \tag{3.5}$$

with the field H evaluated self-consistently by the equation

$$H = \mu \Lambda V \langle \sigma \rangle = \mu \Lambda V \tanh\left(\frac{\mu V H}{k_B T}\right) \tag{3.6}$$

The parameter Λ contains all the constants. It fluctuates in sign and has a magnitude dependent on the density of the embedded SPM nanoparticles. The presence of the dipolar field H alters Eq. (3.2) and yields the following generalized rate constants [23]

$$\lambda_{0\to\pi} = \lambda_0 \exp\left(-\frac{V(\mathrm{K}+\mathrm{H}\mu)}{k_B T}\right) \tag{3.7}$$

$$\lambda_{\pi\to 0} = \lambda_0 \exp\left(-\frac{V(\mathrm{K}-\mathrm{H}\mu)}{k_B T}\right) \tag{3.8}$$

The relative populations of magnetic moments aligned along and against the anisotropy axis (corresponding to $\Phi = 0$ and π) are now described by a master equation for a two-state system [21]

$$\frac{dn_0(\mathrm{t})}{dt} = -\lambda_{0\to\pi} n_0(\mathrm{t}) + \lambda_{\pi\to 0} n_\pi(\mathrm{t}) \tag{3.9}$$

where n_0 and n_π denote the fraction of particles with orientation 0 and π, respectively. Solving Eq. (3.9), we can obtain the time-dependent magnetization

$$M(\mathrm{t}) = V\mu\left[n_0(\mathrm{t}) - n_\pi(\mathrm{t})\right] \tag{3.10}$$

This consequence of the dipolar interactions and the resulting dipolar field is depicted schematically in Figure 3.2.

Consider the above assembly subjected to an oscillating magnetic field $h(t) = h_0 \cos \omega t$ along the anisotropy axis where h_0 and ω ($= 2\pi f$) are the amplitude and frequency of oscillation. In this case, it is necessary to replace the dipolar mean-field H in the exponent of Eqs. (3.7) and (3.8) by a time-dependent field $\tilde{H}(t) = H + h(t)$. All the magnetic fields that are present in this system and the net magnetization M are schematically shown in Figure 3.2. The hysteresis loop area $A = \oint M_z \, dh$ can be computed after a few field cycles to provide sufficient time for the transients to settle down. The area of the loop is a measure of the heat dissipation. More precisely, the volumetric power dissipation is given by $P = f \times A$.

3.2.3 TAILORED HEATING USING SCALING LAWS

It is appropriate to keep a note of the parameter values that are relevant in the context of local heating of cells using magnetic nanoparticles. The temperatures required in hyperthermia and chemotherapy treatments are usually in the range of 42–45 °C [1]. The frequency f of the applied oscillating magnetic field is in the 50–1500 kHz range and the field amplitude usually varies between 1 and 200 Oe (0.001–0.02 T) [1]. The biologically safe limit of magnetic material in the human body is within 100 μg ml^{-1}. We have ensured that our evaluations are within the aforementioned parameter ranges. Further, the problem of tissue cooling due to presence of blood flow is difficult to address due to its mathematical complexity. However, an often-used thumb rule is that a heat deposition rate P of 100 m Wcm^{-3} is sufficient in most situations [1].

We now perform a prototypical calculation using magnetite (Fe_3O_4) nanoparticles having a diameter of 8 nm and an anisotropy constant of 4.68×10^5 ergs cm^{-3} [24]. The particle density that we have taken for these evaluations is in the range 10^{12}–10^{16} particles cm^{-3}, well within the toxicity limit [1]. For body temperature $T = 37°C$, the Ising limit $VK >> k_B T$ that we assume holds for diameters in the range 6–12 nm. Our qualitative observations remain unchanged, even in the case of SPM nanoparticles of alternative compounds, as long as we work in this limit.

In Figure 3.3, we plot the response of the assembly to the applied field $h(t)$ for several values of the volume fraction V_f (i.e., N particles cm^{-3}) (A 1% volume fraction corresponds to around 10^{13} particles cm^{-3} and a 40% volume fraction corresponds to nearly 10^{15}particles cm^{-3}). At low densities (~10^{10}–10^{12} particles cm^{-3}), the supermoments continue to remain noninteracting and the reversal is governed by the Néel relaxation time. The magnetization curve is thus the Langevin function. As the volume fraction is increased, the supermoments begin to interact via dipole–dipole coupling yielding a well-defined loop and consequently, heat dissipation. We quantify the effects of h_o, f, and N on the loop area in Figure 3.4. Our numerical evaluations suggest a power law variation $\pm$where the exponents α, β and δ are 1.5$\pm$0.02, 0.95$\pm$0.02, and 1.3$\pm$0.02 for low and 1.1$\pm$0.02, 0.9$\pm$0.02, and 2.0 $\pm$0.02 for high particle densities, respectively [22]. The clinically useful range of the loop area yielding P~100

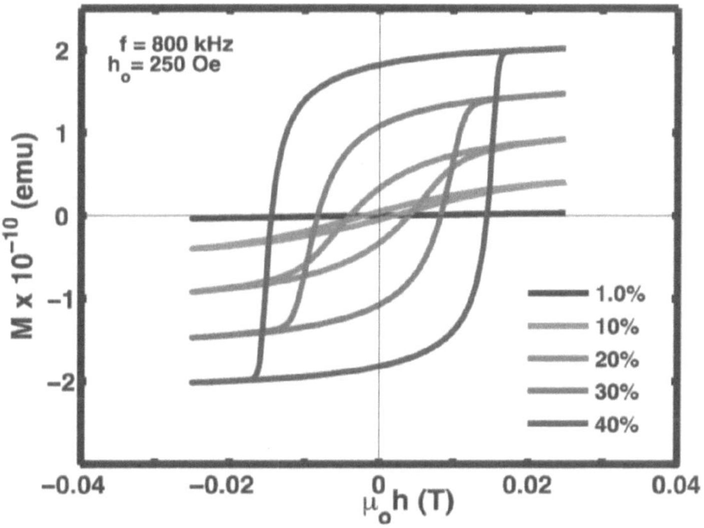

FIGURE 3.3 Typical hysteresis loops for several values of volume fractions. The corresponding number of particles cm^{-3} is specified in the text. (Reprinted with permission from [V. Singh and V. Banerjee, Appl. Phys. Lett. 98, 133702 (2011)]. Copyright [2011], American Institute of Physics).

mW cm^{-3} is encircled in the Figure 3.4. The scaling law can be invoked to select triplets (h_o, f, N) to generate a desired heat deposition rate. Some of them, for the encircled values of A which yield P~100 m Wcm^{-3}, have been provided in Table 3 .1. An appropriate triplet can thus be chosen to suit the conditions in the laboratory.

3.2.4 RETURN POINT MEMORY AND GRADUAL HEATING

A striking phenomenon exhibited by ferromagnets is that of return point memory (RPM). If the field $h(t)$ is made to cycle with a reduced amplitude to generate a subloop, the system returns precisely to the same state from which it left the outer loop. This same memory effect extends to subcycles within cycles. The system thus remembers a hierarchy of states in its past external fields [25, 26]. Figure 3.5 illustrates the return point memory effect for the assembly of magnetite nanoparticles with $N = 10^{14}$ particles cm^{-3}. The inset evaluates the heat deposition rate P as a function of h_o, the maximal field amplitude used to generate the minor loops. We find that return point memory is always observed in standard leaf-shaped loops with saturation. This phenomenon, though unexplored in the context of hyperthermia, seems a promising technique for gradual heating of a local area.

3.3 GUIDELINES FOR COMBINED MH–MRI MODALITY

3.3.1 MEAN FIELD THEORY

A schematic representation indicating the fields experienced by SPM nanoparticles in the combined MH–MRI procedures is shown in Figure 3.6. The dipolar field H is along the z-direction, the transverse magnetic field Γ applied in MRI is in the x-direction, and the

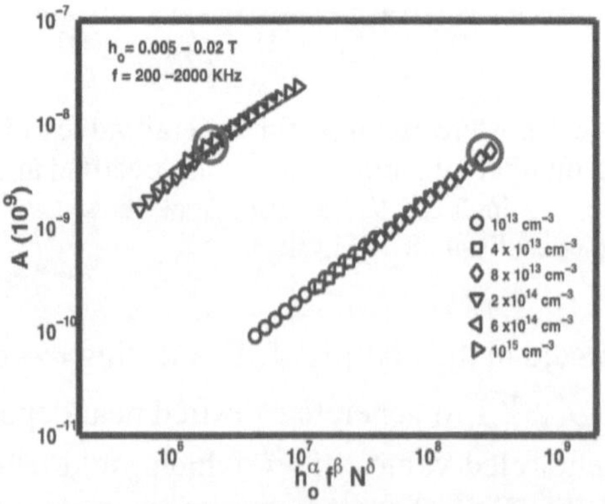

FIGURE 3.4 A scaling plot which demonstrates that the area of the hysteresis loop scales as $A(h_o, f, N) \propto h_o^{\alpha} f^{\beta} N^{\delta}$. The scaling exponents are specified in the text. The encircled values indicate the clinically relevant heat dissipation. (Reprinted with permission from [V. Singh and V. Banerjee, Appl. Phys. Lett. 98, 133702 (2011)]. Copyright [2011], American Institute of Physics).

TABLE 3.1 Sets of triplets (h_o, f, N) corresponding to a heat deposition rate of ~100 m Wcm^{-3} suitable for hyperthermia

h_o (T)	f (kHz)	N (particle cm^{-3})
0.02	1800	4×10^{13}
0.015	1700	8×10^{13}
0.02	1300	8×10^{13}
0.01	900	2×10^{14}
0.015	600	2×10^{14}
0.01	400	6×10^{14}
0.015	300	10^{15}

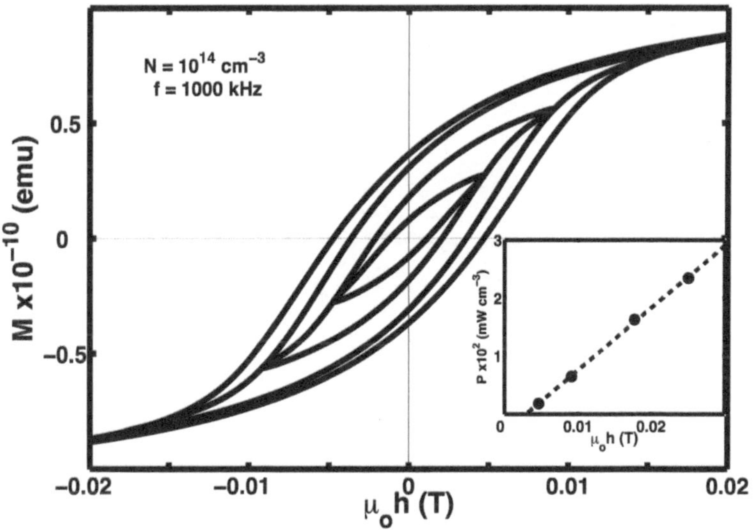

FIGURE 3.5 Hysteresis loop showing return point memory for several minor loops. The inset indicates the heat deposition rate P as a function of h_o, the maximum amplitude of the applied field for the minor loop. (Reprinted with permission from [V. Singh and V. Banerjee, Appl. Phys. Lett. 98, 133702 (2011)]. Copyright [2011], American Institute of Physics).

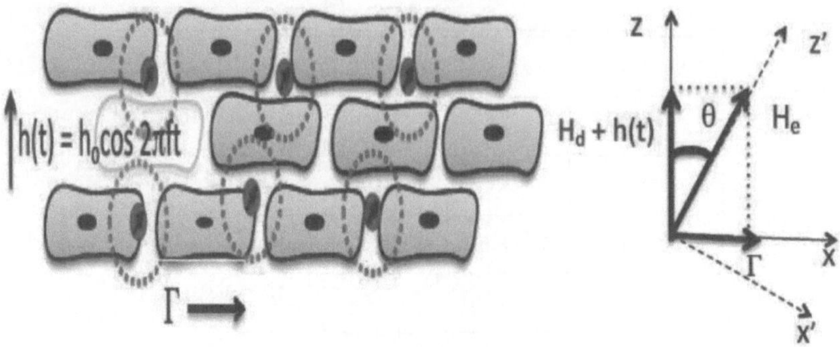

FIGURE 3.6 (a) Super paramagnetic nanoparticles embedded in tissue and subjected to an oscillating magnetic field in the z-direction and a transverse field in x-direction. (b) Schematic representation of the fields and a corresponding rotation of the coordinates along the direction of the effective field H_e. (V.Singh and V. Banerjee, J. Phys. D: Appl. Phys. 46 (2013) 385003 © IOP Publishing. Reproduced by permission of IOP Publishing. All rights reserved).

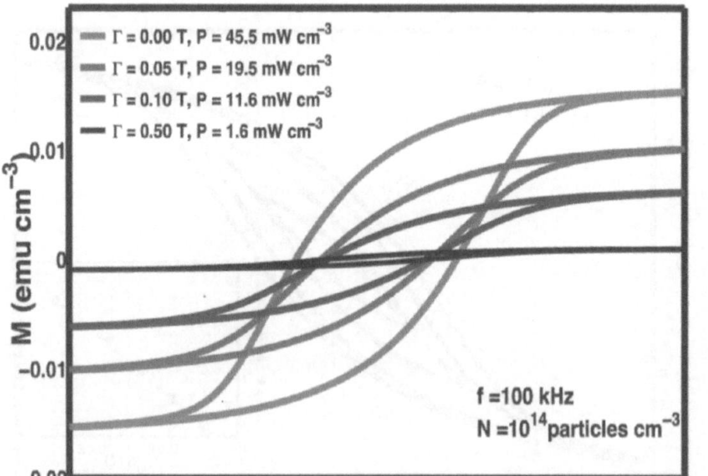

FIGURE 3.7 Typical hysteresis loops corresponding to 10 nm diameter magnetite (Fe_3O_4) nanoparticles for increasing values of Γ, the transverse static field. The loop area shows a marginal change for $\Gamma < h_o$ but a very sharp decrease for $\Gamma \geq h_o$. (V.Singh and V. Banerjee, J. Phys. D: Appl. Phys. 46 (2013) 385003 © IOP Publishing. Reproduced by permission of IOP Publishing. All rights reserved).

oscillating magnetic field $h(t) = h_o \cos(2\pi f t)$ applied for hysteresis in the z-direction. This is a physical realization of the so-called transverse Ising model which holds a special status in statistical mechanics because of its wide applicability to physical systems [27–30]. In the presence of $h(t)$, the Hamiltonian of the TIM is written as [28, 29, 31]

$$H_c = -\sum_{i,j=1}^{N} J_{ij}^{d} \sigma_{zi}\sigma_{zj} - \mu V h(\mathrm{t}) \sum_{i=1}^{N} \sigma_{zi} - \mu V \Gamma \sum_{i=1}^{N} \sigma_{xi} \tag{3.11}$$

Here, σ's are the Ising spins represented by the Pauli matrices and the dipolar interaction J_{ij}^{d} is as defined in Eq. (3.4). As before, mean field theory can be invoked to treat this complicated Hamiltonian. It simplifies to [32]

$$H_c^{MF} = -\mu V H_e \sigma_z \text{ where } H_z = H + h \text{ and } H_z = H + h \tag{3.12}$$

The *effective mean field* H_e along the z-axis experienced by the *effective single spin* is due to all the fields present in the system (see Figure 3.6): (i) the dipolar field H along the z-axis, (ii) the applied oscillating field $h(t)$ along the z-axis, and (iii) the transverse static field Γ along the x-axis. Correspondingly, the components of magnetization are given by

$$M_z(\mathrm{t}) = \frac{H_z}{H_e} M_z^R(\mathrm{t}); \tag{3.13}$$

$$M_x(\mathrm{t}) = \frac{\Gamma}{H_e} M_z^R(\mathrm{t}); \tag{3.14}$$

$$M_y(\mathrm{t}) = 0. \tag{3.15}$$

The loop area $A = \oint M_z \, dh$ can now be evaluated after the initial transients have settled down. It is clear from Figure 3.6, and Eq. (3.13) that the transverse field has the effect of tilting the spin away from the z-axis. In the limit $\Gamma \to \infty$, the spin gets aligned along the x-axis and $M_z \to 0$. The area of the loop is hence zero and there is no heat dissipation or hysteresis!

TABLE 3.2 Set of optimized values of (h_o, f, N, Γ) to obtain heat deposition of ~100–120 mW cm^{-3} suitable for biomedical applications.

h_o (T)	f (kHz)	N (particles cm^{-3})	Γ (T)
0.001	200	5×10^{15}	0.01
0.005	175	2.5×10^{15}	0.03
0.01	120	10^{15}	0.01
0.01	180	2.5×10^{15}	0.03
0.02	90	3×10^{15}	0.05
0.01	200	10^{16}	0.10
0.02	80	2×10^{16}	0.20
0.02	85	3×10^{16}	0.30

3.3.2 SUBVERTING REDUCED HEATING BY PARAMETER SELECTION

Typically, high field MRI machines normally operate at fields in the range 1–10 T, low field machines in the range 0.1–0.4 T, and ultra-low field machines at $\lesssim$0.05 T. Our numerical evaluations of heat dissipation have been made for 10 nm magnetite (Fe_3O_4) nanoparticles with K = 4.68×10^5 ergs cm^{-3} [24]. The chosen particle concentrations are in the range 10^{14} –10^{16} particles cm^{-3}, which is within toxic limits [1]. We again emphasize that qualitative observations remain unchanged for other particle sizes and even for nanoparticles of alternative compounds as long as the Ising limit holds.

In Figure 3.7, we plot hysteresis loops $M_z(t)$ vs. $h(t)$ for different values of the transverse field Γ. The loop area shrinks to zero due to quenching of the *z*-component of the magnetization for reasons discussed in the context of Eq. (3.13). The area is also affected by the field parameters h_o and f and the particle concentration N. We have systematically studied the dependence of A on these parameters. Our numerical data, presented in Figure 3.8, exhibits a power law scaling: $A(h_0, f, N, \Gamma) \sim h_o^{\alpha} f^{\beta} N^{\delta} \Gamma^{\gamma}$, with distinct exponents for ultralow, low, and high values of the

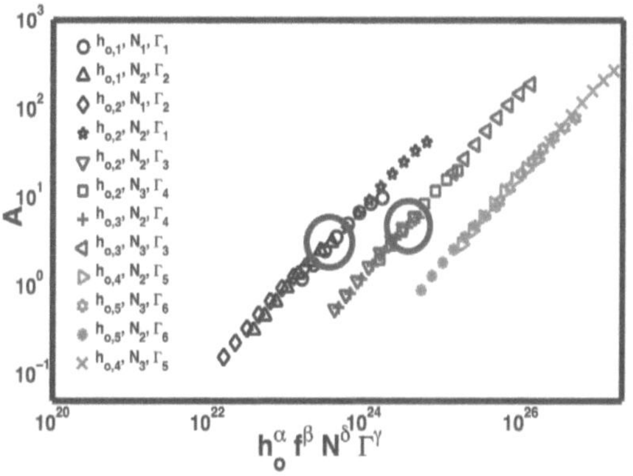

FIGURE 3.8 A scaling plot which exhibits that the area of the hysteresis loop scales as $A(h_0, f, N, \Gamma) \sim h_o^{\alpha} f^{\beta} N^{\delta} \Gamma^{\gamma}$. Here $\mathbf{N}_1 = 10^{14}$, $\mathbf{N}_2 = 10^{15}$, and $\mathbf{N}_3 = 10^{16}$ particles cm^{-3}. $\Gamma_1 = 0.01$ T, $\Gamma_2 = 0.04$ T, $\Gamma_3 = 0.1$ T, $\Gamma_4 = 0.3$ T, $\Gamma_5 = 1.0$ T, and $\Gamma_6 = 5.0$ T. The values of applied field are $h_{o1} = 0.01$ T, $h_{o2} = 0.02$ T, $h_{o3} = 0.03$ T, $h_{o4} = 0.08$ T, and $h_{o5} = 0.1$ T. The frequency $f = 50$–800 kHz. The values of the scaling exponents are given in the text. (V. Singh and V. Banerjee, J. Phys. D: Appl. Phys. 46 (2013) 385003 © IOP Publishing. Reproduced by permission of IOP Publishing. All rights reserved).

transverse field Γ. We find α, β, δ, and γ to be (i) 2.25 ± 0.02, 0.92 ± 0.02, 1.10 ± 0.01, and −0.98 ± 0.01 for ultralow fields, (ii) 2.6 ± 0.02, 0.90 ± 0.01, 1.10 ± 0.01 and −1.00 ± 0:01 for low fields, and (iii) 2.8 ± 0.02, 0.85 ± 0.02, 1.10 ± 0.01, and −1.02 ± 0.01 for high fields.

We have encircled values which correspond to P in the range of ~100 m W cm^{-3}. The scaling laws make it possible to select quadruplets (h_0, f, N, Γ) in many ways to yield the aforementioned range of P. We provide a few of them, chosen from clinically relevant ranges, in Table 3.2. This many-to-one correspondence has important implications: it is possible to undo the effect of changing any one variable on the area Γ by appropriate choices of the others. More contextually, it is possible to mitigate the effect of reduced heating due to the application of Γ by suitable choices of h_o, f, and N! This is demonstrated in Figure 3.9 where we open the loop corresponding to $\Gamma = 0.5$ T of Figure 3.7 by changing h_o, f, and N. The corresponding (significant) rise in heat deposition rate P is also indicated.

3.4 SUMMARY AND DISCUSSION

In this chapter, we have provided a framework to understand heat dissipation in an assembly of SPM nanoparticles. The dipolar interactions between the supermoments play a major role in

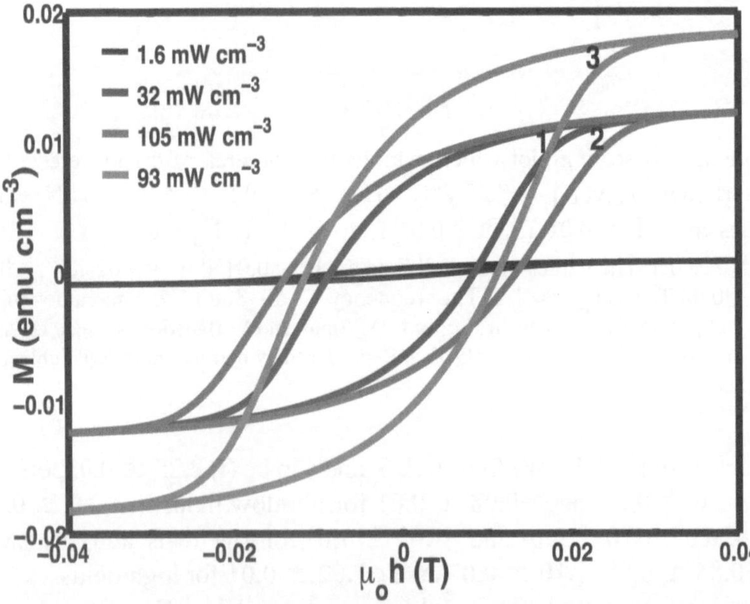

FIGURE 3.9 An increase in hysteresis loop area and hence heat dissipation P (m Wcm^{-3}) by choosing appropriate values of h_o, f and N for a high value of Γ (= 0.5 T) from the scaling law. (V.Singh and V. Banerjee, J. Phys. D: Appl. Phys. 46 (2013) 385003 © IOP Publishing. Reproduced by permission of IOP Publishing. All rights reserved).

creating the effectual ferromagnet. As a consequence, such an assembly, when cycled through an oscillating magnetic field dissipates heat range on account of hysteresis. An important physical application of this phenomenon is in remedial procedures for destroying tumor and cancer cells, usually referred to as MH. Using a mean field approach, we are able to establish precise relationships in the form of scaling laws between heat dissipation and parameters such as particle concentrations, field amplitudes, frequencies, etc. which are easy to alter in a clinical laboratory. It is therefore possible to tailor heat dissipation by a compatible selection of laboratory parameters. And the scaling laws provide not one, but many ways to do it. We hope that the simple mean field framework proves useful in efficient design of experiments in the context of therapeutic applications such as hyperthermia and chemotherapy.

Further, we have also extended our theoretical framework for monitored heat dissipation in MH under surveillance by MRI.
It prevents overheating and damage of collateral tissues. A major obstacle in making it a routine clinical procedure has been decreased heat dissipation due to magnetic fields which are necessarily perpendicular (transverse) in the two protocols. We find that the transverse Ising model (TIM) mimics the environment experienced by magnetic nanoparticles injected in tissues when MH and MRI are performed simultaneously. We derive scaling laws which make it possible to compensate for quenching of heat due to the transverse field by manipulating values of other dependent parameters to ensure significant heat dissipation. We believe that this calculation is an important step toward placing the combined modality of MH and MRI in the mainstream of cancer remedy.

KEYWORDS

- **Dipole**
- **Hyperthermia**
- **MRI**
- **Nanoparticles**
- **Superparamagnetic**

REFERENCES

1. Pankhurst, Q. A.; Connolly, J.; Jones, S. K.; Dobson, J. Applications of Magnetic Nanoparticles in Biomedicine J. Phys. D: Appl. Phys. 2003, 36, R167–R181.
2. Berry, C. C.; Curtis, A. S. G. Functionalisation of Magnetic Nanoparticles for Applications in Biomedicine J. Phys. D: Appl. Phys. 2003, 36, R198–R206.
3. Barreto, J. A.; O'Malley, W.; Kubeil, M.; Graham, B.; Stephan, H.; Spiccia, L. Nanomaterials: Applications in Cancer Imaging and Therapy Adv. Mater. 2011, 23, H18–H40.
4. Hergt, R.; Dutz, S.; Roder, M.; Zeisberger, M. Magnetic Particle Hyperthermia: Nanoparticle Magnetism and Materials Development for Cancer Therapy J. Phys.: Condens. Matter 2006, 18, S2919–S2934.
5. Purushotham, S.; Ramanujan, R. V. Modeling the Performance of Magnetic Nanoparticles in Multimodal Cancer Therapy J. Appl. Phys. 2010, 107, 114701–1-9.
6. Jordan, A.; Scholz, R.; Wust, P.; Fahling, H.; Felix, R. Magnetic Fluid Hyperthermia (MFH): Cancer Treatment with AC Magnetic Field Induced Excitation of Biocompatible Superparamagnetic Nanoparticles J. Magn. Magn. Mater. 1999, 201, 413–419.
7. Bankman, I. N. Handbook of Medical Imaging: Processing and Analysis; Academic Press: USA, 2000.
8. Mehdaoui, B.; Carrey, J; Stadler, M.; Comejo, A.; Nayral, C.; Delpech, F.; Chaudret, B.; Respaud, M. Influence of a Transverse Static Magnetic Field on the Magnetic Hyperthermia Properties and High-frequency Hysteresis Loops of Ferromagnetic FeCo Nanoparticles Appl. Phys. Lett. 2012, 100, 052403–1-3.
9. D´ejardin, P. M.; Kalmykov, Y. P.; Kashevsky, B. E.; Mrabti, H. El.; Poperenchny, I. S.; Raikher, Yu. L.; Titov, S. V. Effect of a DC Bias Field on the Dynamic Hysteresis of Single Domain Ferromagnetic Particles J. Appl. Phys. 2010, 107, 073914–1-6.
10. D´ejardin, P. M.; Kalmykov, Y. P. Effect of a DC Magnetic Field on the Magnetizatio Relaxation of Uniaxial Single Domain Ferromagnetic Particles Driven by a Strong AC Magnetic field J. Magn. Magn. Mater. 2010, 322, 3112–3116.
11. Cantillion-Murphy, P.; Wald, L. L.; Adalsteinsson, E.; Zahn, M. Heating in the MRI Envi- ronment due to Superparamagnetic Fluid Suspensions in a Rotating Magnetic Field J. Magn. Magn. Mater. 2010, 322, 727–733.
12. Tasci, T. O.; Vargel, I.; Arat, A.; Guzel, E.; Korkusuz, P.; Atalar, E. Focused RF Hyperther- mia Using Magnetic Fluids Med. Phys. 2009, 36, 1906–1912.
13. Longmire, M. R.; Ogawa, M.; Choyke, P. L; Kobayashi, H. Biologically Optimized Nanosized Molecules and Particles: More than Just Size Bioconjugate Chem. 2011, 22, 993–1000.
14. Minelli, C.; Lowe, S. B.; Stevens, M. M. Engineering Nanocomposite Materials for Cancer Therapy Small 2010, 6, 2336–2357.
15. Park, J. K.; Jung, J. ; Subramaniam, P.; Shah, B. P.; Kim, C.; Lee, J. K.; Cho, J-H.; Lee, C.; Lee, K-B. Graphite-Coated Magnetic Nanoparticles as Multimodal Imaging Probes and Cooperative Therapeutic Agents for Tumor Cells Small 2011, 7, 1647–1652.
16. Dattagupta, S. Relaxation Phenomena in Condensed Matter Physics; Academic Press: London, UK, 1987.

17. Coffey, W. T.; Kalmykov, Y. P.; Waldron, J. T. The Langevin Equation; World Scientific: Singapore, 1996.
18. Singh, V.; Banerjee, V.; Sharma, M. Dynamics of Magnetic Nanoparticle Suspensions J. Phys. D: Appl. Phys. 2009, 42, 245006–1-9.
19. Singh, V.; Banerjee, V. Ferromagnetism, Hysteresis and Enhanced Heat Dissipation in As- sembly of Superparamagnetic Nanoparticles J. Appl. Phys. 2012, 112, 114912–1-8.
20. Dattagupta, S. A Paradigm Called Magnetism; World Scientific: Singapore, 2008.
21. Chakraverty, S.; Bandyopadhyay, M.; Chatterjee,S.; Duttagupta, S.; Frydman, A.; Sengupta, S.; Sreeram, P. A. Memory in a Magnetic Nanoparticle System: Polydispersity and Interaction Effects Phys. Rev. B 2005, 71, 054401–1-8.
22. Singh, V.; Banerjee, V. Hysteresis in a Magnetic Bead and its Applications Appl. Phys. Lett. 2011, 98, 133702–1-3.
23. Aharoni, A. Effect of a Magnetic Field on the Superparamagnetic Relaxation Time Phys. Rev. 1969, 177, 793–796.
24. Goya, G. F.; Lima, E.; Ardaro, A. D.; Torres, T.; Rechenberg, H. R.; Rossi, L.; Marquina, C.; Ibarra, M. R. Magnetic Hyperthermia with Fe_3O_4 Nanoparticles: The Influence of Particle Size on Energy Absorption IEEE Trans. Magn. 2001, 44, 4444–4447.
25. Sethna, J. P.; Dahmen, K.; Kartha, S.; Krumhansl, J. A.; Roberts, B. W.; Shore, J. D. Hysteresis and Hierarchies: Dynamics of Disorder-Driven First-Order Phase Transformations Phys. Rev. Lett. 1993, 70, 3347–3350.
26. Banerjee, V.; Das, S. K.; Puri, S. Hysteresis and Magnetization Jumps in the T = 0 Dynamics of Spin Glasses Phys. Rev. E 2005, 71, 026105–1-6.
27. de Gennes, P. G. Collective Motion of Hydrogen Bonds Solid State Commun. 1963, 1, 132–137.
28. Stinchcombe, R. B. Ising Model in a Transverse Field. I. Basic Theory J. Phys. C: Solid State Phys. 1973, 6, 2459–2483.
29. Chakrabarti, B. K.; Acharyya, M. Dynamic Transitions and Hysteresis Rev. Mod. Phys. 1999, 71, 847–859.
30. Banerjee, V.; Dattagupta, S. Phase Transition and Relaxation Characteristics of Quantum Magnets and Quantum Glasses Phase Transitions 2004, 77, 525–561.
31. Dattagupta, S.; Puri, S. Dissipative Phenomena in Condensed Matter: Some Applications; Springer: Berlin Heidelberg, 2004.
32. Singh, V.; Banerjee, V. The Transverse Ising Model: A Guide for Combined Modalities of Hyperthermia and Imaging J. Phys. D: Appl. Phys. 2013, 46, 385003–1-5.

CHAPTER 4

SILVER NANOPARTICLES: NEWLY EMERGING ANTIMICROBIALS IN 21ST CENTURY

M. SARAVANAN, VINOY JACOB, K. RAVI SHANKAR, KARTHIK DEEKONDA, JESU AROCKIARAJ, and P. PRAKASH

ABSTRACT

Nanotechnology is the emerging field in the contemporary era, which is the root cause of the next industrial revolution. Bionanoparticles are building blocks of nanobiotechnology as they play a vital role in these applications. Green synthesis of metal nanoparticles renders large-scale production to be free from producing hazardous waste into the environment. Microbial and plant mediated synthesis of nanoparticle from silver and silver derivatives and its biomedical application is the main perspective of this book chapter.

4.1 INTRODUCTION

In contemporary technologies, nanoscience is a fast developing field focusing on a wide spectrum of synthesis and application of different nanomaterials. This emerging field is providing a conclusion for solving many bottlenecks in the field of the sciences [1]. Nanomaterials have received much attention because of their structure and properties varying from those of atoms, molecules and bulk materials [2], and thereby having many potential applications *[3,4]*. Owing to their vast applications, there is immense demand to obtain these nanoparticles in a non-agglomerated, uniform with a well-controlled mean size and narrow distribution *[5,6]*. Nanoparticles referred to as those particles with size upto 100 nm [7]. Silver, gold, copper are some of the noble metals that used for nanoparticle synthesis [8]. Among these, silver nanoparticles (AgNPs) have great applications in various fields, especially in the fields of biological systems, living organisms, medicine and environmental biotechnology *[9,10]*. These have important properties like catalytic activity, surface enhanced Raman scattering effect [11–15], good conductivity, chemical stability, antibacterial activity [16–18], therapeutics [19,20], degrading pesticides [21], filtration and nano-coated medical devices [22]. Therefore, a keen attention focused on synthesis of AgNps in numerous ways.

Considering the fact, the overall scenario of AgNPs among multidisciplinary fields, the production of them in bulk amount has a great criterion in this field of nano-biotechnology. A number of physical, chemical and biological approaches are available for the synthesis of silver nanoparticles [23]. In a broad view, the physical synthesis procedures involve evaporation, condensation, laser ablation, and chemical procedures involve reduction of metal ions in the solution to form small metal clusters or aggregates [24, 25]. Silver is known for its antimicrobial properties, has been used for years in the medical field for antimicrobial applications, and even has shown to prevent HIV binding to host cells. Additionally, silver has been used in water and air filtration to eliminate microorganisms. Silver ions play a crucial role in the antibacterial activity of silver zeolite. Silver nanoparticles synthesized using a variety of methods including, bio-reduction and solvent phase evaporation [26,27]. In one way, the techniques for the synthesis categorized into top-down and bottom-up methods: One involves molding and etching materials into smaller components, and the other involves the assembling structures into useful devices The top down approach use silver in bulk form and reduces its size mechanically to nano scale by lithography and laser ablation [28].

Current trends includes the wide spread use of semiconductor nanocrystals in biological applications as well as key developments in the synthesis and physical properties of semiconductor nanocrystals used in biological environments. In recent years, semiconductor nanocrystals have been synthesized and evaluated for their potential as fluorescent biological labels. Further more, fluorescent semiconductor nanocrystals may be bound to biomolecules to facilitate selective binding of these nanoscale fluorescent structures to specific subcellular structures. Many interesting nanodevices are useful in biomedical field especially for the improved cancer detection, diagnosis and treatment. In case of bottom-up approach, in other words self-assembly method involves the addition of reducing agent into a silver salt solution followed by the addition of a stabilizing agent to prevent agglomeration on synthesized nano particles. Other physical and chemical modes of silver nanoparticles synthesis are of dielectric approach, where magnetron sputtering, ion exchange sol-gel deposition and convective methods followed. Ion implantation is most commonly used fabrication method, which provides the controlled synthesis of metal nanoparticles and reduction of metal salts in the solutions [29–32].

Nanoparticles in their pure or in the form of mixtures help in the formation of sensors, in batteries, in diagnostic kits, in water treatment, also helpful in curing deadly diseases such as cancer. Biological molecules have the qualities by which it can undergo highly controlled and hierarchical assembly for making them suitable for the development of a reliable and Ecofriendly process for metal nanoparticles synthesis. It can also control chemical and photochemical reactions in reversible micelles [33,34], thermal decomposition of silver compounds [35,36] as solvothermal synthesis [37], radiation assisted [38] like ultrasonic radiation, laser irradiation and radio lysis [39,40], electrochemical methods [41–44], sonochemical approach [45,46], facile methods and microwave assisted processes [47]. The surface passivation reagents, including surfactant molecules and polymers, are needed to prevent the nanoparticles from aggregation. Polyethylene glycol has been widely applied as an effective passivation agent in the fabrication of Ag NPs and other metal nanoparticles. Recent reports revealed that PEG 200 was effective for the control of size and shape of AgNPs. The surface modification of these colloidal nanoparticles is very important to facilitate their application to biotechnology and nanocomposites. Nanosilver can be used as liquid form, such as a colloids (used for coating and spraying) or as a shampoo (liquid form) and can also appear embedded in a solid such as a polymer master batch or be suspended in a bar of soap (solid). Nanosilver can also be utilized either in the textile industry by incorporating it into the fiber (spun) or employed in filtration membranes of water purification systems.

Now, there is a great need to develop environmentally safe processes for the synthesis of nanoparticles that avoid toxic chemicals in the synthesis protocols [48]. Biosynthetic methods employing microorganisms and plant extracts have emerged as a simple and viable alternative to chemical synthetic procedures and physical methods [49]. Recently developed method: Biomimetics is a technique in which we use biological systems such as yeast, fungi, bacteria and plants for the synthesis of nanostructures of biocompatible metal and semiconductors (Figure 4.1). Silver toxicity towards wide range of micro-organisms has long been known. Silver can even destroy antibiotic resistant bacteria such as Methicillin resistant *Staphylococcus aureus*. In fact, bacteria are not able to develop resistance against silver like they do with antibiotics. Among all the well known activity of silver ions and silver-based compounds, silver nanoparticles proved to be the material of choice as they kill microbes effectively. Bacterial membrane proteins and DNA make preferential sites for silver nanoparticles interaction as they possess sulfur and phosphorus compounds and silver have higher affinity to react with these compounds. When silver nanoparticles enter the bacterial cell, it forms a low molecular weight region in the center of the bacteria to which the bacteria conglomerates, thus protecting the DNA from the silver ions. The nanoparticles preferably attack the respiratory chain, cell division finally leading to cell death. The nanoparticles release silver ions in the bacterial cells, which enhance their bactericidal activity.

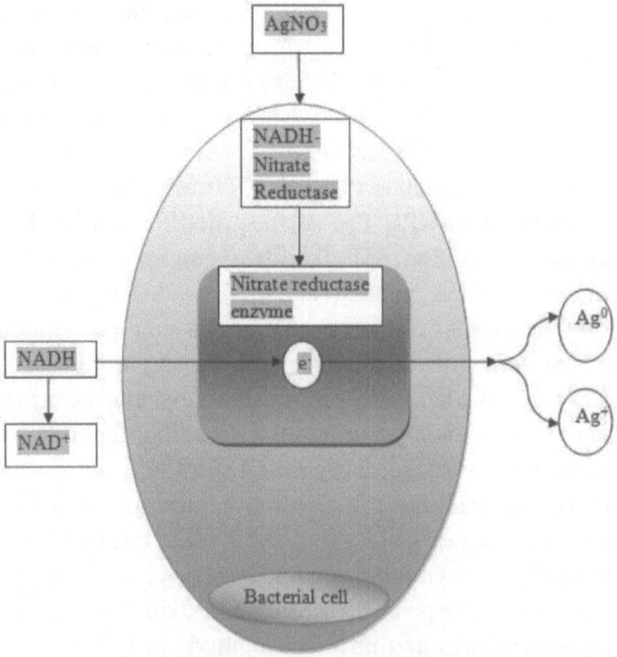

FIGURE 4.1 Bio-reduction of silver nitrate into silver ions in bacterial cell.

4.2 METHODS FOR SILVER NANOPARTICLE SYNTHESIS

The synthesis of metallic nanoparticles is an active area of "Applied research" in nanotechnology. A variety of chemical and physical procedures can be used for synthesis of metallic nanoparticles (Figure 4.2). However, these methods are fraught with many problems including use of toxic solvents, generation of hazardous by-products and high-energy consumption. Accordingly, there is an essential need to develop environmental friendly procedures for the synthesis of metallic bio-nanoparticles. A promising approach to achieve this objective is to exploit the array of biological resources in nature. Indeed, over the past few decades, plants, algae, fungi and bacteria have been widely used for production of low-cost, energy-efficient, and nontoxic metallic bio-nanoparticles. Synthesis of Silver nanoparticles by microbes is due to their resistance mechanism. The resistance caused by the bacterial cell for silver ions in the environment is responsible for its nanoparticles synthesis. The silver ions in nature are highly toxic to the bacterial cells. Therefore, their cellular machinery helps in the conversion of reactive silver ions into stable silver atoms. In addition, temperature and pH plays an important role in their production. At room temperature, the size of nanoparticles is 50 nm; at higher temperature, i.e. at 60°C, the size of nanoparticles reduces to 15 nm. This indicates that with the increase in temperature size decreases. Under alkaline conditions, nanoparticles synthesis by the microbe is more as compared to the acidic conditions. Nevertheless, after pH 10 cell deaths occur. The first evidence of the synthesis comes from *Pseudomonas stutzeri* AG259, a bacterial strain that was originally isolated from silver mine.

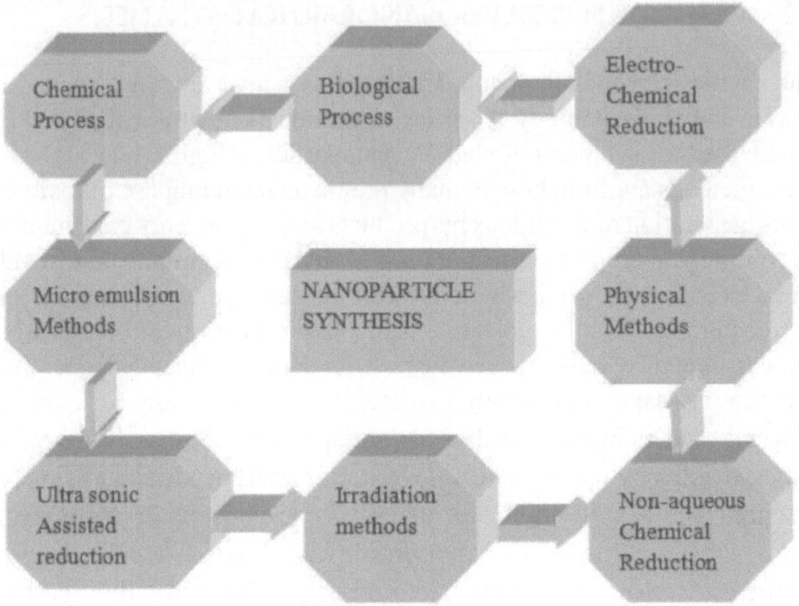

FIGURE 4.2 Various methods for the synthesis of silver nanoparticles.

4.2.1 CHEMICAL SYNTHESIS OF AGNPS

The Chemical reduction method is widely used to synthesize Ag-NPs because of its readiness to generate Ag-NPs under gentle conditions and its ability to synthesize Ag-NPs on a large scale. However, these chemical synthesis methods employ toxic chemicals in the synthesis route which may have an adverse effect in the medical applications and hazard to environment. Therefore, preparation of Ag-NPs by green synthesis approach has advantages over physical and chemical approaches as it is environmental friendly, cost effective and the most significant advantage is that conditions of high temperature, pressure, energy and toxic chemicals are not required in the synthesis protocol [50]. There are lots of research worked carried out in the electro-chemical reduction and non-aqueous chemical reduction. Generally, the chemical synthesis process of the Ag-NPs in a solution usually employs the following three main components: (i) metal precursors, (ii) reducing agents and (iii) stabilizing/capping agents. Polyethylene glycol (PEG) is frequently used in the polymer blends production to improve the biocompatibility of its film due to its wide range of molecular weights, excellent solubility in aqueous medium, low toxicity, chain flexibility and biocompatibility properties. Although PEG has non biodegradability properties, it is readily excreted from the body and forms non-toxic metabolites. Besides that, PEG was able to act both as reducing agent and stabilizer. In several research studies, researchers proposed that long polymer chain of PEG exhibits higher reducing activity and provides higher stability in forming Ag-NPs. These can effectively prevent agglomeration of Ag-NPs. the silver nitrate ($AgNO_3$) in different stirring times of reaction at moderate temperature with sugar and PEG used as green reducing agent and polymeric stabilizer [51]. Plasmonic properties of silver nanoparticles (AgNPs) have been extensively studied for their superior performances that exceed those of other metals with a surface plasmon resonance (SPR) in the visible range like gold or copper.

4.2.2 PHYSICAL SYNTHESIS OF AGNPS

For a physical approach, the metallic NPs can be generally synthesized by evaporation and condensation, which could be carried out by using a tube furnace at atmospheric pressure. However, in the case of using a tube furnace at atmospheric pressure there are several drawbacks such as a large space of the tube furnace, great energy consumption for raising the environmental temperature around the source material and a lot of time for achieving thermal stability. Therefore, various methods of synthesis of Ag-NPs based on the physical approach have been developed. The Ag-NPs were formed by decomposition of a Ag^+-oleate complex, which was prepared by a reaction with $AgNO_3$ and sodium oleate in a water solution, at high temperature of 290°C. The average particle size of the Ag-NPs was obtained of about 9.5 nm with a standard deviation of 0.7 nm [52]. In another work, an attempt to synthesize metal NPs via a small ceramic heater that has a local heating area. The small ceramic heater was used to evaporate source materials [53]. Another work reported that the arc discharge method to fabricate Ag-NPs suspension in deionized water with no added surfactants. In this synthesis, silver wires (Gredmann, 99.99%, 1mm in diameter) were submerged in deionized water and used as electrodes. The experimental results show that Ag-NPs suspension fabricated by means of arc discharge method with no added surfactants contains metallic Ag-NPs and ionic silver [54]. The physical synthesis process of Ag-NPs usually utilizes the physical energies (thermal, AC power, arc discharge) to produce Ag-NPs with nearly narrow size distribution. The physical approach can permit producing large quantities of Ag-NPs samples in a single process. This is also the most useful method to produce Ag-NPs powder. However, primary costs for investment of equipment should be considered.

4.2.3 MICRO EMULSION METHOD, ULTRA SONIC MEDIATED AND IRRADIATION REDUCTION SYNTHESIS

The photo-induced synthetic strategies can be categorized into two distinct approaches, that is the photophysical (top down) and photochemical (bottom up) ones. The former could prepare the NPs via the subdivision of bulk metals and the latter generates the NPs from ionic precursors. The NPs are formed by the direct photoreduction of a metal source or reduction of metal ions using photo-chemically generated intermediates, such as excited molecules and radicals, which is often called photosensitization in the synthesis of NPs [55].

The direct photo-reduction process of $AgNO_3$ in the presence of sodium citrate (NaCit) was carried out with different light sources (UV, white, blue, cyan, green and orange) at room temperature. It was shown that this light-modification process results in a colloid with distinctive optical properties that can be related to the size and shape of the particles. A simple and reproducible UV photo-activation method for the preparation of stable Ag-NPs in aqueous Triton X-100 (TX-100) was reported [56,57].

4.2.4 MICROBIAL SYNTHESIS OF AGNPS

Recently, many studies were conducted to explore the synthesis of AgNPs using microorganisms as a potential bio sources; e.g. *Fusarium oxysporium* and *Verticillium* spp. [58], *Fusarium semitectum* [59]. New enzymatic approaches using bacteria and fungi in the synthesis of nanoparticles by both intra and extracellular, expected to play a key role in many conventional and emerging technologies.

4.2.4.1 MECHANISM OF BIOSYNTHESIS OF AGNPS BY FUNGI

In the biosynthesis of metal nanoparticles by a fungus, the fungus mycelium is exposed to the metal salt solution. That prompts the fungus to produce enzymes and metabolites for its own survival. In this process, the toxic metal ions are reduced to the non-toxic metallic solid nanoparticles through the catalytic effect of the extracellular enzyme and metabolites of the fungus. Fungi can reduce metal ions to nano-sized particles by two different mechanisms. The extracellular synthesis method may possibly involve the NADPH dependent nitrate reductase enzyme that is secreted by the fungi into the reaction medium (Table 4.1). The process of reduction of metal ions to the nano level is accompanied the simultaneous conversion of NADPH to $NADP^+$. It is hypothesized that the hydroxyquinoline shuttles the electrons generated during the enzymatic reaction, involving the conversion of nitrate to nitrite consecutively Ag^+ ions conversion to Ag^0. Absence of the enzyme from the reaction medium leads to the conspicuous disappearance of all the bands thus validating the active enzymatic role in the whole process [60].

The other mechanism involves the intracellular process wherein the fungal cell walls along with sugars present in them play a very important role. The cell wall composition varies during the fungal life cycle and its inner side is associated with a microfibrillar component that is embedded in an amorphous matrix material. The latter usually is made up of water-soluble polysaccharides whereas the cell contain chitin and beta linked glucans. The positively charged groups of cell wall enzymes may responsible for absorbing metals ions from the medium and subsequently these get reduced to nanometals by enzymes in the cell wall. Microscopic studies show aggregates of these particles in the cell wall as well as in the cytoplasmic membrane and cytoplasm. This probably leads to the conclusion that some particles may diffuse through the cell wall and enter the cytoplasm where the reduction takes place with the help of cytoplasmic enzymes also. Production of silver nanoparticles through fungi has several advantages. They include tolerance towards high metal nanoparticle concentration in the medium, easy management in large-scale production of nanoparticles, good dispersion of nanoparticle and much higher amounts of protein expressions. Compared to bacterial broth, fungal broth can be easily filter-by-filter press or similar commonly used equipment, thus saving considerable investment costs for specialized equipment that may be needed for other methods. As a result, for large-scale production of nanoparticles fungi is preferred over other methods [61].

4.2.4.2 MECHANISM OF BIOSYNTHESIS OF AGNPS BY BACTERIA

Silver ions and silver, based compounds are highly toxic to microorganisms and hence show strong bactericidal effects in many common species of bacteria including *Escerichia coli*. AgNPs found to be accumulating in the bacterial membranes, in some way interacting with certain building elements (ribosome and nucleic acid) of the bacterial cells, thus causing structural changes, degradation and finally cell death (Figure 4.3). It is well known that many organisms can provide inorganic materials either intra or extracellularly but it is very important to develop AgNPs in an eco-friendly and easier manner (Table 4.2). Silver bionanoparticles can be produced by physical and chemical methods [62,63], whereas they can also be produced in biological methods specifically by bioreduction [64]. Extracellular synthesis of silver nanoparticles from microorganisms has been proving fruitful, because it possesses antimicrobial activity [65–67]. Another reason to use AgNPs over bulk silver metals is due to its high specific surface area and high fraction of surface atoms; hence, it proved as a good antimicrobial agent for pathogenic microorganisms [68–70].

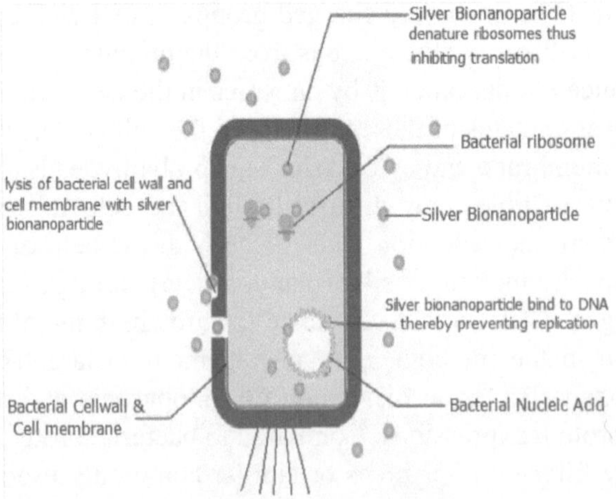

FIGURE 4.3 Mechanism of bacterial lysis by silver bionanoparticle.

4.2.4.3 MECHANISM FOR MICROBIAL SYNTHESIS: EXTRACELLULAR AND INTRACELLULAR

Extracellular enzyme shows an excellent redox properties and it can act as an electron shuttle in the metal reduction. Many other compounds other than extracellular enzymes like hydroquinones, napthoquinones and anthroquinones, act as electron shuttle in metal reduction [71–73]. Studies have indicated that NADH and NADH-dependent enzymes are important factors in the biosynthesis of metal nanoparticles and initiation of metal reduction seemed to be due to the transfer of electron from NADH by NADH- dependent reductase as an electron carrier [1, 65,67,74,75]. Ag^+ ion is a highly active chemical agent, which binds strongly to electron donor groups and this binding with biomolecules like protein could restrict the size of the particle [73]. It is shown that certain fungi have the ability of producing extracellular metabolites that serve as agent, for their own survival when exposed to such environmental stresses like toxic materials (such as metallic ions), predators and temperature variations [76]. To a certain degree, the rate of intracellular particle formation can be controlled by altering the parameters such as pH, temperature, substrate concentration and exposure time to substrate [77]. Even at high metal ion concentrations, microorganisms have tolerance against metal stress by efflux systems, alteration of solubility, toxicity, biosorption, bioaccumulation and extracellular complexation or precipitation of metals [78]. It is also shown that fungi produces wide variety of extracellular enzymes and metabolites such as industrial production of glucosidase, paracelsin, protein, acetyl xylem asterase, cellobiohydrolase D, cellulose, hemicellulase, cell wall lytic enzyme, β-glucosidase, β-1, 3-glucanase, and glucose at industrial scale [79]. It should also be mentioned that *Trichoderma reesei* is the best studied cellulolytic fungus. It is widely used for the large-scale gene transformation and other biotechnology industries dealing with overexpression of extracellular enzymes [80].

TABLE 4.1 Variousfungal strains used for the biosynthesis of AgNPs

Biological entity (Fungal stains)	Size(nm)	References
Aspergillus clavatus	550–650	[71]
Aspergillus flavus	8.92±1.61	[81]

TABLE 4.1 *(Continued)*

Aspergillus fumigates	5–25	[66]
Trichoderma reesei	5–50	[63]
Alternaria solani GS1	5–20	[82]
Penicillium funiculosum GS2	5–10	[82]
Cladosporium cladosporioides	10–100	[83]
Neurospora crassa	11	[84]
Phoma sp.	71–74	[85]
Verticillium sp.	25±12	[86, 87]
Nitrate reductases – Fusarium oxysporium	10–25	[88]
Penicillium sp. K_1	10–100	[89]
Penicillium sp. K_{10}	18–100	[90]
Trichoderma asperellum	13–18	[91]
Trichoderma viride	5–40	[88]
Fusarium oxysporum	5–60	[1, 65, 74, 92, 93, 94]
Fusarium oxysporum PTCC 5115	50	[95]
Fusarium acuminatum	5–40	[96]
Fusarium solani (USM-3799)	5–35	[97]
Cladosporium cladosporioides	10–100	[83]
Volvariella volvacea	15	[97]
Phoma sp. 3.2883	67.6–74.52	[85]
Verticillium sp.	12–37	[98, 59]
Pencillium brevibacterium WA 2315	23–105	[99]
Yeast strain MKY_3	2–5	[62]
Morganella sp.	20±5	[100]
Pencillium fellatunum	5–25	[101]
Penicillium brevicompactum WA 2315	23–105	**[99]**
Fusarium semitectum	35	**[102]**
Phaenerochaete chrysosporium	50–200	**[103]**

TABLE 4.2 Various bacterial strains used for the synthesis AgNPs

Biological entity (Bacteria)	Size(nm)	References
Pseudomonas stutzeri	200	[104]
Staphylococcus aureus	160–180	[105]
Klebsiella pneumonia	5–32	[106]
Escherichia coli	50	[107]
Plectonema boryanum UTEX 485	200	[108]
Brevibacterium casei	10–50	[109]
Bacillus sp.	5–15	[110]
Bacillus lichenformis	50	[111]
Corynebacterium strain SH09	10–15	[112]
Pseudomonas stutzeri AG259	100–200	[112]
Pseudomonas AG4	60–150	[112]
Acetobacter xylinum (strain TISTR 975)	11.34–6.31	**[113]**
Bacillus megaterium (NCIM 2326)	80–98.56	[114]
Bacillus cereus	62.8	[115]

4.2.5 MECHANISM FOR PHYTOSYNTHESIS OF SILVER NANOPARTICLES

Green synthesis of AgNPs follows benign protocols and materials, which is cost effective and involves bulk synthesis. As its procedures, involve no high pressure, energy, temperature and toxic chemicals [116]. Using plant extracts for nanoparticle synthesis can be advantageous over other biological processes by eliminating the elaborate process of maintaining the cell cultures. Phytosynthesis involves preparation of plant extract and then reduction of silver ions into nanoparticles. Extraction procedure is usually by boiling of the air-dried leaf pieces followed by decantation or filtration. To this extract, add silver nitrate and then incubated to produce nanoparticles. The terpenoids believed to be surface-active molecules stabilizing the nanoparticles and the reaction of the metal ions, possibly facilitated by reducing sugars in neem leaf broth [117–119]. Proteins that have amine groups played a reducing and controlling role during the synthesis procedure in *Capsicum annum* [120]. *Desmodium* contain chemically different groups, water-soluble scavenging super-oxide anion radicals and 1, 1-diphenyl-2-picrylhydrazyl (DPHP) radicals present in the plant extract can be responsible for the reduction of silver and the synthesis of nanoparticles [121] and lots of works reported on the silver nanoparticles synthesized from the plant sources [121–123]. The synthesis of AgNPs from plants (Table 4.3) is belonging to basellaceae, asteraceae, poaceae, and their rates of reduction of silver nitrate [8]. Among the plants used the spectrometric analysis of sunflower reaction mixture exhibited strong absorption between 400–500 nm and the particles characterized by XRD. Lingarao and Savitramma accounted that, the phytosynthesis of AgNPs from *Svensonia hyderabadensis* are belongsto Verbenaceae family and listed under the rare taxa [121]. Phytochemical analysis of plants revealed mechniastic aspects of phytosynthesis due to the presence of flavonoids and glycosides. These constituents may play a vital role in the reduction of silver ions and formation of silver nanoparticles. The presence of flavonoids, terpenoids, proteins and reducing sugars in the plant extract are reported to possess stabilizing and capping agents [118,119].

TABLE 4.3 The choice of plant materials used for phyto synthesis of AgNPs

Biological entity (Plant)	Size(nm)	References
Azadirachta indica	50–100	[119]
Aloe vera	15–20	[124]
Emblica officinalis	10–20	[116]
Cinnamomum camphora	55–80	[125]
Parthenium hysterophorus L	40–50	[9]
Ethno-medicinalplant *Gloriosa superba L*	5–20	[82]
Diopyros kaki	15–19	[126]
Camellia sinensis	30–40	[127]
Jatropha curcas(seed)	15–20	[128]
Jatropha curcas(latex)	10–20	[129]
Cinnamomum camphora(dried leaf)	55–60	[130]

TABLE 4.3 *(Continued)*

Pine apple (leaf extract)	15–500	[126]
Persimmon (leaf extract)	15–500	[126]
Ginkgo (leaf extract)	15–500	[126]
Magnolia (leaf extract)	15–500	[126]
Platanus(leaf extract)	15–500	[126]
Geranium(Pelargonium graveolens) leaf extract	16–40	[129]
Phoma glomerata	60–80	[131]
Cinnamomum camphora	5–40	[132]
Macrotyloma uniflorum	12	**[133]**
Anacardium occidentale	15.5	**[134]**
Hibiscus rosa sinensis	14	**[135]**
Cochlospermum gossypium	3	**[136]**
Henna leaf	39	**[137]**
Ocimum tenuiflorum	28	**[138]**
Solanum tricobatum	26.5	**[138]**
Syzygium cumini	65	**[138]**
Centella asiatica	22.3	**[138]**
Citrus sinensis	28.4	**[138]**
Mimusops elengi, L	55–83	**[139]**

4.3 BIOMEDICAL APPLICATIONS OF AG-NPS

Historically, silver compounds and ions extensively used for both hygienic and healing purposes [140, 141]. The first recorded medical use of silver reported during 8th century. Silver vessels used in ancient times to preserve water and wine whereas silver powder used for beneficial purposes like healing and anti-disease properties like ulcer treatment as believed by Hippocrates(father of modern medicine). However, later the uses of silver compounds have emerged in medical practices as well [142]. In the late 1960s, Moyer and Monafo introduced silver nitrate 0.5% solution for burn wound treatment [143] as well in the same period Fox introduced silver sulfadiazine cream for burn wound management [144]. In 1884, Crede German obstetrician introduced 1% silver nitrate as an eye solution for prevention of Gonacoccal opthalmia neonatorum, which is perhaps the first scientifically documented medical use of silver. The disinfectant properties of silver exploited for hygiene and medical purposes, for the treatment of mental illness and infectious diseases like syphilis and gonorrhea [140]. In higher concentrations, silver is toxic to the human beings whereas in low concentration it is non-toxic [29]. Wide ranges of applications are being developed in consumer products, ranging from disinfecting medical devices and home appliances to water treatment [140]. Silver-doped calcium phosphate nanoparticles and silver acetate showed similar effects toward mammalian and prokaryotic cells with toxic silver concentrations in the range of 1–3 $\mu g\ mL^{-1}$ [145].

Silver has been known since ancient times as an effective antimicrobial agent for the treatment of diseases, for food preservation and to keep water safe. With the recent advancements in the field of nanotechnology, AgNPs have been widely used as a novel therapeutic agent extending their use as antibacterial, antifungal, antiviral, anti-inflammatory and anti-cancerous agents [145, 146]. The antimicrobial activity of silver nano particles is comparable or better than the broad spectrum of most prominent antibiotics used worldwide and is dependent on the size of nanoparticle [147]. Silver sulfadiazide as an essential anti-infective tropical medicine listed by World Health Organization (WHO). AgNPS, varied in-vivo and in-vitro applications, as it has the highest bactericidal activity and biocompatibility among all the known antibacterial nanoparticles and it also is more advantageous since it does not require any photocatalytic agent for bactericidal action as required by TiO_2, ZnO, CdSe, ZnS etc [148–151]. Dunn and Edwards-Jones affirmed that some nanosilver applications have received approval from the US FDA [152]. The mechanism of the antimicrobial action of the silver ions is closely related to their interaction with thiol groups in enzymes and proteins that also plays an essential role in its antimicrobial action [153–155]. The biologically synthesized silver nanoparticles could be of immense use in medicinal textiles for their efficient antimicrobial function [156]. The sterile cloth and materials play an important role in hospitals, where often wounds contaminated with microorganisms

[157]. Thus, to reduce or prevent infections, various antibacterial disinfection techniques developed for all types of textiles. Silver ions and silver nano particles have inhibitory and lethal effect on bacterial species such as *E. coli, S. aureus* and even yeast [158,159].The antibacterial activity of Ag NPs against four types of Gram negative bacteria such as *E. coli, V. cholera, P. aeruginosa* and *S. typhus* were also defined [160].

Antimicrobial drug resistance is becoming a major factor, virtually in all hospitals. The acquired infection has become untreatable which has led to a serious public health problem [161]. These concerns have led major research effort to discover alternative strategies for the treatment of bacterial infection [162].Nanobiotechnology is an upcoming and fast developing field with potential application for human welfare. An important area of nanobiotechnology is to develop a reliable and eco-friendly process for synthesis of nanoscale particles through biological systems [163]. Many organisms including unicellular and multicellular microorganisms have been explored as a potential bio-factory for synthesis of metallic nanoparticles (gold, silver, Cadmium sulfide) either intracellularly or extracellularly [62, 65,156,164,165].

Recently few studies, conducted for characterization and antimicrobial effect of silver nanoparticles. The bulk counterparts of AgNPs are an effective antimicrobial agent against various pathogenic microorganisms [92]. The range of 10–15 nm of AgNPshas increased stability and enhanced antimicrobial potency [166]. Bactericidal effect of AgNPs against multidrug resistant bacteria like *Pseudomonas aeroginosa*, ampicillin resistant *E. coli* and erythromycin resistant *Streptococcus pyogenes* was pointed out by some studies [167,168]. The combination effect of AgNPs with different antibiotics described and proved that the antibacterial activities of penicillin G, amoxicillin, erythromycin, clindamycin, and vancomycin increases in the presence of AgNPs against *S. aureus* and *E. coli* [169]. The possibility of free radical involvement in the antibacterial activity of AgNPs reported, but the underlying mechanism and characteristics remain unclear [35]. AgNPs have been used extensively as anti-bacterial agents in the health industry, food storage, textile coatings and a number of environmental applications. It is important to note that despite of decades of use, the evidence of toxicity of silver is still not clear. A range of accredited bodies, including the US FDA, US EPA, SIAA of Japan, Korea's Testing and Research Institute for Chemical Industry and FITI Testing and Research Institute, has approved products made with AgNPs. Moreover, this encouraged the textile industry to use AgNPs in different textile fabrics. In this direction, silver nanocomposite fibres were prepared containing silver nanoparticles incorporated inside the fabric. The cotton fibres containing AgNPs exhibited high anti-bacterial activity against Escherichia coli. The Studies described the interaction between reactive Oxygen Species (ROS) and bacterial cell death. When AgNPs interact with bacterial DNA or mitochondria, it releases ROS such as superoxide anion (O_2^-), hydroxyl radical (OH^-) and singlet oxygen ($^1O_2$) with subsequent oxidative damage.In many studies, direct electron microscopic view determined the structural change of bacterial cell, confirming the cell damage [169]. Studies suggested that the mode of action of silver nanoparticles is similar to that of silver ions, which complex with electron donor groups containing sulfur, oxygen or nitrogen atoms normally present as thiols or phosphates on amino acids and nucleic acids [170]. In addition to their effect on bacterial enzymes, silver ions cause marked inhibition of bacterial growth and they get deposited in the vacuole and cell wall as granules. In the recent times, bioflocculant (polysaccharide) derived from *Bacillus subtilis* MSBN17 was used as a stabilizing agent for AgNP synthesis. They also reported that AgNPs (Core)–chloramphenicol complex inhibits the peptidyl transferase in the 50S ribosomal subunit leading to greater damage to bacterial cells [171].

Candida species is one of the most common fungal pathogens causing hospital acquired sepsis with a mortality rate of about 40%. Prophylaxis with anti fungal materials may lead to the raise of many drug resistant strains. Just a few studies on antifungal efficacy of AgNPs were published [172–174], but the fungicidal effect and mode of action of silver ions remained obscure [175]. By 1980, four classes of antifungal agents- polyenes, azoles, morpholines and allylamines were identified. However, until only oral drug Ketoconazole introduced for the treatment of systemic fungal infections. Studies have also revealed the antifungal activity of AgNPs against Candida.spp, which has very less cytotoxicity effect on human fibroblast at a concentration of 1 mg/L of Ag. It was reported that Sodium dodecyl sulphate (SDS) stabilized nanoparticles penetrate into the cell wall and cytoplasmic membrane to inhibit the activity of various enzymes such as, ATPase activity of P-glycoprotein or Lecitin/cholesterol acyltransferase [23]. Studies have shown the antimicrobial effect of nanosilver on clinical fungal isolates and ATCC strains of *Trichophyton mentagrophytes* and *Candida* species. It also stated that mycelial forms of fungi is responsible for pathogenicity due to the dimorphic transitions from yeast to mycelia,which is found primarily during the invasion of host tissue and he also finally concluded that the its potential activity of nanoAg inhibits the dimorphic transition [175]. Studies have pointed about the biocidal effect of AgNPs on phyto pathogenic fungi *B. sorokiniana* and *M. grisea* in fields and other soil borne sterile fungi that rarely produce spores [176]. Silver chloride nanoparticles (AgCl NPs) were synthesized using *Bacillus subtilis* MTCC 3053 strain with silver nitrate as the precursor salt and sodium chloride in the nutrient broth media. They proved that extracellular nitrate reductase enzyme of *B. subtilis* might be responsible for the silver chloride nanoparticle synthesis. The antifungal activity of AgCl NPs proven to be more susceptible to *Candida albicans, Aspergillus niger* than *Aspergillus flavus* [177]. Application of the silver nanopaticles importance in the scientific research shown in the Figure.4.4

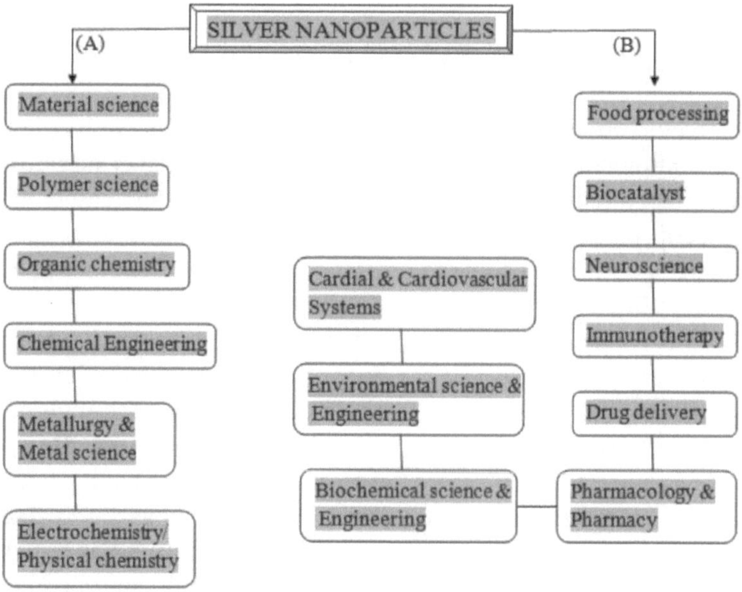

FIGURE 4.4 Application of silver nanoparticles in scientific research: (A) Physical science & (B) Biological science

Furthermore, the electrochemical properties of AgNPs incorporated them in nanoscale sensors that can offer faster response times and lower detection limits. For instance, electrodeposited AgNPs onto alumina plates gold micro-patterned electrode that showed a high sensitivity to hydrogen peroxide. Catalytic activities of nanoparticles differ from the chemical properties of the bulk materials. Furthermore, AgNPs was found to catalyze the chemiluminescence from luminol–hydrogen peroxide system with catalytic activity better than Au and Pt colloid. The optical properties of a metallic nanoparticle depend mainly on its surface plasmon resonance, where the plasmon refers to the collective oscillation of the free electrons within the metallic nanoparticle. It is well known that the plasmon resonant peaks and line widths are sensitive to the size and shape of the nanoparticle, the metallic species and the surrounding medium. For instance, nanoclusters composed of 2–8 silver atoms could be the basis for a new type of optical data storage. Moreover, fluorescent emissions from the clusters could potentially also be used in biological labels and electroluminescent displays.

4.3.1 WOUND DRESSING AND SUTURES

Silver nanoparticles capped with sodium alginate and poly (diallyldimethylammonium chloride) were deposited layer-by-layer on polyamide sutures. The antimicrobial activity against *S. aureus* increases with decreased concentration of alginate due to increased deposition of AgNPs on the suture. This might be due to oozing of silver ions from the capping of Alginate [178].

Recently, Calcium Alginate silver nanoparticles coated surgical gut sutures were synthesized and the release studies at physiological pH of body are found to possess good antimicrobial property for gram positive and negative bacteria with enhanced wound healing ability. Thus, these factors hypothesize for good tissue holding competence paving new pathways for Tissue engineering and biomedical applications [179]. Bacterial cellulose imitative from Acetobacter xylinum TISTR 975 was found to be good means for wound dressing as it can resist the moist environment. The swelling abilities of silver nanoparticles impregnated on them are used for controlling the exudates from the wound followed by further faster healing for prolonged periods by upholding moist environment [180]. Silver/poly(L-lactide) fibres nanocomposite was synthesized by electrospinning process. The uniform distribution of AgNPs in PLL fibres had shown fruitful antibacterial activity against *S. aureus* and *E. coli* with 98.5% and 94.2% respectively. Due to the presence of crystalline silver, the composite fibres tend to be favorable for wound dressing [181]. Poly(vinyl alcohol)/ Silver electrospun nanocomposite fibres were synthesized by heat mediated synthesized which showed PVA fibres form crosslinks with AgNPs forming nonwoven web. This led to increased crystallinity of PVA fibres and also can withstand the dissolution of web by overcoming the gelatinuous and shrinking of web. This had unsolved the moisture environment in the wound areas with continuous release of effectiveness and good antimicrobial [182]. Electrospun Gelatin-AgNPs fibre mats were synthesized by crosslinking **by fusing and shrinking**. This study also demonstrates, with increase in the exposure to crosslinking chamber there is decrease in the weightloss and water retention [183]. Tian et al., reported that silver nanoparticles are not only suppressing the growth of microbes but also modulate the cytokines responsible for scarless wound healing. The crucial cytokine IL-10 and its cascades besides neutrophil apoptosis and decreased matrix metalloproteinases activity directs to anti-inflammatory response thus enhancing the wound healing [184].

4.3.2 SILVER NANOPARTICLES AS POTENTIAL ANTIVIRAL AGENTS

Recently, a study revealed the potential cytoprotective activity of Ag NPs toward HIV-1 infected cells. The activity of AgNPs towards HIV-1 infected Hut/CCR5 cells was investigated using terminal uridyl nucleotide end labeling assay after a three-day treatment. The percentage of apoptotic cells was determined as 49%, 35%, and 19% for vehicle control, 5 and 50 μM Ag, respectively. AgNPs might inhibit the replication in Hut/CCR5 cells causing HIV associated apoptosis. Size dependent interaction of AgNPs with HIV-1 virus has also been demonstrated. AgNPs preferentially bind to gp120 glycoprotein knobs of HIV-1 virus. In the in-vitro studies, it was confirmed that this interaction inhibited the virus from binding with the host cell [185]. List of antiviral silver nanoparicles and their mode of action were listed below in the Figure 4.5 and Table 4.4.

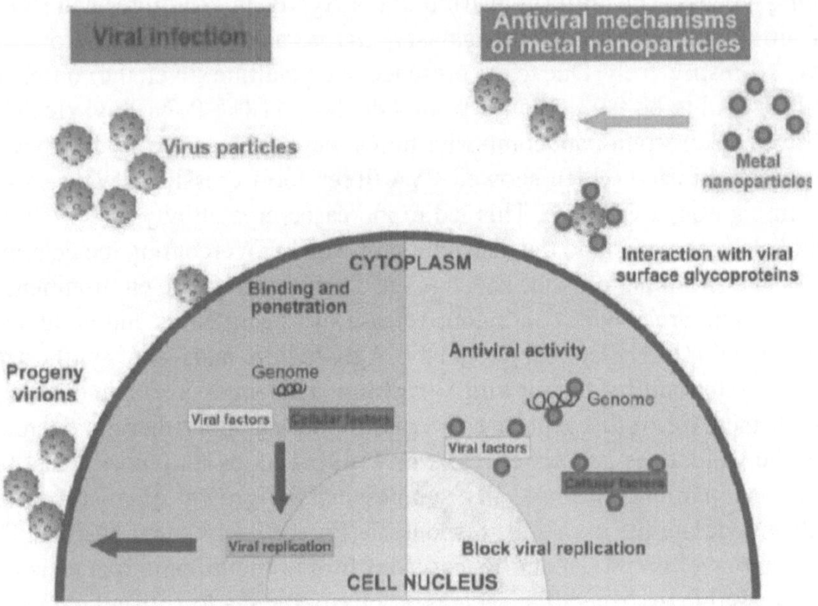

FIGURE 4.5 Antiviral Mechanism of metal nanoparticles

TABLE 4.4 List of antiviral silver nanoparticles and its action

Virus	Metal Nanoparticle Composition (size)	Mechanism of Action
Human immunodeficiency virus type 1 (HIV-1)	PVP-coated silver nanoparticles (1–10 nm)	Interaction with gp120
Herpes simplex virus type 1 (HSV-1)	MES-coated silver and gold nanoparticles (4 nm)	Competition for the binding of the virus to the cell
Respiratory syncytial virus	PVP-coated silver nanoparticles (69 nm +/− 3 nm)	Interference with viral attachment
Monkeypox virus	Silver nanoparticles and polysaccharide-coated Silver nanoparticles (10–80 nm)	Block of virus-host cell binding and penetration
Tacaribe virus (TCRV)	Silver nanoparticles and polysaccharide-coated Silver nanoparticles (10 nm)	Inactivation of virus particles prior to entry
Hepatitis B virus (HBV)	Silver nanoparticles; (10–50 nm)	Interaction with double-stranded DNA and/or binding with viral particles

4.3.3 SILVER DERIVATIVES AND ITS APPLICATION

Mainly silver zeolite and silver zirconium phosphate are of interest for manufacturers aiming to apply antimicrobial compounds to their products. In silver zeolite, alkaline or alkaline earth metal ion complexed with crystal aluminosilicate is partially replaced with silver ion by the ion-exchange method. Since these antimicrobial ceramics are being believed to have low toxicity for humans and the activity of the antimicrobial compound is durable, it is being extensively used for food preservation, disinfection of medical supplies, and decontamination of surfaces of materials such as toys, kitchenwares, and medical supplies and equipment. Silver ion strongly interacts with the ceramics matrix and is minimally released from the matrix in deionized water. Two mechanisms are proposed for the bactericidal action of silver zeolite. One is the action of silver ion itself released from zeolite and the other is that of reactive oxygen species generated from silver in the matrix. While oxygen has been reported to be necessary for the bactericidal activity of silver zeolite by some researchers, silver zeolite has also been reported to be effective on oral bacteria under anaerobic conditions by other investigators [186–192]. Silver ion plays an important role for the bactericidal action of silver zeolite. It is proposed that two possible successive processes may be involved in the action of silver zeolite. First, bacterial cells that make contact with silver zeolite take in silver ion, which inhibits several functions in the cell and consequently damages cells. The second is the generation of reactive oxygen species, which are produced possibly through the inhibition of a respiratory enzymes by silver ion and attack the cell itself. Although we have not yet succeeded in measuring the amount of silver ion transferred from silver zeolite into cells, its measurement is a prerequisite for clarification of the mode of action of silver zeolite. The activity of silver zeolite was inhibited by the addition of L-cysteine, L-methionine, L-histidine, L-tryptophan, bovine serum albumin, and yeast extract but not by glycine, L-alanine, L-leucine, or L-phenylalanine. The presence of L-cysteine indicated a strong inhibition of the bactericidal action of silver zeolite at relatively low concentrations. Sodium chloride at 100 mM substantially inhibited its activity, but neither sodium acetate nor sodium sulfate did. However, an investigation reported a synergistic bactericidal effect against a silver-resistant *E. coli* strain between silver ion and highly concentrated chloride ions and suggested the formation of soluble $AgCl_2$and $AgCl_3$ as the reason for increased activity [193].

4.4 CHARACTERIZATION OF SILVER NANOPARTICLES

Characterization of nanoparticles is important to understand and control nanoparticles synthesis and applications. Characterization is performed using a variety of different techniques such as transmission and scanning electron microscopy (TEM, SEM), atomic force microscopy (AFM), dynamic light scattering (DLS), X-ray photoelectron spectroscopy (XPS), powder X-ray diffractometry (XRD), Fourier transform infrared spectroscopy (FTIR), and UV–Vis spectroscopy. These techniques are used for determination of different parameters such as particle size, shape, crystallinity, fractal dimensions, pore size and surface area. Moreover, orientation, intercalation and dispersion of nanoparticles and nanotubes in nanocomposite materials could be determined by these techniques. For instance, TEM, SEM and AFM could determine the morphology and particle size. The advantage of AFM over traditional microscopes such as SEM and TEM is that AFM measures three-dimensional images so that particle height and volume can be calculated. Furthermore, dynamic light scattering is used for determination of particles size distribution. Moreover, X-ray diffraction is used for the determination of crystallinity, while UV–Vis spectroscopy is used to confirm sample formation by showing the plasmon resonance.

4.5 CONCLUSION

Synthesis of reliable and environmental friendly process of bionanoparticles is an important aspect of nanotechnology in the recent decades. Synthesis, characterization, manipulation and application of nanomaterials are being rapidly used in the development of nanotechnology. The application of nanoscale materials and structures may provide solutions to break barriers inbiomedical fields. With the increased efficiency of pathogenic microorganisms resistant to multiple antimicrobial agents in this urbanized environment, demands have increased for better disinfection methods. The use of nanoparticles nowadays have proved to be a better alternative with antimicrobial efficacy and the properties of Ag^+ ions is known since ancient times and are used widely as bactericide in catheters and wounds. The concept to eclipse the multidrug resistant pathogens is a great deal in the field of nanomedicine. The development of multi-drug resistant clinical pathogens has become a major factor in all hospital acquired infections which are untreatable and inturn causing severe public health problem. These concerns have led to major research efforts to discover alternative strategies using Nano-biotechnology tools for the treatment of multi drug resistant bacterial infections. The regular monitoring of antimicrobial susceptibility pattern of pathogens and formulation of a definite antimicrobial policy may be helpful for reducing the incidence of these infections. The biomedical application of Ag-NPs on selected biosynthetic process in reported in this review. In future nano-biotechnologists will go into the deeper level of understanding on the biochemical and molecular mechanisms of nanoparticles formation and their broad range of applications.

KEYWORDS

- **AgNPs;**
- **Antimicrobials;**
- **NADH;**
- **Nanoparticles;**
- **Nanotechnology**

REFERENCES

1. A. Ahmad, P. Mukherjee, S. Senapati, D. Mandal, M.I. Khan, R. Kumar and M. Sastry, Extracellular biosynthesis of silver nanoparticles using the fungus *Fusarium oxysporum, Colloids Surf. B Biointerf.* 28, 313–318 (2003).
2. F. Rosei, Patterned media for a high density data storage device –patent genius, *J. Phys. Condens. Matter.* 16, 1373–*1436* (2004).
3. S. Papp, R. Patakfalvi and I. Dekany, Metal nanoparticle formation on layer silicate lamellae, *Colloids Polym. Sci.* 286, 3–*14* (2008).
4. D. Chen, X. Qiao, X. Qui and J. Chen, Synthesis of silver nanoparticles on silver flakes to enhance electrical properties in isotropic conductive adhesives, *J. Mater. Sci.* 44, 1076–1081 (2009).

5. M. Brust and C.J. Kiely, Some recent advances in nanostructure preparation from silver particles a short topical review, *Colloids Surf. A: Physicochem. Eng. Aspects*. 202, 175–186 (2002).
6. N.R. Jana, L. Gearheart and C.J. Murphy, Seeding growth for size control, *Langmuir*. 17, 6782–6786 (2001).
7. C.K. Simi and T.E. Abraham, Starch-based completely biodegradable polymer materials, *Bioproc. Biosyst. Eng*. 30, 173–180 (2007).
8. A. Leela and M. Vivekanandan, Plant mediated synthesis of silver nanoparticles using a bryophyte *Fissidens munutus* and its antimicrobial activity, *Afr. J. Biotechnol*. 7, 3162–3165 (2008).
9. I. Hussain, M. Brust, Papworth A.J. Papworth and A.I. Cooper, one-pot synthesis of polyacrylamide-gold nanocomposite, *Langmuir*. 19, 4831–4835 (2003).
10. S.K. Virender, A.Y. Ria and L. Yekaterina, Silver nanoparticles: green synthesis and their antimicrobial activities, *Colloids Interf. Sci*. 145, 83–96 (2009).
11. A.M. Smith, H.W. Duan, M.N. Rhyner, G. Ruan and S. Nie, A systematic examination of surface coatings on the optical and chemical properties of semiconductor quantum dots, *Phys. Chem. Phys*. 8, 3895–3903 (2006).
12. G.J. Kearns, E.W. Foster and J.E. Hutchison, Substrates for direct imaging of chemically functionalized SiO_2 surfaces by transmission electron microscopy, *Anal. Chem*. 78, 298–303 (2006).
13. Z. Li, D. Lee, X.X. Sheng, R.R. Cohen and M.F. Rubner, Two-level antibacterial coating with both release-killing and contact-killing capabilities, *Langmuir*. 22, 9820–9823 (2006).
14. Chen Y, Wang C, Liu H, Qiu J, Bao X, Ag/SiO_2: a novel catalyst with high activity and selectivity for hydrogenation of chloronitrobenzenes, *Chem. Commun*. 42, 5298–5300 (2005).
15. P. Setua, A. Chakraborty, D. Seth, U.M. Bhatta, P.V. Satyam and N. Sarkar, Surface plasmons on metal nanoparticles: the influence of shape and physical environment, *J. Phys. Chem. C*. 111, 3806–3819 (2007).
16. P.M. Tessier, O.D. Velev, A.T. Kalambur, J.F. Rabolt, A.M. Lenhoff and E.W. Kalar, Nanostructured magnetism in living systems, *J. Am. Chem. Soc*. 122, 9554–9555 (2000).
17. Y.C. Cao, R. Jin and C.A. Mirkin, Nanostructured materials prepared with ultrasound, *Science*. 297, 1536–1540 (2002).
18. N.L. Rosi and C.A. Mirkin, Nanostructures in biodiagnostics, *Chem. Rev*. 105, 1547–1562 (2005).
19. M. Rai, A. Yadav and A. Gade, Silver nanoparticles as a new generation of antimicrobials, *Biotechnol. Adv*. 27, 76–83 (2009).
20. J.L. Elechiguerra, J.L. Burt, J.R. Morones, A.C. Bragado, X. Gao, H.H. Lara and M.J. Yacaman, Interaction of silver nanoparticles with HIV-1, *J. Nanobiotechnol.* 3, 6 (2005).
21. C.B. Kuber and S.F. D'Souza, Biomimetic synthesis of nanoparticles: science, technology & applicability, *Colloids Surf. B*, 47, 160–164 (2006).
22. F. Furno, K.S. Morley, B. Wong, B.L. Sharp, P.L. Arnold, S.M. Howdle, R. Bayston, P.D.Brown, P. D. Winship and H.J. Reid, Silver nanoparticles and polymeric medical

devices: a new approach to prevention of infection? *J. Antimicrob. Chemother.* 54, 1019–1024 (2004).
23. A. Panacek, M. Kolar, R. Vecerova et al, Comparison of antifungal activity of silver NPs with those of common antifungals, *Biomaterials*. 30, 6333–6340 (2009).
24. G.B. Khomutov and S.P. Gubin, Interfacial synthesis of noble metal nanoparticles, *Mater. Sci. Eng.* C, 22141–22146 (2002).
25. M. Oliveira, D. Ugarte, D. Zanchet and A.J.G. Zarbin, Influence of synthetic parameters on the size, structure, and stability of dodecanethiol-stablized silver nanoparticles, *J. Colloids Interf. Sci.* 292, 429–435 (2005).
26. G. Canizal, J.A. Ascencio, J. Gardea-Torresday and M.J. Jose-Yacaman, Multiple twinned gold nanorods grown by bioremediation techniques, *J. Nanopart. Res.* 3475–3481 (2001).
27. Y.G. Sun, B. Mayers, T. Herricks and Y.N. Xia, Ag nanowires are formed from twinned nanoparticles, *Nano Lett.* 3, 955–960 (2003).
28. C.H. Bae, S.H. Nam and S.M. Park, Aluminium nanoparticles production by laser ablation in liquids, *Appl. Surf. Sci.* 197, 628–634 (2002).
29. A. Pal, S. Shah and S. Devi, Synthesis of Au, Ag and Au-Ag alloy nanoparticlesin aqueous polymer solution, *Colloids Surf. A Physicochem. Eng. Asp.* 302, 51–57 (2007).
30. M.J. Rosemary and T. Pradeep, Solvothermal synthesis of silver nanoparticles from thiolates, *Colloids Surf. A*. 268, 81–84 (2003).
31. Z.S. Pillai and P.V. Kamat, What factors control the size and shape of silver nanoparticles in the citrate ion reduction method? *J. Phys. Chem. B*. 108, 945–951 (2004).
32. B. Krishna and G.V. Dan, Silver nanoparticles for printable electronics and biological applications, *J. Matter. Res.* 24, 2828–2836 (2009).
33. Y. Xie, R. Ye and H. Liu, Synthesis of silver nanoparticles in reverse micelles stabilized, *Colloids Surf. A Physicochem. Eng. Asp.* 279, 175–178 (2006).
34. M. Maillard, S. Giorgio and M.P. Pileni, Comparative study on bactericidal effect of silver nanoparticles, *Adv. Mater*. 14, 1084–1086 (2002).
35. Y.H. Kim, D.K. Lee and Y.S. Kang, Studies on interactions of thionine with gold nanoparticles, *Colloids Surf. A: Physicoch. Eng.* 273, 257–258 (2005).
36. S. Navaladian, B. Viswanathan, R.P. Viswanath and T.K. Varadarajan, Thermal decomposition as route for silver nanoparticles, *Nanoscale Res. Lett.* 2, 44–48 (2007).
37. M. Starowicz, B. Stypula and J. Banaœ, Electrochemical synthesis of silver nanoparticles, *Electrochem. Commun.* 8, 227–230 (2006).
38. A. Henglein, Article ChemPort, Reduction of $Ag(CN)_2$ on silver and platinum colloidal nanoparticles, *Langmuir.* 17, 2329–2333 (2001).
39. J.P. Abid, A.W. Wark, P.F. Brevet and H.H. Girault, Preparation of silver nanoparticles in solution from a silver salt by laser irradiation, *Chem. Commun.* 792–793 (2002).
40. B. Soroushian, I. Lampre, J. Belloni and M. Mostafavi, Radiolysis of silver ion solutions in ethylene glycol: solvated electron and radical scavenging yields, *Radiat. Phys. Chem.* 72, 111–118 (2005).
41. J. Rodríguez-Sánchez, M.C. Blanco and M.A. López-Quintela, Electrochemical synthesis of silver nanoparticles, *J. Phys. Chem. B* 104, 9683–9688 (2000).
42. Z. Tang, S. Liu, S. Dong and E.Wang, Electrochemical synthesis of Ag nanoparticles on functional carbon surfaces, *J. Elec. Chem.* 502, 146–151 (2001).

43. M. Mazur, Electrochemically prepared silver nanoflakes and nanowires, *Electrochem. Commun.* 6, 400–403 (2004).
44. Z. Jian, Z. Xiang and W. Yongchang, Optical properties of Au/Ag core/shell nanoshuttles, *Microelectron. Eng.* 77, 58–62 (2005).
45. J.J. Zhu, S.W. Liu, O. Palchik, Y. Koltypin and A. Gedanken, Shape-controlled synthesis of silver nanoparticles by pulse sonoelectrochemical methods, *Langmuir.* 16, 6396–6399 (2000).
46. S.R. Esau, S.B. Roberto, J. Ocotlan-Flores and J.M. Saniger, Synthesis of AgNPs by sonochemical induced reduction applications in SERS, *J. Nanopart.* 9, 77 (2010).
47. K. Patel, S. Kapoor, D.P. Dave and T. Murherjee, Synthesis of Pt, Pd, Pt/Ag and Pd/Ag nanoparticles by microwave-polyol method, *J. Chem. Sci.* 117 (4), 311–315 (2007).
48. G.M. Whitesides, Preparation and characterization of IPN microspheres for controlled, *Nat. Biotechnol.* 21, 1161–1165 (2003).
49. N. Revathi and N. Prabhu, Fungal – Silver nanoparticles: preparation and its pH characterization, *Trends Biotechnol.* 8, 27–29 (2009).
50. X.L. Cao, C. Cheng, Y.L. Ma and C.S. Zhao, Preparation of silver nanoparticles with antimicrobial activities and the researches of their biocompatibilities, *J. Mater. Sci. Mater. Med.* 21, 2861–2868 (2010).
51. D.K. Lee and Y.S. Kang, Synthesis of silver nanocrystallites by a new thermal decomposition method and their characterization, *ETRI J.* 26, 252 (2004).
52. J.H. Jung, O.H. Cheol, H. Soo Noh, J.H. Ji and S. Soo Kim, Metal nanoparticle generation using a small ceramic heater with a local heating area, *J. Aerosol Sci.* 37, 1662 (2006).
53. D.C. Tien, K.H. Tseng, C.Y. Liao, J.C. Huang and T.T. Tsung, Discovery of ionic silver in silver nanoparticle suspension fabricated by arc discharge method, *J. Alloys Compounds,* 463, 408 (2008).
54. K. Shameli, M. B. Ahmad, S. D. Jazayeri, P. Shabanzadeh, P. Sangpour, H. Jahangirian and Y. Gharayebi, Investigation of antibacterial properties silver nanoparticles prepared via green method, *Chem. Central J.* 6, 73 (2012).
55. A.J. Christy and M. Umadevi, Synthesis and characterization of monodispersed silver nanoparticles, *Adv. Nat. Sci.: Nanosci. Nanotechnol.* 3, 035013 (2012).
56. M. Sakamoto, M. Fujistuka and T. Majima, Light as a construction tool of metal nanoparticles: synthesis and mechanism, *J. Photochem. Photobiol.* C, 10, 33 (2009).
57. Ulug, B., Turkdemir, M. H., Cicek, A., & Mete, A, Role of irradiation in the green synthesis of silver nanoparticles mediated by fig (Ficus carica) leaf extract. *Spectrochimica Acta Part A: Molecular and Biomolecular Spectroscopy*, 135, 161 (2015).
58. S. Basavaraja, S.D. Balaji, L. Arun kumar, S. Rajasab and A. Venkataraman, Biosynthesis and application of silver and gold nanoparticles, *Mater. Res. Bull.* 43, 1164–1170 (2007).
59. M. Sastry, A. Ahmad, M.I. Khan and R. Kumar, Biosynthesis and application of silver and gold nanoparticles, *Current Sci.* 85, 162–70 (2003).
60. R.K. Mehra and D.R. Winge, Metal ion resistance in fungi: molecular mechanisms and their regulated expression, *J. Cell. Biochem.* 45, 30–40 (1991).
61. K. Vahabi, G. Ali Mansoori and S. Karimi, Biosynthesis of silver nanoparticles by fungus *Trichoderma Reesei, Insci. J.* 1(1), 65–79 (2011); doi:10.5640/insc.010165

62. M. Kowshik, S. Ashtaputre, S. Kharrazi, W. Vogel, J. Urban, S.K. Kulkarni and K.M. Paknikar, Extracellular synthesis of silver nanoparticles by a silver-tolerant yeast strain MKY3, *Nanotechnology,* 14, 95–100 (2003).
63. C.X. Burda, C.R. Narayanan and M.A. El-Sayed, Chemistry and properties of nanocrystals of different shapes, *Chem. Rev*. 105, 1025–1102 (2005).
64. A.E. Porter, K. Muller, J. Skepper, P. Midgley and M. Welland, Uptake of C60 by human monocyte macrophages, its localization and implications for toxicity: studied by high resolution electron microscopy and electron tomography, *Acta Biomater*. 2, 409–419 (2006).
65. A. Ahmad, S. Senapati, M.I. Khan, R. Kumar and M. Sastry, Extracellular biosynthesis of nanoparticle, *Langmuir.* 19, 3550–3553 (2003).
66. K.C. Bhainsa and S.F. D'Souza, Biomimetic synthesis of nanoparticles, *Colloids Surf. B*. 47, 160–164 (2006).
67. P. Mukherjee, S. Senapati, D. Mandal, A. Ahmad, M.I. Khan, R. Kumar and M. Sastry, Extracellular synthesis of gold nanoparticles by the fungus *Fusarium oxysporum, Chem. Biochem*. 3, 461–463 (2002)..
68. K.H. Cho, J.E. Park, T. Osaka and S.G. Park, The study of antimicrobial activity and preservative effects of nanosilver ingredient, *Electrochim. Acta*. 51, 956–960 (2005).
69. K. Paulkumar, S. Rajeshkumar, G. Gnanajobitha, M. Vanaja, C. Malarkodi, and G. Annadurai, Biosynthesis of silver chloride nanoparticles using *Bacillus subtilis* MTCC 3053 and assessment of its antifungal activity, *ISRN Nanomater.* 2013 (2013).70. G. Sathiyanarayanan, G. Seghal Kiran and J. Selvin, *Synthesis of silver nanoparticles* by polysaccharide bioflocculant produced from marine *Bacillus subtilis* MSBN17, *Colloids Surf. B: Biointerf.* (2012).
71. N. Duran, M.P.S. Teixeria, R. De Conti and Esposito, Ecological friendly from fungi, *Crit. Rev. Food Sci*. 42, 53–*66* (2002).
72. A.A. Bell, M.H. Wheeler, J. Liu, R.D. Stipanovic, L.S. Puckhaber and H. Orta, United States Department of Agriculture-Agricultural Research Service studies on polyketide toxins of *Fusarium oxysporum* f sp *vasinfectum*: potential targets for disease control, *Pest Manag. Sci*. 59, 736–747 (2003).
73. D.K. Newman and R. Kolter, A role for excreted quinones in extracellular electron transfer, *Nature*. 405, 94–97 (2000).
74. N. Duran, P.L. Marcato, O.L. Alves and G.I. De Souza, Mechanistic aspects of biosynthesis of silver nanoparticles by several *Fusarium oxysporum* strains, *J. Nanobiotechnol.* 3, 1–7 (2005).
75. S.S. Shankar, A. Rai, A. Ahmad and M. Sastry, Biological synthesis of training gold nanoprisms, *Chem. Mater*. 17, 566 (2005).
76. R.K. Mehra and D.R. Winge, Metal ion resistance in fungi: molecular mechanisms and their regulated expression, *J. Cell. Biochem.* 45, 30–40 (1991).
77. L.S. Devi, D.A. Bareh and S.R. Joshi, Studies on biosynthesis of antimicrobial silver nanoparticles using endophytic fungi isolated from the ethno-medicinal plant *Gloriosa superba* L., *Proc. Natl. Acad. Sci., India, Sect. B Biol. Sci.* DOI:10.1007/s40011-013-0185-7 (2013).
78. M. Gericke and A. Pinches, Biological synthesis of metal nanoparticles, *Hydrometallurgy*. 83, 132–140 (2006).

79. H. Durand, M. Clanet and G. Tiraby, Genetic improvement of *Trichoderma reesei* for large scale cellulase production, *Enzyme Microb. Technol.* 10, 341–346 (1988).
80. R.M. Bruins, S. Kapil and S.W. Oehme, Microbial resistance to metals in thee*environment*, *Ecotox. Environ. Safe*. 45, 198–207 (2000).
81. M. Saravanan and A. Nanda, Extracellular synthesis of silver bionanoparticles from *Aspergillus clavatus* and its antimicrobial activity against MRSA and MRSE, *Colloids Surf. B*. 77, 214–218 (2010).
82. D.B. Archer, D.J. Jeenes and D.A. Mackenzie, Strategies for improving heterologous protein production from filamentous fungi, *Antonie Leeuwenhoek.* 65, 245–250 (1994).
83. N. Vigneshwaran, N.M. Ashtaputrea, P.V. Varadarajana, R.P. Nachanea, K.M. Paralikara and R.H. Balasubramanyaa, Biological synthesis of silver nanoparticles using the fungus *Aspergillus flavus*, *Mater. Lett.* 61, 1413–1418 (2007).
84. D.S. Balaji, S. Basavaraja, R. Deshpande, D. B. Mahesh, B.K. Prabhakar and A. Venkataraman, Extracellular biosynthesis of functionalized silver nanoparticles by strains of Cladosporium cladosporioides fungus, *Colloids Surf. B*. 68, 88–92 (2009).
85. E. Castro-Longoria, A.R. Vilchis-Nestor and M. Avalos-Borja, Biosynthesis of silver, gold and bimetallic nanoparticles using the filamentous fungus *Neurospora crassa*, *Colloids Surf. B*. 83, 42–48 (2011).
86. J.C. Chen, Z.H. Lin and X.X. Ma, Evidence of the production of silver nanoparticles via pretreatment of Phoma sp. 3.2883 with silver nitrate, *Lett. Appl. Microbiol.* 37, 105–108 (2003).
87. P. Mukherjee, A. Ahmad, D. Mandal, S. Senapati, S.R. Sainkar et al, Bioreduction of AuCl4- ions by the fungus, Verticillium sp. and surface trapping of the gold nanoparticles formed, *Angew Chem. Int. Ed.* 40, 3585–3588 (2001).
88. S. Senapati, D. Mandal, A. Ahmad, M.I. Khan, M. Sastry and R. Kumar, Fungus mediated synthesis of silver nanoparticles: a novel biological, *Indian J. Phys.* 78, 101–105 (2004).
89. S.A. Kumar, M.K. Abyaneh, S.W. Gosavi, S.K. Kulkarni, R. Pasricha, A. Ahmad and M.I. Khan, Nitrate reductase-mediated synthesis of silver nanoparticles from $AgNO_3$, *Biotechnol. Lett.* 29, 439–445 (2007).
90. I. Maliszewska and Z. Sadowski, Synthesis and antibacterial activity of of silver nanoparticles, *J. Phys.* Conference Series 146, 1 (2009).
91. P. Mukherjee, M. Roy, B. Mandal, G. Dey, P. Mukherjee and Ghatak, Green synthesis of highly stabilized nanocrystalline silver particles by a non-pathogeni c and agriculturally important fungus Trichoderma asperellum, *Nanotechnology*. 197, 510 (2008).
92. M.F. Amanulla, K. Balaji, M. Girilal, R. Yadav, P.T. Kalaichelvan and R. Venketesan, Silver nanoparticles in*medicine*, *Nanomed. Nanotechol. Biol. Med.* 6, 103–109 (2010).
93. S. Senapati, A. Ahmad, M.I. Khan, M. Sastry and R. Kumar, Extracellular biosynthesis of bimetallic Au-Ag alloy nanoparticles, *Small*. 1, 517–520 (2005).
94. G.I.H. Souza, P.D. Marcato, N. Duran and E. Esposito, Utilization of Fusarium oxysporium in the biosynthesis of silver nanoparticles and its antibacterial activities, In IX national meeting of environmental microbiology, Curtiba, 25 (2004).

95. T. Oksanen, J. Pere, L. Paavilainen, J. Buchert and L. Viikari, Treatment of recycled kraft pulps with Trichoderma reesei hemicellulases and cellulases, *J. Biotechnol.* 78(1), 39–44 (2000).
96. M. Karbasian, S.M. Atyabi, S.D. Siadat, S.B. Momen and D. Norouzian, Optimizing nano-silver formation by *Fusarium oxysporum* PTCC 5115 employing response surface methodology, *Am. J. Agric. Biol. Sci.* 3, 433–437 (2008).
97. A. Ingle, M. Rai, A. Gade and M. Bawaskar, *Fusarium solani*: a novel biological agent for the extracellular synthesis of silver nanoparticles, *J. Nanopart.* 11, 82079–82085 (2008).
98. P. Daizy, Biosynthesis of Au, Ag and Au–Ag nanoparticles using edible mushroom extract, Spectrochimica Acta Part A: Molecular and Biomolecular Spectroscopy Articles, *Spectrochim. Acta A.* 73, 374–381 (2009).
99. P. Mukherjee, A. Ahmad, D. Mandal, S. Senapati, S.R. Sainkar, M.I. Khan, R. Parischa, P.V. Ajayakumar, M. Alam, R. Kumar and M. Sastry, Fungus-mediated synthesis of silver nanoparticles and their immobilization in the mycelial matrix: anovel biological approach to nanoparticle synthesis, *Nano Lett.* 1, 515–519 (2001).
100. S.S. Nikhil, B. Mahesh, B. Rahul, S.S. Rekha, G. Szakacsand and A. Pandey, Biosynthesis of silver nanoparticles using aqueous extract from the compactin producing fungal strain, *Proc. Biochem.* 44, 939–943 (2009).
101. R.Y. Parikh, S. Singh, B.L.V. Prasad, M.S. Patole, M. Sastry and Y.S. Schouche, Extracellular synthesis of crystalline silver nanoparticles and molecular evidence of silver resistance from Morganella sp.: towards understanding biochemical synthesis mechanism, *Chem. Biochem.* 9, 1415–1421 (2008).
102. S. Basavaraja, S.D. Balaji, A. Lagashetty, A.H. Rajasab and A. Venkataraman, Extracellular biosynthesis of silver nanoparticles using the fungus *Fusarium semitectum. Mater. Res. Bull.* 43, 1164–1170 (2008).
103. N. Vigneshwaran, A.A. Kathe, P.V. Varadarajan, R.P. Nachane and R.H. Balasubramanya, Biomimetics of Ag nanoparticles by whiterot fungus *Phaenerochaete chrysosporium. Colloids Surf. B. Biointer.* 53, 55–59 (2006).
104. K. Kathiresan, S. Manivannan, M.A. Nabeel and B. Dhivya, Studies on silver nanoparticles synthesized by a marine fungus, *Penicillium fellutanum* isolated from coastal mangrove sediment, *Colloids Surf. B.* 71, 133–137 (2009).
105. R. Joerger, T. Klaus and C.G. Granqvist, Biologically produced silver-carbon composite materials for optically functional thin-film coatings, *Adv. Mater.* 12, 407–409 (2000).
106. A. Nanda and M. Saravanan, Biosynthesis of silver nanoparticles from *Staphylococcus aureus* and its antimicrobial activity against MRSA and MRSE, *Nanomed. Nanotechol. Biol. Med.* 5, 452 (2009).
107. A.R. Shahverdi, A. Fakhimi, H.R. Shahverdi and M. Sara, Bionanoparticles: synthesis and antimicrobial applications, *Nanomed. Nanotechnol.* 3, 168–171 (2007).
108. S. Gurunathan, K. Kalishwaralal, V. Vaidyanathan, D. Venkataraman, S.R.K. Pandian, J. Muniyandi, N. Hariharan and S.H. Eom, Biosynthesis, purification and characterization of silver nanoparticles using *Escherichia coli, Colloids Surf. B.* 74, 328–335 (2009).

109. M.F. Lengke, M.E. Fleet and G. Southam, Biosynthesis of silver nanoparticles by filamentous cynaobacteria from a silver (I) nitrate complex, *Langmuir*. 23, 2694 (2007).
110. N. Pugazhenthiran, S. Anandan, G. Kathiravan and N.K. Udaya-Prakash, Microbial synthesis of silver nanoparticles by Bacillus sp, *J. Nanopart*. 11, 1811–1815 (2009).
111. S. Silambarasan and J. Abraham, Biosynthesis of silver nanoparticles using the bacteria *Bacillus cereus* and their antimicrobial property, *Int. J. Pharm. Pharmaceu. Sci.* 4(1), 536–540 (2012).
112. Z. Haoran, L. Qingbiao, W. Huixuan and S. Daohua, Biosorption and bioreduction of diamine silver complex by corynebacterium, *J. Chem. Technol. Biotechnol*. 80, 285–290 (2005).
113. D. Wei and W. Qian, Facile synthesis of Ag and Au nanoparticles utilizing, *Colloids Surf B*. 62, 136–142 (2008).
114. T. Maneerunga, S. Tokurab and R. Rujiravanita, Impregnation of silver nanoparticles into bacterial cellulose for antimicrobial wound dressing, *Carbohyd. Polym*. 72, 43–51 (2008).
115. K. Kalishwaralal, V. Deepak and S. Gurunathan, Biosynthesis of silver and gold nanoparticles using *Brevibacterium casei*, *Colloids Surf. B*. 77, 257–262 (2010).
116. M. Saravanan, A.K. Vemu and S.K. Barik, Rapid biosynthesis of silver nanoparticles from Bacillus megaterium (NCIM 2326) and their antibacterial activity on multi drug resistant clinical pathogens, *Colloids Surf. B*. 88, 325–331 (2011).
117. P. Mohanpuria, N.K. Rana and S.K. Yadav, Biosynthesis of nanoparticles: technological concepts and future applications, *J. Nanopart. Res*. 10, 507–517 (2008).
118. J.L. Gardea-Torresdey, J.G. Parsons, E. Gomez, J. Peralta-Videa, H. Troiani, P. Santiago and M. Jose-Yacaman, Formation and growth of Au nanoparticles inside live Alfalfa plants, *Nano. Lett*. 2, 397–401 (2002).
119. S.S. Shankar, A. Rai, A. Ahmad and M. Sastry, Rapid synthesis of Au, Ag, and bimetallic Au core-Ag shell nanoparticles using Neem (*Azadirachta indica*) leaf broth, *J. Colloid Interf. Sci*. 275, 496–502 (2004).
120. S. Li, Y. Shen, A. Xie, X. Yu, L. Qiu, L. Zhang and Q. Zhang, Green synthesis of silver nanoparticles using *Capsicum annuum* L. extract, *Green Chem*. 9, 852–858 (2007).
121. N. Ahmad, S. Sharma, V.N. Singh, S. F. Shamsi, A. Fatma and B.R. Mehta, Biosynthesis of silver nanoparticles from *Desmodium triflorum*: anovel approach towards weed utilization, *Biotechnol. Res. Int*. 8, 1–8 (2011).
122. K. Kalimuthu, R.S. Babu, D. Venkataraman, M. Bilal and S. Gurunathan, Biosynthesis of silver nanocrystals by *Bacillus licheniformis*, *Colloids Surf. B*. 65, 150–153 (2008).
123. M. Linga Rao and N. Savithramma, Antifungal efficacy of silver nanoparticles synthesized from the medicinal plants, *J. Pharm. Sci*. 3, 1117–1121 (2011).
124. S.P. Chandran, M. Chaudhary, R. Pasricha, A. Ahmad and M. Sastry, Synthesis of gold nanotriangles and silver nanoparticles using*Aloevera p*lant extract, *Biotechnol. Prog*. 22, 577–583 (2011).
125. J. Huang, Q. Li, D. Sun, Y. Lu, Y. Su, X. Yang, H. Wang, Y. Wang, W. Shao, N. He, J. Hong and C. Chen, Biosynthesis of silver and gold nanoparticles by novel sun dried Cinnamomum camphora leaf, *Nanotechnol*. 18, 105104–105114 (2007).

126. Y.S. Jae and S.K. Beom, Synthesis of nanoparticles for biomedical applications, *Bioprocess. Eng.* 32, 79–84 (2009).
127. R.A. Vilchis-Nestor, V. Sánchez-Mendieta, A.M. Camacho-López, M.R. Gómez-Espinosa, A.M. Camacho-López and A.J. Arenas-Alatorre, Solventless synthesis and optical properties of Au and Ag nanoparticles using *Camellia sinensis* extract, *Mater. Lett.* 62, 3103–3105 (2008).
128. B. Harekrishna, D.K. Bhui, P.S. Gobinda, S. Priyanka, S. Pyne and A. Misra, Green synthesis of silver nanoparticles using latex of *Jatropha curcas*, *Colloids Surf. A Physicochem. Eng. Asp.* 339, 134–139 (2009).
129. S.S. Shankar, A.M. Ahmad and M. Sastry, Geranium leaf assisted biosynthesis of silver nanoparticles, *Biotechnol. Prog.* 19, 1627–1631 (2003).
130. B. Ankamwar, C. Damle, A. Ahmad and M. Sastry, Biosynthesis of gold and silver nanoparticles using *Emblica Officinalis* fruit extract, their phase transfer and transmetallation in an organic solution, *J. Nano. Sci. Nano. Techno.* 5, 1665–1671 (2005).
131. S.S. Birla, V.V. Tiwari, A.K. Gade, A.P. Ingle, A.P. Yadav and M.K. Rai, Fabrication of silver nanoparticles by *Phoma glomerata* and its combined effect against *Escherichia coli, Pseudomonas aeruginosa* and *Staphylococcus aureus, Lett. Appl. Microbiol.* 48, 173–179 (2009).
132. J. Huang, L. Lin, Q. Li, D. Sun, Y. Wang, Y. Lu, N. He, K. Yang, X. Yang, H. Wang, W. Wang and W. Lin, Continuous-flow biosynthesis of silver nanoparticles by lixivium of sundried *Cinnamomum camphora* leaf in tubular microreactors, *Ind. Eng. Chem. Res.* 47, 6081–6090 (2008).
133. J. Kasthuri, S. Veerapandian and N. Rajendiran, Biological synthesis of silver and gold nanoparticles using apiin as reducing agent. *Colloids Surf. B: Biointer.* 68, 55–60 (2009).
134. P. Logeswari, S. Silambarasan and J. Abraham, Synthesis of silver nanoparticles using plant extract and analysis of their antimicrobial property. *J. Saudi. Chem. Soc.* (2012). http://dx.doi.org/10.1016/j.jscs.2012.04.007
135. P. Prakash, P. Gnanaprakasam, R. Emmanuel, S. Arokiyaraj and M. Saravanan, Green synthesis of silver nanoparticles from leaf extract of *Mimusops elengi*, Linn. for enhanced antibacterial activity against multi drug resistant clinical isolates, *Colloids Surf. B: Biointer.* 108, 255–259 (2013).
136. V.K. Vidhu, S.A. Aromal and D. Philip, Green synthesis of silver nanoparticles using *Macrotyloma uniflorum, Spectrochim. Acta A Mol. Biomol. Spectrosc.* 83, 392–397 (2011).
137. D.S. Sheny, J. Mathew and D. Philip, Phytosynthesis of Au, Ag and Au-Ag bimetallic nanoparticles using aqueous extract and dried leaf of *Anacardium occidentale, Spectrochim, Acta A Mol. Biomol. Spectrosc.* 79, 254–262 (2011).
138. D. Philip, Green synthesis of gold and silver nanoparticles using *Hibiscus Rosa sinensis, Phys. E Low Dimens. Syst. Nanostruct.* 42, 1417–1424 (2010).
139. A.J. Koraa and R.B. Sashidharb, A template for the green synthesis and stabilization of silver nanoparticles with antibacterialapplication, *Carbohyd. Polym.* 82, 670–679 (2010).
140. S.H. Gulbranson, J.A. Hud and R.C. Hansen, Argyria following the use of dietary supplements containing colloidal silver protein, *Cutis.* 66373–66374 (2000).

141. X.Chen and H.J. Schluesener, Nanosilver: a nanoproduct in medical application toxicology, *Toxicol. Lett.* 176, 1–12 (2008).
142. K. Dunn and V. Edwards-Jones, The role of Acticoat™ with nanocrystalline silver in the management of burns, *Burns.* 30(Supp 1), S1–S9 (2004).
143. F. Joy and W. Fiona, Nanocrystalline silver dressings in wound management: a review, *Int. J. Nanomed.* 1(4), 441–449 (2006).
144. A. Peetsch, C. Greulich, D. Braun, C. Stroetges, H. Rehage, B. Siebers, M. Köller and M. Epple, Silver-doped calcium phosphate nanoparticles: synthesis, characterization, and toxic effects toward mammalian and prokaryotic cells. *Colloids Surf. B: Biointerf.* 102, 724–729 (2013).
145. Q. Li, S. Mahendra, D.Y. Lyon, L. Brunet, M.V. Liga, D. Li and P.J.J. Alvarez, Antimicrobial nanomaterials for water disinfection and microbial control: potential applications and implications, *Water Res.* 42, 4591–4602 (2008).
146. A. Melaiye, Z. Sun, K. Hindi, A. Milsted, D. Ely, D. Reneker, C. Tessier and W. Youngs, Silver(I)-imidazole cyclophane gem-diol complexes encapsulated by electrospun tecophilic nanofibers: formation of nanosilver particles and antimicrobial activity, *J. Am. Chem. Soc.* 127, 2285–2291 (2005).
147. R. Roy, M.R. Hoover, A.S. Bhalla, T. Slaweekl, S. Dey, W. Cao, J. Li and S. Bhaskar, Ag-aquasols with extraordinary bactericidal properties: role of the system Ag–O–H_2O, *Mater. Res. Innov.* 11, 3–18 (2008).
148. A.J. Haes and R.P. Van Duyne, A nanoscale optical biosensor:sensitivity and selectivity of an approach based on the localized surface plasmon resonance spectroscopy of triangular silver nanoparticles, *J. Am. Chem. Soc.* 124, 10596–10604 (2002).
149. A.D. McFarland and R.P. Van Duyne, Single silver nanoparticles as real-time optical sensors with zeptomole sensitivity, *Nano Lett.* 3, 1057–1062 (2003).
150. J.A. Kloepfer, R.E.Mielke and J.L. Nadeau, Uptake of CdSe and CdSe/ZnS quantum dots into bacteria via purine-dependent mechanisms, *Appl. Environ. Microbiol.* 71, 2548–2557 (2005).
151. S. Qourzal, M. Tamimi, A. Assabbane, A. Bouamrane, A. Nounah, L. Laanab and Y. Ait-Ichou, Preparation of TiO_2 photocatalyst using $TiCl_4$ as a precursor and its photocatalytic performance, *J. Appl. Sci.* 6, 1553–1559 (2006).
152. K. Dunn and V. Edwards-Jones, The role of Acticoat with nanocrystalline silver in the management of burns, *Burns.* 30, S1–S9 (2004).
153. S. Silver, Bacterial silver resistance: molecular biology and uses and misuses of silver compounds, *Fems. Microbiol. Rev.* 27, 341–353 (2003).
154. B.S. Atiyeh, M. Costagliola, S.N. Hayek and S.A. Dibo, Effect of silver on burn wound infection control and healing: review of the literature, *Burn.* 33, 139–148 (2007).
155. N. Law, S. Ansari, F.R. Livens, J.C. Renshaw and J.R. Lloyd, Formation of nanoscale elemental silver particles via enzymatic reduction by *Geobacter sulfurreducens*, *Appl. Microbiol*, 74, 7090–7093 (2008).
156. N. Vigneshwaran, A.A. Kathe, P.V. Varadarajan, R.P. Nachane and R.P. Balasubramanya, Biomimetics of silver nanoparticles by white rot fungus, *Phaenerochaete chrysosporium*, *Colloids Surf. B.* 53, 55–59 (2006).
157. H.J. Lee, S.Y. Yeo and S.H. Jeong, Influence of halide anion on preparation of silver colloid particles, *J. Mater. Sci.* 38, 2199 (2003).

158. S.K. Gogoi, A. Chattopadhyay and S.S. Ghosh, *Langmuir.* 22(22), 9322–9328 (2006).
159. J.S. Kim, E. Kuk, K.N. Yu, J.H. Kim, S.J. Park and H.J. Lee, Antimicrobial effects of silver nanoparticles, *Nanomed. Nanotechol. Biol. Med.* 3, 95–101 (2007).
160. J.R. Morones, J.L. Elechiguerra, A. Camacho, K. Holt, J.B. Kouri, J.T. Ramírez and M.J. Yacaman, The bactericidal effect of silver nanoparticles, *Nanotechnol.* 16, 2346–2353 (2005).
161. F. Gad, T. Zahra, K.P. Francis, T. Hasan and M.R. Hamblin, Targeted photodynamic therapy of established soft-tissue infections in mice, *Photochem. Photobiol. Sci.* 3, 451–455 (2004).
162. O.V. Salata, Applications of nanoparticles in biology and medicine, *J. Nanobiotechnol.* **2**, 3 (2004).
163. M. Deendayal, E.M. Bolander, D. Mukhopadhyay, G. Sarkar and P. Mukherjee, The use of microorganisms for the formation of metal nanoparticles and their application, *Appl. Microbiol. Biotechnol.* 69, 485–492 (2006).
164. T. Klaus, R. Jeorger, E. Olsson and C. Granqvist, Bacteria as workers in the living factory: metal-accumulating bacteria and their potential for materials science, *Trends Biotechnol.* 19, 15 (2004).
165. B. Nair and T. Pradeep, Coalescence of nanoclusters and formation of submicron crystallites assisted by Lactobacillus strains, *Cryst. Growth Des.* 2293–2298 (2002).
166. S. Shrivastava, T. Bera, A. Roy, G. Singh, P. Ramachandrarao and D. Dash, Applying nanotechnology to human health: revolution in biomedical sciences, *Nanotechnol.* 182, 25103–25111 (2007).
167. H.H. Lara, N.V. Ayala-Nuñez, L. Ixtepan-Turrent and C. Rodriguez-Padilla, Bactericidal effect of silver nanoparticles against multidrug-resistant bacteria. *World J. Microb. Biot.* 26, 615–621 (2010).
168. Y. Matsumura, K. Yoshikata, K. Shin-ichi and T. Tsuchido, Mode of bactericidal action of silver zeolite and its comparison with that of silver nitrate, *Appl. Environ. Microbiol.* 69, 4278–4281 (2003).
169. R.A. Shahverdi, A. Fakhimi, H.R. Shahverdi and S. Minaian, Synthesis and effect of silver nanoparticles on the antibacterial activity of different antibiotics against Staphylococcus aureus and Escherichia coli, *Nanomed. Nanotechol. Biol. Med.* 3, 168–171 (2007).
170. E. Falletta, M. Bonini, E. Fratini, A.L. Nostro, G. Pesavento, A. Becheri, P.L. Nostro, P. Canton and P. Baglioni, Clusters of poly (acrylates) and silver nanoparticles: structure and applications for antimicrobial fabrics, *J. Phys. Chem. C.* 112, 11758–11766 (2008).
171. G. Sathiyanarayanan, G. Seghal Kiran and J. Selvin, Synthesis of silver nanoparticles by polysaccharide bioflocculant produced from marine *Bacillus subtilis* MSBN17, *Colloids Surf. B: Biointer.* 102, 13–20 (2013).
172. D. Roe, B. Karandikar, N.B. Savagel, B. Gibbins and J.B. Roullet, Antimicrobial surface functionalization of plastic catheters by silver nanoparticles, *J. Antimicrob. Chemother.* 61, 869–876 (2008).
173. V.K. Sharma, R.A. Yngard and Y. Lin, Silver nanoparticles green synthesis and their antimicrobial activities, *Adv. Colloids Interf. Sci.* 145, 83–96 (2009).

174. K.J. Kim, W.S. Sung, S.K. Moon, J.S. Choi, J.G. Kim and D.G. Lee, Antifungal effect of silver nanoparticles on dermatophytes, *J. Microbiol. Biotechnol.* 18, 1482–1484 (2008).
175. J. Young-Ki, B.H. Kim and G. Jung, Antifungal activity of silver ions and nanoparticles on phytopathogenic fungi, *Plant Dis.* 93, 1037–1043 (2009).
176. S. Galdiero, A. Falanga, M. Vitiello, M. Cantisani, V. Marra and M. Galdiero, Silver nanoparticles as potential antiviral agents, *Molecules.* 16, 8894–8918 (2011); doi:10.3390/molecules16108894
177. K. Paulkumar, S. Rajeshkumar, G. Gnanajobitha, M. Vanaja, C. Malarkodi, and G. Annadurai, Biosynthesis of silver chloride nanoparticles using *Bacillus subtilis* MTCC 3053 and assessment of its antifungal activity. *ISRN Nanomaterials.* (2013).
178. S.T. Dubas, S. Wacharanad and P. Potiyaraj, Tunning of the antimicrobial activity of surgical sutures coated with silver nanoparticles, *Colloids Surf. A: Physicochem. Eng. Aspects.* 380, 25–28 (2011).
179. R. Augustine and K. Rajarathinam, Synthesis and characterization of silver nanoparticles andits immobilization on alginate coated sutures for the prevention of surgical wound infections and the in vitro release studies. *Int. J. Nano Dim.* 2(3), 205–212 (2012).
180. T. Maneerung, S. Tokur and R. Rujiravanit, Impregnation of silver nanoparticles into bacterial cellulose for antimicrobial wound dressing, *Carbohydrate Polym.* 72, 43–51 (2008).
181. X. Xu, Q. Yang, Y. Wang, H. Yu, X. Chen and X. Jing, Biodegradable electrospun poly(L-lactide) fibers containing antibacterial silver nanoparticles, *Eur. Polym. J.* 42, 2081–2087 (2006).
182. K.H. Hong, Preparation and properties of electrospun poly (vinyl alcohol)/silver fiber web as wound dressings, *Polym. Eng. Sci.* 47(1), 43–49 (2007).
183. P. Rujitanaroj, N. Pimpha and P. Supaphol, Wound-dressing materials with antibacterial activity from electrospun gelatin fiber mats containing silver nanoparticles, *Polymer.* 49, 4723–4732 (2008).
184. J. Tian, K.K.Y. Wong, C.M. Ho, C.N. Lok, W.Y. Yu, C-M. Che, J.F. Chiu and P.K.H. Tam, Topical delivery of silver nanoparticles promotes wound healing, *Chem. Med. Chem.* 2(1), 129–136 (2007).
185. J.L. Clement and P.S. Jarrett, Antimicrobial silver, *Metal-Based Drugs.* 1(5), 467–482 (1994).
186. K. Im, Y. Takasaki, A. Endo and M. Kuriyama, Antibacterial activity of A-type zeolite supporting silver ions in deionized distilled water, *J. Antibact. Antifung. Agents.* 24, 269–274 (1996).
187. Y. Inoue, M. Hoshino, H. Takahashi, T. Noguchi, T. Murata, Y. Kanzaki, H. Hamashima and M. Sasatsu, Bactericidal activity of Ag-zeolite mediatedby reactive oxygen species under aerated condition, *J. Inorg. Biochem.* 92, 37–42 (2002).
188. K. Kawahara, K. Tsuruda, M. Morishita and M. Uchida, Antibacterial effect of silver-zeolite on oral bacteria under anaerobic condition, *Dent. Mater.* 16, 452–455 (2000).
189. H. Kourai, Y. Manabe and Y. Yamada, Mode of bactericidal action of zirconium phosphate ceramics containing silver ions in the crystal structure. *J. Antibact. Antifung. Agents* 22, 595–601 (1994).

190. T. Matsuura, Y. Abe, Y. Sato, K. Okamoto, M. Ueshige and Y. Akagawa, Prolonged antimicrobial effect of tissue conditioners containing silverzeolite, *J. Dent.* 25, 373–377 (1997).
191. H. Miyoshi, H. Kourai, T. Maeda and T. Yoshida, Role of Cl_ adsorbed on silver-loaded zirconium phosphate for the photooxidation of OH_ to OH_, *J. Photochem. Photobiol. A* 113, 243–250 (1998).
192. H. Miyoshi, H. Kourai and T. Maeda, Light-induced formation of 2,5-dihydroxy-*p*-benzoquinone from hydroquinone in photoirradiated silverloaded zirconium phosphate suspension. *J. Chem. Soc. Faraday Trans.* 94, 283–287 (1998).
193. Y. Matsumura, K. Yoshikata, K. Shin-ichi and T. Tsuchido, Mode of bactericidal action of silver zeolite and its comparison with that of silver nitrate, *Appl. Environ. Microbiol.* 69(7), 4278–4281 (2003).

CHAPTER 5

TAILORING PLASMON RESONANCES IN METAL NANOSPHERES FOR OPTICAL DIAGNOSTICS OF MOLECULES AND CELLS

KRYSTYNA KOLWAS, ANASTASIYA DERKACHOVA, and DANIEL JAKUBCZYK

ABSTRACT

The advent of nanotechnology introduced a variety of novel exciting possibilities into diagnostic and sensing applications. Noble metal nanoparticles are of special relevance in that context, because their unusual properties fall mainly in visible, but also near-infrared and near-ultraviolet range, depending on the nanoparticle size and kind of a metal. Nanoparticles interact strongly with light via resonant excitation of localized surface plasmons (LSPs), and so can act as efficient receiving or/and scattering optical nanoantennas. Resonant excitation of LSP give rise to a variety of effects, such as frequency-dependent absorption and scattering (resulting in bright colors of colloids of nanoparticles) and electromagnetic near-field concentration and enhancement. An intense interest in noble-metal nanoparticles is driven by diverse applications in sensing, biomedical diagnostics, and therapy, energy transport and conversion, novel spectroscopic and microscopic techniques, novel (meta-) materials and others.

Direct plasmon characterization versus size greatly facilitates the optimization of plasmon parameters in diverse applications. Functional relationship of plasmon resonance frequencies and particle size provided in this chapter for gold and silver nanoparticles embedded in diverse dielectric media, gives an important practical tool for example predicting spectral performance of single or colloidal nanoparticles, in estimation of the shift of plasmon resonance in colorimetric probes, or adjusting plasmon resonance frequency to the desired molecular transition. Other discussed factors are the maximal enhancement of the local electric near field and its spatial distribution. Predicting the distribution of the field around nanoparticles reveals regions of maximal electric field enhancement and dark areas, which are extremely unfavorable in the context of field enhancement.

5.1 INTRODUCTION

The advent of nanotechnology introduced a variety of novel exciting possibilities into diagnostic and sensing applications. At the nanoscale, optical, electronic, and chemical material properties are significantly different from those seen in bulk. The unique features of nanostructures make them attractive for a wide range of applications. The continuous development of the methodology of synthesis of metal nanoparticles (NPs) (examples in Section 5.4) and of the ability to control their properties by means of size, shape, and composition has opened a variety of novel possibilities for molecular biology, medicine, physics, and chemistry. An intense interest in noble-metal NPs shown by scientists from different domains and by engineers is driven by diverse applications of NPs in sensing, biomedical diagnostics, and therapy, energy transport and conversion, novel spectroscopic and microscopic techniques, novel (meta)materials, and others.

The physical and chemical elementary processes that are vital for life involve energy exchange in the electronvolt range. Visible and near-infrared optical excitations are used to study such processes. Conventional optical investigation methods allow microscopic scale resolution in these spectral ranges. However, routine observation of subwavelength structures with optical microscopes is, according to Abbe's criterion (Born & Wolf 1999), limited to about half the wavelength of light, which is too large to resolve nanometer-sized structures such as biomolecules or cell components. Application of plasmonic structures allows overcoming this limitation.

From the point of view of optical diagnostics, noble-metal nanoparticles are especially interesting because of their unusual optical properties (see Section 5.2) which arise from their ability to resonate with the light field. Nanoparticles interact strongly with light via excitation of localized surface plasmon (LSP), (see Sections 5.2.1 and 5.6.1) and so act as the efficient optical nanoantennas that can capture light (see e.g., Stuart & Haes

2005 for review). Resonant excitation of LSP in nanoparticles gives rise to a variety of effects, such as frequency-dependent absorption and scattering which can be tuned by particle dimensions (Kolwas et al. 2009; Kolwas & Derkachova 2010; Kreibig & Vollmer 1995; Sönnichsen et al. 2002). Another advantage is the near-field concentration and enhancement which can be exploited for a variety of applications (see Aslan et al. 2005b; Quinten 2011; Wang et al. 2011; Willets & Van Duyne 2007 for reviews) such as surface-enhanced Raman scattering (SERS), high-resolution microscopy, solar cells, nondiffraction limited nanoscopic waveguides or nanophotonic devices.

Noble-metal nanoparticles have proven to be useful for biological uses (Aslan et al. 2005b; Khlebtsov & Dykman 2010; Lakowicz 2006; Liao et al. 2006; Sardar et al. 2009; Schuller et al. 2010; Stuart & Haes 2005; Willets & Van Duyne 2007) for reviews). Possibility of tailoring LSP resonance frequency (Sections 5.6 and 5.7) and particle absorbing and scattering properties (Section 5.5) is of fundamental importance to such applications. Another important advantage is the sensitivity of LSP resonance frequency to the changes in refractive index of the immediate environment. Plasmons and the resulting enhanced electric fields, confined in small region adjacent to metal NPs, are the bases of novel, exciting possibilities in optical diagnostics, and sensing. Gold nanoparticles, being resistant to oxidation and photobleaching, are highly potent optical probes, which can replace conventional organic dyes. Gold and silver nanoparticles are amenable to the attachment of biomolecules or ligands. Plasmonic probes making use of absorption and extinction spectra of noble-metal NPs are very attractive for their rapid, real-time performance, and sensitivity. To date, most applications of gold colloids (Aslan & Geddes 2009) have been based on the monitoring of changes in their color (Section 5.2.2) resulting from the spectral shift of plasmon resonance which manifests in the absorption or scattering spectra (Section 5.5). This has led to absorption-based colloid proximity sensors for example DNA, or antibodies (Aslan et al. 2005b and references therein). The aggregation of noble metal nanoparticles, which can be induced by specific bioaffinity reactions, became basis for the development of colorimetric detection of DNA hybridization immunoassays (Aslan et al. 2004; Stuart & Haes 2005 and references therein). The close proximity of plasmonic nanoparticles in an aggregate is known to result in a red-shifted LSP resonance, due to the near-field coupling. In Section 5.3, we give some examples of such applications.

Some applications of LSPs are based on properties of the LSP electric field associated with resonant excitation of plasmons. Others are due to the manner in which these resonances manifest in the light intensity spectra. Understanding the fundamentals of resonant interaction of light with plasmonic nanoparticles is essential for studying LSP properties and for their optimal use. Though the spectacular optical effects in metal nanoparticles are studied systematically since Faraday work (Faraday 1857), the understanding of plasmon basis seems to contain still some confusion.

Plasmonic nanospheres represent the simplest and the most fundamental structures for studying the fundamentals of plasmon phenomena. Surface plasmons (Section 5.2.1) are collective surface charge density waves and the associated surface confined fields, which arise when the incoming light wave resonates with a nanoparticle. Mie theory (Bohren & Huffman 1983; Born & Wolf 1999) (Section 5.5.2), delivers an indispensable formalism enabling to calculate electromagnetic (EM) fields in the near- and far-field regions around a spherical particle of chosen size as well as the resulting field intensities (irradiances) and the scattering, absorption, and extinction spectra (Kolwas et al. 2013). However, the peaks appearing in observed absorption or scattering spectra (Section 5.5) cannot be automatically ascribed to the positions of plasmon resonances. Such presupposition can affect optimization of plasmon parameters for some application (Section 5.2). For example, the enhancement of electric field for surface enhanced Raman scattering (SERS) (Section 5.3.1) requires different optimization than colorimetric methods (Sections 5.6 and 5.7) (which consist in enhancement of intensity of light at certain wavelengths).

The fascinating properties of noble-metal nanospheres emerge when the frequency of the incident light wave coincides with the resonance frequency of the LSP modes (like in case of waves on a string). Plasmon resonance performance can be directly predicted by solving the eigenmode problem of a metal nanosphere, using the formalism of Mie theory (see Section 5.5.2) and realistic modelling of metal optical properties (see Section 5.5.1). Direct dependencies of plasmon resonance frequency and resonance spectral width and strength (defined by plasmon damping rates) versus particle size for Au and Ag NPs embedded in diverse dielectric media follow from the modelling (see Section 5.7.2). Such direct size dependence of LSP resonance frequencies and of corresponding plasmon damping rates make an important tool which allow optimization of plasmon properties for given application. In particular, they provide direct information on the size of gold or silver nanospheres, which must be used to optimize the local electric field enhancement (see Sections 5.7.4 and 5.7.5) by adjusting plasmon resonance frequency to the desired molecular transition. Direct plasmon characterization versus size greatly facilitates the optimization of plasmon parameters (see Section 5.6) by choice of appropriate type of metal and estimation of the shift of plasmon resonance frequency for nanoparticles used as colorimetric probes (see Section 5.3.4). Other factors, which can be optimized by the modelling are the maximal enhancement of the electric near field (see Section 5.7.4) and its spatial distribution (see Sections 5.7.4 and 5.7.5). Predicting the distribution of field around NP reveals regions of maximal electric field enhancement and dark areas on the nanoparticle surface, which are extremely unfavorable for the field enhancement.

Intense research efforts are still underway to exploit new applications of these particles. In view of low toxicity of gold and (functionalized) silver nanoparticles, the large scope of these applications suggests that noble-metal nanoparticles form a whole class of highly promising nanomaterials for new biological and medical applications.

5.2 OPTICAL PROPERTIES OF METALS AT THE NANOSCALE: PLASMON RESONANCES

From everyday experience, we know that metals interact strongly with light. A mirror is the best example: mirrors are usually manufactured by applying a metal coating to the back surface of a substrate (glass) which makes them lustrous. The reason for this is so that metals have high density of freely moving electrons detached from their core atoms. These electrons screen the electric field of light at frequencies below the plasma frequency, causing reflection of light in visible range.

At the nanoscale, metal (in the form of a nanoparticle) possesses unique properties that are astonishingly different from those exhibited by metal in bulk. Nanometals exhibit striking reactive features, unusual electric conductivity, and mechanical properties (including friction and ductility) and spectacular optical effects. The optical effects are the subject of our further interest.

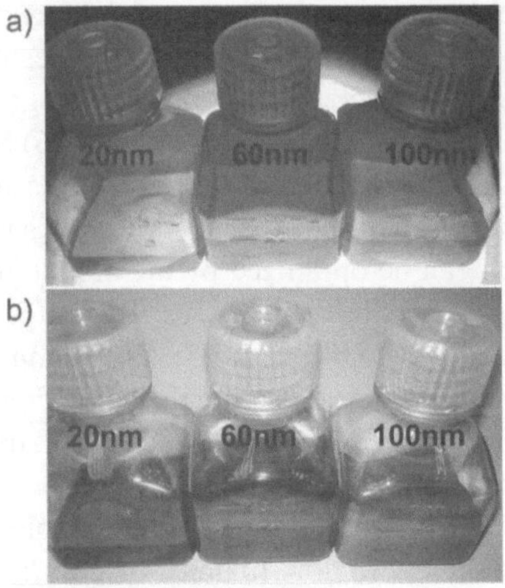

FIGURE 5.1 Suspensions of spherical gold particles with various diameters (20, 60, and 100 nm from right to left) in water. White light illumination from (a) behind; (b) from the front. The color in (a) is due to extinction, in (b) due to diffuse backscattering.

The resonance nanoscale optical effects, which can be observed even with a necked eye, are due to the strong interaction of light with the free electrons of metal nanostructure. Gold, the element known as truly noble, can serve as an example. Gold, which is reduced in size to the nanometer range, loses its well-known color and changes it versus the nanoparticle size (see Figure 5.1).

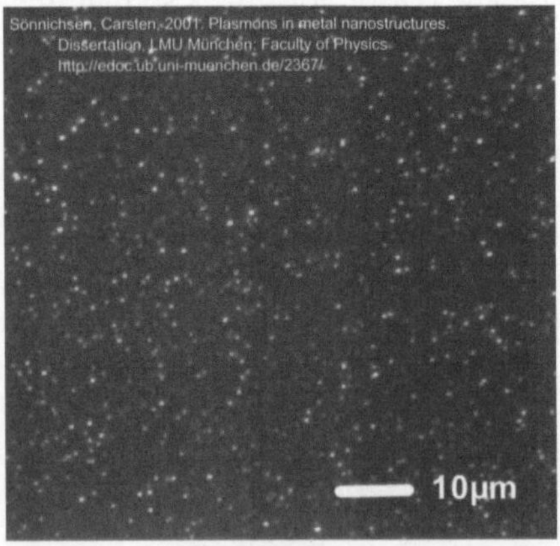

FIGURE 5.2 Image of a sample containing gold and silver nanospheres as well as gold nanorods photographed with a dark field microscope (Sönnichsen 2001).

Figure 5.2 (Sönnichsen 2001) shows a true-color image of a sample containing gold and silver nanospheres as well as gold nanorods observed with a dark field microscope. Each bright dot corresponds to an individual nanoparticle, which scatters light at a different plasmon resonance frequency. The resonance wavelength varies from blue (silver nanospheres) via green and yellow (gold nanospheres) to orange and red (nanorods).

Spectacular optical effects are born in the very proximity of nanometer-sized metal structures (in so-called near-field region), so they belong to the nanoworld. Technically "nano" denotes one billionth of any fundamental unit, while nowadays "nano" is used as an adjective to describe objects, systems, or phenomena with characteristics arising from nanometer-scale structure. Some of these effects can be observed by a far-away observer (in the so-called far-field; at macrodimensions) with a necked eye.

5.2.1 LOCALIZED SURFACE PLASMONS

Spectacular colors of colloidal nanoparticles are due to resonant excitation of localized surface plasmons (LSP) (see e.g., Dragoman & Dragoman 2008; Maier & Atwater 2005; Weiner 2009 for reviews): LSPs are defined as collective oscillations of free-electrons at the metal-dielectric interface and the corresponding fields confined to the surface.

Spherical metal surfaces have an advantage, in contrast to infinite planar interfaces (e.g., Dragoman & Dragoman 2008; Maier & Atwater 2005), of direct excitation of LSP by light. Surface plasmon propagating along an infinite planar metal surface (often called surface plasmon polariton [SPP]) can be excited only under the wave-vector selection rules (see e.g., Dragoman & Dragoman 2008), which limit direct coupling of SP with light. In contrast to SPP waves, LSP are standing waves and do not propagate. Thus, the term "localized" denotes plasmon standing waves on spherical metal/dielectric interface of a nanoparticle, which is comparable to, or smaller than the wavelength λ_{inc} of light used to excite the plasmon.

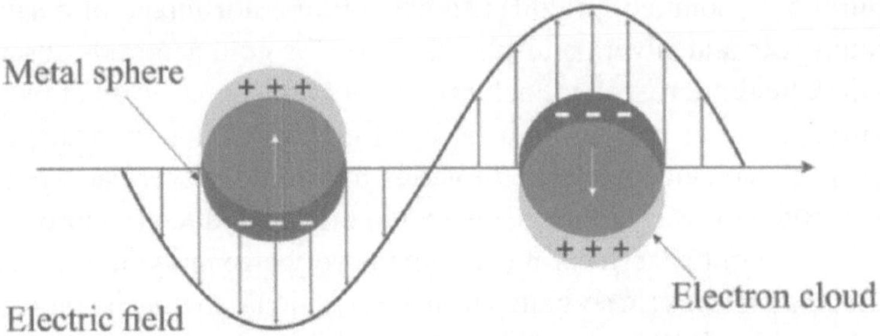

FIGURE 5.3 Schematic diagram illustrating a localized dipole surface plasmon.

Dipole surface plasmons in nanoparticles are usually imagined as back-and-forth normal-mode oscillations of the whole cloud of valence electrons, relative to the rigid, positively charged core (e.g., Kreibig & Vollmer 1995; Willets & Van Duyne 2007). In such a simplified picture, EM field of light drives the collective oscillations of nanoparticles free electrons (see Figure 5.3) at a certain resonance frequency, producing dipole charge density distribution near the surface. Subsequently, the oscillating electrons radiate EM field at the frequency of their oscillation. It is then elastic reradiation of light with the same frequency.

However, in reality, the radiating (scattering) abilities of nanoparticles change with particle size and are weaker for smaller nanoparticles, as discussed in detail in (Kolwas et al. 2009; Kolwas & Derkachova 2010; Kolwas & Derkachova 2013). In general, excited collective charge density waves can be of dipole, quadrupole, hexapole (see Figure 5.4) or higher-polarity type, or can be a combination of multipoles, depending on the particle size. For increasing particle size, higher order plasmons can be effectively excited. Surface charge density waves of LSP are 3D standing waves, which, at the particle circumference, scale with the particle radius. Charge oscillations are coupled to the electric field confined to the surface and the field is strongly enhanced in respect to the incoming field (Section 5.7.4). LSP resonance occurs (Section 5.6) when the frequency of the incident EM wave fits to the resonance frequency of the allowed surface standing waves. Therefore, LSP waves can oscillate only at certain frequencies, which can be tailored by particle size and shape, type of a metal, and of dielectric environment.

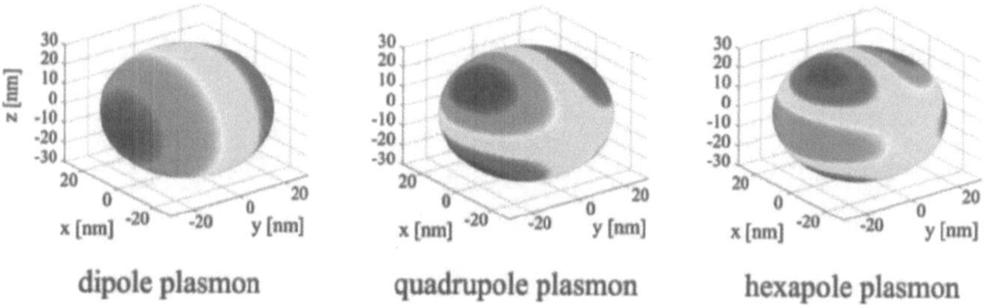

FIGURE 5.4 Examples of charge density distributions of the dipole, quadrupole, and hexapole plasmon at the corresponding resonance frequencies. Illumination from the bottom with light polarized along x-axis (see Figure 5.15a)). Red and blue regions respectively represent maxima and minima of the surface electron densities.

Coherent surface charge oscillations at optical frequencies cannot be observed straightforwardly in any experiment. However, there is evidence of their existence in the near-field optical images (e.g., Esteban et al. 2008; Lin et al. 2010). In principle, SPs are recognized for their intense electromagnetic fields bound to the surface. Therefore, the term "plasmons" refers (Barnes 2006; Barnes et al. 2003; Dragoman & Dragoman 2008; Sambles & Bradbery 1991; Wang et al. 2011; Zakharian et al. 2007), in practice, rather to light confined to the metal-dielectric interface. Their much-desired characteristics are shortened wavelength and enhanced field strength as well as sensitivity of LSP resonance frequency to changes in optical properties of the nanoparticles immediate environment.

5.2.2 COLOR EFFECTS (FAR FIELD)

Plasmon resonances manifest as the intense absorption or scattering of light, depending on the particle size (Section 5.5.2.1), and are the source of beautiful colors which has attracted the interest of artificers and scientists for generations.

The unique properties of nanoparticles were empirically known and utilized since ancient times for coloring ceramics and glasses. The Roman Lycurgus Cup (Figure 5.5) (now at the British Museum, London (Anon n.d.) represents one of the outstanding examples. Under normal lighting, this cup appears green. However, when illuminated from within, it turns to a glowing translucent red. The dichroic effect was achieved by adding tiny proportions of nanometer-sized alloy of silver and gold, which selectively absorbs some wavelengths, but reflects others.

FIGURE 5.5 The Roman Lycurgus Cup viewed in reflected (left) and transmitted (right) light. http://www.britishmuseum.org/explore/online_tours/museum_and_exhibition/the_art_of_glass/the_lycurgus_cup.aspx

In the Middle Ages, gold nanoparticles were used for manufacturing the stained-glass windows of Gothic churches. Stained-glass window makers achieved deep reds by dissolving gold in the molten glass. The stained-glass rose window at Notre Dame de Paris (Figure 5.6) makes a prominent example. When light shines through the glass containing gold nanoparticles, some wavelengths are absorbed within plasmonic mechanism, and others pass through. Introducing different metals to the glass can alter its color. Also, variation of the technological process, that modify the size (and shape) of the nanoparticles embedded in glass, leads to color changes. Metal or metallic oxide nanoparticles are what actually give glass its colour: gold gives a deep ruby red; copper gives blue or green, iron gives green or brown, and so forth.

FIGURE 5.6 Gothic stained glass rose window of Notre-Dame de Paris. The colors were achieved by using colloids of nanoparticles. http://en.wikipedia.org/wiki/File:GothicRayonnantRose003.jpg

However, the scientific research on nanogold seems to have started with Michael Faraday (1857). Faraday reported that the mere variation in size of particles gave rise to a variety of resultant colours. Colloidal suspensions of nanospheres (see Figure 5.1) show bright colors, despite the extremely low concentration of gold particles ($<10^{-2}$ weight %). In addition, the colors in the transmitted light are not a simple complement of the absorbed ones.

Contemporary scattering experiments allow observation of single metal nanoparticles under the dark-field microscopes (see Figure 5.2). The nanoparticles illuminated with white light appear to be of diverse colors depending on particle size and shape. Gold nanoparticles of a diameter of few tens of nanometers (see Figure 5.2) can be easily seen under a microscope using dark field condenser because of intense scattering with plasmonic mechanism (El-Sayed et al. 2005; Jain et al. 2008; Liao et al. 2006). The optical diameter of gold nanoparticles is several orders of magnitude larger than its physical diameter (see Section 5.5.2.1). Taking advantage of the strong LSPR scattering of gold and silver nanoparticles conjugated with specific targeting molecules, allows the molecule-specific imaging and diagnosis of diseases such as cancer (El-Sayed et al. 2005). An additional advantage is that the colors of nanoparticles do not photobleach or blink.

5.2.3 ENHANCEMENT OF THE ELECTRIC FIELD (NEAR-FIELD REGION)

Plasmonic nanoparticles can convert optical radiation into intense, localized EM field distributed near the particle surface. By squeezing light into subwavelength volumes, plasmonic structures have a profound effect on the efficiency of many optical processes, including light–matter interactions (Schuller et al. 2010). The nature of nanometallic light concentrators is distinct from the dielectric lenses used in conventional optics. Such lenses cannot focus light to spots less than about half the wavelength, due to the fundamental laws of diffraction. The ability of strong light concentration makes noble-metal nanoparticles a powerful tool for biological applications, including SERS (see Section 5.3.1) and other surface-enhanced spectroscopic approaches. Plasmonic nanoparticles have been used to enhance the sensitivity of several spectroscopic measurement techniques including fluorescence, Raman scattering, and second harmonic generation. All these methods require optimization of the near-field enhancement factor and of the extent of enhanced field. Such fine-tuning and control of plasmon characteristics is the main key in plasmonic applications and can be achieved by choosing proper nanoparticle parameters (Section 5.6).

5.3 APPLICATIONS (EXAMPLES)

Existing and proposed applications of plasmonic nanoparticles are numerous (e.g., Khlebtsov et al. 2010; Lakowicz 2006; Liao et al. 2006; Sardar et al. 2009; Schuller et al. 2010; Stuart & Haes 2005; Willets & Van Duyne 2007 for reviews). Most applications involve imaging and specimen characterization, however, therapeutic applications are also proposed. Such applications rely on exceptional features of plasmonic nanoparticles. These are enhancement of the electric filed in particle proximity, tunability of plasmon resonance frequency and of absorbing/scattering abilities of nanoparticles, sensitivity of plasmon resonance frequency to the index of refraction of the environment, and to the presence of interactions between nanoparticles In addition, plasmonic nanoparticles inherit an important feature common to all nanostructures: resistance to photobleaching. Furthermore, gold nanospheres are resistant to oxidation while being prone to bioconjugation and biomodification.

5.3.1 SURFACE ENHANCED RAMAN SCATTERING

The prime example of plasmonic specimen characterization technique is SERS (e.g., Mahajan et al. 2010; Schuller et al. 2010; Wang et al. 2007). It is based on the classical Raman spectroscopy (inelastic scattering of light), which enables probing the vibrational states of molecules.

Raman spectroscopy is commonly used to analyse a wide range of materials (gases, liquids, and solids), including highly complex materials such as biological organisms and tissues. However, the extremely weak inelastic scattering intensities limit the applications of this technique. SERS has been used to overcome this limitation. SERS allows detecting trace amount of molecules, microorganisms, or living cells adsorbed at the surface of a metallic nanostructure (see Figure 5.7) illuminated at wavelength corresponding to the plasmon resonance frequency (see Section 5.6). Such detection is possible due to the localization of electromagnetic field and resulting substantial enhancement of the electric field near the plasmonic particle. With the enhancement of the field within plasmonic mechanism, the detection of single molecules adjacent to plasmonic nanoparticles becomes feasible.

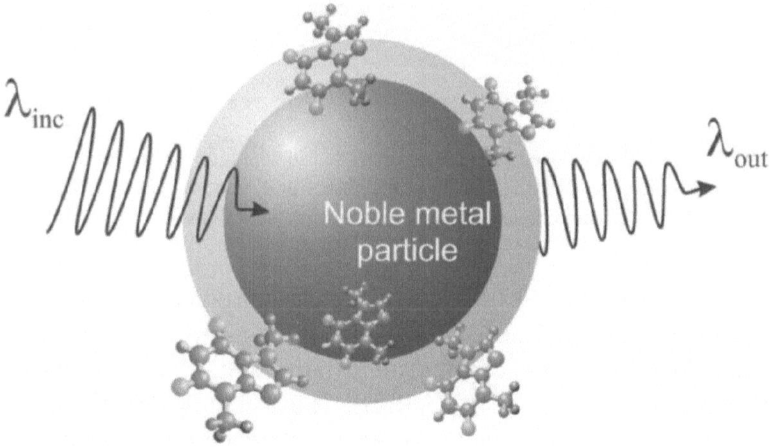

FIGURE 5.7 Schematic drawing illustrating the interaction of molecules with the electric field enhanced in vicinity of a noble-metal nanosphere illuminated with light at plasmon resonance frequency. Nonresonant illumination of a molecule leads to the enhanced Raman emission, while illumination in resonance with absorption or emission band in a molecule can lead to plasmon-controlled fluorescence.

5.3.2 TISSUE MARKERS

Since noble-metal nanoparticles are relatively biocompatible and non-toxic, plasmonic enhancement enables in vivo imaging of organisms and tissues. An important application is detection and diagnostics of specific tissues, such as tumors. When nanospheres are used as a contrast agent. Another example can be detection of specific microorganisms/living cells adsorbed at the surface of nanospheres (see Figure 5.8, e.g., Aslan et al. 2005b; Chen et al. 2005; El-Sayed et al. 2005; Jain et al. 2007, 2008; Khan et al. 2013; Loo et al. 2005).

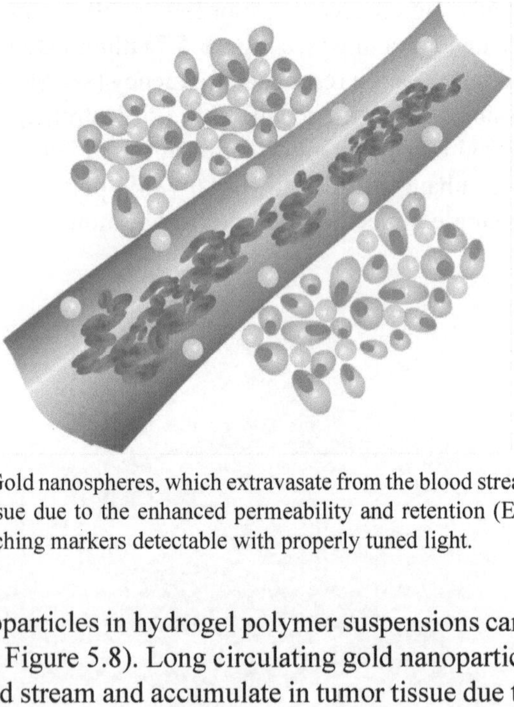

FIGURE 5.8 Gold nanospheres, which extravasate from the blood stream and accumulate in cancerous tissue due to the enhanced permeability and retention (EPR) effect, can be used as nonbleaching markers detectable with properly tuned light.

Gold nanoparticles in hydrogel polymer suspensions can be inserted by injection (see Figure 5.8). Long circulating gold nanoparticles extravasate from the blood stream and accumulate in tumor tissue due to the enhanced permeability and retention (EPR effect). The accumulated gold nanoparticles in cancerous tissue may be used as reference markers for precise localization of tumors in patients during radiation therapy, thereby improving treatment efficiency and reducing side effects due to irradiation of healthy tissue (El-Sayed et al. 2005).

Dark field microscopy enables observation of even single metal nanoparticles, and imaging of biological materials using metal nanoparticles as contrast agents. Dark field microscopy collects scattered light from a sample, whereas in typical transmittance mode microscopy, transmitted light is used to form images. Metal nanoparticles of proper (large enough, see Sections 5.5.2.1 and 5.6.1) size show distinctly under dark-field microscopy because they can efficiently scatter incident light which is in resonance with their own plasmon resonance frequency. Then, even specimen regions with similar refractive indices can be distinguished (see Figure 5.9, El-Sayed et al. 2005, see also Wax & Sokolov 2009).

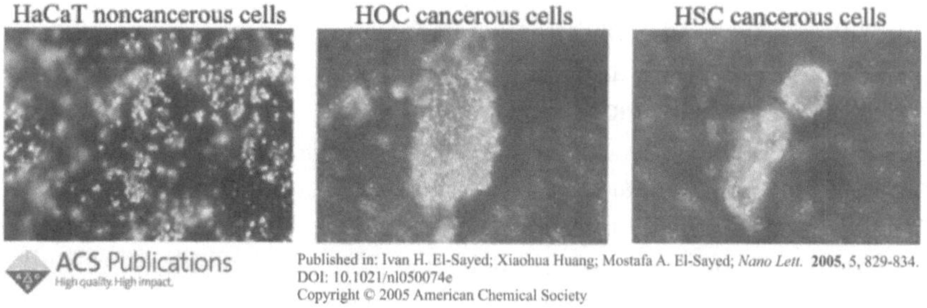

FIGURE 5.9 Dark-field microscope images of specific-antibody-conjugated gold nanospheres used as contrast agent for cancer cell labelling (El-Sayed et al. 2005).

5.3.3 PLASMON-CONTROLLED FLUORESCENCE

Fluorescence detection is the basis of many measurement techniques in biological research (Lakowicz 2006). The sensitivity of fluorescence measurements largely determines progress in biology and medicine. A metallic nanoparticle can modify both the excitation and fluorescence of the molecule. Fluorophores situated in nanoparticle proximity can undergo near-field interactions with plasmonic nanoparticle and vice versa. At very short distances, the nonradiative plasmon excitation decay dominates and the quantum efficiency of the emitter drops (Rogobete et al. 2007). Thus, the apparently very strong enhancements of the excitation field with plasmonic mechanism might not lead to a large enhancement of the molecular emission rate.

Such reciprocal interactions suggest that the novel optical absorption and scattering properties of metallic nanostructures can be used to control the decay rates, location, and direction of fluorophore emission. This opens way to ultrabright single-particle probes that do not photobleach, as well as to probes for selective excitation with decreased light intensities etc (see Lakowicz 2006 for review).

5.3.4 COLORIMETRIC DETECTION

Colorimetric detection methods (Chen et al. 2013; Gartia et al. 2013) are very attractive since they are rapid, sensitive, and real-time. Gold nanoparticles are very efficient colorimetric probes owing to their specific optical properties, which include modification of their plasmon resonance frequency by the index of refraction of the immediate environment, by increasing size resulting from nanoparticle aggregation or by the near-field interaction between nanoparticles.

The close proximity of two nanoparticles is known to result in a red-shift of plasmon resonance, due to near-field coupling. It can be directly utilized as a so called "plasmon ruler" (Jain et al. 2007a; Reinhard et al. 2005) for measuring nanoscale distances between individual molecules. This feature can also be used in sensing techniques, based on looking for changes in plasmon absorption. These changes are caused by the aggregation/disossociation of nanoparticles induced by the presence of a specific agent (detection of DNA damage (see Figure 5.10, Chen et al. 2013), protein conformational changes (Schneider et al. 2013) or heavy-metal ions contamination in water (Pradeep 2009; see also Aslan et al. 2005a).

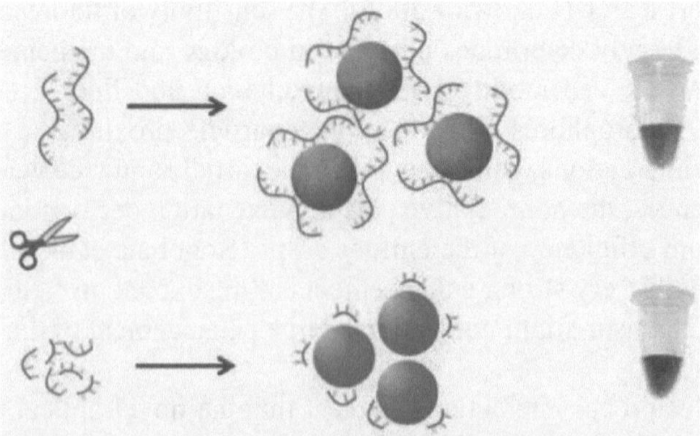

FIGURE 5.10 Binding of DNA fragments to gold nanospheres, influence their absorption spectrum due to the modification of the distance between spheres. The modification is characteristic to the fragment length. Thus the fragment length corresponds to characteristic color of suspension and can be visually detected (Chen et al. 2013).

Functionalized noble-metal nanospheres can act as an assay (immunoassays, aptamer assays, etc.). Size-controlled nanosphere probes can be also used as suspensions exhibiting intense colors.

5.4 NANOPARTICLE SYNTHESIS (EXAMPLES)

With a large variety of potential nanoparticles applications, a large number of fabrication methods was devised to obtain desired particles controllably and reproducibly (see e.g., Heiligtag & Niederberger 2013; Khlebtsov et al. 2010; Shen et al. 2000; Stuart & Haes 2005; Willets & Van Duyne 2007 for reviews). Gold and silver nanoparticles are commonly synthesized using reducing agents such as sodium citrate or sodium borohydride. Highly spherical gold and silver nanoparticles, without harsh reducing agents and with narrow size distribution are offered commercially by for example Sigma-Aldirich, nanoComposix, Cytodiagmostocs or mkNano. Two major groups of methods of nanoparticle synthesis can be distinguished: chemical wet synthesis and lithographic techniques.

5.4.1 CHEMICAL WET SYNTHESIS

Chemical syntheses enable making particles of different shape (triangles, cubes, prisms, tetrahedra, bipyramids, and stars) and size. Though the particles morphology is (partially) controllable by the reaction conditions and the stabilizing surfactant, the effects of polydispersity must be realised and accommodated.

In particular, metal nanoparticles of a variety of shapes and sizes can be produced by the reduction of a metal salt. Metal salts can be also reduced electrochemically in the presence of a surfactant or template. As far as colloidal Au is concerned, it seems that the most popular synthesis protocols involve the reduction of chloroauric acid ($HAuCl_4$) by various reducing agents. For example, a very convenient citrate method (Frens 1973; Turkevich et al. 1954), enables production of relatively monodisperse particles with the controlled average equivolume diameter from 10 to 60 nm. However, for particles larger than 20 nm, the citrate method yields elongated particles (see e.g., Brown et al. 2000). Particles as small as 1 nm can be synthesized using sodium or potassium thiocyanate (Baschong et al. 1985) while only slightly larger (~5 nm) using sodium borohydride/

EDTA (Khlebtsov et al. 1996). Particularly promising spherical nanoparticles are those of a core–shell morphology (Stuart & Haes 2005). Silica/gold (gold forming a shell) particles present an example of nanoparticles with spectral performance which can be tuned over 600–1500 nm range by changing their ratio of the shell thickness to the core radius (Khlebtsov et al. 2010; Oldenburg et al. 1998; Yang et al. 2008; Ye et al. 2009).

5.4.2 LITHOGRAPHIC TECHNIQUES

This class of techniques yields substrate-bound nanostructures. Probably the most straightforward approach for making such structures is electron beam lithography (EBL). In EBL, first, the desired pattern is drawn with a high-energy electron beam on a thin layer of electron-resist, then the resist is chemically developed and the noble metal is deposited on the emerging structure. EBL provides very good control over the nanostructure morphology but is expensive and time-consuming.

An alternative set of methods for making surface-bound nanoparticle arrays is nanosphere lithography (NSL) (Haynes & Van Duyne 2001; Haynes et al. 2002). It is cheap and enables large-scale production, but the control over structure morphology is limited.

5.5 ABSORPTION, SCATTERING, AND EXTINCTION OF NANOPARTICLES

Metal nanoparticles are known to both strongly absorb and scatter incident light, depending on their size, shape, refractive index of the dielectric environment as well as on the proximity to other resonant plasmonic nanostructures. The applicability of nanoparticles as contrast for biomedical imaging and as therapeutic agents relies on their spectral properties in the optical range of wavelengths. For instance, effective photothermal therapy with minimal laser radiation dosage requires high nanoparticle absorption cross section and low scattering losses. On the contrary, cell imaging applications based on light-scattering microscopy require a high-scattering cross section (Jain et al. 2006, 2008). Transmission measurements directly yield the extinction of light, which consists of both absorption and scattering losses (Kreibig & Vollmer 1995). In case of small particles, scattering is negligible with respect to absorption, and the extinction can be directly assigned to the absorption cross section.

If a colloid of such metal nanoparticles is illuminated by light of the frequency, which matches the plasmon resonance frequency characteristic for the nanoparticles size (Derkachova & Kolwas 2013; Kolwas & Derkachova 2010; Kolwas et al. 2009), absorption and/or scattering cross section can be higher than the geometrical cross section of nanoparticles. As compared to fluorophores, a 60 nm gold nanoparticle can produce the same scattering intensity as 3×10^5 fluorescing fluorescein molecules (Aslan et al. 2005b and references therein). The electrons in the metal nanoparticle exposed to the resonant electromagnetic wave can oscillate collectively at the same frequency as the incident wave. Collective oscillation of electrons and the associated EM fields are damped because of radiative and nonradiative processes. The contribution of each of them to the total oscillation damping rate is size dependent (Kolwas & Derkachova 2013). If the electron oscillation is damped mainly due to electron relaxation dissipative processes (the case of nanoparticles much smaller than the incident light wavelength), the absorption cross section is large. With growing particle size, the oscillating electrons effectively reradiate electromagnetic radiation more effectively (see Section 5.6.1).

5.5.1 DESCRIPTION OF MATERIAL PROPERTIES OF METAL NANOPARTICLES

Realistic prediction of scattering and absorbing properties of single plasmonic nanoparticles or of their colloids is of crucial importance in applications. For this purpose, one needs a model describing both: the nanoparticle absorption and scattering spectra, as well as the intrinsic particle properties such as their plasmon resonance frequencies and plasmon damping rates versus size. However, numerical modelling leads to realistic predictions of these quantities only if realistic external parameters of the model are used. One of the crucial parameters is the frequency dependent complex refractive index of a metal nanoparticle $n_{in}(\omega) = n'(\omega) + i{\cdot}n''(\omega)$. It is related to the dielectric function of a metal nanoparticle $\varepsilon_{in}(\omega)$: $n_{in}(\omega) = \sqrt{\varepsilon_{in}(\omega)}$ which has to be supplied by an additional modelling. The recognition of the analytic form of $\varepsilon_{in}(\omega)$ enables the correct description of the dispersion of metal to the overall frequency dependence of absorption and scattering cross-sections as well as to the plasmon dispersion relation.

The simplest dielectric function often used to describe spectral properties of bulk metals is the Drude dielectric function. Drude model of a bulk metal treats free electrons as dimensionless negative charges forming homogeneous plasma uniformly distributed over the bulk and detached from the positively charged ions. In the framework of this model, when an external EM field of frequency ω is applied, the conduction electrons move freely between independent collisions occurring at the average rate of γ. The frequency dependent dielectric function:

$$\varepsilon_D(\omega) = 1 - \frac{\omega_p^2}{\omega^2 + i\gamma\omega} \tag{5.1}$$

explains the optical transparency of perfect metals ($\gamma = 0$) observed above the plasma frequency ω_p (the index of refraction $n(\omega)$ is than imaginary). However, in many plasmonic applications of gold and silver nanoparticles, the function in such form is not satisfactory. The contribution of interband transitions to the optical properties of these metals makes them different from the perfect one. The simplest analytical function often used to describe spectral properties of real metals like gold or silver, results from the Drude–Lorentz–Sommerfeld model including the contribution of the interband electronic transitions accounted for as follows (Bohren & Huffman 1983):

$$\varepsilon_m(\omega) = \varepsilon_{ib} - \frac{\omega_p^2}{\omega^2 + i\gamma\omega} \tag{5.2}$$

ε_{ib} is the phenomenological parameter describing the contribution of bound electrons to the polarizability. If one adopts the effective parameters ω_p = 9.01eV, ε_{ib} = 9.84, and γ = 0.072eV for gold and ω_p = 9.10eV, ε_{ib} = 3.7, γ = 0.18eV for silver (Derkachova & Kolwas 2013; Kolwas et al. 2009). $\varepsilon_m(\omega)$ reproduces the most widely used data sets of refraction indices for pure metals (Johnson & Christy 1972) quite well. The introduction of still another correction to γ over 1.8 eV range for gold enables even better modelling (Derkachova & Kolwas 2015 to be published in Plasmonics).

For a metal nanoparticle of size comparable to the electron mean free path in bulk (42 nm for gold), an additional relaxation mechanism arises due to collisions of electrons with the particle surface (see Kolwas & Derkachova 2013; Kreibig & Vollmer 1995 and references therein). This effect can be accounted by replacing γ in Eq. (5.2) with size dependent relaxation rate:

$$\gamma_R = \gamma_{\text{bulk}} + A\frac{v_F}{R} \tag{5.3}$$

v_F is the Fermi velocity, and A is the theory dependent quantity. In Kolwas & Derkachova (2013), the accepted values are: $A = 0.33$, $v_F = 1.4 \times 10^{-6}$ m/s for gold and $A = 2$, $v_F = 1.4 \times 10^{-6}$ m/s for silver, and $\gamma_{\text{bulk}} = \gamma$ for the corresponding metal. Interface damping has little effect on the position of plasmon resonances in small nanoparticles. However, it has an important impact on the spectral width of plasmon resonance, which manifests in the absorption spectra of smaller particles. It also improves significantly the absorption abilities in comparison to the scattering capacities of nanoparticles of the same size. In all numerical simulations presented in the following, the dielectric function $\varepsilon_{in}(\omega)$ of a metal nanoparticle accounting for the interband transition and surface scattering effect with the effective parameters given above was used.

The optical properties of the nonabsorbing dielectric medium outside the sphere are usually independent on frequency in the optical range.

5.5.2 RIGOROUS MIE SOLUTIONS

Mie scattering theory, which is more than a hundred years old, enables description of many spectacular effects described in Section 5.2. Mie solutions deliver indispensable formalism enabling to calculate EM fields in the near- and far-field regions around a spherical nanoparticle illuminated by light. Based on classical electrodynamics quantities, which are of experimental importance, can be found for chosen particle size and kind for diver optical properties of the environment. These are, for example, the intensities (irradiances) of elastically scattered light, as well as the spectra of absorption and extinction (absorption and scattering cross sections, or absorption and scattering efficiencies). The calculated quantities are very close to the measured ones, if only the external parameters of the theory, such as the dielectric function of a metal or the refractive index of the environment are realistic (Section 5.5.1), and if the radius of the particle is well defined.

Mie theory considers a plane electromagnetic wave illuminating a spherical particle and the divergence-free Maxwell equations (with no external charge or current) supplemented by appropriate boundary conditions. The resulting Helmholtz equation for vector EM fields is solved in spherical coordinates. EM fields are expressed in the form of an infinite sum of partial electromagnetic waves of the "electric" (transverse magnetic [TM]) and "magnetic" (transverse electric [TE]) type, which are reciprocally orthogonal (Bohren & Huffman 1983; Born & Wolf 1999). Mie theory supposes that the nanoparticles and the surrounding medium are homogeneous.

5.5.2.1 ABSORPTION AND SCATTERING SPECTRA OF GOLD AND SILVER NANOPARTICLES

Spectral dependences of absorptive and scattering abilities of a nanosphere follow dependences of absorption and scattering cross sections calculated on the basis of Mie theory for a particle of radius R embedded in a dielectric environment of the refractive index n_{out}

Figure 5.11a) and b), shows the total absorption $C_{abs}(\lambda,R)$ and the scattering $C_{scat}(\lambda,R)$ cross section for gold nanospheres of radii $R = 5$ nm and $R = 75$ nm, respectively. As shown in the figure, the optical properties of spherical gold nanoparticles are highly dependent on the nanoparticle radius. Smaller nanospheres primarily absorb light (with the absorption maximum near 520 nm, (Figure 5.11a), dashed line), while larger nanospheres exhibit increased scattering (Figure 5.11b), solid line). Small silver nanoparticles are much better absorbers than the gold ones (Figure 5.11a) and c)). However, the absorption in small silver nanoparticles falls in the range of wavelength shorter than for the gold ones, which absorb light in the visible range.

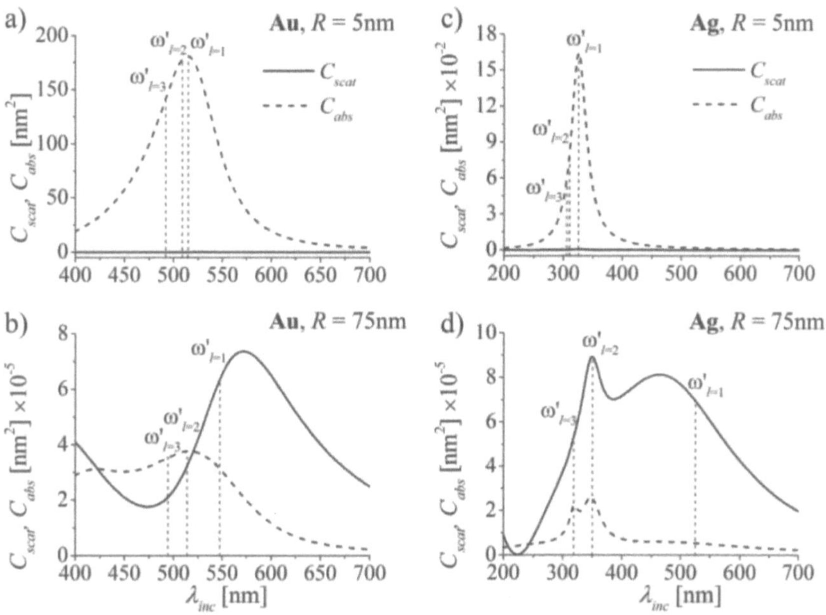

FIGURE 5.11 Absorption and scattering cross sections of gold a), b) and silver c), d) nanoparticles of radius 5 and 75 nm, respectively, $n_{out}=1$. Dashed vertical lines show the positions of surface plasmon resonances ($l = 1, 2, 3$ corresponds to the dipole, quadrupole, and hexapole mode, respectively, see Section 5.6.1) according to the solution of SP dispersion relation.

With increasing nanoparticle size, scattering maxima broaden significantly and shift toward longer wavelengths (red shift). Larger nanospheres scatter much more light than the small ones. Absorption and scattering spectra of both small and large gold nanoparticles fall in the optical range (400–700 nm). It is worth noting that in the infrared spectral range, gold nanoparticles are not optically active, regardless of particle size. Absorption spectra of silver nanoparticles are shifted toward ultraviolet (see Figure 5.11c) and d)). However, large silver nanoparticles, which strongly absorb ultraviolet light, gain scattering abilities in the visible range, as shown in Figure 5.11d). Thus, the increase of the scattering to absorption ratio with increasing size provides a tool for nanoparticle selection for contrast applications.

5.6 DIRECT SIZE CHARACTERISTICS OF SURFACE PLASMONS IN DIVERSE ENVIRONMENTS

Mie scattering theory can correctly reproduce the positions of maxima in the observed spectra of a particle of a given radius. Maxima in the absorption or scattering spectra resulting from Mie theory are usually interpreted as the plasmon resonances. However, spectral positions of peaks in the scattering and absorption spectra are different and are size dependent, as demonstrated and discussed in Section 5.5.2.1 (see Figure 5.11) and in Section 5.6.2. Therefore, the position of a maximum in the spectrum is not a perfect indication of LSP resonance condition. Moreover, Mie scattering theory is not a handy tool for determining the changes of the peaks in the spectrum due to the change of particle size. Effective controlling of spectral properties of plasmonic nanospheres is not possible without the knowledge about the direct dependence of LSP resonance frequencies and spectral widths (defined by plasmon damping rates) on particle size.

For simple geometries and sizes of the order of nanometres, the simplified theory predicts that the resonance frequency is independent of particle size (Kreibig & Vollmer 1995; Schuller et al. 2010). Such approximation is valid when the size of a nanoarticle is significantly smaller than the free-space wavelength of the incident light λ_{inc}, so that the entire particle experiences uniform electric field at any time. Then, spherical nanoparticles exhibit the dipolar plasmon resonance at the frequencies $\omega = 2\pi c/\lambda_{inc}$, where $\varepsilon_{in}(\omega) = -2\varepsilon_{out}(\omega)$, and $\varepsilon_{in}(\omega)$ and $\varepsilon_{out}(\omega)$ are the dielectric functions of a metal nanoparticle and its dielectric environment, respectively, *c* is the velocity of light. When the nanoparticle size approaches the plasmon excitation wavelength, the optical phase varies across the particle and it is necessary to consider retardation effects. These inevitably lead to the dependence of plasmon resonance frequency versus size. Furthermore, not only the dipole plasmon but also higher order plasmons (quadrupole, hexapole, and so on) come into play.

Tailoring of LSP properties by making use of the nanostructure size is a hot topic nowadays due to important applications in a variety of fields. Usually it is supposed that prediction of plasmonic performance versus particle size is not possible within the rigorous modelling, since too many parameters must be taken into account including particle material, size and shape as well as optical properties of particle environment. However, for spherical particles it is feasible (Derkachova & Kolwas 2007, 2013;

Kolwas & Derkachova 2010, 2013). Absorbing and emitting optical properties of a spherical plasmonic nanoantenna can be described in terms of the size dependent resonance frequencies and damping rates of the dipole and higher order LSP. These quantities provide complete intrinsic LSP size characteristics. In (Kolwas & Derkachova 2010; 2013) such characteristics for gold and silver spherical particles were provided up to the large-size retardation regime, where the plasmon radiative damping is significant. The size dependence of both the multipolar LSP resonance frequencies and corresponding damping rates can serve as a convenient tool in tailoring the features of plasmonic nanoparticles in applications. Such characteristics enable to control the operation frequency of a plasmonic nanoantenna and to change the operation range from the spectrally broad to the spectrally narrow and vice versa. It is also possible to switch between particle receiving (enhanced absorption) and emitting (enhanced scattering) abilities. By changing the polarization geometry of observation (Demianiuk & Kolwas 2001; Kolwas et al. 2009) it is possible to separate the dipole and the quadrupole plasmon radiation from all the nonplasmonic contributions to the scattered light.

Strict size characterization of LSP resonances of nanospheres of different metals embedded in diverse dielectric environments results from solving the eigenvalue problem (Kolwas & Derkachova 2010, 2013; Derkachova & Kolwas 2007, 2013). In such description, spectrally selective optical effects are due to the elementary, intrinsic properties of a conducting spherical nanosphere embedded in a dielectric medium that can manifest in the optical response to the external electromagnetic field. The formalism of Mie scattering theory (Bohren & Huffman 1983) is used. However, while the intrinsic plasmonic properties of a nanosphere are studied, there is no illuminating field, as it is in standard formulation of Mie scattering theory. Only TM mode is considered. The dielectric function of a metal, being the external parameter of the modelling, is carefully chosen in order to represent the realistic optical properties of silver and gold (Johnson & Christy 1972) as a function of light wavelength. In addition, the surface electron scattering effect (Kolwas & Derkachova 2013; Kreibig & Vollmer 1995) (also called the interface damping effect, (Section 5.5.1)) is taken into account. This effect causes the substantial modification of electron relaxation rates in particles of sizes comparable or smaller than the electron mean free path in metal. In gold nanoparticle, the additional correction to the electron damping rate above 1.8 eV is added in order to better reproduce the imaginary part of the dielectric function (Derkachova & Kolwas 2015 to be published in Plasmonics).

5.6.1 LSP RESONANCE FREQUENCIES AND DAMPING RATES AS A FUNCTION OF SIZE—RESULTS OF RIGOROUS MODELLING

Plasmon size characteristics result from considering the divergence free Maxwell equations fulfilling continuity relations at the sphere boundary. Conditions for the existence of solutions, which are TM in character, correspond to complex frequencies of the fields. The real parts of these frequencies define frequencies of plasmon oscillation $\omega'_l(R)$ for a given mode l, (l=1 for the dipole plasmon, l=2 for the quadrupole plasmon and so on). Their imaginary parts $|\omega''_l(R)|$ define radiative damping rates (or equivalently: damping times) of plasmon oscillations.

Figure 5.12 illustrates LSP size characteristics of gold spheres in an environment of the index of refraction equal to 1.5 (e.g., the immersing oil). $\omega'_l(R)=\hbar\omega_l(R)$ and $|\omega''_l(R)|=\hbar/T_l(R)$, expressed in units of energy (electronvolts), describe size dependence of plasmon resonance frequency $\omega_l(R)$ and the damping time $T_l(R)$ (from the range of 1–20 fs) of LSP oscillations (Kolwas & Derkachova 2013). The dipole plasmon resonance frequency $\omega'_{l=1}(R)$ is a decreasing function of radius R. The same applies to the quadrupole (l=2) and higher multipolarity plasmon (l>2) frequencies, as illustrated in Figure 5.12a)). Plasmon resonance takes place, when the frequency of the incoming light wave fits to the frequency of plasmon mode in the particle of radius R: $\omega=\omega_l(R)$.

Plasmon oscillations are always damped due to the ohmic losses (absorption) and radiation of electromagnetic energy. With increasing particle size, the damping rate $|\omega''_l(R)|$ (see Figure 5.12b)) is initially governed by the nonradiative damping resulting from electron collisions in a metal nanoparticle (see Eq. 5.3); while for larger particles radiative damping is dominant. Particles of such size are good far-field scattering nanoantennas. Our study, extended toward large particle sizes and plasmon multipolarities, revealed new features of the total plasmon damping rates (Kolwas & Derkachova 2013). In small nanoparticles, the total plasmon damping rate is close to the nonradiative damping rate resulting from absorption and heat dissipation while radiative losses are negligible. For larger particles, the suppression of nonradiative damping takes place when the radiative damping brings the dominant contribution to the total plasmon damping. That corresponds to the range of sizes for which $|\omega''_l(R)|$ is the fast increasing function of R (see Figure 5.12b)). The reduction of nonradiative losses with the increasing contribution of radiation damping is revealed in the absorption spectra of particles; absorptive abilities of large particles are poor (Figure 5.11b) and d)).

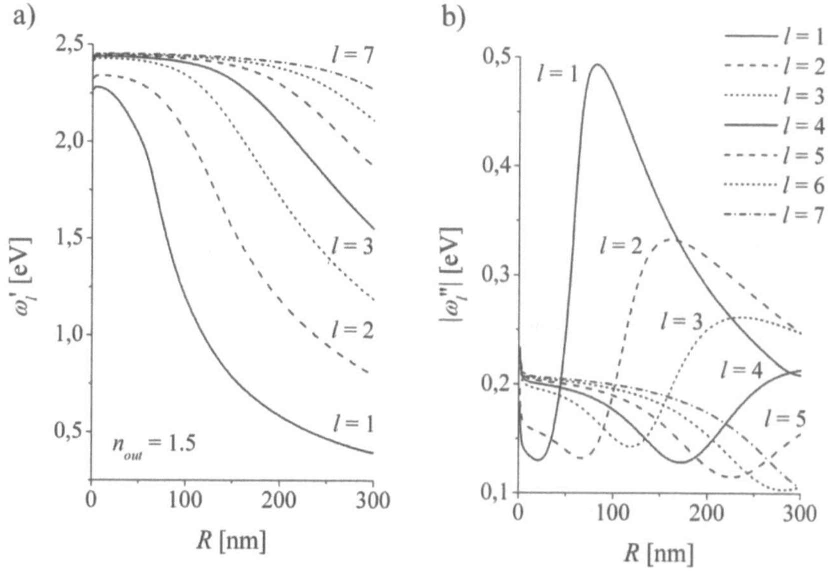

FIGURE 5.12 (a) Multipolar plasmon resonance frequencies ω'_l and (b) plasmon damping rates $|\omega''_l|$ versus particle radius R for $n_{out} = 1.5$.

Figure 5.13a) and b) tells us what wavelength λ_{inc} of the incoming light should be used to excite dipole, quadrupole, and hexapole plasmon resonances in gold nanoparticles of different radii and embedded in different environments: Figure 5.13a) corresponds to air ($n_{out} = 1$), and Figure 5.13b) to water ($n_{out} = 1.3$). Figure 5.13c) and d illustrate the corresponding dependencies for silver nanoparticles.

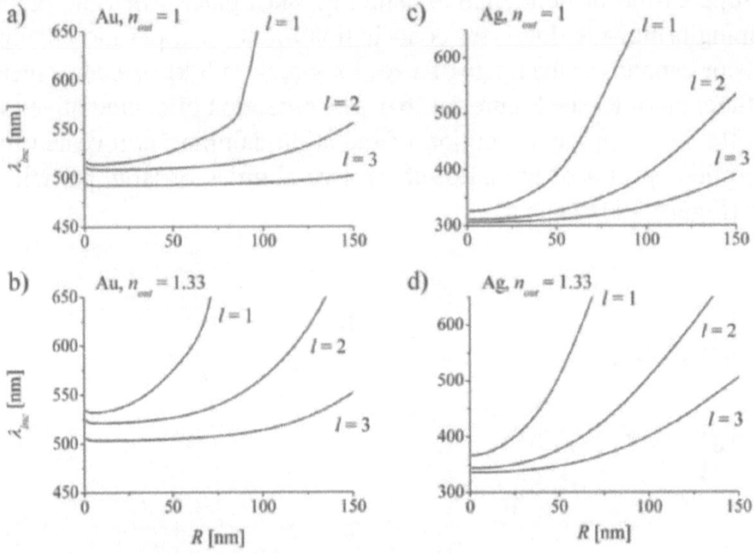

FIGURE 5.13 The wavelength of the incoming light field λ_{inc} corresponding to dipole (l=1), quadrupole (l=2) and hexapole (l=3) plasmon resonance frequency in gold a), (b) and silver c), d) nanoparticles of different radii and in diverse environments: a), c) in air (n_{out}=1), and b), d) in water (n_{out}=1.3).

Figure 5.13 demonstrates that the frequency of plasmon oscillations in gold nanoparticles falls well into the visible range and is not very strong function of particle radius in large range of sizes. Resonance frequencies of the corresponding silver nanoparticles are strongly shifted toward shorter wavelength and for nanoparticles of smaller size fall into near-UV range. Properties of silver nanoparticles are more sensitive to size then the corresponding gold nanoparticles.

5.6.2 MANIFESTATION OF LSP RESONANCES IN THE ABSORPTION AND SCATTERING SPECTRA

Maxima in the optical spectra of nanoparticles are manifestations of plasmon resonances at frequencies $\omega'_l(R)$ which in Figure 5.11 are marked with dotted vertical lines. However, the spectral positions of peaks occurring in the absorption spectrum (red line) and in the scattering spectrum (black line) are different.

In the absorption spectrum of small nanoparticles (R=5nm, dashed line in Figure 5.11), the maximum corresponds to the dipole plasmon resonance frequency $\omega'_{l=1}(R = 5\text{nm})$ for gold (Figure 5.11a)), and for silver (Figure 5.11c)) nanoparticles. However, for larger particles (R=75nm) (Figure 5.11b) and d)), the manifestation of the dipole plasmon at $\omega'_{l=2}(R = 75\text{nm})$ in the absorption spectrum (dashed line) is suppressed (Kolwas & Derkachova 2013). Despite the dipole plasmon resonance frequency $\omega'_{l=1}(R = 5\text{nm})$ (marked with the corresponding horizontal line) shifts toward larger wavelength (red-shift) with increasing particle size, the absorption spectrum shifts toward shorter wavelengths (blue-shift). Though the dipole plasmon manifestation is suppressed in the absorption spectrum of small nanoparticles, in larger particles it participates in formation of the maximum in the scattering spectrum at the frequency $\omega'_{l=1}(R = 75\text{nm})$ marked by dotted vertical line in Figure 5.11b) for gold and Figure 5.11d) for silver nanoparticles. This can be explained by the increased contribution of the radiation damping to the total plasmon damping rate (Figure 5.12b)) in larger particles. Higher order plasmons (l>1) are characterized by lower radiative damping rates for a particles of the same radii. Therefore, as long as the contribution of radiative damping of quadrupole and higher order plasmons is low, absorption in larger nanoparticles is not suppressed. As demonstrated in Figure 5.11b), d), solid line), the position of the maximum in the scattering spectrum for larger particles is shifted in respect to the plasmon resonance position (Kolwas & Derkachova 2013; Kolwas et al. 2009).

We can state, that in general, position of the maximum in the spectrum is not a perfect indication of the plasmon resonance in larger particles. However, the existence of such maxima can be used in applications, for example in the calorimetric techniques. With the increasing size of a particle, not only a substantial down-shift of plasmon frequencies (longer wavelengths), but also a significant broadening of the plasmon resonance take place (Kolwas & Derkachova 2010, 2013; Sönnichsen et al. 2002). This feature can be predicted by our plasmon size characteristics (resonance frequencies $\omega'_l(R)$ and plasmon damping rates $|\omega''_l(R)|$ presented in Figure 5.12)), which give the plasmon resonance position and the spectral width of plasmon resonance as a function of particle radius R. Despite the plasmon resonance in larger particles shifts toward longer wavelengths (see Figure 5.12a), the absorption spectrum shifts toward shorter wavelengths (see Figure 5.11b) and d)) due to suppression of the absorptive particle abilities in the dipole mode. This refers to the existing controversy in interpretation of the observed plasmon optical properties. Modelling presented in this section allows explaining the observed changes in spectral performance of metal nanoparticles as well as tailoring plasmon properties with understanding of the basis of the observed phenomena.

The interpretation of changes in colors of colloids, used in colorimetric detection (Section 5.3.4) of oxidative DNA damage caused by peroxynitrite ($ONOO^-$) can serve as a practical example. The long, single stranded DNA adsorbed onto gold nanoparticles prevents the aggregation of nanoparticles, and their optical response is the same as those of single, well separated nanoparticles with the dipole plasmon maximum in the red part of the spectrum (see Figure 5.11a)). If DNAs are cleaved by $ONOO^-$ to form small fragments, the gold nanoparticles aggregate forming large gold particles. One could expect, that due to the red-shift of plasmon resonance frequencies with increasing size (Figure 13a) and b)), the color of the colloid will change toward the same spectral range. However, it is not the case. Large particles lose their absorbing abilities in the dipole mode due to the domination of radiative damping over all energy dissipation processes leading to absorption in nanoparticles (see the discussion above and Section 5.6.1). As the result, the spectrum is blue shifted, as confirmed by Figure 5.11b), dotted line.

5.7 OPTIMIZATION OF PLASMON FEATURES FOR DIVERSE APPLICATIONS OF NANOPARTICLES: GOLD OR SILVER?

Possibility of tailoring plasmon properties of nanostructures is the key to their applications and their development. Direct size dependencies of LSP resonance frequencies and of corresponding plasmon damping rates provide an important tool for optimization of plasmon properties in diverse application. Plasmon resonance frequencies, damping rates of plasmon oscillations and the number of plasmon models manifesting in optical spectra define the spectral performance of a plasmonic nanoparticle and can be controlled by the particle size, the type of metal, and the optical properties of dielectric environment according to the model predictions (Section 5.6). The numerical tools based on electrodynamical description allow calculating the dependence of plasmon resonance frequencies and damping rates of plasmon oscillations (see Section 5.6.1) of metal nanoparticles embedded in diverse environments. The results of such modelling give direct information on the size of a gold or silver nanosphere which must be used to optimize the plasmon local electric field strength or scattered\ absorbed light intensity spectra in the desired spectral range. Plasmon size characteristics help to decide what kind of metal nanoparticles should be used for a given application. In the following, we discuss the usefulness of the results of our modelling of plasmon features for optimization of performance of plasmonic nanoparticles in diverse applications. We consider such features as the shift of plasmon resonance frequency with size, the size dependence of nanoparticles absorbing and scattering abilities, the spectral sensitivity of plasmon features to the index of refraction of the environment, the strength of near-field enhancement, its spatial distribution and the extent of the plasmon enhanced field.

5.7.1 SIZE DEPENDENT PLASMON RESONANCE FREQUENCY

The explicit description of size dependence of plasmon resonances is very useful in numerous plasmon applications utilising for example the effective enhancement of the local near-field with plasmonic mechanism (e.g., in SERS (see Section 5.3.1)). If we intend to strengthen the fluorescence of the only few molecules which we have at our disposal, we should decide what size of plasmonic nanoparticle should be used in order to enhance the electric field around plasmonic nanoparticle. If, for instance, a molecular excitation wavelength $\lambda = 514$ nm, and we intend to use a gold nanoparticles, their radius should fall in the range up to 30 nm (in air) (as predicted by modelling, see Figure 5.13a)). Such wide range of sizes makes using gold nanoparticles convenient. Moreover, the quadrupole plasmon frequency is in spectral proximity of the dipole plasmon, so we can expect excitation of both and thus an additional near-field enhancement. However, if gold nanoparticles are immersed in water, neither dipole nor quadrupole plasmon can be used for this purpose (see Figure 5.13b)). What will happen if we replace gold by silver? In water, dipole plasmon resonance can be excited, but silver nanoparticle must be large, of radius larger than 50 nm (Figure 5.13d)). Such large silver nanospheres are hardly available with the contemporary nanofabrication techniques.

5.7.2 SIZE DEPENDENT SPECTRAL ABSORBING AND SCATTERING ABILITIES

The ratio of light scattering to light absorption can be tailored by the nanoparticle size (see Section 5.5.2.1, Figure 5.11). This provides a tool for nanoparticle selection for example contrast agent applications. For instance, larger nanoparticles are more suitable for light-scattering-based imaging techniques, while smaller nanoparticles mainly absorb radiation and turn it into heat what can be of potential use in tumor therapy.

Changes in absorbing and scattering cross sections with size can be explained by the size dependence of the total damping rates (Kolwas & Derkachova 2013) (see Section 5.6.1). If the dissipative, nonradiative damping contribution to the total damping rate prevails, absorption cross section C_{abs} is large and scattering cross section C_{scat} is negligible (see Figure 5.11). As long as the contribution of the radiative decay is negligible, the particle is not able to couple to the incoming light field effectively and has weak radiative (scattering) abilities. If the contribution of the radiative damping is large, the particle is able to emit light within plasmonic mechanism efficiently. Figure 5.12b) gives an example of $|\omega''_l(R)|$ dependence for gold nanoparticles. This dependence allows distinguishing the size ranges in which efficient transfer of radiation energy into heat takes place (smaller nanoparticles) and those in which particles are good radiating antennas (larger nanoparticles with the fast increasing $|\omega''_l(R)|$ dependence).

Plasmonic absorption activity of silver nanoparticle falls in UV spectral range (see Figure 5.11c) and d)). However, by increasing particle size it is possible to make them plasmonically active also in the visible range, where their scattering cross section C_{scat} (see Figure 5.11d)) is large.

Due to the plasmon enhancement mechanism, absorption and scattering cross sections of gold and silver nanoparticles are significantly larger than their geometrical cross section. These plasmonic particles are also superior to the absorbing and fluorescing dyes conventionally used in biological and biomedical imaging (Jain et al. 2006, 2008). Note that in the infrared region gold and silver nanoparticles are optically inactive, regardless to particle size. Therefore, they are not able to enhance the infrared radiation (e.g., in cancer therapy), as sometimes expected. However, properly designed gold nanoshell particles are used in experimenting with photothermal tumor therapy (Gobin et al. 2007; Loo et al. 2005).

5.7.3 SENSIBILITY TO THE INDEX OF REFRACTION OF THE ENVIRONMENT

An important advantage of LSP is the sensitivity of plasmon resonance frequency to changes in the refractive index of the immediate environment. Thereby, noble-metal nanospheres can act as efficient contrast agents (see Section 5.3.2).

The explicit size dependence of LSP resonance frequencies of nanoparticle in diverse environment is presented in Figure 5.14a) for the dipole and in Figure 5.14b for the quadrupole LSP of gold and silver nanoparticles correspondingly. The dipole LSP resonance frequency $\omega'_{l=1}(R)$ of both silver and gold nanoparticles exhibits a shift toward lower frequencies (longer wavelengths) with the increasing index of refraction n_{out} of the environment (see Figure 5.14a)). Up to about 100 nm, plasmon resonance frequency of silver nanoparticles is much more sensitive to changes in optical properties of particle environment than that of gold. Therefore silver nanoparticles are more suitable in such applications as plasmonic biosensors. However, their plasmonic activity falls in UV. A similar conclusion can be drawn from the size dependence of the quadrupole resonance frequency, illustrated in Figure 5.14b). For very large nanoparticles $\omega'_l(R)$ for gold and silver is similar when embedded in the same environment.

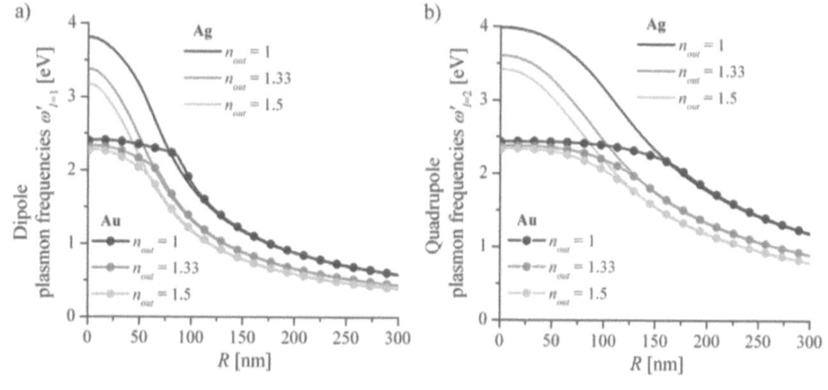

FIGURE 5.14 a) Dipole and b) quadrupole resonance frequencies as the functions of radius for silver and gold nanoparticles in air (n_{out}=1), water (n_{out}=1.33), and immersion oil (n_{out}=1.5).

5.7.4 THE ELECTRIC NEAR-FIELD STRENGTH AND SPATIAL DISTRIBUTION

Plasmon collective charge oscillations are coupled to the electric fields on (and near) the nanoparticle surface. These surface confined fields are strongly enhanced in respect to the incoming field of the light wave. The enhancement factor is not uniform in space and time. Its maximal value changes with particle size and for metal nanoparticles (if realistic modelling of their material properties is used) the enhancement is not largest for the smallest nanoparticles.

An important optimization factor results from the spatial distribution of total electric near field strength around the nanosphere. The local enhancement factor $\eta(R,\theta,\phi,t)$ of the incoming electric wave with the amplitude E_0 is:

$$\eta(R,\theta,\varphi,t) = |\mathbf{E}^{out}(R,\theta,\varphi,t)| / E_0, \tag{5.4}$$

where $|E^{out}| = |E^{inc} + E^{scat}|$ is the strength of the total electric field at the particle surface, illuminated by the incoming plane wave E^{inc}, (r,θ,ϕ) are spherical coordinates (see Figure 5.15a)).

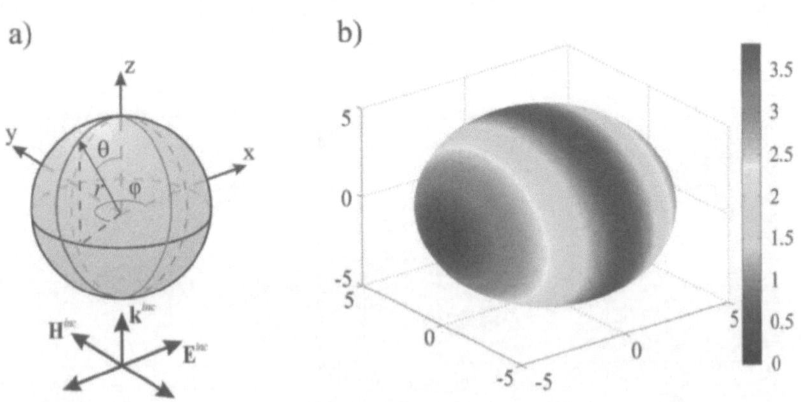

FIGURE 5.15 a) A scheme of plasmon excitation. b) 3D image of the enhancement factor $\eta(R,\theta,\phi,t)$, for the dipole plasmon resonance frequency $\omega'_{l=1}(R=5\text{nm})$.

Spatial distribution of the enhancement factor displays a characteristic dynamic pattern, which follows the temporal dependence of the incident field. Analysis of the distribution pattern of $|E^{out}(R,\theta,\phi,t)|$ field allows distinguishing these areas on (or near) the nanoparticle surfaces, which are the most favorable for the electric field enhancement, and those which should be avoided. Figure 5.15a), red circle, shows the favorable regions lying in the plane parallel to the plane of incidence on the meridian circumference. Exceptions are the south and north poles of a nanosphere, which are dark. The plane of incidence is defined by the polarization and the propagation direction of the incident light. The dark locations (dark blue colour) and the locations of maximally enhanced field (red brown) are presented in Figure 5.15b). If used for example in SERS application (see Section 5.3.1), the molecules adjacent to the nanoparticle dark regions will not be excited at all; they will be totally lost from detection.

The maximal enhancement factor of small gold nanoparticles is lower than that of larger particles (Figure 5.16). It is partially due to the enhanced electron relaxation γ_R (see Eq. (5.3) in Section 5.5.1) in small nanoparticles, which increases the absorption of light by a nanoparticle. Also, while smaller particles are not good scattering nanoantennas (see Section 5.5.2.1) the strength of the total near field, must be smaller than for larger particles, as the numerical model simulations prove.

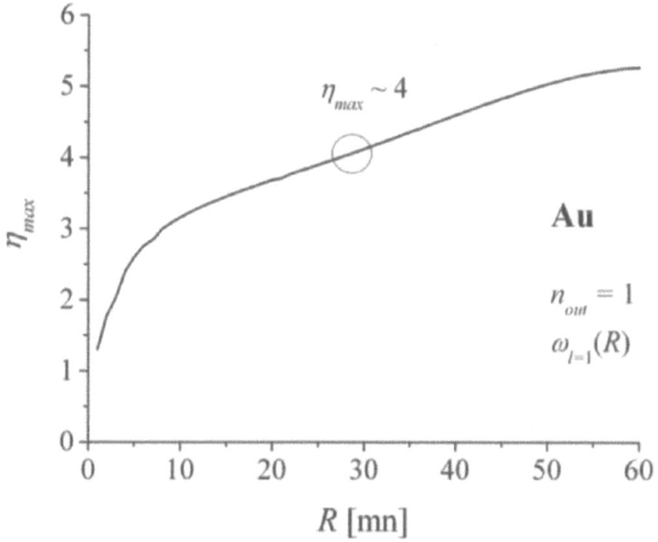

FIGURE 5.16 Maximal enhancement factor η_{max} of the near electric near-field strength versus radius at the dipole resonance frequency of gold nanoparticle.

5.7.5 EXTENT OF THE PLASMON ENHANCED FIELD

Plasmonic nanoparticles are of particular interest in biological applications because they are of the same scale size as biological macromolecules, proteins, and nucleic acids. Plasmon enhanced electric fields are known to be strongly localized near the surface. The extent of the enhanced near field around a plasmonic particle plays crucial role in applications. The penetration depth of the field into the particle environment should be in reasonable relation with the size of the object adjacent to the particle, and subjected to interaction with the field enhanced within plasmonic mechanism.

Figure 5.17 illustrates the maximal enhancement factor η_{max} versus distance d_0 form the surface of gold nanoparticles of diverse radii (R=10, 30, and 100 nm) at the corresponding dipole resonance frequency $\omega'_{l=1}(R)$. In larger nanoparticles, there is more room on the particle surface, and the maximal enhancement is not smaller. Therefore, it is worth to consider the extent of the enhanced plasmon field around a nanoparticle in comparison with the size of the studied object, which can be the atom, molecule or bacteria, in order to optimise their relative relation.

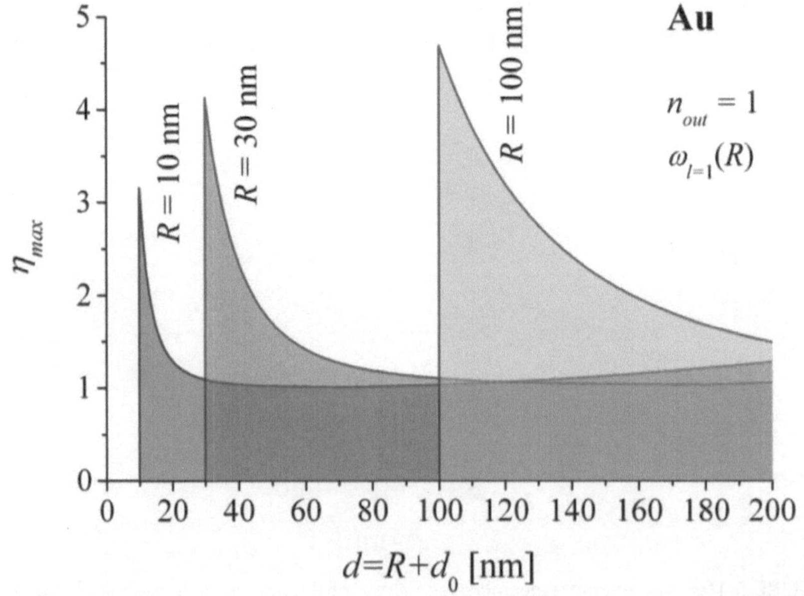

FIGURE 5.17 Maximal enhancement factor η_{max} for the dipole resonance frequency versus distance d_0 form the surface of a gold nanoparticle of diverse radii.

5.8 SUMMARY AND CONCLUSIONS

The main aim of the chapter was to present some practical tools for tailoring plasmon characteristics to improve the optical diagnostics of molecules and cells, as seen by physicists. In the introductory sections (Sections 5.1–5.4) we review those features of metals at the nanoscale which make them suitable in applications. These exceptional features result from ability of metal nanoparticles to resonate with electromagnetic radiation in the optical range, which leads to excitation of localized surface plasmon waves. Plasmonic nanoparticles can convert optical radiation into intense, localized EM field distributed near the particle surface. By squeezing light into a subwavelength volume in particle proximity, plasmonic structures have a profound effect on the efficiency of many optical processes being the basis of numerous applications. Some of them are reviewed in the introductory sections. Plasmon resonances manifest as the intense absorption or scattering of light, depending on the particle size, shape, and optical properties of particle environment. The sensitivity of plasmon resonances to these parameters is the basis of color effects, constituting a large group of colorimetric detection methods. We do not pretend to make a full survey of applications of nanoparticles in biology and medicine. A choice of examples of the optical applications of plasmons is expected to point to further possibilities.

In Sections 5.5–5.7, we present a review of practical tools for tailoring plasmon characteristics to improve the optical diagnostics of molecules and cells. As we believe, improvement in applications is possible only, if the physical basis of the phenomena involved is well understood. Fundamentals of resonant interaction of light with plasmonic nanoparticles are essential for studying LSP properties and for their optimal use. Practically, optimization is possible, if one has the information concerning how the plasmon properties can be conveniently tailored, which parameters should be optimized and according to which formula it should be done. The presented study, which is an original contribution to the domain of physics of metal nanoparticles, delivers direct modelling of plasmon parameters as a function of nanosphere size, kind of metal used (gold and silver) and index of refraction of the particle environment. Language of classical electrodynamics was used and the appropriate solutions of Maxwell equations were found. Using the obtained plasmon characteristics it is possible for example to predict the spectral performance of colloids and explain the observed colors and interpret their changes. In addition, it is possible to control the changes in performance of plasmonic nanoantenna from good absorbing to good scattering device, and to optimize its near-field strength, spatial field distribution, and extent of the plasmon-enhanced field. We hope that such study can be useful to a broad scientific community.

ACKNOWLEDGMENTS

We acknowledge the financial support for this work by the Polish Ministry of Science and Higher Education (Grant No. N N202 126837).

KEYWORDS

- **Dipole plasmon**
- **Hexapole plasmon**
- **Nanoparticles**
- **Nanoscale**
- **Localized Surface Plasmon**
- **Quadrupole plasmon**
- **Raman scattering**

REFERENCES

Anon, The Lycurgus Cup. Available at: http://www.britishmuseum.org/explore/online_tours/museum_and_exhibition/the_art_of_glass/the_lycurgus_cup.aspx.

Aslan, K. & Geddes, C.D., 2009. Wavelength-ratiometric plasmon light scattering-based immunoassays. Plasmonics, 4(4), pp. 267–272.

Aslan, K., Lakowicz, J.R. & Geddes, C.D., 2004. Nanogold-plasmon-resonance-based glucose sensing. Analytical Biochemistry, 330(1), pp. 145–155.

Aslan, K., Lakowicz, J.R. & Geddes, C.D., 2005a. Nanogold plasmon resonance-based glucose absorption bands due to electron oscillations induced by. Analytical Chemistry, 77(7), pp. 2007–2014.

Aslan, K., Lakowicz, J.R. & Geddes, C.D., 2005b. Plasmon light scattering in biology and medicine: new sensing approaches, visions and perspectives. Current Opinion in Chemical Biology, 9(5), pp. 538–544.

Barnes, W.L., 2006. Surface plasmon–polariton length scales: a route to sub-wavelength optics. Journal of Optics A: Pure and Applied Optics, 8(4), pp. S87–S93.

Barnes, W., Dereux, A. & Ebbesen, T., 2003. Surface plasmon subwavelength optics. Nature, 424(August), pp. 824–830.

Baschong, W., Lucocq, J.M. & Roth, J., 1985. "Thiocyanate gold": small (2–3 nm) colloidal gold affinity cytochemical labeling in electron microscopy. Histochemistry, 83, pp. 409–411.

Bohren, C.F. & Huffman, D.R., 1983. Absorption and scattering of light by small particles, Wiley.

Born, M. & Wolf, E., 1999. Principles of Optics, Cambridge University Press.

Brown, K.R., Walter, D.G. & Natan, M.J., 2000. Seeding of colloidal Au nanoparticle solutions. 2. Improved control of particle size and shape. Chemistry of Materials, 12(2), pp. 306–313.

Chen, J. et al., 2005. Gold nanocages: bioconjugation and their potential use as optical imaging contrast agents. Nano Letters, 5(3), pp. 473–477.

Chen, L. et al., 2013. Colorimetric detection of peroxynitrite-induced DNA damage using gold nanoparticles, and on the scavenging effects of antioxidants. Microchimica Acta, 180(7–8), pp. 573–580.

Demianiuk, S. & Kolwas, K., 2001. Dynamics of spontaneous growth of light-induced sodium droplets from the vapour phase. Journal of Physics B: Atomic, Molecular and Optical Physics, 4075(01), pp. 1651–1671.

Derkachova, A. & Kolwas, K., 2007. Size dependence of multipolar plasmon resonance frequencies and damping rates in simple metal spherical nanoparticles. The European Physical Journal Special Topics, 144(1), pp. 93–99.

Derkachova, A. & Kolwas, K., 2013. Simple analytic tool for spectral control of the dipole plasmon resonance frequency of gold and silver nanoparticles. Photonics Letters of Poland, 5(2), pp.69–71.

Derkachova, A. & Kolwas, K., 2014, *to be published.*

Dragoman, M. & Dragoman, D., 2008. Plasmonics: applications to nanoscale terahertz and optical devices. Progress in Quantum Electronics, 32(1), pp. 1–41.

El-Sayed, I. H.; Huang, X.; El-Sayed, M. A., 2005. Surface plasmon resonance scattering and absorption of anti-EGFR antibody conjugated gold nanoparticles in cancer diagnostics: applications in oral cancer. Nano Letters, 5, pp. 829–834.

Esteban, R. et al., 2008. Direct near-field optical imaging of higher order plasmonic resonances. Nano Letters, 8(10), pp. 3155–3159.

Faraday, M., 1857. The Bakerian lecture: experimental relations of gold (and other metals) to light. Philosophical Transactions of the Royal Society of London, 147, pp. 145–181.

Frens, G., 1973. Controlled nucleation for the regulation of the particle size in monodisperse gold suspensions. Nature Physical Science, 241, pp. 20–22.

Gartia, M.R. et al., 2013. Colorimetric plasmon resonance imaging using nano lycurgus cup arrays. Advanced Optical Materials, 1(1), pp. 68–76.

Gobin, A.M. et al., 2007. Near-infrared resonant nanoshells for combined optical imaging and photothermal cancer therapy. Nano Letters, 7(7), pp. 1929–1934.

Haynes, C.L. & Van Duyne, R.P., 2001. Nanosphere lithography: a versatile nanofabrication tool for studies of size-dependent nanoparticle optics. Journal of Physical Chemistry B, 105, pp. 5599–5611.

Haynes, C.L. et al., 2002. Angle-resolved nanosphere lithography: manipulation of nanoparticle size, shape, and interparticle spacing. Journal of Physical Chemistry B, 106(8), pp. 1898–1902.

Heiligtag, F. & Niederberger, M., 2013. The fascinating world of nanoparticle research. Materials Today, 16(7/8), pp. 262–271.

Jain, P.K. et al., 2006. Calculated absorption and scattering properties of gold nanoparticles of different size, shape, and composition: applications in biological imaging and biomedicine. Journal Of Physical Chemistry B, 110(14), pp. 7238–7248.

Jain, P.K. et al. 2007a. On the universal scaling behavior of the distance decay of plasmon coupling in metal nanoparticle pairs: a plasmon ruler equation. Nano Letters, 7(7), pp. 2080–2088.

Jain, P.K., El-Sayed, I. & El-Sayed, M., 2007. Au nanoparticles target cancer. Nanotoday, 2(1), pp. 18–29.

Jain, P.K. et al., 2008. Noble metals on the nanoscale: optical and photothermal properties and some applications in imaging, sensing, biology, and medicine. Accounts of Chemical Research, 41(12), pp. 1578–1586.

Johnson, P.B. & Christy, R.W., 1972. Optical constants of the noble metals. Physical Review B, 6(12), pp. 4370–4379.

Khan, M.S., Vishakante, G.D. & Siddaramaiah H., 2013. Gold nanoparticles: a paradigm shift in biomedical applications. Advances in Colloid and Interface Science, 199–200, pp. 44–58.

Khlebtsov, N.G. & Dykman, L.A., 2010. Optical properties and biomedical applications of plasmonic nanoparticles. JQSRT, 111(1), pp. 1–35.

Khlebtsov N.G. et al., 1996. Spectral extinction of colloidal gold and its biospecific conjugates. Journal of Colloid Interface Science, 180(1), pp. 436–445.

Khlebtsov, B.N., Khanadeev, V.A. & Khlebtsov, N.G., 2010. Attenuation, scattering, and depolarization of light by gold nanorods with silver shells. Optics and Spectroscopy, 108(1), pp. 59–69.

Kolwas, K. & Derkachova, A., 2010. Plasmonic abilities of gold and silver spherical nanoantennas in terms of size dependent multipolar resonance frequencies and plasmon damping rates. Opto-Electronics Review, 18(4), pp. 429–437.

Kolwas, K. & Derkachova, A., 2013. Damping rates of surface plasmons for particles of size from nano- to micrometers; reduction of the nonradiative decay. JQSRT, 114, pp. 45–55.

Kolwas, K., Derkachova, A. & Shopa, M., 2009. Size characteristics of surface plasmons and their manifestation in scattering properties of metal particles. JQSRT, 110(14–16), pp. 1490–1501.

Kolwas, K, Derkachova, A. & Shopa, M., 2013. Practical tools for optimising plasmon enhanced near fields for optical diagnostic of molecules and cells. Journal of Biomaterials and Tissue Engeneering, pp. 1–9, *in print*.

Kreibig, U. & Vollmer, M., 1995. Optical properties of metal clusters. Heidelberg: Springer.

Lakowicz, J.R., 2006. Plasmonics in biology and plasmon-controlled fluorescence. Plasmonics (Norwell, Mass.), 1(1), pp.5–33.

Liao, H., Nehl, C.L. & Hafner, J.H., 2006. Biomedical applications of plasmon resonant metal nanoparticles. Nanomedicine, 1(2), pp. 201–8.

Lin, H.-Y. et al., 2010. Direct near-field optical imaging of plasmonic resonances in metal nanoparticle pairs. Optics Express, 18(1), pp. 165–172.

Loo, C. et al., 2005. Immunotargeted nanoshells for integrated cancer imaging and therapy. Nano Letters, 5(4), pp. 709–711.

Maier, S.A. & Atwater, H.A., 2005. Plasmonics: Localization and guiding of electromagnetic energy in metal/dielectric structures. Journal of Applied Physics, 98(1), p. 011101.

Mahajan, S. et al., 2010. Understanding the surface-enhanced raman spectroscopy "background". The Journal of Physical Chemistry C, 114(16), pp. 7242–7250.

Oldenburg, S.J. et al., 1998. Nanoengineering of optical resonances. Chemical Physics Letters 288, pp. 243–247

Pradeep, T., 2009. Noble metal nanoparticles for water purification: a critical review. Thin Solid Films, 517(24), pp. 6441–6478.

Quinten, M., 2011. Optical properties of nanoparticle systems, Wiley-VCH Verlag GmbH & Co.

Rogobete, L. et al., 2007. Design of plasmonic nanoantennae for enhancing spontaneous emission. Optics Letters, 32(12), pp. 1623–1625.

Sambles, J. & Bradbery, G., 1991. Optical excitation of surface plasmons: an introduction. Contemporary Physics, 32(3), pp. 173–183.

Sardar, R. et al., 2009. Gold nanoparticles: past, present, and future. Langmuir: The ACS Journal of Surfaces and Colloids, 25(24), pp. 13840–13851.

Schneider, T. et al., 2013. Localized surface plasmon resonance (LSPR) study of DNA hybridization at single nanoparticle transducers. Journal of Nanoparticle Research, 15(4), p. 1531.

Schuller, J.A. et al., 2010. Plasmonics for extreme light concentration and manipulation. Nature Materials, 9(3), pp. 193–204.

Shen, Y. et al., 2000. Nanophotonics: interactions, materials and applications. Journal of Physical Chemistry B, 104, pp. 7577–7587.

Sönnichsen, C., 2001. Plasmons in metal nanostructures. Ludwig-Maximilians-Universität München. http://edoc.ub.uni-muenchen.de/2367/1/Soennichsen_Carsten.pdf.

Sönnichsen, C. et al., 2002. Plasmon resonances in large noble-metal clusters. New Journal of Physics, 4, pp. 93–93.

Stuart, D. & Haes, A., 2005. Biological applications of localised surface plasmonic phenomenae. IEE Proceedings of Nanobiotechnology, 152(1), pp. 13–32.

Turkevich, J., Garton, G. & Stevenson, P., 1954. The color of colloidal gold. Journal of Colloid Science, 9, Supplement 1, pp. 26–35.

Wang, Y. et al., 2007. SERS opens a new way in aptasensor for protein recognition with high sensitivity and selectivity. Chemical Communications (Cambridge, UK), 48, pp. 5220–5222.

Wang, Y., Plummer, E.W. & Kempa, K., 2011. Foundations of plasmonics. In Advances in physics. Taylor & Francis.

Wax, A. & Sokolov, K., 2009. Molecular imaging and darkfield microspectroscopy of live cells using gold plasmonic nanoparticles. Laser & Photonics Review, 3(1-2), pp. 146–158.

Weiner, J., 2009. The physics of light transmission through subwavelength apertures and aperture arrays. Reports on Progress in Physics, 72(6), p. 064401.

Willets, K. & Van Duyne, R., 2007. Localized surface plasmon resonance spectroscopy and sensing. Annual Reviw of Physical Chemistry, 58, pp. 267–297.

Yang, S.-M. et al., 2008. Synthesis and assembly of structured colloidal particles. Journal of Material Chemistry, 18(19), pp. 2177–2190.

Ye, J. et al., 2009. Fabrication, characterization, and optical properties of gold nanobowl submonolayer structures. Langmuir, 25(3), pp. 1822–1827.

Zakharian, A.R., Moloney, J. V. & Mansuripur, M., 2007. Surface plasmon polaritons on metallic surfaces. Optics Express, 15(1), pp. 183–197.

CHAPTER 6

RECENT ADVANCES IN NANOPARTICULATE DRUG DELIVERY SYSTEM FOR ANTIVIRAL DRUGS

DIPALI M. DHOKE, RUPESH V. CHIKHALE, AMIT M. PANT, SUNIL MENGHANI, NILESH RAROKAR, and PRAMOD B. KHEDEKAR

ABSTRACT

Nanomedicines has given an innovative way to prevent the diseases caused by viruses by giving various therapeutic technologies that target the virus-infected site in the body and has achieved the success by curing the diseases. To achieve this target specificity nanoparticles may be formulated using different biocompatible and biodegradable polymers. These polymers aim at targeting the antiviral drug to the virus-infected cells or tissues by forming a ligand–receptor complex. However, it has been observed that some of the polymers like poly (D,L-lactide-co-glycolide) (PLGA) copolymer have limited types of functional groups available on the surface for conjugation to targeting ligands. To overcome this drawback the Interfacial Activity Assisted Surface Functionalization (IAASF) technique can be used to incorporate reactive functional groups such as maleimide onto the surface of PLGA nanoparticles which can be further conjugated with different peptides to achieve target specificity. Thus the functionalized nanoparticulate drug delievery system might increase the efficacy, reduce drug- and formulation related side effects, change the release kinetics of antivirals, increase the target specificity, increase their bioavailability, improve the pharmacokinetics, reduce treatment costs, and improve patient compliance. These features are particularly functional in viral diseases where high drug doses are needed, the disease is chronic and fast remedy is required to overcome the illness, drugs are expensive, and the success of a therapy is associated with a patient's adherence to the administration procedure.

6.1 INTRODUCTION

Nanoparticles are particulate dispersions or solid particles with a size of 1–100 nm. The major goals in designing nanoparticles as a delivery system are to control particle size, surface properties, and release of pharmacologically active agents in order to achieve the site-specific action of the drug at therapeutically optimal rate and dose regimen (1). Nanoparticles can be prepared by different methods which include nanoprecipitation technique, solvent diffusion technique, and emulsion polymerization technique (2). After preparation, nanoparticles are usually dispersed in liquids. Such a system can be administered to humans by parentral route, oral route, or used in ointments and ocular products. Alternatively, nanoparticles can be dried to a powder, which allows pulmonary delivery or further processing to tablets or capsules (3). After intravenous administration, nanosized particles are mainly taken up by the macrophages of the mononuclear phagocyte system (MPS) (4) and, thus, can be localized in the liver, spleen, and lungs (5). By modifying particle surface, by coating, defence mechanisms of the body can be avoided to some extent leading to longer circulation times of nanoparticles in the blood (6). Tailored coating also enables another promising application of nanoparticles in drug delivery across the blood–brain barrier (BBB) (7). After oral administration, intracellular uptake may occur from the intestine and prior to that, the nanoparticles can adhere to the mucosa (bioadhesion) and thus improve pharmacokinetics of the drug (8). Nanoparticles in pharmaceutical applications have gained plenty of research attention during recent decades. Although the research concerning formulation of nanoparticles into drug delivery devices has been extensive, only a few nanoparticulate products have reached the market, for example, Lupron, Nutropin, Sandostatin, LAR, Trelstar, Suprecur MP, Somatuline LA, Arestin, Risperdal Consta, and others (9).

Targeted drug delivery has the potential to increase the fraction of administered dose reaching the disease site while minimizing nonspecific drug distribution. Delivery systems used for targeting most often comprise a drug carrier attached to a ligand that binds with a specific target over expressed at the disease site (10).

In drug delivery, nanoparticles should be biocompatible (not harmful for humans) and biodegradable (deteriorate and expulse in the body conditions). These properties can be affected by nanoparticle material selection and by surface modification. Materials such as synthetic polymers, proteins, or other natural macromolecules are used in the preparation of nanoparticles. In recent years, biodegradable, biocompatible polymeric nanoparticles, particularly those coated with hydrophilic polymer such as poly(ethylene glycol) (PEG), PLGA known as long-circulating particles, have been used as potential drug delivery devices because of their ability to circulate for a prolonged period time, target a particular organ, as carriers of DNA in gene therapy, and their ability to deliver proteins, peptides, genes, and other drugs at the site of their action (11) (Figure 6.1).

In the current review, it is demonstrated that the Interfacial Activity Assisted Surface Functionalization (IAASF) method may be used to incorporate reactive functional groups onto the surface of PLGA nanoparticles to achieve target specificity for antiviral drugs (12). The antiviral developed against some viruses for example, HSV, HBV, herpes virus, and HIV are unable to reach to the site of actual dense viral hub as a result they are able to treat the acute diseases but do not cure the latent infection. This results in recurrent or chronic diseases that require treatment for longer duration. These and other issues such as low efficacy of antiviral drugs, low bioavailability, short half-life, low solubility, side effects, and long duration therapy represent a major challenge in antiviral research and development. Further, it is difficult to design safe and effective antiviral drug because viruses uses the host's cell to replicate. As a result it becomes difficult to find targets for the drug that would interfere with the virus without harming the host cells. Moreover, the major difficulty in developing vaccines and antiviral drugs is due to variations in the genetic makeup of the virus.Therefore, a deep and expanded knowledge of the genetic and molecular function of organisms is required for the biomedical researchers to understand the structure and function of viruses to develop an antiviral. Various advanced technique are in progress for finding new drugs (13).

Further improvements to the therapy can also be obtained through the use of novel drug delivery system delivery system (NDDS) for antiviral administration. NDDS has proved to be a better approach to enhance the effectiveness of the antivirals and improve the patient compliance and decrease the adverse effect. The NDDS have reduced the dosing frequency and shorten the duration of treatment, thus, which could lead the treatment more cost-effective. The development of NDDS for antiviral therapy aims to deliver the drug devoid of toxicity, with high compatibility and biodegradability, targeting the drug to specific sites for viral infection, and in some instances it also avoids the first pass metabolism effect. This chapter aims to discuss the usefulness of novel delivery approaches of antiviral agents such as the use of nanotechnology has led to the development of nanoparticles (14). Nanotechnologial approaches can be used to improve the design, formulation, and delivery of antiviral drugs as shown in Figure 6.1 (15). This review describes the use of nanoparticles as a carrier for the transport of antiviral drugs.

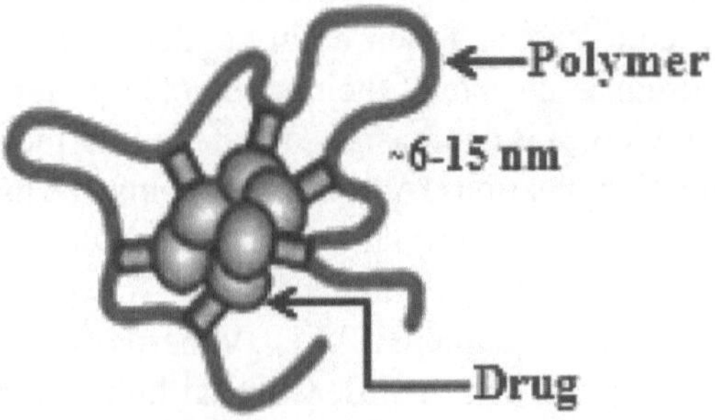

FIGURE 6.1 A schematic representation of polymer–drug conjugate.

6.2 NANOTECHNOLOGIES TO IMPROVE THE DELIVERY OF ANTIVIRAL AGENTS

Over the past two decades, nanotechnology solutions have been developed to improve the delivery of active molecules, especially in cancer to target the cancerous tissues and avoid the the damage of normal cells. It is the creation and utilization of materials and systems on the nanometre scale (a nanometer is one-billionth of a meter) (see Figure 6.2).

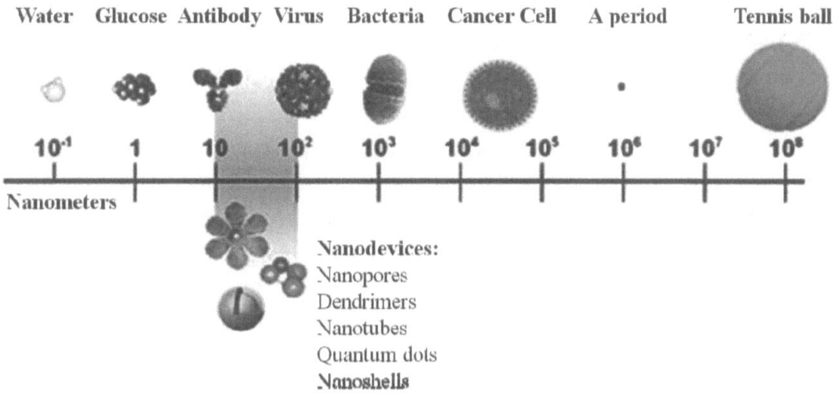

FIGURE 6.2 A schematic representation of nanometer scale

Nanostructures confer novel properties because of their size. These have applications in disease diagnosis, treatment, and prevention. Nanomedicine is defined as the application of nanotechnology within medicine, that is, the monitoring, repair, construction, and control of human biological systems at the molecular level using nanodevices and nanostructures. Nanoparticles may be effective delivery systems for drug and gene therapy, and can alter the distribution of drugs and other therapeutic agents within the patient's body. Nanoparticles can also be used for the efficient delivery of antiviral drugs which aim at obtaining higher potency and lower toxicity in the patient. Until now nanotechnology has been used to modify existing drugs to improve their biopharmaceutic and pharmacokinetic (16). Currently, the researches on nanoparticle drug delivery system focus on (1) to obtain suitable drug release speed using the selective and favorable combination of carrier materials, (2) to improve the targeting ability by proper surface modification of the nanoparticle, (3) to increase their drug delivery capability, their application in clinics, and the possibility of industrial production by optimizing the preparation of nanoparticles, and (4) the study of in vivo dynamic process to reveal the interaction of nanoparticles with blood and targeting tissues and organs (17).

6.2.1 MARKET VIEW

Reports by Global Market for nanocarriers (18) have estimated that by 2012 nano-enabled DDS has capture over a 15% share of the market from its current portion of 4%. Nanocarriers will account for 40% of a $136 billion nanotechnology which enables drug delivery market by 2021. They forecasted the total market size in 2021 to be US$136 billion, with a 60/40 split between nanocrystals and nanocarriers, respectively. Thus, developing new targeted delivery mechanisms may allow more value to be created for companies and entrepreneurs. It was recently estimated that the drug delivery industry is currently worth approximately US$80 billion and a major component of this sum is devoted to the design of controlled release and targeting systems. Thus, the development of new methods for achieving controlled release is a very attractive research area, both in terms of the need to improve healthcare and from the perspective of pharmaceutical companies to maintain revenue and to ensure patent positions in both existing and new drugs. Some forecasts have predicted the nanotechnology market to reach close to a trillion dollars by 2015, presenting investors with unique opportunities. However, the market for applications of nanotechnology is complex, multidisciplinary, and highly segmented.

Different types of nanocarriers or nanoparticulates (including microspheres, nanocapsules, nanoparticles vesicles, dendrimers, micelles liposomes, and inorganic nanomaterials), have been designed by the nanodelivery systems to deliver small molecular weight drugs, but they can also be exploited for the delivery of macromolecules and biological therapeutics such as oligonucleotides (19–20).

The miniaturization of materials often imparts novel physicochemical properties. Specifically, as a particle's size decreases, there is an increase in the effective surface area, often rendering the particle more reactive. Greater effective surface area leads to more intimate contact between the solid surface and the aqueous solvent giving higher dissolution rate. Nanoparticles can be synthesized by various methods, such as self-assembly, vapor and electrostatic deposition, nanoprecipitation (see Figure 6.3), solvent diffusion and solvent evaporation techniques, coacervation, nanomanipulation polymerization method, supercritical fluid technology, and particle replication in nonwetting templates (PRINT) (21).

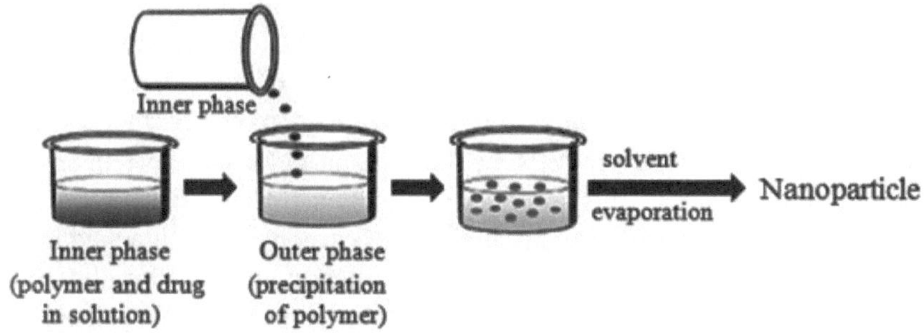

FIGURE 6.3 Schematic illustration of the nanoprecipitation process.

By using different types of nanocarriers it might be possible to overcome many conventional dosage forms related problems of antiviral drugs. It might increase drug bioavailability, control solubility and dissolution rates (improvement in BCS score), reduce side effects, improve tissue drug tolerance, protect sensitive drugs from degradation, overcome anatomical and cellular barriers increase the efficacy of the drug, decrease the emergence of drug resistance, and target the drug at specific biological sites either passively or actively. The size and lipophilicity of the nanoparticles enable them to target drugs to specific tissues or organs, such as the liver and brain. Modifying the surface of the nanoparticle enables them to reach particular sites and deliver the drug to specific cellular targets (22).

Antiviral drugs can be delivered through local or systemic route using nanodelivery systems. During the past three decades the pharmaceutical industry has developed and marketed several nanoparticulate pharmaceuticals with major emphasis on intravenous products, for example, intravenous (IV) nutritional fat emulsion (Intralipid®), and liposomal products (AmBisome®, Doxil®). But these IV products were unable to achieve high drug loading, the cost of ingredients and processing, and the restricted number of suitable excipients has until now limited the broader use of these formulation approaches. Elan's NanoCrystal® Technology, which focuses on poorly water-soluble drugs, has addressed many of these major concerns and has successfully expanded the scope and use of nanoparticulates or nanosuspensions to include the oral, inhalation, intravenous, subcutaneous (SubQ), intramuscular (IM), and ocular routes of delivery. In order to deliver the drug through intravenous route it is essential that the nanodelivery system must be in the nanometer range so that it can circulate in the bloodstream without being retained by the pulmonary capillaries (23). Specific strategies have been designed to overcome their uptake by the reticuloendothelial system (RES). The longevity of nanocarriers and the avoidance of RES uptake in the bloodstream can be increased by (1) modify the surface property of the nanoparticle with certain hydrophilic polymers/surfactants, such as polyethylene glycol (PEG); (2) The ligand (antibody, another protein, peptide, or carbohydrate) attached to the carrier surface may increase the rate of its uptake; and (3) formulation of nanoparticles with biodegradable copolymers with hydrophillic segments such as PEG, polyethylene oxide, polyoxamer, poloxamine, and polysorbate 80 (Tween 80).

Studies show that PEG conformation at the nanoparticle surface is of utmost importance for the opsonin repelling function of the PEG layer. PEG surfaces in brush-like and intermediate configurations reduced phagocytosis and complement activation, whereas PEG surfaces in mushroom-like configuration were potent complement activators and favored phagocytosis (24).

6.3 POLYMER-BASED NANOPARTICULATE DRUG DELIVERY

Natural or synthetic polymers are used to formulate polymeric nanoparticles. These polymers are biodegradable and have high level of biocompatibility to reduce cytotoxicity and maximize tissue compatibility. Polymers that have been approved by the US Food and Drug Administration (USFDA) for human use are poly-D,L-lactic acid (PLA), polyglycolic acid (PGA), poly(lactic-co-glycolic acid) (PLGA), poly(methylmethacrylate), and poly (ε-caprolactone). Nanoparticles made of poly (D,L-lactide) and poly (ε-caprolactone) containing dexamethasones were prepared by nanoprecipitation process to evaluate the anti-inflammatory activity of dexamethasone (25). In vitro experiments were carried out to compare the ocular pharmacokinetics and pharmacodynamics of pilocarpine-loaded nanospheres with an aqueous suspension of the free drug to treat glaucoma.The nanoparticles were prepared using polybutylcyanoacrylate by an emulsion polymerization pocess. The polybutylcyanoacrylate-coated nanoparticle showed higher ophthalmic bioavailability as compared to the normal eye drop (26). Verdun et al. demonstrated in mice treated with doxorubicin incorporated into poly (isohexylcyanoacrylate) nanopsheres that higher concentrations of doxorubicin manifested in the liver, spleen, and lungs than in mice treated with free doxorubicin (27).

6.3.1 FUNCTIONALIZATION OF POLYMERIC NANOPARTICLES

Polymeric nanoparticles, especially those formulated using poly(D,L-lactide-coglycolide) (PLGA) copolymer, have emerged as promising carriers for targeted delivery of a wide variety of payloads (28). PLGA nanoparticles have the advantages of biocompatibility, biodegradability, ease of formulation, and tunable sustained release properties (29). An important drawback with PLGA nanoparticles is the limited types of functional groups available on the surface for conjugation to targeting ligands (30). To overcome this limitation, PLGA with specific end terminal groups (such as carboxyl) have been used, the rationale being that some of these functional groups could be available on the surface for chemical reaction (31–33). Other reports have utilized the hydroxyl groups of residual polyvinyl alcohol present on the surface of PLGA nanoparticles after fabrication (34).

Other surfactants with the required functional groups have also been used in the fabrication of nanoparticles, which enables ligand attachment to the surfactant adsorbed on the surface of the particles to achieve targeted drug delivery (35). Danhier et al. (36) used the diblock copolymer polycaprolactone–polyethylene glycol dissolved along with PLGA to prepare PLGA nanoparticles with PEG on the surface. These nanoparticles were then conjugated to cRGD peptide for tumor targeting (see Figure 6.4).

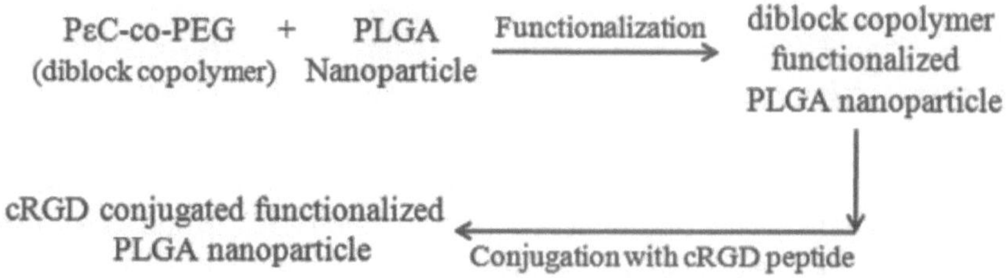

FIGURE 6.4 The IAASF technique and the synthetic scheme for conjugation of cRGD peptide to nanoparticles.

Recently, it is reported that the use of the interfacial activity assisted surface functionalization (IAASF) technique to introduce PEG molecules and targeting ligands such as folic acid on the surface of PLGA nanoparticles in a single step. This technique involves the partitioning of polylactide (PLA)-PEG–ligand conjugate at the oil/water interface formed during nanoparticle formulation, resulting in the formation of nanoparticles with PEG-conjugated ligand on the surface of nanoparticles. However, this procedure involves an oil/water interface, and may, therefore, not be suitable for ligands that are sensitive to organic solvents (e.g., peptides and proteins) (37) (Figure 6.5).

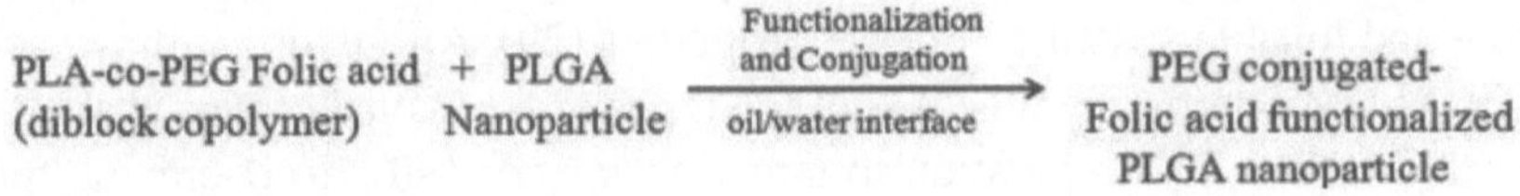

FIGURE 6.5 The IAASF technique for conjugation of PEG peptide and folic acid to PLGA nanoparticles in single step.

Thus, studies show that the polymeric composition of nanoparticles such as type, hydrophobicity, and biodegradation profile of the polymer along with the associated drug's molecular weight, its localization in the nanospheres, and mode of incorporation technique, adsorption, or incorporation, have a great influence on the drug distribution pattern in vivo. Some of the promising compounds shown to have antiviral effects in vitro, but because of solubility and bioavailability problems they are not currently being administered in vivo. Mostly, the peptides and the nucleic acids fall in this category. Therefore, these compounds could be successfully administered using nanoparticles.

6.4 PEPTIDE-NANOPARTICLE MEDIATED DRUG DELIVERY

Peptides and proteins can be successfully delivered using certain nanodelivery systems protecting them from degradation (38). Hood et al. (39) investigated whether cytolytic melittin peptides could inhibit human immunodeficiency virus-1 (HIV-1) infectivity when carried in a nanoparticle construct which was used as a topical vaginal virucide. Free melittin and melittin-loaded nanoparticles were prepared and compared for cytotoxicity and their ability to inhibit infectivity by CXCR4 and CCR5 tropic HIV-1 strains. The conclusion obtained illustrated the first proof-of-concept for therapeutic and safe nanoparticle-mediated inhibition of HIV-1 infectivity through melittin peptides.

6.5 SIRNA NANOPARTICLE DELIVERY

Ocular, nasal, and pulmonary administration routes could be targeted through the nanoparticle drug delivery system. Nanoparticles could be useful for the selective delivery of antiviral drugs or small interfering RNA (siRNA) to the nasal epithelia and lungs in order to target viruses that infect the respiratory tract, such as parainfluenza virus, influenza viruses, corona vius, measles and human metapneumovirus, respiratory syncytial virus, and rhinoviruses.

In the past few years, the gene expression of human viral pathogens, including that of influenza viruses, severe acute respiratory syndrome virus, corona vius, flavivirus, human metapneumovirus, HIV, HCV, and HBV has been silenced by RNA interference (RNAi) which has emerged as a promising antiviral strategy (40–43). There are two distinct pathways of RNAi such as (1) siRNA pahway and (2) microRNA pathway. Drugs affecting RNA offer several potential advantages over conventional small molecule dug development approaches. Very recently, a study by DeVincenzo et al. (44) provided a unique proof-of-concept for an RNAi-based therapy in humans directed against respiratory syncytial virus. But there are many obstacles that prevent the translation of RNAi into a potential therapeutic platform, and the most important obstacle is the delivery of siRNA in vivo. It has been observed that siRNA that are not chemically modified are rapidly degraded by the endo or exonucleases. Thus, the half-life of siRNA in the blood is short. Several methods are under preclinical study to overcome the shortcomings of siRNA and achieve safe and systemic delivery of siRNA. Thus, nanotechnology-based many delivery strategies are currently under development to address the challenges for siRNA delivery (45).

6.6 CELLULAR TRAFFICKING OF NANOPARTICLES

The submicron size range of nanoparticles can provide intracellular uptake and transport of active compounds possible. The delivery of macromolecules like proteins, peptides, and nucleic acids into the cytoplasm is limited by their low membrane permeability and their degradation in the endosomal environment after uptake by endocytosis. But their incorporation into a nanoparticulate system could promote cell internalization and protect the molecules from degradation (46). There are various mechanisms of nanocarrier cell internalization that are dramatically influenced by nanoparticles' physicochemical properties. This is an important feature to promote the delivery of antiviral drugs because most antiviral drugs, like nucleoside analogs, target viral functions that are carried out within a cell. There are various mechanisms that govern the entry of nanoparticulates into cells: phagocytosis, macropinocytosis, caveolae-mediated endocytosis, clathrin-mediated endocytosis, and other clathrin- and caveolae-independent cytosis. To study particle uptake by cells and their cellular trafficking, fluorescent-labelled nanoparticles can be used (47).

6.7 NANOPARTICLES AND PHYSIOLOGICAL BARRIERS

Nanoparticles are able to overcome the physiological barriers. They are also found to delivery the drugs to the central nervous system. Different nanocarriers have reported enhanced in vitro and in vivo blood–brain barrier (BBB) permeability and drug accumulation in the brain in various studies. Nanocarriers have been found (1) to increase the local drug concentration gradients, (2) facilitate drug transport into the brain via endocytotic pathways and (3) inhibit the adenosine triphosphate (ATP)-binding cassette (ABC) transporters expressed at the barrier sites. Recent studies show that the specificity and efficiency of antiviral drug delivery can be further enhanced by using nanocarriers with specific brain targeting, cell penetrating ligands, or ABC-transporters inhibitors. By selecting the components and the formulation parameters it is possible to prepare nanoparticles with physicochemical properties that allow delivery to the brain (48).

Antiretrovirals can be delivered to the central nervous system using four different categories of nanocarriers. They are micelle-based, polymer-based, dendrimer-based, and lipid-based systems (see Figure 6.6) (49). The endothelial barrier, cell membrane barrier, BBB, cerebrospinal fluid barrier, placental barrier, blood–testis barrier, blood–retinal barrier could also be overcome using nanocarriers. Moreover, both physiological and anatomical barriers, cellular viral reservoirs that are not easily accessible to drugs in their current dosage form could be easily reached using nanotechnology-based nanocarriers. For instance, the central nervous system, the cerebrospinal fluid, the lymphatic system, the macrophages, hematopoietic progenitor cells (HPCs), memory CD4+ T cells and the semen are almost completely inaccessible to drugs. As a result, these are the compartments where HIV proliferates and evolves independently despite of a successful highly active antiretroviral therapy. However, the treatment of HIV infection and the eradication of viral reservoirs from the patient become complicated due to suboptimal drug administration (50–52).

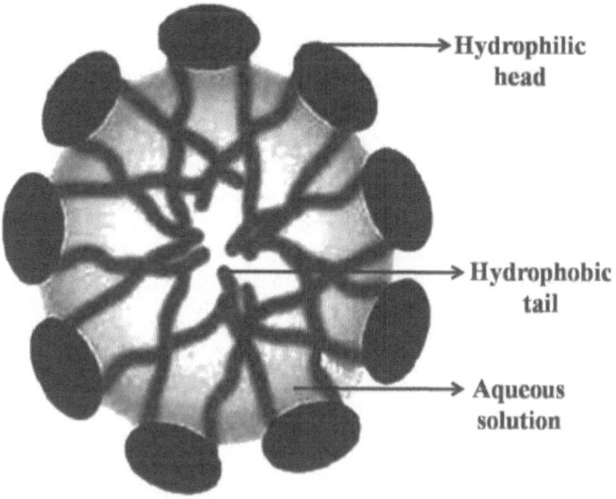

a) Self-assembling nanosized colloidal particles with a hydrophobic core and hydrophilic shell currently used for the solubilization of various nanoparticles and poorly soluble pharmaceuticals.

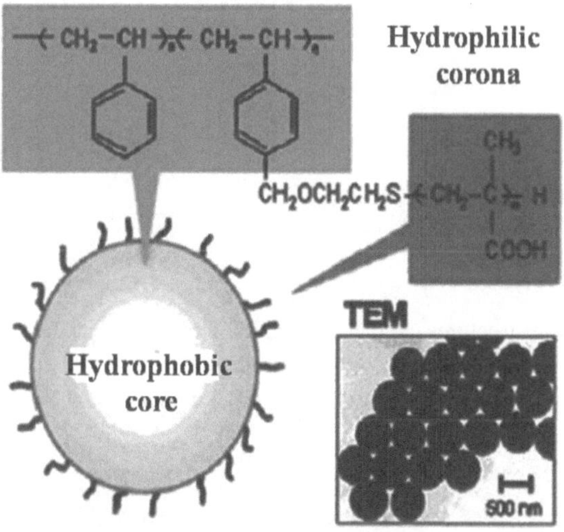

b) Polymeric nanoparticles are prepared from polymers. The drug is dissolved, entrapped, encapsulated, or attached to a nanoparticle and depending upon the method of preparation, nanospheres or nanocapsules are obtained.

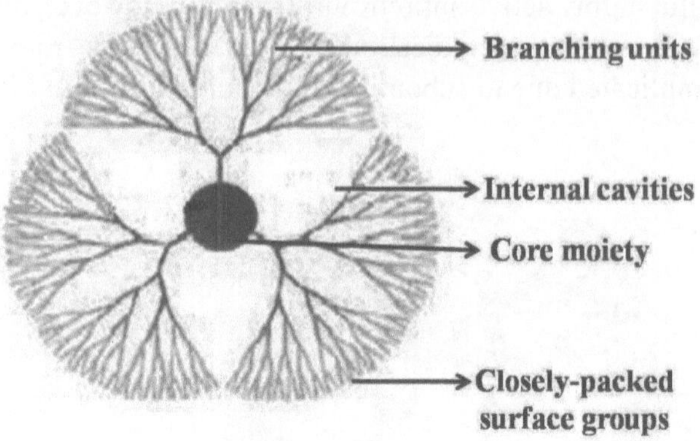

c) Synthetic polymeric macromolecule of nanometer dimensions, which is composed of multiple highly branched monomers that emerge radially from the central core.

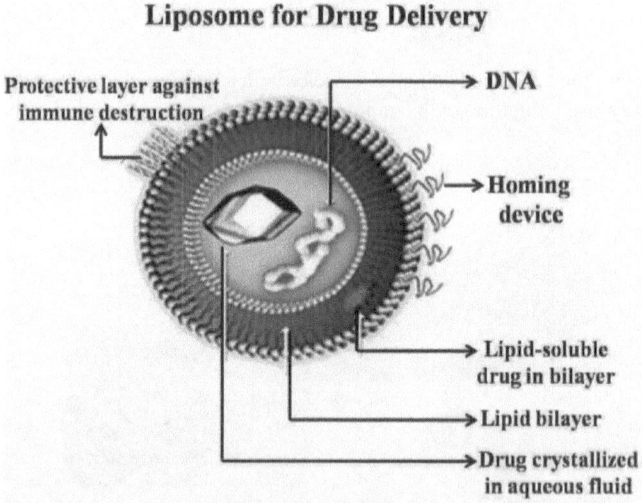

d) Self-assembling structures composed of lipid bilayers in which an aqueous volume is entirely enclosed by a membranous lipid bilayer.

FIGURE 6.6 Images of (a) micelle, (b) polymeric nanoparticle, (c) dendrimer, and (d) liposome.

Efflux transporters are inhibited by the administration of antivirals in nanoparticles thus affecting the therapeutic efficacy. Many orally administered drugs must overcome several barriers before reaching their target site. The first major obstacle to cross is the basolateral membrane in the intestinal epithelium. Efflux proteins located at the apical membrane, which include P-glycoprotein (Pgp), are MDR1 and MRP2. These proteins may drive compounds from inside the cell back into the intestinal lumen, preventing their absorption into blood. Thus, it leads to the subtherapeutic drug concentrations. P-gp is found to develop resistance for the anticancer drugs in cancer cells (53). The activity of efflux transporters is found to expel drugs from cells which leads to subtherapeutic drug concentrations. Indeed, P-gp inhibition represents one potential strategy for the improvement of antiviral intestinal absorption. Salama et al. (54) and Yang et al. (55) have demonstrated that the absorption of acyclovir in vitro is increased in the presence of P-gp-specific inhibitors without affecting the paracellular or transcellular transport, but this inhibition can increase side effects. However, administering the drug through nanoparticulate delivery system can enhance the drug absorption through the cell.

6.8 MULTIFUNCTIONAL PHARMACEUTICAL NANOCARRIERS

Engineering the surface of nanoparticles can give rise to multifunctional systems. These results in many advantageous characteristics that include longevity, targetability, intracellular penetration, stimuli sensitivity, and contrast loading. Thus, the engineering of multifunctional pharmaceutical nanocarriers combine to produce multifunctional nanoparticle that can simultaneously perform more than one useful function (56). Such multifunctional nanocarriers could significantly enhance the efficacy of many therapeutic and diagonostic protocols.

The idea of multifunctional nanosystem can be utilized for specific target recognition. Targeting of nanoparticulate drug delivery system with the aid of specific ligands selective to certain cell-surface receptors allows for the selective drug delivery to those namely cells. It has been proposed that conjugating an albumin peptide ligand on the surface of maleimide attached diblock copolymer functionalized PLGA nanoparticle loaded with antiviral drug for acting on hepatitis B virus would target the drug to the specific target site, that is, the asialoglycoprotein receptor. The asialoglycoprotein receptor (ASGP-R) is a glycoprotein present in large amount only on hepatocytes where it binds and internalizes a broad range of molecules exposing galactose or *N*-acetyl-galactosamine residues. The albumin conjugates containing galactose residues bind to the ASGP-R and selectively enterhepatocytes, where the lysosomal enzymes split the bond between the carrier and the drug and the drug becomes concentrated in liver cells (57, 58). Thus using this technique an antiviral drug of nucleoside analogs used in chronic viral hepatitis could be targeted to a specific site and extrahepatic side effects such as neurotoxicity related to it could be avoided (see Figure 6.7).

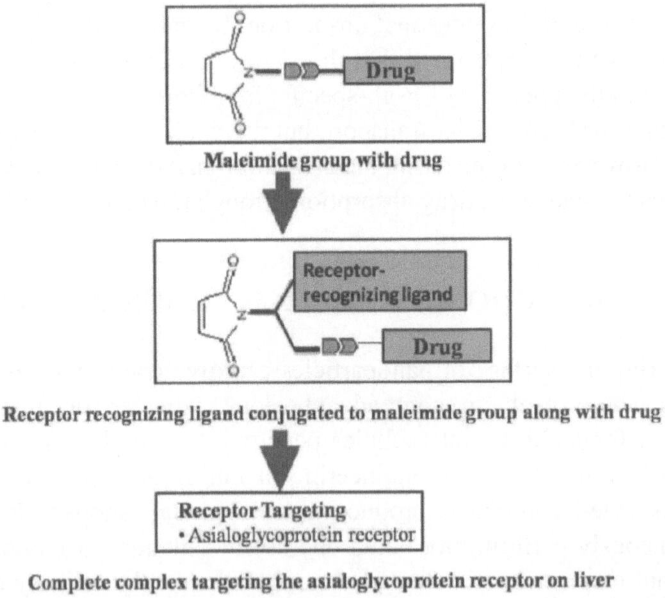

FIGURE 6.7 Novel concept for receptor targeting via malemide functionalized drug loaded PLGA nanoparicle.

Stefano et al. (59) developed a novel technique in order to reduce the extrahepatic toxicity of anticancer agents in the treatment of liver tumor. A (6-maleimidocaproyl) hydrazone derivative of doxorubicin (DOXO-EMCH) was coupled to a thiolated form of lactosaminated human albumin (L-HSA). The resulting conjugate L-HSA–DOXO bind to the hepatocyte ASGP-R with subsequent internalization in the liver and degradation of the carrier in lysosomes. The conjugate L-HSA–DOXO achieved a very efficient targeting of the drug to the liver of treated mice reaching the doxorubicin concentrations upto the level of 7–20 times higher than those raised in extrahepatic tissues thus facilitating the use of the drug in hepatocarcinoma treatment (see Figure 6.8).

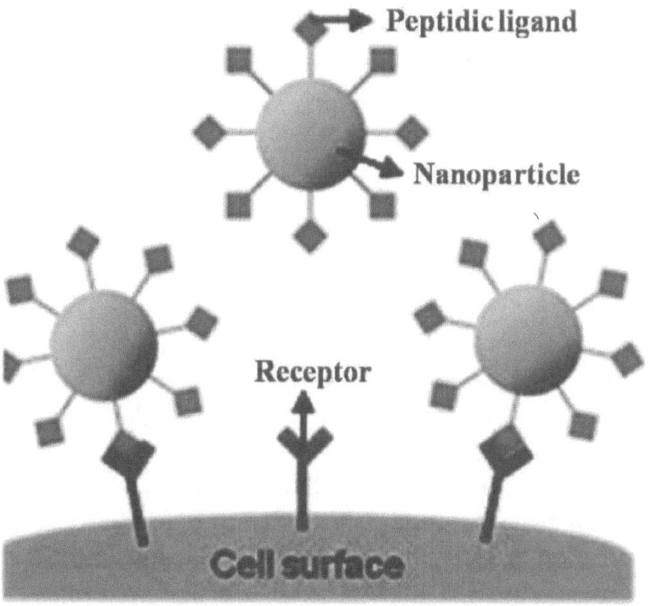

FIGURE 6.8 Schematic illustration of targeted nanoparticles with peptidic ligand.

Development of integrated multifunctional nanosystems is another emerging area of research for diagnosis and therapy. These novel systems, called theranostics, are designed to form a class of agents which can serve diagnostic and therapeutic functions simultaneously. In the current state of technology, tumor detection and therapy are mostly performed separately. In this condition the theranostic nanoparticles can help to achieve a more efficient and effective method. They would integrate the efforts for diagonosis, treatment, and monitoring of tumor response, and assist in the decision-making process for the need for further treatment. This concept has drawn interest in the cancer research community and it is moving ahead in order to develop an innovative nanoparticle-based technology for cancer diagnosis and treatment (60). In the future this technique can be further extended for the molecular diagnosis, treatment, and monitoring of viral infections at the cellular level.

The advancement in nanotechnology has enabled us to utilize particles in the size of the nanoscale. This has created new therapeutic horizons, and in the case of silver, the currently available data only reveals the surface of the potential benefits and the wide range of applications. Interactions between viral biomolecules and silver nanoparticles suggest that the use of nanosystems may contribute importantly for the enhancement of current prevention of infection and antiviral therapies. Recently, it has been suggested that silver nanoparticles (Ag-NPs) bind with external membrane of lipid-enveloped virus to prevent the infection. Viruses have almost the same dimensions as nanoparticles. This feature provoked the researchers to study the interaction between the viruses and the nanoparticle. Further, this interaction could be exploited to study the antiviral activity. Interactions between viral biomolecules and silver nanoparticles which fall in the same range of 10–50 nm, suggest that the use of silver nanosystems may contribute importantly for the enhancement of current prevention of infection and antiviral therapies. Indeed, silver nanoparticles have been shown to inhibit infection by various viruses including HIV-1, HBV, respiratory syncytial virus, herpes simplex virus type 1, monkeypox virus, influenza virus, and Tacaribe virus (61–65). All the silver nanoparticles are found to be in the mean diameter range of 10–50 nm. As a result, it is suggested that direct interaction between the virus and the nanopaticle is responsible for the antiviral activity. Thus, it is suggested that silver nanoparticles exert antiviral activity at an early stage of viral replication as a virucidal agent or as an inhibitor of viral entry. Sun et al. (66) have utilized silver nanoparticles conjugated to various proteins to study the inhibition of respiratory syncytial virus RSV infection in HEp-2 cell culture. In their study, they conjugated the siver nanoparticle with different capping agents, such as (1) poly(N-vinyl-2-pyrrolidone) (PVP); (2) bovine serum albumin (BSA); and (3) a recombinant F protein from RSV (RF 412). The results suggested that PVP-coated silver nanoparticles were able to bind to the viral surface where they might have interacted with G proteins that are evenly distributed on the envelope of the RSV virion and thus inhibited the RSV infection by 44%. The hypothesized interpretation for the interaction of PVP-nanoparticles with G proteins is that their small size and uniformity

(4–8 nm), compared to the other (BSA and RF 412) coated nanoparticles (3–38 nm) may contribute to the effectiveness of the binding.

6.9 TARGETED DELIVERY OF ANTIVIRAL AGENTS

Paul Ehrlich in 1906 first suggested the concept of targeted drug delivery. Targeted drug delivery by functionalized nanoparticle has become one of the most attractive, innovative, and promising areas of research in nano-medicine one century after this intuition. However, there are some fundamental laws which cannot be denied in our attempts to target carriers to anatomically distant targets. Tumors are the standard target experiencing most of the barriers which prevent quantitative carrier and hence drug uptake (67). Site-specific drug delivery could be obtained with different types of nanoparticles.

Various researches have been carried out to develop systems that could improve the biodistribution of anticancer drugs and their accumulation in specific tissues. Drug targeting could be achieved using three different ways: direct injection to a specific site, passive targeting, and active targeting.

6.10 STRATEGIES FOR PASSIVE TARGETING

Passive targeting means the nanoparticulate carrier can reach a given organ by the virtue of its intrinsic properties, such as particle size or lipophilicity, whereas active targeting involves the presence of a "homing device" that guides the carrier to its target site. Maeda et al. (68) demonsrated that passive targeting permits the penetration of nanoparticles into tumor tissues because of the presence of leaky vasculature and the nanocarrier's size facilitates its penetration. This effect is referred to as the "enhanced permeability and retention (EPR) effect" which results in nanoparticle accumulation within the tumors (see Figure 6.9). Because of their small sizes and surface characteristics, nanoparticles can be taken up by the lymphatic tissue in the gut (i.e., the Peyer's patches containing M cells) after oral administration (69).

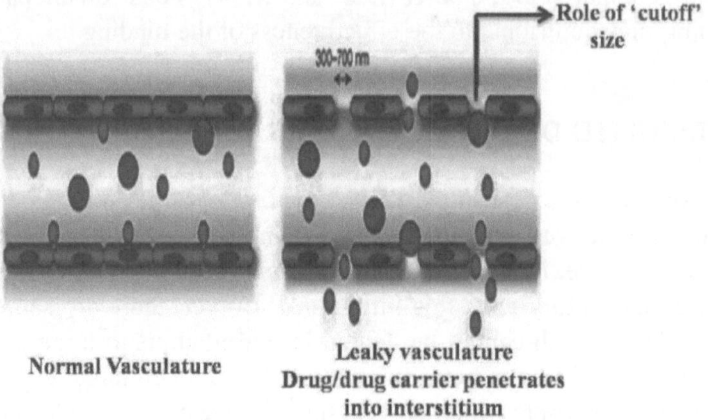

FIGURE 6.9 Schematic representation of passive targeting of tumor tissues associated with the EPR effect.

6.11 STRATEGIES FOR ACTIVE TARGETING

Active targeting can be accomplished by surface modifications, that is, by attaching targeting ligand at the surface of the nanocarrier for binding to appropriate receptors expressed at the target site. Thus, it could be said that active targeting requires homing device. The primary strategy uses monoclonal antibodies, antibody fragments, or nonantibody ligands raised against specific cells or tissues. Other homing devices used are sugars, polymers, proteins, vitamins, lectins, and aptamers molecules (70). Danihier et al. (71) suggested that the ligand is chosen to bind to a receptor overexpressed by tumor cells or tumor vasculature and not expressed by normal cells. Moreover, targeted receptors should be expressed homogeneously on all targeted cells. The binding affinity of the ligands influences the tumor penetration because of the binding-site barrier. The tumor vasculature in which the cells are readily accessible because of the dynamic flow environment of the bloodstream their high affinity binding appears to be preferable.

PEG chains act as stealth agents targeting molecules attached to the surface of the functionalized nanoparticle. Image contrast agent illuminates the interaction of the nanoparticle with a target cell. Magnetic probe permits the nanoparticle localization using an external magnetic field (see Figure 6.10).

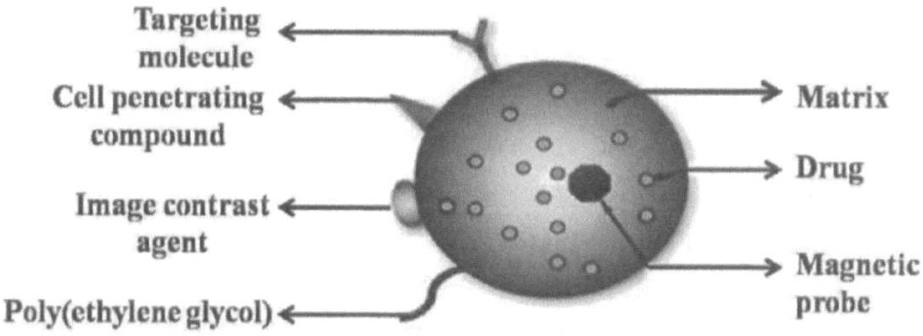

FIGURE 6.10 Schematic representaion of a functionalized nanoparticle.

6.12 INTEFACIAL ACTIVITY-ASSISTED SURFACE FUNCTIONALIZATION TECHNIQUE

Active targeting via specific ligand–receptor like mechanism often comprise a drug carrier attached to a ligand that binds with a specific target overexpressed at the disease site. The intefacial activity-assisted surface functionalization (IAASF) technique may be used to form a ligand–carrier conjugate which can further target the receptors present over the affected cells. However, as described in Figures 6.4 and 6.5 this procedure involves an oil/water interface, and may, therefore, not be suitable for ligands that are sensitive to organic solvents (e.g., peptides and proteins). To overcome this drawback, in the current review it is demonstrated that the IAASF method can be used to incorporate reactive functional groups onto the surface of PLGA nanoparticles. In addition to enabling the introduction of new functional groups, this approach lends itself to more chemoselective bioconjugation approaches (72). The IAASF strategy differs from previously reported surface functionalization techniques in the fact that nanoparticles formed may not be micellar but are polymeric matrix-type devices (73, 74). This enables the incorporation and sustained release of a wide variety of therapeutic agents including proteins, nucleic acids, and antiviral drugs (75–77). Using the IAASF technique PLGA nanoparticles may be fabricated with various functional groups which can then be used to attach various peptides. These peptides can then help to target the drug at a particular site. Fluorescence microscopy can be used to verify that nanoparticles were internalized into the cells and not simply bound to the cell surface. A dose–response study can be performed to evaluate if functionalization enabled greater cell uptake of nanoparticles at different doses.

In vitro studies carried out by Alex et al. (78) showed that cRGD-functionalized nanoparticle with maleimide group using IAASF technique resulted in greater internalization and longer retention of nanoparticles within tumor epithelial and microvascular endothelial cells. These interactions could have resulted in increased accumulation and retention of cRGD-functionalized nanoparticles in the affected area. No such interactions were possible for nonfunctionalized nanoparticles, resulting in slow accumulation and rapid decline of control nanoparticles, and an overall decrease in nanoparticle exposure in the tumor tissue. Considering the fact that PLGA nanoparticles release their payload over a period of weeks, the longer residence time of cRGD-functionalized nanoparticles in tumor will help increase the anticancer efficacy of the payload (see Figure 6.11).

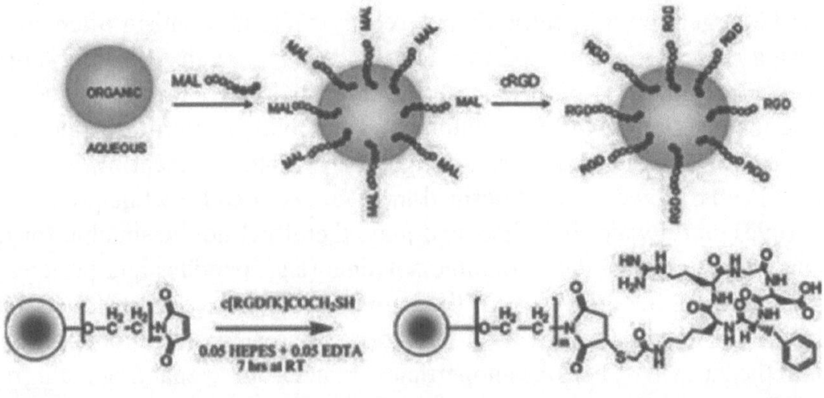

FIGURE 6.11 The IAASF technique (top panel) and the synthetic scheme for conjugation of cRGD peptide to PLGA nanoparticles that are surface functionalized with maleimide groups (bottom panel). PLA-PEG-MAL block copolymer orients itself at the oilwater interface such that PEG molecules are present on the surface of PLGA nanoparticles. Maleimide is then reacted with the cRGD peptide to create cRGD-functionalized nanoparticles.

Source: Udaya T.; Bharath G.; Alex G.; Jayanth P. Interfacial activity assisted surface functionalization: A novel approach to incorporate maleimide functional groups and cRGD peptide on polymeric nanoparticles for targeted drug delivery. Molecular Pharmaceutics, 2010, 7(4), 1108–1117.

IAASF technique has the potential to enable the incorporation of reactive functional groups and the conjugation of peptide ligands on the surface of PLGA nanoparticles. Studies investigating the anticancer efficacy of drug-loaded, cRGD-functionalized nanoparticles in different tumor models are in progress. However, this technology can be used for the identification of molecules or functions that are differentially expressed or carried out by a virus-infected cell. Ideally, this knowledge should be provided for each virus against which an antiviral drug is available.

6.13 STIMULI-SENSITIVE NANOCARRIERS

Stimuli-sensitive nanocarriers are another novel approach to target specific body areas or intracellular compartments. This strategy utilizes the advantage of certain abnormal functions in the body which includes either intrinsically abnormal pH, redox and temperature values of intracellular organelles (i.e., the endosomes) and pathological sites or externally applied stimuli, such as a magnetic field, temperature, and ultrasounds. All of these stimuli are expected to guide, dissolve, or modify the sensitive nanocarriers, resulting in the release of the loaded drug in a particular region, such as tumors, inflammation sites, infarcts, or endosomes (79).

6.13.1 PH-SENSITIVE NANOCARRIERS

pH-sensitive nanocarriers are one of the form of stimuli-sensitive nanocarriers and are of particular interest in the area of therapeutic applications. pH-sensitive systems emerged from the knowledge that (1) certain enveloped viruses (e.g., the influenza virus) lose their envelope in the acidic environment of the endosomal lumen thereby infecting the cells, and (2) some pathological tissues as in case of tumors, inflammations, and infections, exhibit a relatively more acidic environment than normal tissues. These concepts lead to the emergence of pH-sensitive nanocarriers. Hydrogels, nanogels, polymeric micelles, and liposomes are the different classes of pH-sensitive systems (80). These pH-sensitive carriers can promote the intracellular release of the encapsulated drug when the pH changes. Since the pH changes throughout the gastrointestinal tract, the drug release from the pH-sensitive nanocarriers also changes subsequently. The mechanisms of drug release are the result of carriers' dissolution, swelling or both of them at specific pH. Advantages of pH-sensitive nanoparticles over conventional nanoparticles includes (1) the rapid dissolution or swelling of carriers at specific pH results in quick drug release and high drug concentration gradient, which is helpful for absorption and (2) at the specific pH, carriers dissolve or swell, and the bioadhesion of carriers to mucosa becomes high because nanoparticles turn from solid to gel, which can facilitate drug absorption (81). Lu et al. (82) prepared pH-sensitive nanoparticle drug delivery system derieved from natural polysaccharide pullulan for doxorubicin release.

6.13.2 EXTERNAL STIMULI SENSITIVE NANOCARRIERS

External stimuli in the form of magnetic field, temperature and ultrasound can be used in combination with labelled nanocarriers for specific- and targeted drug delivery. Magnetic field applied from outside the body can help the magnetic-labeled nanocarriers to reach the target site while in case of physical stimulus, such as temperature or ultrasounds outer guided temperature and ultrasound variation can help to target the drug.

6.14 MAGNETIC DRUG DELIVERY

In magnetic drug delivery, the drug-loaded nanoparticles having a magnetic probe in it are guided by an external magnet to the targeted organ. The carrier is thus magnetically concentrated in the target organ. The magnetic probe-containg nanocarrier is found to release the drug by passive process affected by the properties of the particulate system. Currently, some superparamagnetic iron oxide nanoparticles (SPION) are in early clinical trials and several formulations have been approved for medical imaging, such as Combidex® or Ferumoxytol®, Feridex® (83, 84). Magnetic microspheres containing interferon exhibited rapid localization to the immunization-related tissues on application of external magnetic field. This interferon loaded magnetic microsperes have shown better antiviral activity as compared to the conventional polymeric protein loaded microspheres (85).

6.15 ULTRASOUND DRUG DELIVERY

External ultrasound can trigger the drug delivery or drug release from fom the nanocarriers. Ultrasound induces either thermal or mechanical effect and permits the activation of the drug release at the site of action.The thermal effect can be induced through high-intensity focused ultrasound (HIFU), inducing phase transition of the polymer causing drug release while the mechanical effect can be induced through the transient cavitation (86).

The targeted delivery system developed for antiviral drugs is listed in Table 6.1 (87–98) and mainly concerned with liposomes or nanoparticles designed for the HIV treatment.

TABLE 6.1 Targeted delivery system developed for antiviral drugs

Drug	Virus	Nanodevice	Targeting	Targeted tissue	In vivo studies	References
siRNA	HCV	Cationic liposomes	Apolipoprotein A1	Liver	Yes	Kim *et al.* (87)
AZT	HIV	Albumin nanoparticles	Transferrin	Brain	Yes	Mishra *et al.* *(88)*
Protease inhibitor	HIV	Pegylated liposomes	Monoclonal antibody against gp120	HIV-positive cells	No	Clayton *et al.* (89)
siRNA	HIV	Immunoliposomes	Antibody against LFA1	Lymphocytes	Yes	Kim *et al.* (90)
Nosiheptide	HBV	Recombinant HDL	Recombinant HDL	Liver	Yes	Feng *et al.* (91)
Acyclovir	HBV	Recombinant HDL	Recombinant HDL	Liver	Yes	Feng *et al.* (92)
Interferon		Magnetic microspheres	External magnetic field	–	No	Zhou *et al.* (93)
Saquinavir	HIV	Nanoparticles	Transferrin	Brain	No	Mahajan *et al.* (94)
gp120 Folding inhibitor	HIV	Liposomes	CD4 antigen	HIV-positive cells	No	Pollock *et al.* (95)

Interferon-α		Nanoparticles	Digalactosyl diacyl glycerol	Liver	No	Chiellini *et al.* (96)
Indinavir	HIV	Immunoliposomes	Antibodies against human and murine HLA-DR and CD4 antigen	Lymphoid tissues	Yes	Gagne *et al.* (97)
Protease inhibitor	HIV	Liposomes	CD4 antigen	Lymphocytes	Yes	Duzgunes *et al.* (98)

AZT, zidovudine; siRNA, small interfering RNA; HDL, high density lipoprotein; LFA1, lymphocyte funcion-associated antigen1; HLA, human leukocyte antigen

Drug could be very beneficially targeted by modifying the surface of the nanocarriers. This surface modification can help the delivery of many peptides and proteins inside the cell which usally get degraded due to the lysosomal degradation. Thus, intracellular distribution of nanocarriers could be controlled by coupling transactivating transcriptional activator (TAT) peptide or cell penetrating peptides to the nanoparticle surface. This is particularly important for drugs that act within the cytosol or that must reach the nucleus (99).

Torchilin et al. (100) used cell-penetrating peptides (CPPs) and TAT peptide (TATp) for intracellular delivery of various nanoparticulate pharmaceutical carriers (liposomes, micelles, and nanoparticles). Here, he carried out the TATp-mediated delivery of liposomes and DNA. They also developed the "smart" stimuli-sensitive nanocarriers, where cell-penetrating function can be activated by the decreased pH only inside the biological target minimizing thus the interaction of drug-loaded nanocarriers with nontarget cells (101).

The various nanoparticle surface modification strategies used for targeting purposes are presence of surface charge, surface coating, PEG coating, antibody binding, antibody fragment conjugation, CD4-derived peptide conjugation, mannose conjugation, galactose conjugation, transferrin, apolipoprotein and recombinant HDL (high-density lipoprotein).

Out of all the targeted delivery systems described above the active targeting accomplished by surface modification, in particular via IAASF technology can be useful to target the virus-infected cells or tissues by ligand–receptor bonding mechanism and thus will reduce the drug-related toxicity in other cells.

6.16 CONCLUSION

The nanomedicine approach opens new therapeutic strategies for attacking viral diseases and for improving treatment success rates. Innovative nanomedicine solutions are expected to have great effects in the treatment as well as the eradication of infectious diseases. Their role could be important in prevention, early diagnosis, more effective drug delivery systems, specific targeting, and personalized therapy. In antiviral therapy nanomedicines may lead to increased bioavailability, improved antiviral delivery, control of adverse side effects, monitoring of antiviral therapy, nanomicrobicides in prevention therapy, and patient compliance.

In the current review, we evaluated the use of the IAASF technique to incorporate reactive maleimide groups on the surface of PLGA nanoparticles. The IAASF technique depends only on the interfacial activity of the block copolymer and the presence of oil/water interface. Further, this approach is suitable for encapsulation and sustained release of antiviral drugs, proteins, and nucleic acids. Importantly, the IAASF technique can be used to incorporate multiple functional groups and the conjugation of peptide ligands on the surface of PLGA nanoparticle which will further expand the usefulness of PLGA nanoparticles in targeted drug delivery since the ligand will help to target the drug to the virus-infected cells and achieve cell internalization and thus will reduce the drug-related toxicity in other tissues.

In particular, nanoparticulate-based systems could improve the effectiveness of the antivirals, restrict adverse drug side effects, reduce treatment costs, and improve the patient compliance. This nanoparticulate drug delivery system can reduce the dosing frequency and shorten the duration of treatment, which could lead the treatment more cost-effective. These features are particularly useful in viral diseases where high drug doses are needed, drugs are expensive, and the success of a therapy is associated with patient adherence to the administration protocol. The complicated and chronic regimens required in viral treatments could be overcome by using this technology. Nanotechnology can also enhance the effectiveness of approved antiviral drugs by targeting the drug to a specific site, increasing their bioavailability, preventing drug-related toxicity in other tissues, crossing the cellular barriers effectively, and intracellular drug distribution. However, several important issues must be addressed before the potential of nanotechnology can be translated into safe and effective antiviral formulations for clinical use. Thus, the objective of the review is to highlight the potential of nanotechnology in delivering the antiviral drugs.

6.17 FUTURE PERSPECTIVES

From a technological point of view, the main objectives of future antiviral therapy research will be the identification of new technologies for the characterization of nanoscale materials, the development of nanodelivery systems devoid of cytotoxicity, and with high biocompatibility and biodegradability. The functionalization of nanocarriers for effectively targeting specific sites of viral infection in order to reduce drug-related toxicity in other tissues is proving to be a new ray of hope. The selective targeting of a nanocarrier to infected tissues requires the identification of molecules or functions that are differentially expressed or carried out by a virus-infected cell, so that a particular ligand–receptor combination could be developed. This concept will be independent for each virus against which an antiviral drug is available. A targeted delivery of various antiviral agents, infected organs, or cells will be possible.

ACKNOWLEDGMENTS

We are sincerely thankful to all the authors and editors who have granted us the permission to refer their articles and allowed us to make use of text or figures from their articles. We specially thank Office of Cancer Nanotechnology Research, National Cancer Institute, Prof. Ross Brian of Center for Molecular Imaging, Department of Radiology, University of Michigan, Ann-Arbor, USA, Prof. Jayant Panyam, Department of Pharmaceutics, College of Pharmacy, University of Minisota, USA, and Prof. Roberta Cavalli of Department of Science and Pharmaceutical Technology, University of Turin, Italy for their permission and suggestions in preparing this chapter.

KEYWORDS

- **Antiviral drug**
- **IAASF technique**
- **Nanomedicine**
- **Nanoparticles**
- **NDDS**
- **PLGA**

REFERENCES

1. Vila, A.; Sanchez, A.; Tobio M.; Calvo, P.; Alonso J. Design of biodegradable particles for protein delivery. J. Control. Rel. *2002*, 78, 15–24.
2. Stephanie, S.; Joseph, D; Ulrich, S. Nanoprecipitation and nanoformulation of polymers: from history to powerful possibilities beyond poly (lactic acid). Soft Matter. *2011*, 7, 1581–1588.
3. Kreuter J. Nanoparticles. In Encyclopedia of Pharmaceutical Technology; Swarbrick, J; Boylan J., Ed.; Marcel Dekker: New York, 1994; Vol. 10; p165.
4. Illum, L.; Davis, S.; Wilson, G.; Thoma,s W.; Frier, M.; Hardy, G. Blood and organ deposition of intravenously administered colloidal particles – The effects of particle size, nature and shape. Int. J. Pharm. *1982*, 12, 135–146.
5. Couvreur, P.; Kante, B.; Lenaerts, V.; Scailteur, V.; Roland, M.; Speiser P. Tissue distribution of antitumor drugs associated with polyalkylcyanoacrylate nanoparticles. J. Pharm. Sci. *1980*, 69, 199–202.
6. Storm, G.; Belliot, O.; Daemen, T.; Lasic, D. Surface modification of nanoparticles to oppose uptake by the mononuclear phagocyte system. Adv. Drug Deliv. Rev. *1995*, 17, 31–48.
7. Garcia-Garcia, E.; Andrieux, K.; Gil, S.; Couvreur, P. Colloidal carriers and blood-brain barrier (BBB) translocation: a way to deliver drugs to the brain. Int. J. Pharm. *2005*, 298, 274–292.
8. Florence, T. The oral absorption of micro- and nanoparticulates: neither exceptional nor unusual. Pharm. Res. *1997*, 14, 259–266.
9. Namita, G.; Priti, T.; Vijai, K.; Ravi, S. Targeted novel surface-modified nanoparticles for interferon delivery for the treatment of hepatitis B. Acta Biochim Biophys Sin. *2011*, 43, 877–883.
10. Suri, S.; Fenniri, H.; Singh, B. Nanotechnology-based drug delivery systems. J. Occup. Med. Toxicol *2007*, 2, 16.
11. McCarron, A.; Marouf, M.; Quinn, J.; Fay, F.; Burden, E.; Olwill, A.; Scott, J. Antibody targeting of camptothecin-loaded PLGA nanoparticles to tumor cells. Bioconjugate Chem. *2008*, 19, 1561–1569.

12. Beaudette, T.; Cohen A.; Bachelder, M.; Broaders, E.; Cohen, L.; Engleman, G.; Frechet, J. Chemoselective ligation in the functionalization of polysaccharide-based particles. J. Am. Chem. Soc. *2009*, 131, 10360–10361.
13. Antiviral drug, *Wikipedia, the free encyclopedia*, en.wikipedia.org/wiki/Antiviral_drug
14. Sharma, P.; Chawla, A.; Arora, S.; Pawar, P. Novel drug delivery approaches on antiviral and antiretroviral agents. J. Adv. Pharm. Technol. Res. *2012*, 3, 147–159.
15. David, L.; Roberta, C. Review – Nanoparticulate delivery system for antiviral drugs. Antiviral Chem. Chemother. *2010*, 21, 53–70.
16. Thomas, T.; Foster, G. Nanomedicines in the treatment of chronic hepatitis C – Focus on pegylated interferon alpha-2a. Int. J. Nanomed. *2007*, 2, 19–24.
17. Zonghua, Liu, Z.; Jiao, Y.; Wang, Y.; Zhou, C.; Zhang, Z. Polysaccharides-based nanoparticles as drug delivery systems. Adv. Drug Deliv. Rev. *2008*, 60, 1650–1662.
18. Nanotechnology for Drug Delivery: Global Market for Nanocarriers, Cientifica Ltd., http://www.cientifica.com/research/market-reports/nanotechnology-for-drug-delivery-global-market-for-nanocarriers/
19. Chen, M.; Zhong, Z.; Tan, W.; Wang, S.; Wang, Y. Recent advances in nanoparticle formulation of oleanolic acid. *Chin. Med. 2011,* **6**(20), 1–4.
20. Mishra, B.; Patel, B.; Tiwari, S. Colloidal nanocarriers: a review on formulating technology, types and applications toward targeted drug delivery. Nanomedicine, *2010*, 6, 9–24.
21. Mohanraj, J.; Chen, Y. Nanoparticles–A review. Trop. J. Pharm. Res. *2006*, 5(1), 561–573.
22. Nanoparticulate Drug Delivery System, Thassus, D.; Deleers, M.; Pathak, Y. Ed.; Drugs and the pharmaceutical sciences series 166; Informa Healthcare: USA, 2007.
23. Elaine, M.; Merisko, L.; Gary, G. Drug nanoparticles: formulating poorly water-soluble compounds. Toxicol. Pathol. *2008*, 36, 43–48.
24. Bhadra, D.; Bhadra, S.; Jain, P.; Jain, K. Pegnology: a review of PEG-ylated systems. Pharmazie *2002*, 57, 5–29.
25. Beck, R.; Guterres, S.; Freddo, R.; Machalowski, C.; Barcellos, I.; Funck, J. Nanoparticles containing dexamethasone: physicochemical properties and anti-inflammatory activity. Acta Farm. Bonaerense *2003*, 22, 11–5.
26. Zimmer, A.; Mutschler, E.; Lambrecht, G.; Mayer, D.; Kreuter, G. Pharmacokinetic and pharmacodynamic aspects of an ophthalmic pilocarpine nanoparticle delivery system. Pharm. Res. *1994*, 11, 1435–1442.
27. Verdun, C.; Brasseur, F.; Vranckx, H.; Couvreur, P.; Roland M. Tissue distribution of doxorubicin associated with polyhexylcyanoacrylate nanoparticles. Cancer Chemother. Pharmacol. *1990*, 26, 13–18.
28. Lu, M., Wang, X.; Wang, H.; Lin, H.; Yao, Q.; Chen, C. Current advances in research and clinical applications of PLGA-based nanotechnology. Expert Rev. Mol. Diagn. *2009*, 9, 325–341.
29. Panyam, J.; Labhasetwar, V. Sustained cytoplasmic delivery of drugs with intracellular receptors using biodegradable nanoparticles. Mol. Pharmaceutics *2004*, 1, 77–84.
30. Mohamed, F.; Van der Walle, F. Engineering biodegradable polyester particles with specific drug targeting and drug release properties. J. Pharm. Sci. *2008*, 97, 71–87.

31. McCarron, A.; Marouf, M.; Donnelly, F.; Scott, C. Enhanced surface attachment of protein-type targeting ligands to poly(lactide-co-glycolide) nanoparticles using variable expression of polymeric acid functionality. J. Biomed. Mater. Res. A. *2008*, 87, 873–884.
32. McCarron, A.; Marouf, M.; Quinn, J.; Fay, F.; Burden, E.; Olwill, A.; Scott, J. Antibody targeting of camptothecin-loaded PLGA nanoparticles to tumor cells. Bioconjugate Chem. *2008*, 19, 1561–1569.
33. Kocbek, P.; Obermajer, N.; Cegnar, M.; Kos, J.; Kristl, J. Targeting cancer cells using PLGA nanoparticles surface modified with monoclonal antibody. J. Control. Rel. *2007*, 120, 18–26.
34. Sahoo, K.; Labhasetwar, V. Enhanced antiproliferative activity of transferrin-conjugated paclitaxel-loaded nanoparticles is mediated via sustained intracellular drug retention. Mol. Pharmaceutics *2005,* 2, 373–383.
35. Keegan, E.; Royce, M.; Fahmy, T.; Saltzman, M. *In vitro* evaluation of biodegradable microspheres with surface-bound ligands. J. Control Rel. *2006*, 110, 574–580.
36. Danhier, F.; Vroman, B.; Lecouturier, N.; Crokart, N.; Pourcelle, V.; Freichels, H.; Jerome, C.; Marchand-Brynaert, J.; Feron, O.; Preat, V. Targeting of tumor endothelium by RGD-grafted PLGAnanoparticles loaded with Paclitaxel. J. Control Rel. *2009*, 140, 166–173.
37. Patil, B.; Toti, S.; Khdair, A.; Ma, L..; Panyam, J. Singlestep surface functionalization of polymeric nanoparticles for targeted drug delivery. Biomaterials *2009*, 30, 859–866.
38. Ketzinel-Gilad, M.; Shaul, Y.; Galum, E. RNA interference for antiviral therapy. J. Gene Med. *2006*, 8, 933–950.
39. Hood, L.; Jallouk, P.; Campbell, N.; Ratner, L.; Wickline, A. Cytolytic nanoparticles attenuate HIV-1 infectivity. Antivir. Ther. *2013*, 18(1), 95–103.
40. Kumar, P.; Sood, V.; Vyas, R.; Gupta, N.; Banerjea, C.; Khanna, M. Potent inhibition of influenza virus replication with novel siRNA-chimeric-ribozyme constructs. Antivir. Res. *2010*, 87, 204–212.
41. Subramanya, S.; Kim, S.; Abraham, S. Targeted delivery of small interfering RNA to human dendritic cells to suppress dengue virus infection and associated proinflammatory cytokine production. J. Virol. *2010*, 84, 2490–2501.
42. Singh, K. RNA interference and its therapeutic potential against HIV infection. Expert Opin. Biol. Ther. *2008*, 8, 449–461.
43. Huang, D. The potential of RNA interference-based therapies for viral infections. Curr. HIV/AIDS Rep. *2008*, 5, 33–39.
44. De, Vincenzo, J.; Williams, R.; Wilkinson, T. A randomized, double-blind, placebo-controlled study of an RNAi-based therapy directed against respiratory syncytial virus. Proc. Natl. Acad. Sci. U S A. *2010*, 107, 8800–8805.
45. Lee, K.; Jeon, H.; Ryoo, R.; Hyeon, C.; Min, D. Efficient functional delivery of siRNA using mesoporous silica nanoparticles with ultralarge pores. Small, *2012*, 8, 1752–1761.
46. Tiemann, K.; Rossi, J. RNAi-based therapeutics – Current status, challenges and prospects. EMBO Mol. Med. *2009*, 1, 142–151.
47. Hillaireau, H.; Couvreur, P. Nanocarriers' entry into the cells: relevance to drug delivery. Cell Mol. Life Sci. *2009*, 66, 2873–2896.

48. Wong, L.; Chattopadhyay, N.; Wu, X.; Bendayan R. Nanotechnology applications for improved delivery of antiretroviral drugs to the brain. Adv. Drug Deliv. Rev. *2010*, 62, 503–517.
49. Strain, C.; Letendre, S.; Pillai, K. Genetic composition of human immunodeficiency virus type 1 in cerebrospinal fluid and blood without treatment and during failing antiretroviral therapy. J. Virol. *2005*, 79, 1772–1788.
50. Emeje, M..; Obidike, I.; Akpabio, E.; Ofoefule, S. Nanotechnology in Drug Delivery, Ali Demier Sezer, Ed.; ISBN 978-953-51-0810-8, Intech Open Science, 2012; 69–106.
51. Pillai, K.; Good, B.; Pond, K. Semen-specific genetic characteristics of human immunodeficiency virus type 1 env. J. Virol. *2005*, 79, 1734–1742.
52. Alexaki, A.; Liu, Y.; Wigdahl, B. Cellular reservoirs of HIV and their role in viral persistence. Curr. HIV Res. *2008*, 6, 388–400.
53. Chan, S.; Lowes, S.; Hirst, H. The ABCs of drug transport in the intestine and liver: efflux proteins limiting drug absorption and bioavailability. Eur. J. Pharm. Sci. *2004*, 21, 25–51.
54. Salama, N.; Scott, R.; Eddington, D. DM27, an enaminone, modifies the *in vitro* transport of antiviral therapeutic agents. Biopharm. Drug Dispos. *2004*, 25, 227–236.
55. Yang, G.; Meng, H.; Zhang, X. Effect of quercetin on the acyclovir intestinal absorption. Beijing Da Xue Xue Bao *2004*, 36, 309–312.
56. Yu, M.; Park, J.; Sangyong, J. Targeting strategies for multifunctional nanoparticles in cancer imaging and therapy. Theranostics *2012*, 2, 3–44.
57. Trere, D.; Fiume, L.; Badiali, G.; Stefano, G.; Migaldi, M.; Derenzini, M. The asialoglycoprotein receptor in human hepatocellular carcinomas: its expression on proliferating cells. British J. Canc. *1999*, 81, 404–408.
58. Kratz, F. Albumin as a drug carrier: design of prodrugs, drug conjugates and Nanoparticles. J. Control. Rel. *2008*,132, 171–183.
59. Stefano, G.; Kratz, F.; Lanza, M.; Fiume, L. Doxorubicin coupled to lactosaminated human albumin remains confined within mouse liver cells after the intracellular release from the carrier. Digest. Liver Disease *2003*, 35, 428–433.
60. Riehemann, K.; Scheider, W.; Luger, A.; Godin, B.; Ferrari, M.; Fuchs, H. Nanomedicine – Challenge and perspectives. Angew Chem. Int. Ed. Engl. *2009*, 48, 872–897.
61. Lara, H.; Ayala-Nunez, V.; Ixtepan-Turrent, L.; Padilla, C. Mode of antiviral action of silver nanoparticles against HIV-1. J. Nanobiotechnol. *2010*, 8, 1–10.
62. Elechiguerra, L.; Burt, L.; Morones, R. Interaction of silver nanoparticles with HIV-1. J. Nanobiotechnol. *2005*, 3, 6–12.
63. Lu, L.; Sun, W.; Chen, R. Silver nanoparticles inhibit hepatitis B virus replication. Antivir. Ther. *2008*, 13, 253–262.
64. Mori, Y.; Ono, T.; Miyahira, T.; Nguyen, V.; Matsui, T.; Ishihara, M. Antiviral activity of silver nanoparticle/chitosan composites against H1N1 influenza A virus. Nanoscale Res. Lett. *2013*, 8(93), 1–6.
65. Rogers, V.; Parkinson, V.; Choi, W.; Speshock, L.; Hussain, L. A preliminary assessment of silver nanoparticles inhibition of monkeypox virus plaque formation. Nanoscale Res. Lett. *2008*, 3, 129–133.
66. Sun, L.; Singh, K.; Vig, K.; Pillai, R.; Singh, R. Silver nanoparticles inhibit replication of respiratory syncytial virus. J. Biomed. Nanotechnol. *2008*, 4, 149–158.

67. Ruenraroengsak, P.; Cook, M.; Florence, T. Nanosystem drug targeting: facing up to complex realities. J. Control. Rel. *2010*, 141, 265–276.
68. Maeda, H.; Wu J.; Sawa, T.; Matsumura, Y.; Hori, K. Tumour vascular permeability and the EPR effect in macromolecular therapeutics: a review. J. Control. Rel. *2000*, 65, 271–284.
69. Nanotechnology in drug delivery; Gupta, U.; Kompella, U., Eds.; Biotechnology: pharmaceutical aspects; 10 New York: Taylor & Francis, 2006.
70. Sanvicens, N.; Marco, M. Multifunctional nanoparticles-properties and prospects for their use in human medicine. Trends Biotechnol. *2008*, 26, 425–432.
71. Danhier, F.; Feron, O.; Preat, V. To exploit the tumor microenvironment: passive and active tumor targeting of nanocarriers for anti-cancer drug delivery. J. Control. Rel. *2010*, 148, 135–146.
72. Beaudette, T.; Cohen, A.; Bachelder, M.; Broaders, E.; Cohen, L.; Engleman, G.; Frechet, J. Chemoselective ligation in the functionalization of polysaccharide-based particles. J. Am. Chem. Soc. *2009*, 131, 10360–10361.
73. Farokhzad, C.; Jon, S.; Khademhosseini, A.; Trant, T.; LaVan, A.; Langer, R. Nanoparticle-aptamer bioconjugates: a new approach for targeting prostate cancer cells. Cancer Res. *2004*, 64, 7668–7672.
74. Gu, F.; Zhang, L.; Teply, A.; Mann, N.; Wang, A.; Radovic-Moreno, F.; Langer, R.; Farokhzad, C. Precise engineering of targeted nanoparticles by using self-assembled biointegrated block copolymers. Proc. Natl. Sci. U.S.A. *2008*, 105, 2586–2591.
75. Panyam, J.; Zhou, Z.; Prabha, S.; Sahoo, K.; Labhasetwar, V. Rapid endo-lysosomal escape of poly(DL-lactide-co-glycolide) nanoparticles: implications for drug and gene delivery. FASEB J. *2002*, 16, 1217–1226.
76. Sahoo, K; Ma, W.; Labhasetwar, V. Efficacy of transferring conjugated paclitaxel-loaded nanoparticles in a murine model of prostate cancer. Int. J. Cancer. *2004,* 112, 335–340.
77. Sun, B.; Ranganathan, B.; Feng, S. Multifunctional poly(D,Llactide-co glycolide)/montmorillonite (PLGA/MMT) nanoparticles decorated by trastuzumab for targeted chemotherapy of breast cancer. Biomaterials *2008*, 29, 475–486.
78. Alex, G.; Udaya, T.; Bharath, G.; Jayanth, P. Interfacial activity Assisted surface functionalization: a novel approach to incorporate maleimide functional groups and cRGD peptide on polymeric nanoparticles for targeted Drug delivery. Mol. Pharmaceutics *2010*, 7(4), 1108–1117.
79. Torchilin, V. Multifunctional and stimuli-sensitive pharmaceutical nanocarriers. Eur. J. Pharm. Biopharm. *2009*, 71, 431–444.
80. Smart nanoparticles in nanomedicine, Arshady, R., Kono, K., Eds.; MML series 8; Kentus Books: London, 2006.
81. Wang, Q.; Zhang, Q. pH-sensitive polymeric nanoparticles to improve oral bioavailability of peptide/protein drugs and poorly water-soluble drugs. Eur. J. Pharm. Biopharm. *2012*, 82, 219–229.
82. Lu, X., Wen, T., Liang, J.,Zhang, X., Gu, Z.; Fan, J. Novel pH sensitive drug delivery system based on natural polysaccharide for doxorubicin drug release. Chiene J. Polymer Sci. *2008*, 26, 369–374.
83. Veiseh, O.; Gunn, J.; Zhang, M. Design and fabrication of magnetic nanoparticles for targeted drug delivery and imaging. Adv. Drug Deliv. Rev. *2010*, 63, 284–304.

84. Mahoudi, M.; Sant, S.; Wang, B.; Laurent, S.; Sen, T. Superparamagnetic iron oxide nanoparticles (SPIONs): development, surface modification and applications in chemotherapy. Adv. Drug Deliv. Rev. *2011,* 63, 24–46.
85. Zhou, S.; Sun, J.; Sun, L. Preparation and characterization of interferon-loaded magnetic biodegradable microspheres. J. Biomed. Mater Res. B. Appl. Biomater *2008*, 87, 189–196.
86. Schroeder, A.; Kost, J.; Barenholz, Y. Ultrasound, liposomes, and drug delivery: principles for using ultrasound to control the release of drugs from liposomes. Chem. Phys. Lipids, *2009*, 162, 1–16.
87. Kim, I.; Shin, D.; Lee, H. Targeted delivery of siRNA against hepatitis C virus by apolipoprotein A-I-bound cationic liposomes. J. Hepatol. *2009*, 50, 479–488.
88. Mishra, V.; Mahor, S.; Rawat, A. Targeted brain delivery of AZT via transferrin anchored pegylated albumin nanoparticles. J. Drug Target *2006*, 14, 45–53.
89. Clayton, R.; Ohagen, A.; Nicol, F. Sustained and specific in vitro inhibition of HIV-1 replication by a protease inhibitor encapsulated in gp120-targeted liposomes. Antiviral Res. *2009*, 84, 142–149.
90. Kim, S.; Peer, D.; Kumar, P. RNAi-mediated CCR5 silencing by LFA-1-targeted nanoparticles prevents HIV infection in BLT mice. Mol. Ther. *2010*, 18, 370–376.
91. Feng, M.; Cai, Q.; Shi, X., Huang, H.; Zhou, P.; Guo, X. Recombinant high-density lipoprotein complex as a targeting system of nosiheptide to liver cells. J. Drug Target *2008*, 16, 502–508.
92. Feng, M.; Cai, Q.; Huang, H.; Zhou P. Liver targeting and anti-HBV activity of reconstituted HDL-acyclovir palmitate complex, Eur. J. Pharm. Biopharm. *2008*, 68, 688–693.
93. Zhou, S.; Sun, J.; Sun, L. Preparation and characterization of interferon-loaded magnetic biodegradable microspheres. J. Biomed. Mater Res. B. Appl. Biomater. *2008*, 87, 189–196.
94. Mahajan, D.; Roy, I.; Xu, G. Enhancing the delivery of anti retroviral drug saquinavir across the blood brain barrier using nanoparticles. Curr. HIV Res. *2010*, 8, 396–404.
95. Pollock, S.; Dwek, A.; Burton, R.; Zitzman, N. N-Butyldeoxynojirimycin is a broadly effective anti-HIV therapy significantly enhanced by targeted liposome delivery. AIDS, *2008*, 22, 1961–1969.
96. Chiellini, E.; Chiellini, F.; Solaro, R. Bioerodible polymeric nanoparticles for targeted delivery of proteic drugs. J. Nanosci. Nanotechnol. *2006*, 6, 3040–3047.
97. Gagne, F.; Désormeaux, A.; Perron, S.; Tremblay, J.; Bergeron, G. Targeted delivery of indinavir to HIV-1 primary reservoirs with immunoliposomes. Biochim. Biophys. Acta. *2002*, 1558, 198–210.
98. Duzgunes, N.; Pretzer, E.; Simoes, S. Liposome-mediated delivery of antiviral agents to human immunodeficiency virus-infected cells. Mol. Membr. Biol. *1999*, 16, 111–118.
99. Zhang, X.; Jin, X.; Plummer, R. Endocytosis and membrane potential are required for HeL a cell uptake of R.I.-CKT at 9, a retro-inverso Tat cell penetrating peptide. Mol. Pharmaceutics *2009*, 6, 836–848.
100. Torchilin, V. Intracellular delivery of protein and peptide therapeutics. Protein Therap. *2008*, 5, 95–103.

101. Torchilin, V. Cell penetrating peptide-modified pharmaceutical nanocarriers for intracellular drug and gene delivery. Biopolymers *2008*, 90, 604–610.

CHAPTER 7

TRIGGERABLE LIPOSOMES: NEWER APPROACH IN CYTOPLASMIC DRUG DELIVERY

NEERAJ K. SHARMA and VIMAL KUMAR

ABSTRACT

Delivery of anticancer agents, genes, and others having cytoplasmic, nuclear, or other subcellular compartments as a site of action, is usually affected by presence of endocytic pathway. This is the most common pathway for intracellular drug delivery. Besides this, drugs suffer with basic problems such as toxicity, instability, and improper biodistribution. One of the several methods to overcome these problems is encapsulation of the drug in a suitable carrier. However, problems like infavorable biodistribution, high risk of toxicity and instability can be solved by delivery carriers but proper delivery to the site of action should also be considered while selecting a carrier. Several nanocarriers are developed and explored so far to release drugs at desired sites. Among these, exhaustive research has established liposomes as an effective carrier to trigger the release of loaded cargo to site of action. Liposomes are one of the well recognized and effective drug delivery vehicles of the present era that protects the drug from the body milieu as well as improves its stability, safety, and effectiveness. This chapter covers various aspects of triggering modalities found and tested to date such as temperature, pH, enzymes, and light, using liposomes.

7.1 INTRODUCTION

The endocytic pathway represents one of the primary uptake pathways of a cell. This pathway uses vesicles known as early endosomes, late endosomes, and lysosomes. Endosomes having an internal pH more on the acidic side (around 5) and they convert from early endosomes to late endosomes during maturation. After this stage, they fuse with digestive organelle of cells called lysosomes [1]. Hence, delivered therapeutics enters into the cells via this endolysocytic pathway and eventually trapped in the lysosomes. Digestive enzymes degrade these materials and result in a limited delivery of biologicals to the cytoplasmic targets. Many compounds due to this reason cannot be exploited for in vivo study, in spite of having tremendous in vitro activity. Hence, attempts were made to select suitable method amongst several approaches available to promote the early release of payload from the endosomal pathway into the cytosol. First challenge was to identify mechanisms for endosomal escape and effective methods like pores formation in the endosomal membrane, the pH buffering effect, and conformational changes in endosomal escape enhancers were found to be most suitable. These methods exploit viral and bacterial originated proteins and several synthetic agents called as endosomal-releasing agents. A suitable agent to promote endosomal escape should be safe as well as effective at the same time. So far, different researchers have made many attempts to select such type of agents for escaping the endocytic mechanism [2–4]. Several synthetic peptides, synthetic polymers with pH-sensitive properties have been tested as endosomal escaping agents. However, these agents have their own limitations because of immunogenicity, low stability, and toxicity. These problems have inspired researchers to design and test newer approaches for this endosomal escape process [5]. The successful nanomedicines solely depend on the use of suitable and effective drug delivery system [6, 7]. Efforts are already being made to develop efficient and selective drug carrier to deliver therapeutic agents to the target areas but this field can be explored more to solve the problems associated with existing therapies [8, 9]. After conventional drug delivery system researchers moved to more selective drug delivery aspect called targeted delivery methods [10]. Nanovectors have been developed to minimize toxicity and immunogenicity associated with the use of viral vectors. They enter into the cells by receptor-mediated mechanism and in this way they copy the cell entry pattern of viruses. This was an early breakthrough in delivery of biologicals but less efficient delivery due to capture by endosomal pathway was observed [11, 12]. Amongst these carrier systems like particulate, dendrimers, and vesicular drug delivery systems, for the last 35 years liposomes have been studied for effective clinical applications like cancer and other infective disease treatment, plasmid/oligonucleotide delivery, and imaging purposes. They have ability and advantages to (1) incorporate a therapeutic moiety for appropriate response, (2) provide long circulation time to drug, (3) deliver drug at the target area, (4) release their content at a therapeutically sufficient level, and (5) alter bioavailability of the drug for improved action. Various specific receptors can be targeted with proper design and modification of these lipid based nanocarriers [13, 14]. Market is now full of these types of approved therapeutic products due to advanced research in this area during the past 15 years [15]. Conventional drug delivery systems distribute drugs in whole body hence target site does not get a therapeutic effective dose. Due to delivery to nonspecific organs unwanted and nonavoidable side effects arises. Hence, an appropriate drug delivery approach with suitable and effective carrier is required for such type of delivery to combine the advantages of targeting, enhanced drug uptake, and release for improved therapeutic response. This chapter covers various aspects of triggering modalities examined to date such as temperature, pH, enzymes, and light using liposomes.

7.2 LIPOSOMES AS A EFFECTIVE VEHICLE FOR CYTOSOLIC DELIVERY

Some of the very important targets are situated in the cytoplasm (glucocorticoid receptors) or in nucleus (deoxy ribo nucleic acid [DNA], antisense oligonucleotides, targets of some most valuble, and effective agents such as 5-fluorouracil and doxorubicin). Some subcellular compartments are also involved in this category like mitochondria (example is antioxidants). Cytosolic delivery is also challenged by the presence of multidrug resistance proteins and P-glycoproteins. They work as some efflux transporters of cell; hence, effective cytoplasmic delivery is desired to get the best therapeutic response [16]. In case of plasmid DNA that has its site of action in nucleus, success is dependent on how effectively gene delivery occurs. For therapeutic benefits a gene has to localize and integrate with the nuclear or mitochondrial DNA. Enzyme degradation presents another obstacle. These enzymes are usually located in cytosol and nucleus. Beside this the plasmid DNA suffers from a short half-life of only 60–90 min, hence protection is required for enhanced gene expression. A suitable and effective delivery system is required to solve these above mentioned problems [17]. Liposomes have been established themselves as a valuable drug delivery system for cytoplasmic delivery due to their structural characteristics and compositions (Figure 7.1). A drug has to pass through a series of membrane barriers in order to get the target site and valuable thearpeutic quantity of the drugs is lost at each subsequent steps. These obstacles are represented by cellular association and internalization within endolysosomal compartments. Endocytic pathway represents the commonest mechanism for liposomes to enter into the cells (Figure 7.2) [18, 19]. Steric stabilizers such as polyethylene glycol (PEG) and lipids improve the circulation half-life and biodistribution of liposomes and promote the accumulation of drug into target areas like tumors by enhanced permeation and retention (EPR) effect. This passive delivery can be improved by replacing it with active one like targeting with suitable ligands. Binding to a targeting ligand, for selective recognition of receptors present on cell membranes, is an attaractive strategy. However, triggering of delivered vehicle is essential to release the entrapped payload. This process provides numerous opportunities to research in this developing area to enhance the endosomal escape of drug carriers [20]. These triggering mechanisms should have characteristics like minimum or no toxicity, effectiveness, affordable cost, and instant reactiveness within the early endosome upon getting stimulation [13].

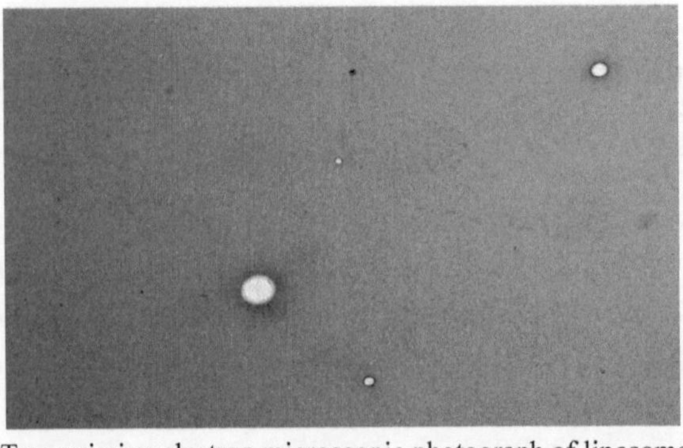

FIGURE 7.1 Transmission electron microscopic photograph of liposomes

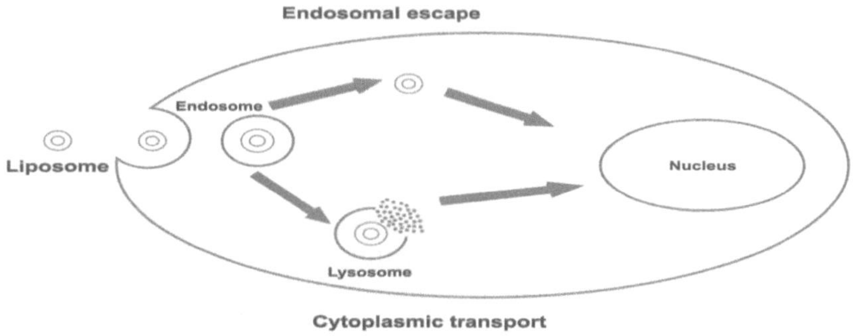

FIGURE 7.2 Liposomal entry in to the cell and endosomal pathway with escaping stage

7.3 DIFFERENT ENDOSOMAL ESCAPE MECHANISMS

The virus and bacteria have their own mechanisms for endosomal escape. These can be explored for the effective delivery to the cytoplasm and nucleus. Under enveloped viral approach fusion occurs between viral envelope and lipid bilayer of the cell membrane, whereas in case of nonenveloped viruses transfer of their genetic material takes place either through lysis or creation of pore in the membrane [21, 22]. The common most mechanism followed by bacteria is pore formation using bacterial exotoxins [23]. This acidifies the endosomal environment and eventually causes interactions between the peptides and the lipid bilayer of the endosomes. Some domains of amino acids gain conformational changes which favors these interactions and eventually promote formation of pores, membrane fusion, and/or lysis [24]. There are several possible approaches to perform cytosolic delivery (Table 7.1).

TABLE 7.1 List of commonly used components for triggered liposomal formulation

S. No.	Type of liposomes	Examples of components
1.	pH sensitive	PE, DOPE, HSPC, CHEMS, DSPE, DPPC
2.	Fusogenic	DTA, DOPE, Sendai virus, PC, DOPE
3.	Photsensitive	Bis-azo PC, Bis-sorb PC, CHOL, DPPC, DSPC, HSPC
4.	Thermosensitive	DSPE, PC, CHOL, DOPE
5.	Ultrasound triggered	PC, CHOL, DOPE

7.3.1 pH-BUFFERING MECHANISM

The pH-buffering effect also known as proton sponge effect is driven by high buffering capacity and swelling property when gets protonated. Endosomes face an influx of ion followed by excessive inflow of water, eventually lead to rupture of its own membrane, result in release of entrapped materials into cytoplasm. pH-sensitive liposomes (PSLs) use this buffering effect to improve the delivery of entrapped material into cytoplasm. After endocytic internalization, liposomes destabilize and/or fuse with the endosomal membrane. This process takes place due to acidic environment of the endosome and results in release of liposomal contents into the cytoplasm [25]. PSLs are the typical examples of this class.

7.3.1.1 pH-SENSITIVE LIPOSOMES

On the primary basis PSLs have been developed with the help of unsaturated phosphatidyl ethanolamine (PE) and amphipathic stabilizers. They are effective but suffer from stability problems compared with conventional liposomes in terms of effective cytoplasmic delivery [26]. Role of 1,2 dioleoyl-sn-glycero-3-phosphoethanolamine (DOPE), as a component in their composition, was investigated later. Experiments on the cells pretreated with metabolic inhibitors or lysosomotropic agents revealed that internalization of these DOPE-containing liposomes principally occur via endocytosis. It was also found that besides acidification of the endosomes there could be another mechanism involved in the process of destabilization of the liposomes [27]. The most explored class of PSLs contains unsaturated PE along with mildly acidic amphiphiles, like oleic acid or cholesterylhemisuccinate (CHEMS). They show poor in vivo applicability due to their low stability and/or rapid removal by the mononuclear phagocyte system after intravenous administration. Hence, pH-responsive behavior can be provided to PEGylated liposomes by attaching hydrophobically modified copolymers of N-isopropylacrylamide (NIPAM) and methacrylic acid by surface modification [28–31]. This pH-triggered ability of liposomes to promote cytoplasmic delivery of therapeutic moieties can be specifically a target for the treatment of tumors [32].

7.3.1.1.1 pH-SENSITIVE LIPOSOMES FOR CANCER CHEMOTHERAPY

One of the several formulations, DOPE and CHEMS in the molar ratio of 3:2 were used to formulate PSLs. N-acetylglucosamine derivative of bovine serum albumin (N-Ac-BSA) was anchored to target the asialoglycoprotein receptors. Studies with fluorescence-activated cell sorter and confocal microscope were revealed the association of liposomes with the chicken hepatoma, hence showed exceptional effectiveness in receptor-mediated targeting to chicken hepatoma cells [33]. Examples of pH sensitive stealth liposomes (PSSLs) with doxorubicin (DOX) were found. They were explored to enhance the potency of DOX in vitro for the treatment of B-cell lymphoma. In order to improve stability of PSLs, formulations with different molar ratio of the hydrogenated soy phosphatidylcholine (HSPC) and/or cholesterol (CHOL), were used to the lipid materials and mixture of DOPE/HSPC/CHEMS/CHOL/mPEG2000-DSPE at a molar ratio of 4:2:2:2:0.3 and DOPE/HSPC/CHEMS/CHOL at a molar ratio of 4:2:2:2. This showed the best drug retention along with good pH sensitivity [34]. In another work, PSSLs containing cisplatin (CDDP) were developed and evaluated for tissue distribution in comparison with free CDDP using solid Ehrlich tumor-bearing mice. Higher accumulation of CDDP was observed with prepared formulation [35]. Cancer chemotherapy struggles with multiple phenomenon like multidrug resistance and drug toxicity. CDDP in the form of PSSLs were formulated in order to solve these problems. This formulation was administrated intraperitoneally in male and female mice with a single administration of free CDDP (5, 10, and 20 mg/kg) or prepared formulation (7, 12, 30, 45, and 80 mg/kg). Formulation of reduced CDDP-induced toxicity can be used as a promising carrier for intraperitoneal chemotherapy [36]. High-molecular weight polymers like poly(styrene-co-maleic acid) can be used to impart pH-responsive property to liposomes. It follows conformational transition from a charged extended structure to an uncharged globule below its pK1 value. Cytosolic delivery of bioactive molecules through endosomal destabilization was mediated by spherocyte formation. Using this appraoch 5-FU was delivered to colon cancer cells (HT-29) with better cytocompatibility compared to conventional liposomes. Results showed enhanced apoptosis due to increased bioavailability of the drug to the target site and favored the clinical applicability of styrene-co-maleic acid-based vesicles [37]. Immunoliposomes (ILs) based on the use of monoclonal antibodies can be used as a promising strategy for tumor targeting. Approach is exemplified by PSSLs with EGFR antibody and delivered to human nonsmall cell lung cancer cell line (A549) and BALB/c-nu/nu mouse tumor model. Results suggested that tumor growth inhibition by gemcitabine was a result of increased apoptosis [38]. Another work was represented by PSLs having a terminally alkylated copolymer of NIPAM in the liposomal bilayer and anchored with the anti-CD33 monoclonal antibody to target leukemic cells. Flow cytometry and confocal microscopy analysis advocated the content release from the endosomes after receptor-mediated internalization. The formulation exhibited the highest cytotoxicity against HL60 cells, hence supported the NIPAM copolymer mediated endosomal escape of intact arabinoside [39].

7.3.1.1.2 pH-SENSITIVE LIPOSOMES FOR GENE DELIVERY

Liposomes have potential application in gene delivery applications. Novel PSLs using various combinations of cationic/anionic lipids were formulated and explored for gene delivery. PSLs for targeted delivery to folate receptor (FR) were formulated with dimethyldioctadecylammonium bromide/CHEMS/folate-poly-PEG-PE, combined with polylysine-condensed plasmid DNA. Formualtion was delivered to KB cells in the presence of 10% serum for FR-mediated delivery of a luciferase reporter gene. On the basis of excellent findings, these cationic lipid-containing PSLs can be exploited as an effective vehicle for intracellular gene delivery [40]. Dendritic cells (DCs) are representative of potent antigen-presenting cells and being used for cancer immunotherapy in several aspects. PSLs with fusogenic properties and complexed with lipoplexes were formulated and transfected to a murine DC line DC2.4. They contain polymers based on poly(glycidol) with carboxyl groups. In addition, no effects were observed on transfection or cell association after anchoring with ligands such as transferrin and mannan. It was observed that endosomal escape is a critical step in transfection of DC2.4 cells. The prepared formulation exhibited improved transfection toward DC2.4 cells compared to some commercially available agents [41]. Some viral infections, cancer, and inflammatory diseases can be effectively treated with antisense oligonucleotides. They have ability to inhibit gene expression. However, problems like poor stability in biological medium and weak intracellular penetration show obstacles in this delivery. Therefore, liposomes having pH-sensitive properties and stability in plasma can be explored for cytoplasmic delivery of such biologicals [42]. In another experiment, env mRNA complementary oligonucleotide encapsulated in liposomes was used to inhibit proliferation of the friend retrovirus and results were compared with the same oligonucleotide incubated freely. There was lack of antiviral activity observed in a later case. Virus inhibition was estimated with focus immunoassay or reverse transcriptase assay [43].

7.3.1.1.3 pH-SENSITIVE LIPOSOMES FOR PEPTIDE/PROTEIN DELIVERY

Long circulating conventional liposomes or PSSLs were employed to improve circulation half-life and impart intracellular delivery characteristics for delivery of therapeutics to nucleus. Three cell lines, Hs578t human epithelial cells from breast carcinoma, MDA-MB-231 human breast carcinoma cells, and WI-26 human diploid lung fibroblast cells, were used to show cellular uptake of peptide-loaded liposomes. Confocal microscopy and flow cytometry were used in this regard. Results advocated the usefulness of PSLs over the conventional formulation in terms of delivery of hydrophilic materials to the cytoplasm [44]. PSLs modified with 3-methylglutarylated poly(glycidol) of linear (MGlu-LPG) or hyperbranched structure (MGlu-HPG) were formulated and delivered via subcutaneous or nasal routes. Ovalbumin (OVA) specific cytotoxic T cells were generated abundantly compared with conventional liposomes. Furthermore, delivery of polymer-modified OVA-loaded liposomes reduced tumor burden remarkably after administration into mice bearing E.G7-OVA tumor. Results suggested them to be used as a vehicle of antigens for efficient cancer immunotherapy [45].

7.3.2 FUSION IN THE ENDOSOMAL MEMBRANE

Another approach to get endosomal escape is the destabilization of the endosomal membrane by fusogenic agents such as peptides. They can be combined with a number of carriers like liposomes and others [46, 47].

7.3.2.1 FUSOGENIC LIPOSOMES

This approach can be implemented in different areas like drug delivery, vaccine development, gene delivery, and cancer treatment with the help of liposomes [48].

7.3.2.1.1 FUSOGENIC LIPOSOMES FOR VACCINE DEVELOPMENT AND GENE DELIVERY

In the vaccine field, liposomes that are associated with the functional viral envelope proteins are termed as virosomes. They represent an excellent delivery vehicle with good immunogenicity and safety profile in terms of tolerability [49, 50]. To deliver therapeutic agents based on DNA or RNA oligonucleotides, nanoparticles were encapsulated into liposome and further fused with ultraviolet (UV)-inactivated Sendai virus to formulate fusogenic liposomes. Results showed an excellent delivery of high amount of nanoparticles into the cytoplasm without any cytotoxicity using this fusogenic liposomal formulation [51].

7.3.2.1.2 FUSOGENIC LIPOSOMES FOR CANCER CHEMOTHERAPY

Application of the fusogenic liposomes in cancer treatment is very attractive. One of the several approaches is represented by fusogenic liposomes containing fragment A of diphtheria toxin (DTA). In vitro study against sarcoma-180 (S-180) cells revealed the fact that highest cytotoxcity was achieved with this formulation as compared to conventional liposomes containing DTA or empty fusogenic liposomes [52]. Current examples of drug delivery by fusogenic liposomes especially in cancer chemotherapy were represented by octaarginine (R8)-modified fusogenic DOPE-liposomes (R8-DOPE-BLM) and DOX-loaded PEGylated liposomes. Both formulations delivered the drug to the nucleus and did not end their journey in lysosomes. Remarkable increase in cytotoxcity was observed using this approach. Fusogenic nature of these formulations improved their delivery efficacy as well as effectiveness against cancer [53, 54].

7.3.3. PHOTOCHEMICAL DESTABILIZATION OF ENDOSOMAL MEMBRANE

Light is considered as one of the effective approaches to achieve cytoplasmic release of compounds from liposomes due to its ability to exihibit spatial and temporal control over radiation. After exposure to light, generation of reactive singlet oxygen species occur from these photosensitive agents that eventually induce destruction of the endosomal/lysosomal membrane, whereas the contents of the organelles deliver to the cytosol unharmed [55, 56]. This destabilization is mediated through different mechanisms like light-induced isomerization, cleavage, or polymerization [57]. This technique has been combined with several delivery systems like such as lipid carriers and polymers-based formulations [58–61].

7.3.3.1 PHOTOSENSITIVE LIPOSOMES

Azobenzene in the form of 1,2-bis(4-n-butylphenylazo-4′-phenylbutyroyl)-L-α-phosphatidylcholine (Bis-azo PC) has been used with photoisomerizable phenomenon in the light-mediated release of liposomal contents [62]. Photoisomerizable liposomes can be formulated using retinoyl-phospholipids. Liposomal content release can be achieved via photoisomerization of spiropyran [63, 64]. Naturally occurring lipids called plasmalogens undergo photoinduced cleavage with light treatment [65]. Liposomes composed of 1,2-bis[10-(2′,4′-hexadienoyloxy)-decanoyl]-sn-phosphatidylcholine (Bis-sorbPC) in combination with cholesterol, 1,2-dioleoyl-sn-glycero-3-phosphatidylcholine (DOPC), and PEG2000-DOPE were formulated. This formulation caused about 100-fold increase in the permeability of encapsulated fluorescent marker [66, 67]. In a very recent work, a novel photosensitive liposome composed of egg yolk phosphatidylcholine (PC) was developed and surface modified with hydrophobically modified Poly(vinyl alcohol)–epoxypropoxy coumarin conjugate (HmPVA-EPC). Further, 5(6)-carboxy fluorescein was consumed as a fluorescence marker and decanoyl chloride (DC) as a hydrophobic pendant. Fluorescence marker was released for 60 min upon UV irradiation from prepared formulation because of photodimerization of EPC residues [68]. Numerous drug delivery systems have been examined carefully for the intracellular delivery of small molecular chemotherapeutic drugs to potentiate their therapeutic efficacies as well as to minimize associated side effects . Variety of anticancer drugs exhibit toxicity to nuclear DNA or its associated enzymes to execute their cytotoxic effect on cancer cells [69]. After entering into tumor cells, there should a system to deliver them further to the nucleus to get therapeutic benefits.

7.3.3.1.1 PHOTOSENSITIVE LIPOSOMES FOR ANTICANCER DRUG DELIVERY

In the search to design a vehicle capable in delivery of biologicals to cytoplasm and ultimately up to nucleus, recently a novel formualtion of phototriggerable liposomes was developed using dipalmitoyl phosphatidylcholine (DPPC) and photopolymerizable diacetylene phospholipid ($DC_{8,9}PC$). For in vivo applications this system was triggered using light to assess cellular toxicity for anticancer effect. Calcein-loaded liposomes containing various ratios of DPPC:$DC_{8,9}PC$ and 4 mol% distearoyl phosphoethanolamine (DSPE)-PEG2000 released remarkable amount of calcein. After this laser-mediated release (514 nm) of DOX was estimated, and cellular toxicity was examined. Stable liposomes were found and after 514 nm laser treatment wavelength-specific release of DOX was observed. Cocultures of DOX-loaded liposomes and cells (Raji and MCF-7) after laser treatment resulted in at least two-to-threefold enhanced cell killing as compared to untreated samples. To the best of our knowledge, this was the first reported study of improved cell killing after light treatment of an encapsulated anticancer agent from photosensitive liposomes [25, 70].

7.3.3.1.2 PHOTOSENSITIVE LIPOSOMES FOR PHOTODYNAMIC THERAPY

Photodynamic therapy (PDT) represents a therapeutic approach for the treatment of certain types of skin cancers. This treatment included delivery of photosensitive drugs followed by targeted light exposure. Reactive oxygen species generated in this process causes cell death. PDT is now being used for the successful treatment of actinic keratosis and nonmelanoma skin cancers [71, 72]. These photosensitizers usually suffer from poor penetration through biological barriers of skin and cell membranes. Long circulatory liposomes are more effective in this respect because of their improved plasma half-life compared to conventional liposomes in delivering sufficient amount of therapeutics for tumor uptake [73]. Hence, aminolevulinic acid (ALA) and its esterified derivatives ALA-Hexyl ester (He-ALA) and ALA-Undecanoyl ester (Und-ALA) were delivered using liposomal formulations. Good entrapment with stability was shown by the vesicles upon storage for 1 week at 4°C [74]. PDT is an innovative rising technique for the treatment of neovascular diseases of the eye [75]. One recent example in this field was represented by verteporfin encapsulated in cationic liposomes (CL-VTP). Results are compared between CL-VTP and visudyne® in the case of choroidal neovascularization. CL-VTP caused less retinal damage compared to visudyne® [76].

7.3.3.1.3 GOLD-EMBEDDED PHOTOSENSITIVE LIPOSOMES FOR DRUG DELIVERY

Gold nanoparticles encapsulated within liposomes responsive to light irradiation, represent a unique class of drug delivery system for the controlled release of drugs in specific target areas including eye and skin. Deeper layer of body can be approached by proper selection and adjustment of laser wavelength used. In three different formulations, hydrophobic nanoparticles embedded into the lipid bilayer, negatively charged hydrophilic nanoparticles encapsulated in the core of liposomes, and lipid functionalized gold particles localized on the inner and the outer surface of the liposomes were formulated [77]. Recently, liposomes embedded with gold nanoparticles were studied to identify the mechanism behind light-induced changes and functionality. Phase transitions analysis in distearoyl phosphatidylcholine (DSPC)/DPPC liposomes upon heating was studied using real time small angle X-ray scattering. Gold nanoparticles absorb light energy and then convert it into heat. This cause lipid phase transition from gel to fluid state. Delivery of gold nanoparticle-loaded liposomes to human retinal pigment epithelial cell line ARPE-19 cells results in internalization of liposome. Sequence is followed by light-triggered release of hydrophilic fluorescent probe (calcein) and fluorescence-activated cell sorting was used to demonstrate it [78].

7.3.4 TEMPERATURE-MEDIATED RELEASE

Early liposomes have been employed as conventional as well as targeted delivery vehicles but after this, the technology has shifted toward stimuli–responsive liposomal approach, in order to get more successful delivery toward target site. With the use of these systems site-specific chemotherapy became possible with triggered drug release at the target site; hence, greater control was provided in terms of spatial and temporal therapy. Liposomal chemotherapy can be improved in the context of increased vascular permeability in solid tumors with the use of mild hyperthermia (HT) as a stimuli. In this way liposomal accumulation, and release at the target site can be enhanced by triggering with the help of hyperthermia [79, 80].

7.3.4.1 THERMOSENSITIVE LIPOSOMES

Temperature-sensitive liposomes (TSLs) were one of those innovative approaches, formulated by Yatvin and coworkers first time [81]. More recently, several thermosensitive vesicular formulations were developed by including phospholipids having phase transition temperature (T_m) between 41°C and 42°C in the preparation or with leucine zipper sequence peptide which has dissociative, unfolding properties to increase drug release under mild hyperthermia. Long circulating thermosensitive liposomes can be formulated using PEG with usual components or by novel lipid 1.2-dipalmitoyl-sn-glycero-3-phosphoglyceroglycerol [82–86].

7.3.4.1.1 THERMOSENSITIVE LIPOSOMES FOR THE DELIVERY OF CHEMOTHERAPEUTIC DRUGS

In vitro release study for the amount of DOX was carried out by incubating LTSLs over a range of temperatures and durations, and compared to conventional formulations. In vivo experiments were performed on murine adenocarcinoma tumors by administration of liposomes and free DOX. Mice were supplied pulsed-high intensity focused ultrasound after 0 and 24 h. Growth inhibition in the tumors was evaluated for doxorubicin concentration and after this combination of exposure, viable clinical results were obtained applying this strategy [87]. In another case, novel

TSLs (40°C) containing DOX were combined with local HT to evaluate the efficiency against the treatment of solid growing rat rhabdomyosarcomas. Comparisons were made with free DOX with or without HT in terms of tumor inhibition and systemic toxicity. Repeated treatments with DOX-liposomes along with HT exhibited a statistically significant ($p<0.05$) tumor growth delay. No sign of systemic toxicity was observed as compared to free DOX [88]. Radiofrequency ablation combined with lysothermosensitive liposomal DOX was employed to cure for large tumors in hepatocellular carcinoma. Unfortunately unclear tumor margins show limiting factor in the curative efficacy of this therapy [89]. Copolymers containing temperature-responsive nisopropylacrylamide (NIPAAm) and pH-responsive propylacrylic acid (PAA) were combined to develop novel polymer-modified thermosensitive liposomes for the delivery of DOX for cancer. In comparison to conventional thermosensitive formulations, this formulation improved the release profile and showed significant ($p<0.0001$) lowered thermal dose threshold [90]. A thermosensitive formulation composed of DPPC and Brij78 loaded with DOX remarkably altered the stability in serum at 37°C. Confirmation was made by cell-based assays, and overall enhanced drug release rates were observed and compared to lysolipid temperature sensitive liposomes. Brij78-liposomes and low TSLs displayed mild hemolytic activity with comparable blood compatibility [91, 92]. In another study, temperature-dependent drug release was compared for adriamycin (ADM)-encapsulated TSLs and conventional liposomes. Most of the loaded ADM (>90%) was released from thermosensitive formulation within a period of 30 min at 42°C temperature. However, conventional liposomes released negligible amount with the same condition as revealed by in vitro study. C6 glioma bearing mice model was explored to evaluate body distribution and antitumor efficacy for the prepared formulation. The maximum concentration of ADM in the brain with enhanced survival time was noted compared to ADM solution and liposomes-ADM, respectively [93]. Semisynthetic therapeutic agent like vinorelbine bitartrate TSLs gained advantage of targeting and antitumor effect when delivered with TSLs. Lung tumor model was employed to deliver a stable formulation that quickly releases drug at 42°C. Overall inhibition of tumor was higher in a group treated with TSLs compared to the group treated with normal injection [94]. In another study low TSLs and non-TSLs were developed. Formulations were dual-labeled using ^{3}H-cholesteryl hexadecyl ether lipid and loaded with ^{14}C-DOX. DOX showed ultrafast release at 42°C immediately after HT that results in highest drug accumulation in tumors [95]. DOX concentrations in heated tumors were found 26.7 times higher than in unheated tumors (p=0.017) when, magnetic resonance imaging controlled focused ultrasound HT was combined with thermosensitive liposomes [96]. Some of the critical advantages like less frequent dosing, lower accumulation in healthy tissues, and enhanced concentration in tumor periphery due to the EPR effect, can be achieved by delivery of chemotherapeutic agents encapsulated in PEGylated liposomes [97]. Based on this approach, combinations of long circulating liposomes with thermosensitive properties were formulated to deliver epirubicin hydrochloride (EPI) for cancer chemotherapy. According to the in vitro results more than 90% of loaded EPI was released within 4 min at 43°C. The EPI-low TSLs formulation prolonged the circulation time and in vivo performance compared to nonthermal formulation and EPI solution [98]. Another recent example was represented by PEGylated cationic TSLs, a combination of stealth liposomal technology with hyperthermia-mediated release [99].

7.3.4.1.2 GEL LOADED THERMOSENSITIVE LIPOSOMES FOR THE DELIVERY OF ANTICANCER DRUGS

Cytarabine-loaded liposomes were formulated in order to reduce dosing frequency and sustaining the drug action. This formulation was embedded in biodegradable and biocompatible chitosan-beta-glycerophosphate thermosensitive solution. The advantage associated is that of gel formation at body temperature. Prepared formulation shows sustained release of encapsulated cargo for more than 60 h as compared to drug-loaded liposomal suspension which show up to 48 h only. Pharmacokinetic studies in rats revealed the fact that a higher $t1/2$ (28.86 h) and AUC 2526.88 mug/mL h was achieved with this formulation compared with cytarabine-loaded liposomal suspension and free cytarabine. Thus, combination of effects like gelling at body temperature and sustained release were obtained using this formulation [100]. Thermoreversible gel (Pluronic® F127 gel) in the form of in situ gel system was formulated for liposomes-containing paclitaxel (PTX) to get controlled release and improved antitumor drug efficiency. High viscosity of liposomal gel creates a drug reservoir effect and hence the extent of the drug release compared with liposomes, general gel, and commercial formulation Taxol®. Treatment with PTX-loaded liposomal 18% Pluronic F127 resulted into intercellular fluorescence intensity that revealed the presence of drug cargo into cytosol. This formulation showed much higher cytotoxicity and drug concentration in KB cells compared to conventional liposomes [101].

7.3.4.1.3 TARGETED THERMOSENSITIVE LIPOSOMES FOR THE DELIVERY OF ANTICANCER DRUGS

Targeted temperature sensitive magnetic liposomes were employed with the property of magnetic hyperthermia-triggered drug release for thermochemotherapy [102]. Recently, thermosensitive magnetic liposomes were formulated to target FRs. The aim was to explore the benefits of combination of drug targeting and magnetic hyperthermia-triggered drug release [103]. Human epidermal growth factor receptor 2 (HER2) specific affibody (ZHER2:342-Cys) was conjugated to thermosensitive liposomes in order to improve DOX accumulation in human breast adenocarcinoma cells (SK-BR-3). Results showed two-to-threefold increase with prepared formulations compared to control liposomes. Brief exposure of heat up to 45°C was made for liposomes–cell complexes and resulted in improved cytotoxicity for affisomes and control liposomes, but Doxil®, however, showed lower toxicity under the same conditions. Results advocated the usefulness of formulation in both targeting and triggering potential for the safe and effective treatment of breast cancer [86, 104].

7.3.5 MISCELLANEOUS APPROACHES

7.3.5.1 PORE FORMATION IN THE ENDOSOMAL MEMBRANE

Pore formation is controlled by a balance between a membrane tension and a line tension. Effects are counterbalancing in nature. Some materials like peptides have a high binding capacity toward the rim of the pore. It has been shown that some agents like cationic amphiphilic peptides upon binding to the lipid bilayer produce internal stress or internal membrane tension that eventually lead to formation of pores in the lipid bilayer [105].

Similarly, binding of some peptides to the membrane has an effect on the line tension negatively which ultimately keep the pore radius stable [106].

7.3.5.2 ULTRASOUND TRIGGERED LIPOSOMES

Nanosterically stabilized liposomes were evaluated on BALB/c mice with C26 colon adenocarcinoma tumors in a footpad using low frequency ultrasound (LFUS) to trigger the release of therapeutic agent. Combination of formulation with LFUS produced the best therapeutic result compared to free cisplatin with or without LFUS, or liposomal CDDP without LFUS, or LFUS alone, or no treatment [107]. A different kind of work was reported for liposomes with attached with microbubble with the help of covalent thiol-maleimide linkages to prepare a self-assembled system [108]. These loaded complexes carried a high drug payload and unload their content with the help of US [109]. Another system explored against breast cancer with possible ultrasound-triggered targeted chemotherapy was examplified by liposomes–microbubble complexes loaded with PTX. Due to remarkable release of payload from complex, it was concluded that complex plus US were effectively inhibited tumor growth compared to complex without US [110]. Cell-impermeable dye (TO-PRO-3) was delivered using TSLs and was followed by evaluation of intracellular uptake. To promote the release combination of two consecutive steps—heating and sonication—were taken into account. This strategy can be used to deliver cell impermeable drugs those have usually low clearance and/or degradation in blood and no uptake by cells [111].

7.4 CONCLUSION

The development of delivery vehicles for endosomal escape, represent a rising field for research. Different mechanisms for endosomal escape use different procedure for this effect. A safe endosomal escape agent should possess safe, effective, low cost, and patient compliance characteristics for application. Pharmaceutical liposomal formulations are a fully developed and equipped area. However, formulation-based research approaches have the goal to deliver drugs effectively to cytoplasmic targets. Recent research with new classes of materials approaches to a newer hope for the development of rational delivery mechanisms for the cytoplasmic delivery of materials using liposomes.

LIST OF ABBREVIATIONS

Adriamycin (ADM)
Aminolevulinic acid (ALA)
ALA-Hexyl ester (He-ALA)
ALA-Undecanoyl ester (Und-ALA)
1,2-bis(4-n-butylphenylazo-4′- phenylbutyroyl)-L-α-phosphatidylcholine (Bis-azo PC)
1,2-bis[10-(2′,4′-hexadienoyloxy)-decanoyl]-sn-phosphatidylcholine (Bis-sorbPC)
Cholesterol (CHOL)
Cholesterylhemisuccinate (CHEMS)
Cisplatin (CDDP)
1,2-dioleoyl-sn-glycero-3-phosphoethanolamine (DOPE)
1,2-dioleoyl-sn-glycero-3-phosphatidylcholine (DOPC)
Dipalmitoyl phosphatidylcholine (DPPC)
Diphtheria toxin (DTA)
Distearoyl phosphatidylcholine (DSPC)
Distearoyl phosphatidylethanolamine (DSPE)
Decanoyl chloride (DC)
Dendritic cells (DCs)
Doxorubicin (DOX)
Epidermal growth factor receptor (EGFR)
Epirubicin hydrochloride (EPI)
Enhanced permeation and retention (EPR)
5-Fluorouracil (5-FU)
Folate receptor (FR)
Human oral carcinoma cells (KB)
Hyperthermia (HT)
Immunoliposomes (ILs)
Low frequency ultrasound (LFUS)
N-isopropylacrylamide (NIPAM)
Ovalbumin (OVA)
Paclitaxel (PTX)

Phosphatidylcholine (PC)
Phosphatidyl ethanolamine (PE)
Photodynamic therapy (PDT)
Photopolymerizable diacetylene phospholipid ($DC_{8,9}PC$)
pH-sensitive liposomes (PSLs)
pH-sensitive stealth liposomes (PSSLs)
Polyethylene glycol (PEG)
Poly(vinyl alcohol)–epoxypropoxy coumarin conjugate (HmPVA-EPC)
Propylacrylic acid (PAA)
Soy phosphatidylcholine (SPC)
Thermosensitive liposomes (TSLs)
Transition temperature (T_m)
Verteporfin loaded cationic liposomes (CL-VTP)

KEYWORDS

- **Anticancer**
- **Cytoplasmic delivery;**
- **Encapsulation**
- **Endosomal escape**
- **Liposomes**

REFERENCES

1. Gruenberg, J.; van der Goot, F. G. Mechanisms of pathogen entry through the endosomal compartments. *Nat. Rev. Mol. Cell Biol.*, **2006**, 7, 495–504.
2. Chakrabarti, R.; Wylie, D. E.; Schuster, S. M. Transfer of monoclonal antibodies into mammalian cells by electroporation. *J. Biol. Chem.*, **1989**, 264, 15494–15500.
3. Arnheiter, H.; Haller, O. Antiviral sTATe against influenza virus neutralized by microinjection of antibodies to interferon-induced Mx proteins. *EMBO J.*, **1988**, 7, 1315–1320.
4. Stevenson, D. J.; Gunn-Moore, F. J.; Campbell, P.; Dholakia, K. Single cell optical transfection. *J. R. Soc. Interface.*, **2010**, 7, 863–871.
5. Hoffman, A. S.; Stayton, P. S.; Press, O.; Murthy, N.; Lackey, C. A.; Cheung, C.; Black, F.; Campbell, J.; Fausto, N.; Kyriakides, T. R.; Bornstein, P. Design of "smart" polymers that can direct intracellular drug delivery. *Polym. Adv. Technol.*, **2002**, 13, 992–999.

6. Nishikawa, M.; Huang, L. Nonviral vectors in the new millennium: delivery barriers in gene transfer. *Hum. Gene Ther.*, **2001**, 12, 861–870.
7. Tokatlian, T.; Segura, T. siRNA applications in nanomedicine, *Wiley Interdiscip. Rev. Nanomed. Nanobiotechnol.,* **2010**, 2, 305–315.
8. Pfeifer, A.; Verma, I. M. Gene therapy: promises and problems. *Ann. RevGenomics Hum. Genet.*, **2001**, 2, 177–211.
9. Kay, M. A.; Glorioso, J. C.; Naldini, L. Viral vectors for gene therapy: the art of turning infectious agents into vehicles of therapeutics. *Nat. Med.*, **2001**, 7, 33–40.
10. Medina-Kauwe, L. K.; Maguire, M.; Kasahara, N.; Kedes, L. Nonviral gene delivery to human breast cancer cells by targeted Ad5 penton proteins. *Gene Ther.*, **2001**, 8, 1753–1761.
11. Fominaya, J.; Wels, W. Target cell-specific DNA transfer mediated by a chimeric multidomain protein—novel non-viral gene delivery system. *J. Biol. Chem.* **1996**, 271, 10560–10568.
12. Gottschalk, S.; Sparrow, J. T.; Hauer, J.; Mims, M. P.; Leland, F. E.; Woo, S. L. C.; Smith, L. C. A novel DNA-peptide complex for efficient gene transfer and expression in mammalian cells. *Gene Ther.*, **1996**, 3, 448–457.
13. Gerasimov, O. V.; Boomer, J. A.; Qualls, M. M.; Thompson, D. H. Cytosolic drug delivery using pH- and light-sensitive liposomes. *Adv. Drug Deliv. Rev.,* **1999**, 38, 317–338.
14. Ta, T.; Porter, T. M. Thermosensitive liposomes for localized delivery and triggered release of chemotherapy. *J. Control Release.,* **2013**, 169, 112–125.
15. Torchilin, V. P. Recent advances with liposomes as pharmaceutical carriers. *Nat. Rev. Drug Discov.,* **2005**, 4, 145–160.
16. Panyam, J.; Labhasetwar, V. Targeting intracellular targets. *Curr. Drug. Deliv.*, **2004**, 1, 235–247.
17. Pathak, A.; Patnaik, S.; Gupta, K. C. Recent Trends in Non-viral Vector-mediated Gene Delivery. *Biotechnol. J.,* **2009**, 4, 1559–1572.
18. Düzgüne&scedil, N.; Nir, S. Mechanisms and kinetics of liposome-cell interactions. *Adv. Drug Deliv. Rev.,* **1999**, 40, 3–18.
19. Varkouhi, A. K.; Scholte, M.; Storm, G.; Haisma, H. J. Endosomal escape pathways for delivery of biologicals. *J. Control Release.,* **2011**, 151, 220–228.
20. Vasir, J. K.; Labhasetwar, V. Biodegradable nanoparticles for cytosolic delivery of therapeutics. *Adv. Drug Deliv. Rev.,* **2007**, 59, 718–728.
21. Meier, O.; Greber, U. F. Adenovirus endocytosis. *J. Gene Med.,* **2003**, 5, 451–462.
22. Hogle, J. M. Poliovirus cell entry: common structural themes in viral cell entry pathways. *Annu. Rev. Microbiol.*, **2002,** 56, 677–702.
23. Mandal, M.; Lee, K. D. Listeriolysin O-liposome-mediated cytosolic delivery of macromolecule antigen in vivo: enhancement of antigen-specific cytotoxic T lymphocyte frequency, activity, and tumor protection. *Biochim. Biophys. Acta.*, **2002**, 1563, 7–17.
24. Parente, R.A.; Nir, S.; Szoka, F.C. pH-dependent fusion of phosphatidylcholine small vesicles. *J. Biol. Chem.,* **1988**, 263, 4724–4730.
25. Miller, D. K.; Griffiths, E.; Lenard, J.; Firestone, R. A. Cell killing by lysosomotropic detergents. *J. Cell Biol.* **1983**, 97, 1841–1851.

26. Kono, K.; Igawa, T.; Takagishi, T. Cytoplasmic delivery of calcein mediated by liposomes modified with a pH-sensitive poly(ethylene glycol) derivative. *Biochim. Biophys. Acta.*, **1997**, 1325, 143–154.
27. Simões, S.; Slepushkin, V.; Düzgünes, N.; Pedroso de Lima, M. C. On the mechanisms of internalization and intracellular delivery mediated by pH-sensitive liposomes. *Biochim. Biophys. Acta.*, **2001**, 1515, 23–37.
28. Connor, J.; Norley, N.; Huang, L. Biodistribution of pH-sensitive immunoliposomes. *Biochim. Biophys. Acta.*, **1986**, 884, 474–481.
29. Ellens, H.; Bentz, J.; Szoka, F. C. pH-induced destabilization of phosphatidylethanolamine-containing liposomes: role of bilayer contact. *Biochemistry.*, **1984**, 23, 1532–1538.
30. Senior, J. H. Fate and behavior of liposomes in vivo: a review of controlling factors. *Crit. Rev. Ther. Drug Carrier Syst.*, **1987**, 3, 123–193.
31. Connor, J.; Norley, N.; Huang, L. Biodistribution of pH-sensitive immunoliposomes. *Biochim. Biophys. Acta.*, **1986**, 884, 474–481.
32. Fonseca, C.; Moreira, J. N.; Ciudad, C. J.; Pedroso de Lima, M. C.; Simões, S. Targeting of sterically stabilised pH-sensitive liposomes to human T-leukaemia cells. *Eur. J. Pharm. Biopharm.*, **2005**, 59, 359–366.
33. Skalko, N.; Peschka, R.; Altenschmidt, U.; Lung, A.; Schubert, R. pH-sensitive liposomes for receptor-mediated delivery to chicken hepatoma (LMH) cells. *FEBS Lett.*, **1998**, 434, 351–356.
34. Ishida, T.; Okada, Y.; Kobayashi, T.; Kiwada, H. Development of pH-sensitive liposomes that efficiently retain encapsulated doxorubicin (DXR) in blood. *Int. J. Pharm.*, **2006**, 309, 94–100.
35. Júnior, A. D.; Mota, L. G.; Nunan, E. A.; Wainstein, A. J.; Wainstein, A. P.; Leal, A. S.; Cardoso, V. N.; Oliveira, M. C. De Tissue distribution evaluation of stealth pH-sensitive liposomal cisplatin versus free cisplatin in Ehrlich tumor-bearing mice. *Life Sci.*, **2007**, 80, 659–664.
36. Leite, E. A.; Giuberti Cdos, S.; Wainstein, A. J.; Wainstein, A. P.; Coelho, L. G.; Lana, A. M.; Savassi-Rocha, P. R.; Oliveira, M. C. De Acute toxicity of long-circulating and pH-sensitive liposomes containing cisplatin in mice after intraperitoneal administration. *Life Sci.*, **2009**, 84, 641–649.
37. Banerjee, S.; Sen, K.; Pal, T. K.; Guha, S. K. Poly(styrene-co-maleic acid)-based pH sensitive liposomes mediate cytosolic delivery of drugs for enhanced cancer chemotherapy. *Int. J. Pharm.*, **2012**, 436, 786–797.
38. Kim, I. Y.; Kang, Y. S.; Lee, D. S.; Park, H. J.; Choi, E. K.; Oh, Y. K.; Son, H. J.; Kim, J. S. Antitumor activity of EGFR targeted pH-sensitive immunoliposomes encapsulating gemcitabine in A549 xenograft nude mice. *J. Control Release.*, **2009**, 140, 55–60.
39. Simard, P.; Leroux, J. C. pH-sensitive immunoliposomes specific to the CD33 cell surface antigen of leukemic cells. *Int. J. Pharm.*, **2009**, 381, 86–96.
40. Shi, G.; Guo, W.; Stephenson, S. M.; Lee, R. J. Efficient intracellular drug and gene delivery using folate receptor-targeted pH-sensitive liposomes composed of cationic/anionic lipid combinations. *J. Control Release.*, **2002,** 80, 309–319.
41. Yuba, E.; Kojima, C.; Sakaguchi, N.; Harada, A.; Koiwai, K.; Kono, K. Gene delivery to dendritic cells mediated by complexes of lipoplexes and pH-sensitive fusogenic polymer-modified liposomes. *J. Control Release.*, **2008**, 130, 77–83.

42. Fattal, E.; Couvreur, P.; Dubernet, C. "Smart" delivery of antisense oligonucleotides by anionic pH-sensitive liposomes. *Adv. Drug Deliv. Rev.*, **2004**, 56, 931–946.
43. Oliveira, M. C.; Fattal, E.; Ropert, C.; Malvy, C.; Couvreur, P. Delivery of Antisense Oligonucleotides by Means of pH-sensitive Liposomes. *J. Control Release.*, **1997**, 48, 179–184.
44. Ducat, E.; Deprez, J.; Gillet, A.; Noël, A.; Evrard, B.; Peulen, O.; Piel, G. Nuclear delivery of a therapeutic peptide by long circulating pH-sensitive liposomes: benefits over classical vesicles. *Int. J. Pharm.*, **2011**, 420, 319–332.
45. Yuba, E.; Harada, A.; Sakanishi, Y.; Watarai, S.; Kono, K. A liposome-based antigen delivery system using pH-sensitive fusogenic polymers for cancer immunotherapy. *Biomaterials.*, **2013**, 34, 3042–3052.
46. Marsh, M.; Helenius, A. Virus entry into animal cells. *Advan. Virus Res.*, **1989**, 36, 107–151.
47. Horth, M.; Lambrecht, B.; Khim, M. C. L.; Bex, F.; Thiriart, C.; Ruysschaert, J.M.; Burny, A.; Brasseur, R. Theoretical and functional analysis of the SIV fusion peptide. *EMBO J.*, **1991**, 10, 2747–2755.
48. Wiley, D. C.; Skehel, J. J. The structure and function of the hemagglutinin membrane glycoprotein of Influenza virus. *Ann. Rev. Biochem.*, **1987**, 56, 365–394.
49. Felnerova, D.; Viret, J. F.; Glück, R.; Moser, C. Liposomes and virosomes as delivery systems for antigens, nucleic acids and drugs. *Curr. Opin. Biotechnol.*, **2004**, 15, 518–29.
50. Kunisawa, J.; Nakagawa, S.; Mayumi, T. Pharmacotherapy by intracellular delivery of drugs using fusogenic liposomes: application to vaccine development. *Adv. Drug Deliv. Rev.*, **2001**, 52, 177–86.
51. Kunisawa, J.; Masuda, T.; Katayama, K.; Yoshikawa, T.; Tsutsumi, Y.; Akashi, M.; Mayumi, T.; Nakagawa, S. Fusogenic liposome delivers encapsulated nanoparticles for cytosolic controlled gene release. *J. Control Release.*, **2005**, 105, 344–53.
52. Mizuguchi, H.; Nakanishi, M.; Nakanishi, T.; Nakagawa, T.; Nakagawa, S.; Mayumi, T. Application of fusogenic liposomes containing fragment A of diphtheria toxin to cancer therapy. *Br. J. Cancer.*, **1996**, 73, 472–476.
53. Koshkaryev, A.; Piroyan, A.; Torchilin, V. P. Bleomycin in octaarginine-modified fusogenic liposomes results in improved tumor growth inhibition. *Cancer Lett.*, **2013**, 334, 293–301.
54. Biswas, S.; Dodwadkar, N. S.; Deshpande, P. P.; Parab, S.; Torchilin, V. P. Surface functionalization of doxorubicin-loaded liposomes with octa-arginine for enhanced anticancer activity. *Eur. J. Pharm. Biopharm.*, **2013**, 84, 517–525.
55. Berg, K.; Selbo, P. K.; Prasmickaite, L.; Tjelle, T. E.; Sandvig, K.; Moan, D.; Gaudernack, G.; Fodstad, O.; Kjolsrud, S.; Anholt, H.; Rodal, G. H.; Rodal, S. K.; Hogset, A. Photochemical internalization: a novel technology for delivery of macromolecules into cytosol. *Cancer Res.*, **1999**, 59, 1180–1183.
56. Lou, P. J.; Lai, P. S.; Shieh, M. J.; MacRobert, A. J.; Bergs, K.; Bown, S. G. Reversal of doxorubicin resistance in breast cancer cells by photochemical internalization. *Int. J. Cancer*, **2006**, 119, 2692–2698.
57. Leung, S. J.; Romanowski, M. Light-activated content release from liposomes. *Theranostics.*, **2012**, 2, 1020–1036.

58. Fretz, M. M.; Hogset, A.; Koning, G. A.; Jiskoot, W.; Storm, G.; Cytosolic delivery of liposomally targeted proteins induced by photochemical internalization. *Pharm. Res.*, **2007**, 24, 2040–2047.
59. Oliveira, S.; Fretz, M. M.; Hogset, A.; Storm, G.; Schiffelers, R.M. Photochemical internalization enhances silencing of epidermal growth factor receptor through improved endosomal escape of siRNA. *Biochim. Biophys. Acta.*, **2007**, 1768, 1211–1217.
60. Cabral, H.; Nakanishi, M.; Kumagai, M.; Jang, W. D.; Nishiyama, N.; Kataoka, K. A photoactivated targeting chemotherapy using glutathione sensitive camptothecin-loaded polymeric micelles. *Pharm. Res.*, **2009**, 26, 82–92.
61. Bonsted, A.; Wagner, E.; Prasmickaite, L.; Hogset, A.; Berg, K. Photochemical enhancement of DNA delivery by EGF receptor targeted polyplexes. *Methods Mol. Biol.*, **2008**, 434, 171–181.
62. Bisby, R. H.; Mead, C.; Morgan, C. G. Photosensitive liposomes as "cages" for laser-triggered solute delivery: the effect of bilayer cholesterol on kinetics of solute release. *FEBS Lett.*, **1999**, 463, 165–168.
63. Pidgeon, C.; Hunt, C. A. Light Sensitive Liposomes. *J. Photochem. Photobiol.*, **1983**, 37, 491–494.
64. Ohya, Y.; Okuyama, Y.; Fukunaga, A.; Ouchi, T. Photo-sensitive lipid membrane perturbation by a single chain lipid having terminal spiropyran group. *Supramol. Sci.*, **1998**, 5, 21–29.
65. Anderson, V. C.; Thompson, D. H. Triggered release of hydrophilic agents from plasmalogen liposomes using visible light or acid. *Biochim. Biophys. Acta.*, **1992**, 1109, 33–42.
66. O'Brien, D. F.; Armitage, B.; Benedicto, A.; Bennett, D. E.; Lamparski, H. G.; Lee, Y. S.; Srisiri, W.; Sisson, T. M. Polymerization of Preformed Self-organized Assemblies. *Acc. Chem. Res.*, **1998**, 31, 861–868.
67. Clapp, P. J.; Armitage, B. A.; O'Brien, D. F. Dimensional Polymerization of Lipid Bilayers:Visible-light-sensitized Photoinitiation. *Macromolecules.*, **1997**, 30, 32–41.
68. Seo, H. J.; Cha, H. J.; Kim, T. S.; Kim, J. C. Photo-responsive liposomes decorated with hydrophobically modified poly(vinylalcohol)–coumarin conjugate. *J. Ind. Eng. Chem.*, **2013**, 19, 310–315.
69. Sui, M.; Liu, W.; Shen, Y. Nuclear drug delivery for cancer chemotherapy. *J. Control Release.*, **2011**, 155, 227–236.
70. Yavlovich, A.; Singh, A.; Blumenthal, R.; Puri, A. A novel class of photo-triggerable liposomes containing DPPC:DC(8,9)PC as vehicles for delivery of doxorubcin to cells. *Biochim. Biophys. Acta.*, **2011**, 1808, 117–126.
71. Lopez, R. F.; Lange, N.; Guy, R.; Bentley, M. V. Photodynamic therapy of skin cancer: controlled drug delivery of 5-ALA and its esters. *Adv. Drug Deliv. Rev.*, **2004**, 56, 77–94.
72. Yano, S.; Hirohara, S.; Obata, M.; Hagiya, Y.; Ogura, S.; Ikeda, A.; Kataoka, H.; Tanaka, M.; Joh, T. Current states and future views in photodynamic therapy. *J. Photochem. Photobiol.*, **2011**, 12, 46–67.
73. Derycke, A. S.; Witte, P. A. de Liposomes for photodynamic therapy. *Adv. Drug Deliv. Rev.*, **2004**, 56, 17–30.
74. Gabriela, Di V.; Hermida, L.; Batlle, A.; Fukuda, H.; Defain, M. V.; Mamone, L.; Rodriguez, L.; MacRobert, A.; Casas, A. Characterisation of liposomes containing aminolevulinic acid and derived esters. *J. Photochem. Photobiol. B, Biol.*, **2008**, 92, 1–9.

75. Christie, J. G.; Kompella, U. B. Ophthalmic light sensitive nanocarrier systems. *Drug Discov. Today.*, **2008**, 13, 124–134.
76. Gross, N.; Ranjbar, M.; Evers, C.; Hua, J.; Martin, G.; Schulze, B.; Michaelis, U.; Hansen, L. L.; Agostini, H. T. Choroidal neovascularization reduced by targeted drug delivery with cationic liposome-encapsulated paclitaxel or targeted photodynamic therapy with verteporfin encapsulated in cationic liposomes. *Mol. Vis.*, **2013**, 19, 54–61.
77. Paasonen, L.; Laaksonen, T.; Johans, C.; Yliperttula, M.; Kontturi, K.; Urtti, A. Gold nanoparticles enable selective light-induced contents release from liposomes. *J. Control Release.*, **2007**, 122, 86–93.
78. Paasonen, L.; Sipilä, T.; Subrizi, A.; Laurinmäki, P.; Butcher, S. J.; Rappolt, M.; Yaghmur, A.; Urtti, A.; Yliperttula, M. Gold-embedded photosensitive liposomes for drug delivery: triggering mechanism and intracellular release. *J. Control Release.*, **2010**, 147, 136–143.
79. Hossann, M.; Syunyaeva, Z.; Schmidt, R.; Zengerle, A.; Eibl, H.; Issels, R. D.; Lindner, L. H. Proteins and cholesterol lipid vesicles are mediators of drug release from thermosensitive liposomes. *J. Control Release.*, **2012**, 162, 400–406.
80. Koning, G. A.; Eggermont, A. M.; Lindner, L. H.; Hagen, T. L. ten Hyperthermia and thermosensitive liposomes for improved delivery of chemotherapeutic drugs to solid tumors. *Pharm. Res.*, **2010**, 27, 1750–1754.
81. Yatvin, M. B.; Weinstein, J. N.; Dennis, W. H.; Blumenthal, R. Design of liposomes for enhanced local release of drugs by hyperthermia. *Science.*, **1978**, 202, 1290–1293.
82. Chiu, G. N.; Abraham, S. A.; Ickenstein, L. M.; Ng, R.; Karlsson, G.; Edwards, K.; Wasan, E. K.; Bally, M. B. Encapsulation of doxorubicin into thermosensitive liposomes via complexation with the transition metal manganese. *J. Control Release.*, **2005**, 104, 271–288.
83. Al-Ahmady, Z. S.; Al-Jamal, W. T.; Bossche, J. V.; Bui, T. T.; Drake, A. F.; Mason, A. J.; Kostarelos, K. Lipid-peptide vesicle nanoscale hybrids for triggered drug release by mild hyperthermia in vitro and in vivo. *ACS Nano.*, **2012,** 6, 9335–9346.
84. Li, L.; Hagen, T. L. ten; Schipper, D.; Wijnberg, T. M.; Rhoon, G. C. van; Eggermont, A. M.; Lindner, L. H.; Koning, G. A. Triggered content release from optimized stealth thermosensitive liposomes using mild hyperthermia. *J. Control Release.*, **2010**, 143, 274–279.
85. Lindner, L. H.; Eichhorn, M. E.; Eibl, H.; Teichert, N.; Schmitt-Sody, M.; Issels, R. D.; Dellian, M. Novel temperature-sensitive liposomes with prolonged circulation time. *Clin. Cancer Res.*, **2004,** 10, 2168–2178.
86. Sen, K.; Mandal, M. Second generation liposomal cancer therapeutics: Transition from laboratory to clinic. *Int. J. Pharm.*, **2013**, 448, 28–43.
87. Dromi, S.; Frenkel, V.; Luk, A.; Traughber, B.; Angstadt, M.; Bur, M.; Poff, J.; Xie, J.; Libutti, S. K.; Li, K. C.; Wood, B. J. Pulsed-high intensity focused ultrasound and low temperature-sensitive liposomes for enhanced targeted drug delivery and antitumor effect. *Clin. Cancer Res.*, **2007**, 13, 2722–2727.
88. Morita, K.; Zywietz, F.; Kakinuma, K.; Tanaka, R.; Katoh, M. Efficacy of doxorubicin thermosensitive liposomes (40° C) and local hyperthermia on rat rhabdomyosarcoma. *Oncol. Rep.*, **2008**, 20, 365–372.

89. Poon, R. T.; Borys, N. Lyso-thermosensitive liposomal doxorubicin: a novel approach to enhance efficacy of thermal ablation of liver cancer. *Expert Opin Pharmacother.*, **2009,** 10, 333–343.
90. Ta, T.; Convertine, A. J.; Reyes, C. R.; Stayton, P. S.; Porter, T. M. Thermosensitive liposomes modified with poly(N-isopropylacrylamide-co-propylacrylic acid) copolymers for triggered release of doxorubicin. *Biomacromolecules.*, **2010**, 11, 1915–1920.
91. Tagami, T.; Ernsting, M. J.; Li, S. D. Optimization of a novel and improved thermosensitive liposome formulated with DPPC and a Brij surfactant using a robust in vitro system. *J. Control Release.*, **2011**, 154, 290–297.
92. Tagami, T.; May, J. P.; Ernsting, M. J.; Li, S. D. A thermosensitive liposome prepared with a Cu2+ gradient demonstrates improved pharmacokinetics, drug delivery and antitumor efficacy. *J. Control Release.*, **2012**, 161, 142–149.
93. Gong, W.; Wang, Z.; Liu, N.; Lin, W.; Wang, X.; Xu, D.; Liu, H.; Zeng, C.; Xie, X.; Mei, X.; Lu, W. Improving efficiency of adriamycin crossing blood brain barrier by combination of thermosensitive liposomes and hyperthermia. *Biol. Pharm. Bull.*, **2011**, 34, 1058–1064.
94. Zhang, H.; Wang, Z. Y.; Gong, W.; Li, Z. P.; Mei, X. G.; Lv, W. L. Development and characteristics of temperature-sensitive liposomes for vinorelbine bitartrate. *Int. J. Pharm.*, **2011**, 414, 56–62.
95. Al-Jamal, W. T.; Al-Ahmady, Z. S.; Kostarelos, K. Pharmacokinetics & tissue distribution of temperature-sensitive liposomal doxorubicin in tumor-bearing mice triggered with mild hyperthermia. *Biomaterials.*, **2012**, 33, 4608–4617.
96. Staruch, R. M.; Ganguly, M.; Tannock, I. F.; Hynynen, K.; Chopra, R. Enhanced drug delivery in rabbit VX2 tumours using thermosensitive liposomes and MRI-controlled focused ultrasound hyperthermia. *Int. J. Hyperthermia.*, **2012**, 28, 776–787.
97. Agarwal, A.; Mackey, M. A.; El-Sayed, M. A.; Bellamkonda, R. V. Remote triggered release of doxorubicin in tumors by synergistic application of thermosensitive liposomes and gold nanorods. *ACS Nano.*, **2011**, 5, 4919–4926.
98. Wu, Y.; Yang, Y.; Zhang, F. C.; Wu, C.; Lü, W. L.; Mei, X. G. Epirubicin-encapsulated long-circulating thermosensitive liposome improves pharmacokinetics and antitumor therapeutic efficacy in animals. *J. Liposome Res.*, **2011**, 21, 221–228.
99. Dicheva, B. M.; Hagen, T. L. M.; Li, L.; Schipper, D.; Seynhaeve, A. L.; Rhoon, G. C.; Eggermont, A. M.; Lindner, L. H.; Koning, G. A. Cationic thermosensitive liposomes: a novel dual targeted heat-triggered drug delivery approach for endothelial and tumor cells. *Nano Lett.*, **2012**, 13, 2324–2331.
100. Mulik, R.; Kulkarni, V.; Murthy, R. S. Chitosan-based thermosensitive hydrogel containing liposomes for sustained delivery of cytarabine. *Drug Dev. Ind. Pharm.*, **2009,** 35, 49–56.
101. Nie, S.; Hsiao, W. L.; Pan, W.; Yang, Z. Thermoreversible Pluronic F127-based hydrogel containing liposomes for the controlled delivery of paclitaxel: in vitro drug release, cell cytotoxicity, and uptake studies. *Int. J. Nanomedicine.*, **2011,** 6, 151–166.
102. Pradhan, P.; Jyotsnendu, G.; Samanta, G.; Sarma, H. D.; Mishra, K. P.; Bellare, J.; Banerjee, R.; Bahadur, D. Preparation and characterization mangenese ferrite based magnetic liposome for hyperthermia treatment of cancer. *J. Biomed. Mater. Res. B.*, **2007**, 81, 12–22.

103. Pradhan, P.; Giri, J.; Rieken, F.; Koch, C.; Mykhaylyk, O.; Döblinger, M.; Banerjee, R.; Bahadur, D.; Plank, C. Targeted temperature sensitive magnetic liposomes for thermo-chemotherapy. *J. Control Release.,* **2010**, 142, 108–121.
104. Smith, B.; Lyakhov, I.; Loomis, K.; Needle, D.; Baxa, U.; Yavlovich, A.; Capala, J.; Blumenthal, R.; Puri, A. Hyperthermia-triggered intracellular delivery of anticancer agent to HER2(+) cells by HER2-specific affibody (ZHER2-GS-Cys)-conjugated thermosensitive liposomes (HER2(+) affisomes). *J. Control Release.,* **2011**, 153, 187–194.
105. Jenssen, H.; Hamill, P.; Hancock, R.E.W. Peptide antimicrobial agents. *Clin. Microbiol. Rev.*, **2006**, 19, 491–511.
106. Miller, D. K.; Griffiths, E.; Lenard, J.; Firestone, R A. Cell killing by lysosomotropic detergents. *J. Cell Biol.* **1983**, 97, 1841–1851.
107. Schroeder, A.; Honen, R.; Turjeman, K.; Gabizon, A.; Kost, J.; Barenholz, Y. Ultrasound triggered release of cisplatin from liposomes in murine tumors. *J. Control Release.,* **2009**, 137, 63–68.
108. Yan, F.; Li, L.; Deng, Z.; Jin, Q.; Chen, J.; Yang, W.; Yeh, C.K.; Wu, J.; Shandas, R.; Liu, X.; Zheng, H. Paclitaxel-liposome-microbubble complexes as ultrasound-triggered therapeutic drug delivery carriers. *J. Control Release.,* **2013**, 166, 246–255.
109. Geers, B.; Lentacker, I.; Sanders, N. N.; Demeester, J.; Meairs, S.; De Smedt, S. C. Selfassembled liposome-loaded microbubbles: The missing link for safe and efficient ultrasound triggered drug-delivery. *J. Control Release.,* **2011**, 152, 249–256.
110. Lentacker, I.; Wang, N.; Vandenbroucke, R. E.; Demeester, J.; De Smedt, S. C.; Sanders, N.N. Ultrasound exposure of lipoplex loaded microbubbles facilitates direct cytoplasmic entry of the lipoplexes. *Mol. Pharm.,* **2009**, 6, 457–467.
111. Yudina, A.; Smet, M. de; Lepetit-Coiffé, M.; Langereis, S.; Ruijssevelt, L. Van.; Smirnov, P.; Bouchaud, V.; Voisin, P.; Grüll, H.; Moonen, C. T. Ultrasound-mediated intracellular drug delivery using microbubbles and temperature-sensitive liposomes. *J. Control Release.,* **2011**, 155, 442–448.

CHAPTER 8

P-gp INHIBITORS: A POTENTIAL TOOL TO OVERCOME DRUG RESISTANCE IN CANCER CHEMOTHERAPY

ANKIT JAIN and SANJAY K. JAIN

ABSTRACT

P-glycoprotein (P-gp; also known as MDR1 and ABCB1) being the first identified ABC (ATP [adenosine triphosphate]-binding cassette) transporter utilizes the energy of ATP hydrolysis to transport a broad range of substrates, including anticancer agents, across biological membranes against concentration gradients. It has the ability to confer multidrug resistance (MDR) in various types of cancer cells. Its expression is often associated with adverse prognosis of patients and is supposed to be a major obstacle limiting the therapeutic efficacy of cancer chemotherapy. Hence, intensive efforts have been made to search for specific and high-affinity inhibitors that could antagonize their action and overcome MDR in the clinic. Development of the first-generation (G_1) P-gp inhibitors (e.g., verapamil, cyclosporine A etc.), second-generation (G_2) P-gp inhibitors (e.g., Valspodar and Biricodar etc.), and third-generation (G_3) P-gp inhibitors (e.g., Tariquidar and CBT-1 etc.) have been limelighted as cancer chemotherapeutics at clinic outset. This chapter provides a better understanding and advances in the development of P-gp inhibitors to reverse MDR. The problems highlighted by off-site effects, the inhibition of the normal function of ABCB1 transporters in healthy tissue, and the need for combinatorial therapies, have prompted the development of nanotechnology-based targeting strategies.

8.1 INTRODUCTION

ABCB1 was the first human ABC transporter identified. It was cloned in 1986 and is the best characterized in multidrug resistance (MDR) (Chen et al., 1986). P-glycoproteins (P-gp; also known as MDR1 and ABCB1), are integral membrane proteins, which are a type of the ATP (adenosine triphosphate)-binding cassette (ABC) transporters, have the ability to confer MDR in various types of cancer cells similar to multidrug resistance-associated protein 1 (MRP1; also known as ABCC1), and breast cancer resistance protein (BCRP) (Gillet & Gottesman 2011). They utilize the energy of ATP hydrolysis to transport neutral and cationic hydrophobic compounds, such as the natural products vinblastine, vincristine, doxorubicin, daunorubicin, actinomycin D, etoposide, and paclitaxel, across biological membranes against concentration gradients. Being the first identified ABC transporter, P-gp is the product of the human ABCB1 gene, localized to chromosome 7q21. It is a membrane-bound glycoprotein consisting of 1,280 amino acids and is composed of two homologous half-transporters with each one containing transmembrane domains, that is, TMD (or transmembrane helicles, i.e., TMH) and nucleotide bind domain, that is, nucleotide-binding domain (NBD). The two TMDs and two NBDs of P-gp are arranged in the sequence of TMD1-NBD1-TMD2-NBD2 with a linker region connecting the two half-transporters. The linker region plays a critical role in ensuring proper interaction of the two subunits and it provides communication between the two ATP sites of NBD (Sarkadi et al., 2006).

Many cytotoxic anticancer drugs are transported by P-gp, which was first identified because it was overexpressed in cell lines made resistant to such cytotoxic drugs. Owing to the broad substrate specificity of P-gp, the cells displayed cross-resistance to many different cytotoxic drugs, thus named as MDR. There are few common structural denominators for transported P-gp substrates. They are usually organic molecules ranging in size from less than 200 Da to almost 1,900 Da. Many contain aromatic groups, but nonaromatic linear or circular molecules are also transported. Most of the efficiently transported molecules are uncharged or (weakly) basic in nature, but some acidic compounds (e.g., methotrexate, phenytoin) can also be transported, albeit at a low rate. The only common denominator identified so far in all P-gp substrates is their amphipathic nature. This may have to do with the mechanism of drug translocation by P-gp: it has been postulated that intracellular P-gp substrates first have to insert into the inner hemileaflet of the cell membrane, before being "flipped" to the outer hemileaflet, or perhaps being extruded directly into the extracellular medium by P-gp (Higgins & Gottesman, 1992). Only amphipathic molecules would have the proper membrane insertion properties. Both MDR1 P-gp and the highly related MDR3 P-gp can transport intrinsic (amphipathic) membrane components such as phosphatidylcholine and analogs thereof, which support the notion that substrates are taken from the inner hemileaflet. As most P-gp substrates are quite hydrophobic, in principle they can diffuse passively across biological membranes at a reasonable rate. For cell-biological and pharmacological studies, this means that in the absence of active transport, P-gp substrates will cross membranes and in vivo penetrate into tissues and pharmacological compartments. It also means that a contribution of active transport by P-gp will only result in noticeable distribution effects if the rate of active transport for a certain compound is substantial relative to the passive diffusion rate. If not, the pump activity will be overwhelmed by the passive diffusion component (Smith et al., 1994; van Helvoort, 1996).

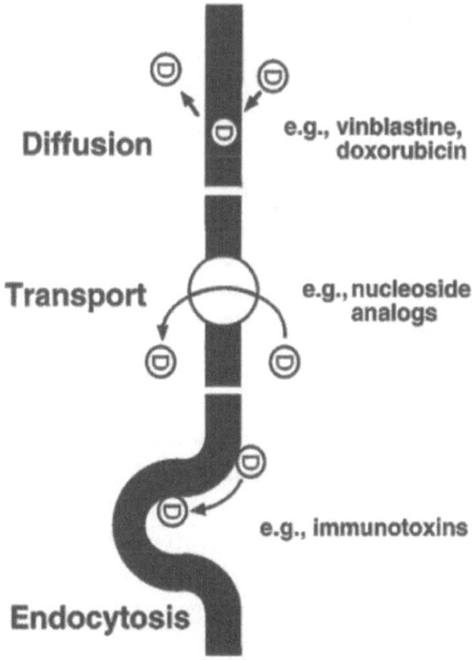

FIGURE 8.1 Ways for drugs to get into cells.

Three major routes (Figure 8.1) for drugs to get into cells are listed hereunder.

- Diffusion across the plasma membrane,
- Piggy-backing onto a receptor or transporter, and
- Endocytosis.

P-gp now becomes the primary focus of biomedical research to overcome cancer drug resistance. One option to overcome MDR is to identify or develop an efficacious drug that is not a substrate for the transporter but a minority of drugs currently in use falls under this category, such as alkylating agents, antimetabolites, and anthracycline derivatives. The more suitable approach is to develop an effective ABC inhibitor, modulator, or MDR-reversal agent. Ideally, this should be a nontoxic compound, potent and specificity, and efficacious so that it will not adversely affect the pharmacokinetics of the therapeutic drug that would be coadministered to kill the cancer cells (Darby et al., 2011).

8.2 P-gp AND ITS GENETIC EXPRESSION

P-gp expression is regulated by various factors. Mutation of the p53 gene and overexpression of the p63 gene and/or the p73 gene in certain tumors may facilitate P-gp expression. ABCB1 promoter activation by a nuclear protein, MDR1 promoter-enhancing factor, or a component of the multiprotein complex, RNA helicase A, may upregulate P-gp expression. Epigenetic methylation, in contrast, results in silence of ABCB1. Even after three decades of P-gp research, there is still no clear therapeutic strategy to overcome the classical drug efflux pump in tumors. Recently, it was found that the expression of hyaluronan (HA) and its receptor, CD44, are tightly linked to MDR and tumor progression. In breast and ovarian cancer cell lines, HA-CD44 interaction may activate the stem cell marker. HA-CD44 binding forms a complex with ankyrin, the downstream effector of CD44 that results in an efflux of chemotherapeutic drugs so anti-CD44 antibody not only blocks HA-CD44 binding but also inhibits ABCB1-mediated efflux pump thereby enhancing chemosensitivity (Bourguignon et al., 2008). Numerous genetic polymorphisms have been identified in ABCB1 gene, the majority of which are single nucleotide polymorphisms (SNPs) (Pauli-Magnus & Kroetz, 2004). Stein et al. (1996) showed the first evidence for genetic variants in ABCB1: a +103T>C variant in the ABCB1 promoter region which was identified in human osteosarcoma cells. After treatment with ABCB1 substrate drugs, the +103T>C leads to increased transcritional activity in an in vitro reporter assay compared with the wild type controls. Another SNP (+8T>C) was found in the ABCB1 promoter from hematological malignancies, but the effect on gene function was unknown because it was similarly detected in normal controls (Rund et al., 1999). Mfold (a software for RNA secondary structure prediction) analysis on the ABCB1 mRNA sequence surrounding the three SNPs revealed that all three SNPs changed singlestrandedness count (ss-count) compared to the reference sequence, whereas changes in ss-count occurred more globally across the entire mRNA with 3435C>T than 1236C>T and 2677G>T. These results support the notion that the SNP, even the synonymous polymorphism, may affect the secondary structure and turnover of ABCB1 mRNA (Hoffmeyer et al., 2000). Further, 1236C>T and 2677G>T, being in strong linkage disequilibrium with 3435C>T, were used as additional marker SNPs and revealed similar results as 3435C>T, suggesting that the differences did not result from analytical artifacts. Cancer cells grown under selection pressure may quire some mutations in the ABCB1 gene, resulting in changes in the pattern of substrate specificity. Colchicine-selected cells exhibited Gly185Val substitution, resulting in increased resistance to colchicine but not to other drugs. A multidrug-resistant sarcoma cell line isolated by coselection with doxorubicin and PSC-833 (ABCB1 inhibitor) had a deletion of a phenylalanine at amino acid residue 335, resulting in increased resistance to doxorubicin and paclitaxel, but decreased resistance to vinca alkaloids, and loss of resistance to dactinomycin (Kioka et al., 1989). These results suggest that these sites may be important for finding of the ABCB1 substrates and inhibitors. For the majority of reported ABCB1 polymorphisms, the associations with the gene expression and function are not always reproducible (Bonhomme-Faivre et al., 2004; Lepper et al., 2005; Puisset et al., 2004). However, a majority of clinical studies support the notion that ABCB1 polymorphisms can affect clinical pharmacokinetics and response to cancer therapy (Wu et al., 2006). For example, ABCB1 haplotypes including 236C>T, 2677G>T and 3435C>T reduced renal clearance irinotecan and its metabolites (Huang, 2007).

8.3 STRUCTURE OF P-GP AND PHYSIOLOGICAL ROLES

P-gp is a single polypeptide made up of two homologous halves. Each half comprises six transmembrane helices (TMHs) and one NBD located on the cytoplasmic side of the membrane, which are connected by a cytoplasmic linker (Figure 8.2).

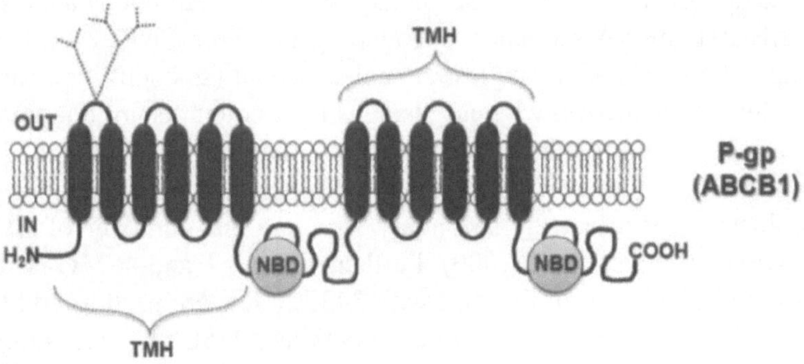

FIGURE 8.2 Structure of P-gp

P-gp appears to bind multiple drugs by having a large highly flexible binding cavity which can accommodate many compounds in different locations by an "induced fit" type of mechanism. Biochemical cross-linking and fluorescence studies had already pointed to a substrate-binding region with these properties (Loo & Clarke, 2005). P-gp is expressed abundantly at the apical surface of epithelial cells of the large and small intestines, liver bile ductules, and kidney proximal tubules in most humans but it is also found in the adrenal gland, the placenta, and blood–brain barrier (BBB) which plays a rate-limiting role for entry of drug to the brain (Schinkel et al., 1999; Thiebaut et al., 1987) (Figure 8.3). Similarly, it protects the fetus from the toxic endogenous and exogenous molecules. P-gp is found to affect the disposition of number of clinically administered anticancer drugs and thereby manifests pharmacokinetic changes. In the intestine, it extrudes out drugs into the lumen, thus reducing their absorption and oral bioavailability.

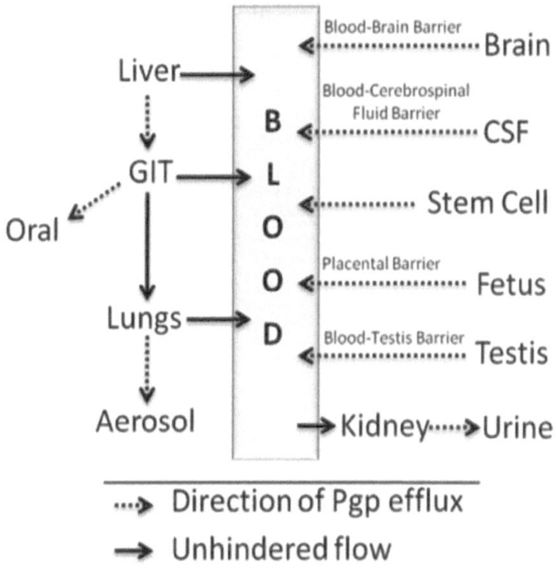

FIGURE 8.3 Rate-limiting role of P-gp at various sites of body

P-gp actively transports the drug back out of the cell and acts as a "flippase" to eject the drug from the membrane (Homolya et al., 1993). Figure 8.4 shows the vaccum cleaner and flippase modes of P-gp. The polyspecific nature of the P-gp-binding pocket and its ability to bind more than one drug molecule simultaneously makes the rational design of specific high-affinity inhibitors a challenging problem.

P-gp expression is known to be upregulated in the MDR phenotype in cancer cells and as such is a key target for improving the efficacy of chemotherapy (Ferreira et al., 2005). P-gp inhibitors, such as guggulsterone and CJX1, are capable of resensitising cells to drugs (Xu et al., 2009). They are known to achieve this through inhibiting P-gp drug expulsion mechanisms in addition to preventing expression of the channel protein in the lipid bilayer. The P-gp theory looks adequate on its own but there is one point that needs further investigation. Why would drugs interact with a transporter like P-gp and be expelled efficiently? Remind that by definition MDR violates the law of enzyme specificity. The question is to envision a mechanism whereby chemical efficiency without the need for chemical affinity occurs. Recent developments in pharmacology, such as the introduction of HTS technology and "screen-friendly" synthetic chemical libraries, combined with improved understanding of substrate–protein interactions should enable rational planning and de novo synthesis of novel Pgp modulators (Pleban & Ecker, 2005; Seelig & Gatlik-Landwojtowicz, 2005). These strategies to engage, evade, or even exploit efflux-based resistance mechanisms are illustrated in Figure 8.5.

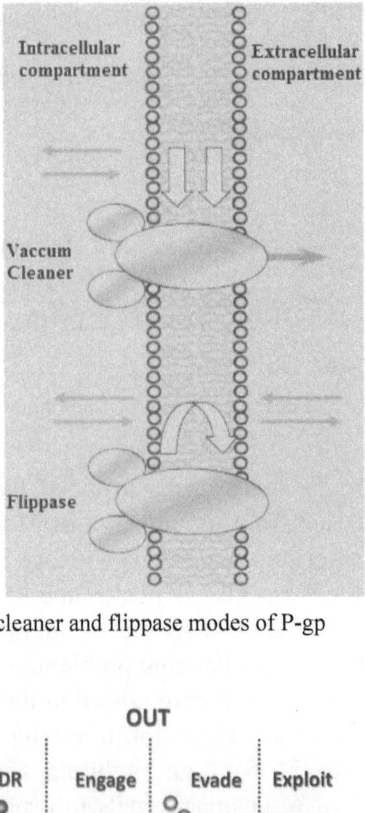

FIGURE 8.4 Vaccum cleaner and flippase modes of P-gp

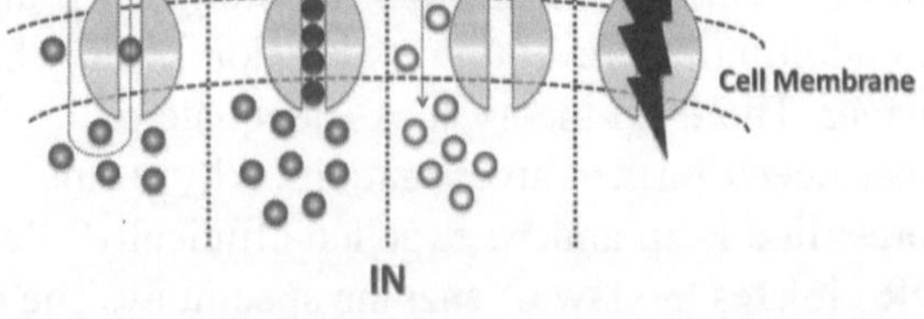

FIGURE 8.5 Know-how of targeting multidrug-resistant cancer. P-glycoprotein (P-gp) actively extrudes many types of drugs from cancer cells, keeping their intracellular levels below a cell-killing threshold.

Strategies that circumvent P-gp-mediated multidrug resistance (MDR) include the following:

- **Engage** – The coadministration of pump inhibitors and cytotoxic agents.
- **Evade** – The use of cytotoxic agents that bypass P-gp-mediated efflux.
- **Exploit** – Taking advantage of the collateral sensitivity of MDR cells.

The studious efforts of cancer researchers to understand MDR in cancer have resulted in the identification of a limited number of distinct, clinically proven mechanisms. Overexpression of ABC transporters, particularly P-gp, has consistently been implicated as a cause for MDR both in vitro and in vivo. Recent strategies to engage, evade, or exploit this transporter to improve cancer treatment reflect both the creativity and hopefulness of cancer researchers that at least this cause of MDR can be vanquished.

8.4 DRUG PUMPING BY P-GP AND RESISTANCE TO CHEMOTHERAPY

8.4.1 THE P-GP CATALYTIC AND TRANSPORT CYCLE

In the symmetric state, both halves of the dimer interface are open, with each binding an ATP molecule loosely (ATP_l). The catalytically active pump rapidly progresses to the asymmetric state in which one ATP molecule is tightly bound (ATP_t) in one NBD, where the dimer interface is closed. The tightly bound ATP molecule is committed to enter the transition state and undergoes hydrolysis, which provides the energy for movement of drug from the binding pocket within the transporter to the extracellular space. Drug transport involves a conformational change from an inward-facing to an outward-facing protein conformation. After ATP hydrolysis, ADP is loosely bound (ADP_l), resulting in opening of the dimer interface. The other catalytic site simultaneously switches to the high-affinity state, resulting in tight binding of the second ATP molecule and closure of the dimer interface in the other NBD. P_i then dissociates from the catalytic site of the first NBD, followed by ADP, which is replaced by another molecule of loosely bound ATP (nucleotide exchange) to achieve the asymmetric state again. A second round of ATP hydrolysis and drug transport then takes place at the other NBD (Siarheyeva et al., 2010).

P-gp exists in an asymmetric state at all stages of the catalytic cycle, thus requiring that the two NBDs alternate in hydrolyzing ATP. P-gp alters not only their absorption but also tissue distribution (Zhou, 2008). Interaction of a drug with P-gp can cause poor uptake in the intestine, thus reducing oral bioavailability, and prevent delivery of drugs to the brain, which is a serious problem in the treatment of brain diseases. The coadministration of two drugs having asymmetric ATP site affinity switches the state (Frances, 2011).

8.4.2 ATP HYDROLYSIS AND DRUG TRANSPORT

ATP site affinity switch asymmetric states that are both P-gp substrates can also lead to major pharmacokinetic effects as they compete for the transporter. Plasma drug level stays higher for longer, and a reduction in drug dose is often necessary to avoid toxicity. Similar effects are observed when a P-gp substrate drug is consumed with foods or herbal supplements containing natural products that are also P-gp substrates (e.g., plant fiavonoids, St John's wort) (Borrelli & Izzo, 2009; Staud et al., 2010). For these reasons, drug testing for P-gp interactions is now recommended by the United States Food and Drug Administration (US FDA) as part of its approval process. Such tests are usually carried out using polarized monolayers of epithelial cell lines either transfected with human P-gp, or expressing P-gp naturally. Movement of the test compound from the medium on the basolateral side of the monolayer to the medium on the apical side is compared with movement in the opposite direction. This is a time-consuming and expensive assay, and there is currently a pressing need for new rapid high-throughput methods (Giacomini et al., 2010). Over 50 polymorphisms (single nucleotide polymorphisms and insert ions/deletions) in the ABCBI gene are known, and some of them appear to change the mRNA expression, protein expression, and function of P-gp (Choudhuri & Klaassen, 2006). Polymorphisms may be responsible for the variation in drug responses observed between different individuals and populations, and, in recent years, there has been considerable interest in how they might affect the outcome of drug therapy in people carrying them. However, there have been many conflicting reports in this field, and no clear associations between genotype and altered response to drug treatment have emerged. Similarly, polymorphisms have been reported to alter the susceptibility of certain individuals to various disease states, including colon cancer, renal cancer, inflammatory bowel disease, and Parkinson's disease (Sharom, 2008). This is a rapidly developing field, and will require substantial further investigation before firm conclusions can be reached.

8.5 P-gp SUBSTRATES AND MODULATORS

P-gp can interact with a vast array of compounds that are structurally unrelated, including natural products, chemotherapeutic drugs, steroids, linear and cyclic peptides, fluorescent dyes, and ionophores. Most are weakly amphipathic and relatively hydrophobic, often containing aromatic rings and a positively charged N atom. However, direct measurement of transport has been carried out for only a few of these putative substrates, and most have been identified on the basis of resistance of P-gp-overexpressing cell lines to their cytotoxic effects. Table 8.1 lists some of the different classes of P-gp substrates, and Figure 8.6 shows representative structures of P-gp modulators and substrates.

TABLE 8.1 Anticancer drugs that interact with P-gp substrates

Category	Examples
Anthracyclines	Doxorubicin, daunorubicin
Camptorhecins	Topotecan
Epipodophyllotoxins	Etoposide, teniposide
Taxanes	Paclitaxel, docetaxel
Vinca alkaloids	Vinbiastine, vincristine

A second group of compounds can also interact with P-gp, known as modulators (also called inhibitors or chemosensitizers). Modulators are able to reverse MDR in intact cells by interfering with the ability of P-gp to efflux drugs (Robert & Jarry, 2003). Like substrates, P-gp modulators are structurally diverse, and many have been identified over the years. Modulators may inhibit P-gp transport function in a number of ways, and their mode of action at the molecular level is not well understood. Some are transported themselves (e.g., cyclosporin A), and thus act as competing substrates, whereas others (e.g., T.Y335979) bind tightly to the protein's drug-binding pocket for prolonged periods of time without being transported (Dantzig et al., 2001; Saeki et al., 1993). A few modulators appear to interact with the NBDs (e.g., steroids or disulfiram) and interfere with ATP hydrolysis. Several membrane P-gp substrates, P-gp modulators vinblastine GF120918 fluidizers, and surfactants (e.g., benzyl alcohol, Nonidet P40, or Cremaphor El.) do not appear to directly interact with P-gp, and probably act as nonspecific modulators at the level of the membrane bilayer (Loo & Clarke, 2000). P-gp modulators may be valuable for their ability to increase delivery of therapeutic drugs to the brain, which is desirable in the treatment of diseases such as cancer, acquired immunodeficiency syndrome (AIDS), Alzheimer's disease, Parkinson's disease, schizophrenia, and epilepsy (Breedveld et al., 2006).

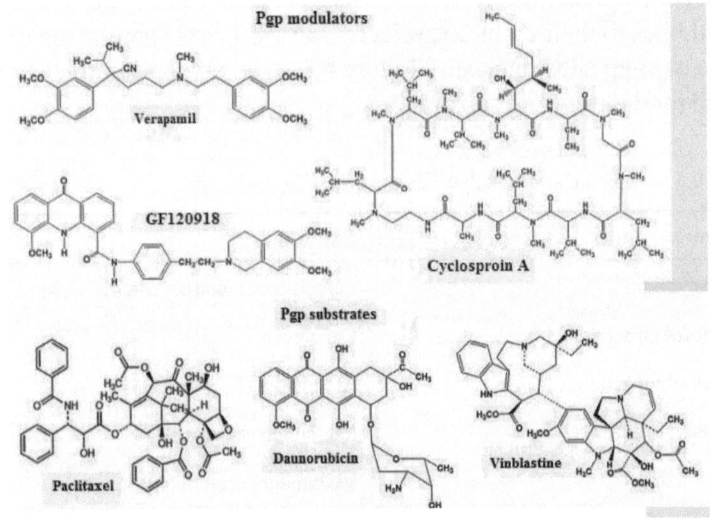

FIGURE 8.6 Representative structures of P-gp modulators and substrates

8.6 P-gp IN DRUG DISCOVERY AND RESISTANCE TO CANCER CHEMOTHERAPY

The presence of P-gp in the intestinal epithelium is a serious problem in drug discovery, since new drug candidates may be poorly absorbed, making them ineffective clinically. It is especially important to screen out P-gp substrates when developing drugs targeted to the brain, since their efficacy depends on their ability to cross the BBB. Many pharmaceutical companies now routinely include testing for interactions with P-gp as part of their drug discovery process. Methods that are employed range from measurements of drug uptake in intact cells to in vitro assays that assess the effect of a compound on P-gp ATPase or transport activity in membrane vesicles (Sharom & Siarheyeva, 2008). Predictive methods have been of limited value in determining whether a compound is likely to be a substrate for P-gp, although this may change as we gain a better understanding of its drug-binding pocket.

Resistance of tumors to the cytotoxic action of chemotherapy drugs is the single most common cause of cancer treatment failure. Some cancers, including tumors of the kidney, liver, breast, and ovary, are drug resistant at the outset (intrinsic resistance), whereas others, typically leukemia, lymphomas, and multiple myeloma, develop resistance after one or more rounds of chemotherapy treatment (acquired resistance). For several cancers, including acute myelogenous leukemia (AML), acute lymphoblastic leukemia (ALL), and ovarian tumors, high P-gp expression levels are strongly linked to a weak response to chemotherapy treatment and poor overall prognosis (Steinbach & Legrand, 2007). However, in other cases, it has proved difficult to link MDR in cancer to P-gp expression, probably because there are multiple mechanisms by which some tumors can develop drug resistance. There has been much interest in combining modulators with chemotherapy drugs to improve the outcome of cancer treatment (Szakacs et al., 2006). However, only a few of the hundreds of compounds identified as modulators in vitro proved suitable for use in clinical trials. G_1 modulators already in use for other medical conditions (e.g., verapamil) were tested in the 1980s and generally proved to be highly toxic to patients. They were modified to produce G_2 compounds, which often showed adverse pharmacokinetic interactions in which decreased clearance of anticancer agents led to toxicity. In recent years, more rationally designed G_3 modulators of low toxicity have been produced. These are highly selective and potent inhibitors of P-gp transport function. One notable success has been the combined use of modulators and chemotherapy to cure childhood retinoblastoma (Coley, 2010). However, the results of clinical trials using P-gp modulators in cancer treatment have generally been disappointing (Polgar & Bates, 2005), possibly because these studies suffered from serious experimental limitations such as poor patient selection criteria. There is current interest in exploring the use of natural components of foods plants as P-gp modulators (e.g., flavonoids, curcuminoids etc.), and they could serve as useful leads for the development of G_4 agents (Coley, 2010).

8.7 P-gp INHIBITORS IN CANCER CHEMOTHERAPY

It was soon recognized that transport of cytotoxic and other substrates for P-gp can also be inhibited by certain compounds (Tsuruo et al., 1981). These so-called "reversal agents" or "P-gp blockers" are discussed in greater depth elsewhere in this chapter, but we present some general considerations here. Many of the initially identified inhibitors, like the calcium channel blocker verapamil or the immunosuppressive agent cyclosporin A, turned out to be themselves transported substrates of P-gp, suggesting that they act as competitive inhibitors. For other inhibitors no significant transport by P-gp could be demonstrated, indicating that they probably work through other mechanisms. Initial thoughts on clinical application of P-gp inhibitors were focused on reversing MDR in chemotherapy-resistant tumor cells that contain significant amounts of P-gp, but later insights indicated that such inhibitors might also be useful to modulate the general pharmacological behavior of drugs in the body. The P-gp inhibitors that were initially recognized, such as verapamil, are actually relatively poor P-gp inhibitors in vivo, and they frequently have their own pharmacodynamic effects that put severe restrictions on the plasma levels that can be safely achieved in patients. We therefore briefly discuss so-called G_2 or G_3 P-gp inhibitors, that were selected specifically for their high P-gp-inhibiting capacity, and for lack of other pharmacodynamic effects. Major ABCB1 (MDR1) chemotherapy substrates and their inhibitors are given in Table 8.2.

TABLE 8.2 Major ABCB1 (MDR1) chemotherapy substrates and their inhibitors

Substrates	Inhibitors
Adriamycin	Anthranilamide
Actinomycin-D	Cyclosporine D
Bisantrene	NSC-38721 (mitotane)
Daunorubicin	Pipecolinate
Docetaxel	Quinoline

TABLE 8.2 *(Continued)*

Doxorubicin	OC-144-093
Etoposide	PSC-833 (valspodar)
Epirubicin	MS-209
Homoharringtonine	LY-335979 (zosoquidar)
Mitoxantrone	XR-9576 (tariquidar)
Paclitaxel	R-101933 (laniquidar)
Teniposide	VX-710 (biricodar)
Topotecan	GF-120918 (elacridar)
Vinblastine	ONT-093
Vincristine	Isothiocyanates
Vinorelbine	Diallyl sulfide
VP-16	PK11195
Miscellaneous inhibitors: Amooranin, siRNA, tRA 98006, Agosterol A, flavonoids etc.	

SDZ PSC 833 (or PSC 833) is a cyclosporin A analog that does not have the immunosuppressive effect of cyclosporin A, and can be given at quite high dosages to patients (Boesch et al., 1991). PSC 833 is a high affinity, but slowly transported substrate for MDR1 P-gp, which probably acts as an effective inhibitor because its release from P-gp is very slow (Smith et al., 1998). Although PSC 833 is an efficient P-gp inhibitor, it does have the complication that it is also an inhibitor of cytochrome P450 3A4 (CYP3A4), one of the main drug-metabolizing enzymes in the body (Fischer et al., 1998). Consequently, when administered to patients, next to inhibiting P-gp, it may have additional effects on the clearance of drug substrates that are degraded by CYP3A4. Many cytotoxic anticancer drugs that are P-gp substrates, such as etoposide and doxorubicin, are also extensively degraded by CYP3A4. Therefore, coadministration with PSC 833 can intensify the toxic side effects of these anticancer drugs, necessitating a dose reduction for safe treatment of the patient (Sikic, 1999). The main dose-limiting toxicity of PSC 833 itself in patients is ataxia. PSC 833 is currently tested in a number of Phase III clinical trials for reversal of drug resistance in tumors. GF120918 (or GG918) is a highly effective P-gp inhibitor and it can be given at very high oral dosages to mice and patients without obvious toxic effects. In mice, GF120918-treatment improves the response of implanted P-gp-containing tumors to chemotherapy. Currently running clinical trials with GF120918 are amongst others testing inhibition of P-gp in the intestine. LY335979 is another specifically developed highly effective P-gp inhibitor (Dantzig et al., 1996). Like GF120918, it improves chemotherapy response in mice with transplanted P-gp-containing tumors, and it does not clearly affect the plasma clearance of intraperitoneally administered doxorubicin or etoposide. However, this was only tested with intravenously or intraperitoneally administered LY335979. In line with the absence of pronounced plasma pharmacokinetic interactions, LY335979 has much lower affinity for CYP3A than for P-gp (Dantzig et al., 2001) LY335979 is currently being tested in clinical trials. Two other compounds, XR9576 and OC144-093 appear to be very promising P-gp inhibitors as well (Mistry et al, 2001). Both have high affinity for P-gp, they can be given both orally and intravenously to improve the chemotherapy response of transplanted P-gp-containing tumors in mice, and they do not affect the plasma pharmacokinetics of intravenously administered paclitaxel. The latter point again suggests that XR9576 and OC144-093 do not have extensive interaction with paclitaxel-metabolizing enzymes (Schinkel & Jonker, 2012).

8.7.1 FIRST-GENERATION (G_1) P-GP INHIBITORS

The G_1 inhibitors include the use of drugs developed for other conditions such as verapamil (an antihypertensive), quinine (an antimalarial), and cyclosporine A (an immunosuppressant). Despite their efficacy in inhibiting ABCB1-dependent drug efflux in vitro (Tsuruo et al., 1981), these inhibitors failed to provide a positive effect in the clinical setting and were prone to pharmacokinetic complications (and also cardiac toxicity in the case of verapamil). The inhibitors lacked specificity and potency and as a result they were characterized by high toxicity (Ozols et al., 1987). Therefore, it is not surprising that clinical trials using these G_1 MDR inhibitors mostly failed, largely because of undesired side effects (Tiwari et al., 2011).

8.7.2 SECOND-GENERATION (G_2) P-GP INHIBITORS

Since the major reason for the failure of the G_1 MDR-reversal agents resided in their low specificity, some of the G_2 ABCB1 inhibitors were designed to specifically target this transporter. Valspodar (formerly PSC833), for example, is a derivative of cyclosporine A. In vitro, it is characterized by higher specificity (it lacks the immunosuppressive effects of its forerunner) and higher potency than cyclosporine A and other G_1 inhibitors. However, it also failed to improve outcome in Phase III clinical trials when coadministered with the anticancer drugs vincristine and doxorubicin, or daunorubicin and etoposide (Baer et al., 2002; Friedenberg et al., 2006) in patients with AML. Indeed, patients on the combinatorial treatment that included Valspodar had a poorer outcome. Despite the promising preclinical evidence, it became apparent that Valspodar also inhibited CYP450s, with the ensuing pharmacokinetic impact of elevated systemic concentrations of inhibitor and therapeutic drug leading to toxicity. The piperidine derivative Biricodar (VX-710) was also identified at this time as a more potent inhibitor of ABCB1 than the G_1 compounds. Biricodar was considered a promising drug for recalcitrant breast, ovarian, small-cell lung, and prostate carcinomas. However, in Phase II clinical trials with doxorubicin, vincristine or paclitaxel, no patient benefit was evident, and the significant complication of neutropenia was observed alongside several more moderate side effects (nausea, asthenia, paresthesia, and headache) (Gandhi et al., 2007).

8.7.3 THIRD-GENERATION (G_3) P-GP INHIBITORS

The G_3 inhibitors were specifically designed to overcome the limitations of the G_2, and so inhibitor development focused on compounds that lacked inhibition of the CYP450s and did not alter the pharmacokinetics of the anticancer drugs. In addition, inhibitors were rationally designed based on quantitative structure–activity relationships (QSAR) to target ABCB1 specifically. Currently, Tariquidar (an anthranilamide, XR9576), Elacridar (an acridone caroxamide), Zosuquidar (LY335979), and CBT-1 (both quinolone derivatives) and Laniquidar (a piperidine) are at various stages of clinical trial. They are characterized by higher potency and lower toxicity than their predecessors. After promising early reports (Abraham et al.,

2009), both Tariquidar and Zosuquidar have entered Phase II/III trials in combination with vinorelbine and doxorubicin, respectively, for a variety of advanced malignancies or AML. Zosuquidar, perhaps the most specific of the G_3 inhibitors of ABCB1, has also run into difficulty with neurotoxicity and drug–drug interactions reported for doxorubicin and vinorelbine (Sandler et al., 2004). Clearly, therefore, there remain difficulties with this strategy that may or may not be overcome. Indeed it could be argued that inhibition of ABCB1 in the target cancer cell is also likely to inhibit the transporter on the canalicular membrane and inevitably result in interaction with the therapeutic drug, as this is likely to be the clearance route for an ABCB1 transport-substrate. Careful stratification of patients to understand the mechanism of MDR in order to tailor a personalized therapy, and careful dosing to minimize side effects associated with the likelihood of raised inhibitor and therapeutic drug concentrations will surely help, and studies are ongoing with these inhibitors (Kelly et al., 2011).

8.8 ALTERNATIVE APPROACHES TO TARGETING MDR

8.8.1 TARGETED DOWNREGULATION OF MDR GENES

Selective downregulation of resistance genes in cancer cells is an emerging approach in therapeutics. Although in cell lines MDR is often a result of the amplification of the MDR1 gene, the overexpression of the protein has transcriptional components as well. Regulation of P-gp expression is amazingly complex, and could include different mechanisms in normal tissues compared with cancer cells (Scotto, 2003). If mechanisms governing expression of P-gp in malignant cells were mediated through tumor-specific pathways, cancer-specific approaches to circumvent P-gp overexpression could be developed with minimal effect on constitutive expression of normal cells (Kang et al., 2004). Using peptide combinatorial libraries, Bartsevich et al. designed transcriptional repressors that selectively bind to the MDR1 promoter. Expression of the repressor peptides in highly drug-resistant cancer cells resulted in a selective reduction of P-gp levels and a marked increase in chemosensitivity (Bartsevich & Juliano, 2000; Xu et al., 2004). Similarly, antagonists of the nuclear steroid and xenobiotic receptor (SXR), which coordinately regulate drug metabolism and efflux, can be used in conjunction with anticancer drugs to prevent the induction of P-gp (Synold et al., 2001). Using technologies that enable the targeted regulation of genes—antisense oligonucleotides, hammerhead ribozymes, and short-interfering RNA (siRNA)—has produced mixed results. Sufficient downregulation of Pgp has proved difficult to attain and the safe delivery of constructs to cancer cells in vivo remains a challenge (Pichler et al., 2005). However, transcriptional repression is a promising new strategy that is not only highly specific but also enables the prevention of P-gp expression during the progression of disease.

8.8.2 NOVEL ANTICANCER AGENTS DESIGNED TO EVADE EFFLUX

Several novel anticancer drugs are exported by ABC transporters, including irinotecan (also its metabolite SN-38), depsipeptide, imatinib (Gleevec; Novartis), flavopiridol etc. Moreover, the NCI60 screen suggests that a significant portion of the compounds in the drug development pipeline are substrates of ABC transporters. Epothilones are novel microtubule-targeting agents with a paclitaxel-like mechanism of action that are not recognized by P-gp, providing proof of the concept that new classes of anticancer agents that do not interact with the multidrug transporters can be developed to improve response to therapy. As most anticancer agents subject to efflux are currently irreplaceable in chemotherapy regimens, an attractive solution would be to chemically modify their susceptibility to being transported while retaining antineoplastic activity. Although such modifications frequently decrease the bioavailability or efficacy of drugs, some new agents have been developed using this approach (Perego et al., 2001). The intracellular concentration of drugs can also be elevated by increasing the rate of influx. This apparent circumvention of P-gp-mediated efflux can be achieved by increasing the lipophilicity of compounds (positive charge and degree of lipophilicity dictate, or at least influence, whether compounds are recognized by MDR1) or by stealth formulations. For example, highly lipophilic anthracycline analogs, such as annamycin and idarubicin, were shown to elicit a high remission rate in P-gp-positive AML cases with primary resistance to chemotherapy (Byrne et al., 1999; Lampidis et al., 1997). The efficacy of these drugs is currently being evaluated in the MRC AML15 trial. Encapsulation of doxorubicin in polyethylene glycol-coated liposomes (PLD) might be safer and occasionally more effective than conventional doxorubicin (Vail et al., 2004). PLD was found to cross the BBB, and seemed to overcome the MDR of tumors in preclinical models. The combination of this formulation with PSC-833 suppressed tumor growth to an even greater degree in mouse xenograft models, providing proof-of-principle for Phase I studies (Fracasso et al., 2005). Theoretically, the simplest way to counter efflux mechanisms is to increase drug exposure of cancer cells through prolonged or higher-dose chemotherapy. Indeed, it could well be that the benefit of classical inhibitors was derived solely from the augmented dose intensity of the concomitantly administered chemotherapeutics, as opposed to the pharmacodynamic modulation of target cells106. Unfortunately, the therapeutic window of anticancer agents is very narrow, as even a slight increase in chemotherapy dosages results in potentially lethal side effects.

8.8.3 EXPLOITING DRUG RESISTANCE BY TARGETING MDR CELLS WITH PEPTIDES AND ANTIBODIES

P-gp-mediated drug resistance can be reversed by hydrophobic peptides that are high-affinity P-gp substrates. Such peptides, showing high specificity to P-gp, could represent a new class of compounds for consideration as potential chemosensitizers (Sharom et al., 1999). Small peptides corresponding to the transmembrane segments of P-gp act through a different mechanism. Peptide analogs of TMDs are believed to interfere with the proper assembly or function of the target protein, as was shown in experiments aimed at the in vitro or in vivo inhibition of G-protein-coupled receptors (George et al., 2003). Small peptides designed to correspond to the transmembrane segments of P-gp act as specific and potent inhibitors, suggesting that TMDs of ABC transporters can also serve as templates for inhibitor design (Tarasova et al., 2005). Studies suggest that immunization could be an alternative supplement to chemotherapy. A mouse monoclonal antibody directed against extracellular epitopes of P-gp was shown to inhibit the in vitro efflux of drug substrates. Similarly, immunization of mice with external sequences of the murine gene mdr1 elicited antibodies capable of reverting the MDR phenotype in vitro and in vivo, without eliciting an autoimmune response (Pawlak-Roblin et al., 2004). Ideally, therapy is directed against specific target cells. MDR cancer cells are eminent targets for destruction, and the high surface expression of P-gp could be exploited in strategies that use antibodies to bridge effector molecules and cells. Anti-P-gp antibodies have been successfully used to destroy P-gp-expressing cells in antibody-mediated cytolysis experiments, and have also been used as immunotoxins (Heike et al., 1990). More recently, Morizono et al. have used a mouse melanoma model engineered to express the human ABCB1 gene to show that metastatic cells can be successfully targeted with a vector linked to an anti-P-gp monoclonal antibody. Immune response to the anti-P-gp immunoglobulins and the toxic side effects expected in normal tissues expressing P-gp are the concerns that have to be addressed before the widespread clinical use of these strategies (Morizono et al., 2005). Future enhancements of the technology, such as the replacement of the monoclonal antibodies with peptide fragments, will be important for successful clinical applications.

8.8.4 EXPLOITING THE PARADOXICAL SENSITIVITY OF MDR CELLS

Gene expression studies have shown that MDR cells can be profoundly different from their sensitive counterparts31. Perhaps as a result of these differences, MDR cells that are cross-resistant to structurally and functionally unrelated drugs can simultaneously show paradoxical hypersensitivity to certain compounds. MDR cells were collaterally sensitive to membrane-active agents such as the calcium-channel blocker verapamil; inhibitors such as PSC-833 or LY294002 (Lehne et al., 2002); and various stress-inducing compounds, including 2-deoxy-D-glucose (Bell et al., 1998), tunicamycin and 5-fluorouracil (Warr et al., 2002). In an effort to catalog compounds against which MDR cells might show collateral sensitivity, the expression profile of the 48 ABC transporters were characterized in the NCI60 cancer cell panel 25. The NCI60 cell panel was set up by the Developmental Therapeutics Program of the NCI to screen the toxicity of chemical compound repositories (Monks et al., 1991). The relationship between ABC transporter expression levels and sensitivity to drugs or drug candidates were explored in asking which of the transporters confer resistance or sensitivity to various classes of agents. In particular, statistical correlations between the cell lines' sensitivity to cancer drugs and the expression of ABC transporters were searched. Using this pharmacogenomic approach, strongly correlated "drug–gene" pairs in which the expression of MDR1/P-gp correlated with increased sensitivity to a drug were identified. This correlation suggested that the toxicity of several compounds can be potentiated, rather than antagonized, by the MDR1 multidrug transporter. Follow-up studies have verified that cells become hypersensitive to "MDR1-inverse" compounds, such as NSC 73306, in proportion to their P-gp function. The physiological function of P-gp includes transmembrane transport of a broad spectrum of endogenous substrates, some of which have a role in regulation of cell growth. Recent observations support the possibility that P-gp can promote cell survival by efflux-independent pathways, including the inhibition of caspase-dependent apoptosis or the reduction of ceramide levels through either the reduction of inner leaflet sphingomyelin pools or the modulation of the glucosylceramide synthase pathway (Turzanski et al., 2005). In view of these findings, it can be speculated that downstream changes in the apoptosis-inducing pathways in MDR cells might be responsible for the preferential susceptibility to MDR1-inverse compounds. Cells expressing other ABC transporters could become similarly sensitive. For example, increased MRP1 expression could be accompanied by the intracellular depletion of important molecules, such as GSH, resulting in an increased susceptibility to oxidative stress (Trompier et al., 2004).

8.9 CLINICAL EFFORTS TO OVERCOME MDR WITH P-GP INHIBITORS

The clinical importance of P-gp might also be determined through clinical trials designed to abrogate P-gp function. Toward this end, less than 10 years after the discovery of P-gp-mediated MDR, Phase I and II clinical trials began to test the clinical potential of P-gp inhibitors. Initial trials used G_1 P-gp inhibitors, including verapamil, quinine, and cyclosporine (also known as cyclosporin A), which were already approved for other medical purposes. In general, these compounds were ineffective or toxic at the doses required to attenuate P-gp function. Despite these problems, a randomized Phase III clinical trial exhibited the benefit of addition of cyclosporine to treatment with cytarabine and daunorubicin in patients with poor-risk AML (List et al., 2001). Similarly, quinine was shown to increase the complete remission rate as well as survival in P-gp-positive MDS cases treated with intensive chemotherapy, suggesting that successful P-gp modulation is feasible (Daenen et al., 2004).

Promising early clinical trials encouraged further development. The G_2 inhibitors were devoid of side effects related to the primary toxicity of the compounds. For example, the R-enantiomer of verapamil and the cyclosporin D analog PSC-833 (Valspodar) antagonized P-gp function without blocking calcium channels or immunosuppressive effects, respectively (Hollt et al., 1992). PSC-833 has been tested most frequently in clinical trials with little success rate. Characteristic of the failures of G_2 inhibitors, PSC-833 induced pharmacokinetic interactions that limited drug clearance and metabolism of chemotherapy, thereby elevating plasma concentrations beyond acceptable toxicity. To preserve patient safety, empirical chemotherapy dose reductions were necessary; however, because pharmacokinetic interactions were generally unpredictable, some patients were probably underdosed, whereas others were overdosed. Similarly, development of another G_2 inhibitor showing initial promise (VX-710; biricodar) has been curtailed (Goldman, 2003). G_3 inhibitors are designed specifically for high transporter affinity and low pharmacokinetic interaction. Inhibition of cytochrome P450 3A, which is responsible for many adverse pharmacokinetic effects with previous-generation inhibitors, has generally been avoided with the latest generation of inhibitors, including laniquidar (R101933), oc144-093 (ONT-093), zosuquidar (LY335979), elacridar (GF-120918)75, and tariquidar (XR9576) (Guns et al., 2002). Tariquidar has the added benefit of extended P-gp inhibition, as a single intravenous dose inhibited efflux of rhodamine from CD56+ cells (biomarker lymphoid cells that express P-gp) for at least 48 h (Stewart et al., 2000).

Biricodar (VX-710) and GF-120918 bind P-gp. Although affinity for multiple drug transporters might extend the functionality of these inhibitors to P-gp-negative tumors showing MDR, the scope of possible side effects also increases. In 2002, Phase III clinical trials began using tariquidar as an adjunctive treatment in combination with first-line chemotherapy for patients with NSCLC. Despite the promising characteristics mentioned above, the studies were stopped early because of toxicities associated with the cytotoxic drugs (Minderman et al., 2004). This study also illustrates a defect in experimental design, as there is no strong evidence to suggest that NSCLC expresses P-gp to a significant extent. Zosuquidar has been evaluated in patients with AML. Preliminary analysis indicates that zosuquidar can be safely given without chemotherapy dose reductions (L. D. Cripe, personal communication); trial endpoints have not yet been analysed. Although P-gp is clearly established as a prognostic marker in adult

AML, after more than three decades of research, the clinical benefit of modulating P-gp-mediated MDR is still in question. This is, in part, due to limitations of candidate inhibitors, and the inadequate design of the trials (van Zuylen et al., 2000). Clearly, the inhibitors used today are much improved from those used in the past, with greater substrate specificity, lower toxicity, and improved pharmacokinetic profiles. Results from Phase III trials using G_3 inhibitors will be pivotal in determining whether inhibition of P-gp, or other ABC transporters, can result in improved patient survival. The perfect reversing agent is efficient, lacks unrelated pharmacological effects, shows no pharmacokinetic interactions with other drugs, tackles specific mechanisms of resistance with high potency, and is readily administered to patients. This might be too much to ask from a cancer drug that targets a network of transporters with a pivotal role in ADMET. In more realistic terms, the ideal inhibitor should restore treatment efficiency to that observed in MDR-negative cases. Nevertheless, modulators are unlikely to improve the therapeutic index of anticancer drugs unless agents that lack significant pharmacokinetic interactions are found. The search for such "G_4" inhibitors is ongoing, and there is no shortage of compounds showing in vitro sensitization of MDR cells. Similar to their predecessors, some of the emerging candidates are "off the shelf" compounds (old drugs with new tricks), such as disulfiram, used to treat alcoholism, or herbal constituents shown to inhibit P-gp function in vitro in concentrations that are compatible with clinical applicability (Zhou et al., 2004).

8.9.1 POTENTIAL REASONS FOR A LACK OF CLINICAL BENEFIT

8.9.1.1 INADEQUATE PRECLINICAL EVALUATION

Regardless of the underlying mechanisms that led to the limited therapeutic success in the clinic, it is evident that many of the MDR inhibitors were introduced into clinical trials. However, all these trials were sought to evaluate their ability to improve the therapeutic index in patients with advanced cancer, and most of the agents tested were not investigated in rodents with established macroscopic tumors. It is ethically questionable to embark on hugely expensive clinical trials without preclinical evaluation in appropriate animal models. No future strategy should advance to clinical testing without thorough preclinical evaluation.

8.9.1.2 ENHANCED TOXICITY AT NONTARGET SITES

A common explanation for early termination or clinical failure of MDR modulators is their inherent pharmacological action. This was particularly true of the G_1 P-gp inhibitors that were developed for other medical indications: these agents cause unacceptable toxicity (such as cardiac toxicity, ataxia, and immunosuppression) at doses required for P-gp inhibition (Baer et al., 2002; Kolitz et al., 2004). While these effects are less for the G_2 and G_3 inhibitors, all of them inhibit ABC transporters in normal tissues, which can lead to toxicity. For example, P-gp is expressed in some early hematopoietic precursor cells and in cells lining various tissue barriers, including the BBB, and its inhibition in these sites could lead directly to hematological or neurological toxicity and/or to increased toxicity of administered chemotherapy.

8.9.1.3 PHARMACOKINETIC INTERACTIONS BETWEEN MDR INHIBITORS AND ANTICANCER DRUGS

Some inhibitors of the MDR proteins are substrates for the cytochrome P450 system and may compromise the action of these biotransformation enzymes, which are necessary for the clearance and/or metabolism of co-administered anticancer drugs. As a result, there is a need to reduce doses of anticancer drugs, which may contribute to the lack of benefit as compared to control treatments with standard dose chemotherapy (Goldman, 2003).

8.9.1.4 LACK OF PRESELECTION OF PATIENTS AND MULTIPLE MECHANISMS OF DRUG RESISTANCE

The expression of MDR proteins was evaluated retrospectively in a minority of the clinical studies, and in none of the trials was patient selection based on prospective evaluation of the expression of any of these markers. Most of the patients in the clinical studies were enrolled after multiple lines of therapy, so resistance to chemotherapy was likely to be multifactorial. Furthermore, the chemotherapy usually included drugs that were and were not substrates of ABC transporters. The action of an individual drug transporter cannot account for the full spectrum of MDR found in the clinic. Some causes of MDR (such as enhanced ability to repair DNA and increased survival pathways) are unrelated to ABC transporters. Also, most anticancer drugs are not recognized by a single efflux transporter and, when the action of one drug transporter is inhibited, others may cause drug resistance in cancer cells by mediating active outward efflux of the drugs. For example, in clinical trials that evaluated P-gp inhibitors in non-small cell lung cancer, other mechanisms of resistance were probably due to the expression of MRP1 and BCRP (Guns et al., 2002; Minderman et al., 2004).

8.9.1.5 INTERACTIONS BETWEEN INHIBITORS AND MULTIPLE ABC TRANSPORTERS

Suppressing the activity of a given transporter may impact on the function of nontarget transporters and such effects can result in unexpected side effects. For example, high doses of G_3 P-gp inhibitors, such as tariquidar (>1 μM), cross-react with BCRP. This may cause the inhibition of BCRP and resultant toxicities in healthy tissues since physiological levels of BCRP expression exert protective roles in the intestine, blood–testis barrier, BBB, and hematopoietic progenitor cells. A better understanding of the complex interplay among the transporters and their relative roles in drug resistance will be essential to develop modulators that can simultaneously inhibit the function of multiple ABC protein targets known to confer drug resistance, without leading to unacceptable toxicity (Xenova Group Limited Tariquidar, 2006).

8.9.1.6 GENETIC POLYMORPHISMS OF ABC TRANSPORTER GENES

The prevalence of many single-nucleotide polymorphisms (SNPs) within the ABC transporter genes may have prevented optimal results from the clinical evaluation of their inhibitors. The impact of these genetic variations on drug bioavailability and resistance is often unknown, but some SNPs and haplotypes (combinations of SNPs) of MDR1 have been shown to modify P-gp expression, its conformational structure at target sites, and its function. These effects probably contribute to differences in outcome when using MDR modulators in different individuals and ethnic groups

(Seelig & Gatlik-Landwojtowicz, 2005). Genetic polymorphisms in drug transporters can influence the absorption, metabolism, and excretion of administered anticancer drugs, thereby causing interindividual variability in the pharmacokinetics of the drugs, as well as in drug response (Pleban & Ecker, 2005). For example, patients homozygous for the T allele of the MDR1 C1236T polymorphism correlated with increased exposure to irinotecan and its active metabolite SN-38. Another study showed that the response to preoperative chemotherapy in patients with breast cancer was increased in women with the C3435T polymorphism (a silent mutation) in exon26 of MDR1 as compared to women with the wild-type CC genotype (George et al., 2003; Tarasova et al., 1999). Women with ovarian cancer who have the MDR1 G1199A SNP (amino acid substitution Ser400Asn) had significantly shorter PFS (~2 months) following chemotherapy than those expressing the wild-type allele (~19 months) (Tarasova et al., 2005). Further research will be needed to address whether some of these SNPs also influence the specificity and effectiveness of MDR inhibitors. Selection of patients with expression of forms of ABC transporters that are known to be inhibited by candidate agents will be important if future clinical trials are to be more successful.

8.9.1.7 EFFECTS TO ADVERSELY MODIFY DRUG DISTRIBUTION IN SOLID TUMORS

Anticancer drugs access solid tumors via the bloodstream and must penetrate tumor tissues to reach the cancer cells. It has shown that limited penetration of tissue by anticancer drugs is an important and neglected cause of drug resistance, independent of the sensitivity of the constituent cells: if a drug does not reach all of its target population, then even drug-sensitive cells may be spared. Our previous studies using multilayered cell cultures have shown that augmented P-gp expression is associated with better penetration of tissue by doxorubicin, as might be expected if there is minimal uptake of drug in proximal cells, while P-gp inhibitors reduce drug penetration (Pawlak-Roblin et al., 2004). Recent in vivo experiments demonstrate that both verapamil and valspodar inhibit penetration of doxorubicin in P-gp overexpressing murine mammary sarcoma and human breast tumor xenografts (Scotto, 2003). This effect leads to improved uptake of doxorubicin only in tumor cells within a restricted radius around functional blood vessels. In contrast, drug uptake remains largely unchanged in cells at intermediate distances from the blood vessel and may even decrease in more distal cells. These data emphasize that, in solid tumors, there is a trade-off between uptake of anticancer drugs into proximal cells and penetration to distal cells, and the therapeutic effectiveness of inhibitors of ABC transporters may be limited because they inhibit the penetration of substrate drugs in solid tumors. Repopulation of tumors may originate from cells in regions far from blood vessels (Kang et al., 2004), and it is important to consider factors such as the tumor microenvironment and poor distribution of chemotherapeutic drugs in solid tumors when developing novel treatment options. This points once again to the importance of evaluating the effectiveness of inhibitors of ABC transporters in appropriate animal models.

8.9.1.8 HIGH CELL CONCENTRATION LIMITS THE ACTIVITY OF P-GP INHIBITORS

Most inhibitors of ABC transporters are tested initially for their effects against MDR cells in dilute tissue culture (e.g., at $\sim 10^5$ cells/ml). ABC transporters are expressed at high concentration on drug-resistant cells, raising the question as to whether inhibitors can be delivered in sufficient concentration to provide effective inhibition in dense tissue such as would occur within solid tumors (cell concentration of 108–109 cells/ml). The impact of cell concentration in tissue culture on the ability of P-gp inhibitors, verapamil and cyclosporin A, to potentiate the sensitivity of MDR cells to doxorubicin was investigated. The effects of these MDR-modulating agents on the uptake and cytotoxicity of doxorubicin declined substantially at higher concentration of cultured cells (Bartsevich & Juliano, 2000). In addition, the cell concentration effects cannot be overcome by continuous replenishment of doxorubicin and verapamil to maintain a stable extracellular concentration. Several mechanisms might explain these observations. High cell concentration may lead to increased nonspecific binding of MDR inhibitors to cellular components, with a reduction in their effective concentration. Suppression of drug transport across cell membranes by an inhibitor also relies, under nonequilibrium conditions that are likely to apply in vivo, on the number of drug transporters present in different tissue types. For example, if P-gp molecules are expressed on drug-resistant tumor cells at a concentration of $\sim 10^5$ molecules/cell, as has been reported, there will be $\sim 10^{14}$ target P-gp molecules/ml in solid tissues containing ~10^9 tumor cells/ml. This is comparable to the number of molecules per milliliter at a 1 μM concentration of MDR inhibitors. Under such situations, the concentration of a reversal agent may be too low to inhibit efficiently the activity of target P-gp molecules, even if it is fully active in dilute tissue culture (Xu et al., 2002).

8.9.2 POSSIBLE REASONS FOR FAILURE IN PHASE III TRIALS TARGETING P-GLYCOPROTEIN

Possible reasons for the failure of targeting Pgp in Phase III trials include the following:

- ✓ Alternative mechanisms of resistance
- ✓ Unfavorable pharmacological properties of the inhibitors
  - o Low affinity (ineffective inhibition)
  - o Poor specificity (unrelated pharmacological activity)
  - o Low bioavailability at tumor site
- ✓ Toxicity of the inhibitors
  - o Primary toxicity of the G_1 and G_2 reversing agents (e.g., hypotension, ataxia, and immunosuppression)
  - o Secondary toxicity due to inhibition of P-gp in physiological sanctuaries such as bone marrow stem cells
- ✓ Pharmacokinetic interactions
  - o P-gp modulators can decrease the systemic clearance of anticancer drugs, thereby increasing exposure to normal and malignant cells and so potentially increasing the severity and/ or incidence of adverse effects associated with the anticancer therapy (Relling, 1996)
  - o There is a considerable overlap in the substrate specificities and regulation of cytochrome P450 3A (CYP3A) and Pgp. CYP3A, the major Phase I drug-metabolizing enzyme, and Pgp have complementary roles in intestinal drug metabolism, where, through repeated extrusion and reabsorption, P-gp ensures elongated exposure of the drugs to the metabolizing enzyme (Benet et al., 2004).
  - o Inhibition of P-gp can interfere with CYP3A-mediated intestinal or liver metabolism, resulting in reduced drug clearance.

  - o Interaction with other ATP-binding cassette (ABC) transporters, such as ABCB4 and ABCB11, which results in compromised biliary flow (Bohme et al., 1993).
- ✓ Empirical dose-modification of chemotherapy
  - o To accommodate expected elevations in systemic drug exposure, some patients might have been overdosed or undertreated.

The relevance of ABCB1 in clinical MDR in cancer is unambiguous and other ABC transporters, such as ABCC1 and, less convincingly, ABCG2 have now been implicated (List et al., 2001). The search for inhibitors or reversal agents has concentrated mostly on ABCB1-mediated MDR. Potent and specific inhibitors of ABCB1 have been developed to avoid off-site side effects but this has also highlighted a number of competing risks as ABCB1 has important roles in normal physiology including as a major determinant of the permeability barrier of sanctuary sites. Drug–drug interactions have also become evident as inhibition of ABCB1in the gut and the liver can have a strong influence on the pharmacokinetics of coadministered chemotherapeutic agents. In future clinical trials of combinatorial therapies, it will be important to reevaluate the safety and efficacy of the coadministered anticancer therapeutic in light of likely changes to its pharmacokinetic behavior following inhibition of ABCB1 (or other polyspecific drug efflux pumps). The key question is where the field should go from here. Despite the negative results from clinical trials conducted with ABC transporter inhibitors, the field retains great potential especially considering the emerging novel concepts. Indeed, in recent years it has become increasingly apparent that other ABC transporters may play roles in MDR and thereby in cancer progression is emerging (Dantzig et al., 2003). In other words, ABC transporters appear to contribute to tumor biology independently of their ability to efflux cytotoxic drugs. ABC transporters also appear to be important in cancer stem cells and this represents another area of investigation and development for ABC transporter inhibitors. Recent work on cancer stem cells has revealed that they are protected against widely used chemotherapeutic agents by a variety of mechanisms, such as a increased efficacy in DNA damage repair, hyperactivation of signaling pathways such as PI3K and Wnt, low division rate, and last but not the least, the expression of ABC transporter drug-efflux pumps (Pleban & Ecker, 2005; Sharom, 2008). The ultimate goal of this field of biomedical research is to specifically inhibit or down regulate ABC transporters in cancer and cancer stem cells that are relevant to MDR and/or disease progression. There are promising developments, not least in the biochemical characterization of the different ABC transporters, the classification of their transport substrates, and identification of novel inhibitors. Among these is the surprising discovery that MDR cancer cells can display hypersensitivity to other drugs, a phenomenon described as "collateral sensitivity" (Mechetner & Roninson, 1992). Although the prospect that drug-resistant cells may be hypersensitized to an alternative cytotoxic agent is attractive at present, it remains characterized only in vitro and more studies are needed to translate this to the clinic. There is also a problem relating to preclinical studies of small molecule inhibitors in cancer cell lines in vitro.

8.10 CONCLUSION

Obviously, this discussion of P-gp inhibitors cannot be comprehensive, but it is clear that by now there are many P-gp inhibitors with increasingly suitable properties for clinical use. Present studies tell us that they have a marked impact on clinical multidrug resistance. The problems highlighted by off-site effects, the inhibition of the normal function of ABC transporters in healthy tissue, and the need for combinatorial therapies, have prompted the development of targeting strategies and the use of nanoparticles technology. These remain in the early stages of development but as more cancer biomarkers are being identified, the promise of targeted delivery of a particle containing a potent and specific inhibitor (or siRNA) for the ABCB1 transporter relevant to the cancer type, in combination with a cytotoxic drug, becomes closer.

KEYWORDS

- **ATP;**
- **Cancer cells;**
- **Chemosensitizers;**
- **Chemotherapy;**
- **MDR;**
- **P-gp**

REFERENCES

Abraham, J.; Edgerly, M.; Wilson, R. et al. A phase I study of the P-glycoprotein antagonist tariquidar in combination with vinorelbine. Clin Cancer Res. *2009*, 15, 3574–3582.

Baer, M.R.; George, S.L.; Dodge, R.K. et al. Phase III study of the multidrug resistance modulator PSC-833 in previously untreated patients 60 years of age and older with acute myeloid leukemia: cancer and Leukemia Group B Study 9720. Blood *2002*, 100, 1224–1232.

Bartsevich, V.V. & Juliano, R.L. Regulation of the MDR1 gene by transcriptional repressors selected using peptide combinatorial libraries. Mol Pharmacol. *2000*, 58, 1–10.

Bell, S.E.; Quinn, D.M.; Kellett, G.L.; Warr, J.R. 2-Deoxy-D-glucose preferentially kills multidrug-resistant human KB carcinoma cell lines by apoptosis. Br J Cancer *1998*, 78, 1464–1470.

Benet, L.Z.; Cummins, C.L.; Wu, C.Y. Unmasking the dynamic interplay between efflux transporters and metabolic enzymes. Int J Pharm. *2004*, 277, 3–9.

Boesch, D.; Gavériaux, C.; Jachez, B.; Pourtier-Manzanedo, A.; Bollinger, P.; Loor, F. In vivo circumvention of P-glycoprotein-mediated multidrug resistance of tumor cells with SDZ PSC 833, Cancer Res. *1991*, 51, 4226–4233.

Bohme, M.; Buchler, M.; Muller, M.; Keppler, D. Differential inhibition by cyclosporins of primary-active ATP-dependent transporters in the hepatocyte canalicular membrane. FEBS Lett. *1993*, 333, 193–196.

Bonhomme-Faivre, L.; Devocelle, A.; Saliba, F.; Chatled, S.; Maccario, J.; Farinotti, R. et al. MDR-1 C3435T polymorphism influences cyclosporine a dose requirement in liver-transplant recipients. Transplantation *2004*, 78, 21–25.

Borrelli, F. & Izzo, A.A. Herb-drug interactions with St John's wort (Hypericum perforatum): an update on clinical observations. AAPS J. *2009*, 11, 710–727.

Bourguignon, L.Y.W.; Peyrollier, K.; Xia, W.; Gilad, E. Hyaluronan-CD44 interaction activates stem cell marker, Nanog, Stat-3-mediated MDR1 gene expression, and ankyrin-regulated multidrug efflux in breast and ovarian tumor cells. J Biol Chem. *2008*, 283, 17635–17651.

Breedveld, P.; Beinen, J.H.; Schellens, J.H. Use of P-glycoprocein and BCRP inhibitors to improve oral bioavailabilicy and CNS penetration of anticancer drugs. Trends Pharmacol Sci. *2006*, 27, 17–24.

Byrne, J.L. et al. Early allogeneic transplantation for refractory or relapsed acute leukaemia following remission induction with FLAG. Leukemia *1999*, 13, 786–791.

Chen, C.J.; Chin, J.E.; Ueda, K. et al. Internal duplication and homology with bacterial transport proteins in the MDR1 (P-glycoprotein) gene from multidrug-resistant human cells. Cell *1986*, 47, 381–389.

Choudhuri, S. & Klaassen, C.D. Structure, function, expression, genomic organization, and single nucleotide polymorphisms of human ABCB1 (MDR1), ABCC (MRP), and ABCG2 (BCRP) efflux transporters. Int J Toxicol. *2006*, 25, 231–259.

Coley, H.M. Overcoming multidrug resistance in cancer: clinical studies of P−glycoprotein inhibitors. Methods Mol Biol. *2010*, 596, 341–358.

Daenen, S. et al. Addition of cyclosporin A to the combination of mitoxantrone and etoposide to overcome resistance to chemotherapy in refractory or relapsing acute myeloid leukaemia; a randomised phase II trial from HOVON, the Dutch-Belgian Haemato-Oncology Working Group for adults. Leuk Res. *2004*, 28, 1057–1067.

Dantzig, A.H.; Shepard, R.L.; Cao, J.; Law, K.L. ; Ehlhardt, W.J.; Baughmann, T.M.; Bumol, T.F.; Starling, J.J. Reversal of P-glycoprotein-mediated multidrug resistance by a potent cyclopropyldibenzosuberane modulator, LY335979. Cancer Res. *1996*, 56, 4171–4179.

Dantzig, A.H.; Law, K.L.; Cao, J.; Starling, J.J. Reversal of multidrug resistance by the P-glycoprotein modulator, LY335979, from the bench to the clinic. Curr Med Chem. *2001*, 8, 39–50.

Dantzig, A.H.; de Alwis, D.P.; Burgess, M. Considerations in the design and development of transport inhibitors as adjuncts to drug therapy. Adv Drug Deliv Rev. *2003*, 55, 133–150.

Darby, R.A.; Callaghan, R.; McMahon, R.M. P-glycoprotein inhibition: the past, the present and the future. Curr Drug Metab. *2011*, 12, 722–731.

Ferreira, M.J.; Gyemant, N.; Madureira, A.M.; Tanaka, M.; Koos, K.; Didziapetris, R.; Molnar, J. The effects of jatrophane derivatives on the reversion of MDR1- and MRP-mediated multidrug resistance in the MDA-MB-231 (HTB-26) cell line. Anticancer Res. *2005*, 25, 4173–4178.

Fischer, V.; Rodríguez-Gascon, A.; Heitz, F.; Tynes, R.; Hauck, C.; Cohen, D.; Vickers, A.E.M. Themultidrug resistance modulator valspodar (PSC 833) is metabolized by human cytochrome P450 3A. Drug Met Disp. *1998*, 26, 802–811.

Fracasso, P.M. et al. Phase I study of pegylated liposomal doxorubicin and the multidrug-resistance modulator, valspodar. Br J Cancer *2005*, 93, 46–53.

Frances J. The Authors Journal compilation. Chapter 9. Sharom. The P−glycoprotein multidrug transporter. Biochem Soc Essays Biochem. *2011*, 50, 161–178; doi:10.1042/BSE0500161.

Friedenberg, W.R.; Rue, M.; Blood, E.A. et al. Phase III study of PSC-833 (valspodar) in combination with vincristine, doxorubicin, and dexamethasone (valspodar/VAD) versus VAD alone in patients with recurring or refractory multiple myeloma (E1A95): a trial of the Eastern Cooperative Oncology Group. Cancer *2006*,106, 830–838.

Gandhi, L.; Harding, M.W.; Neubauer, M. et al. A phase II study of the safety and efficacy of the multidrug resistance inhibitor VX-710 combined with doxorubicin and vincristine in patients with recurrent small cell lung cancer. Cancer *2007*, 109, 924–932.

George, S.R. et al. Blockade of G protein-coupled receptors and the dopamine transporter by a transmembrane domain peptide: novel strategy for functional inhibition of membrane proteins in vivo. J Pharmacol Exp Ther. *2003*, 307, 481–489.

Giacomini, K.M.; Huang, S.M.; Tweedie, D.J.; Benet, L.Z.; Brouwer, K.L.; Chu, X.; Dahlin, A.; Evers, R.; Fischer, V.; Hillgren, K.M. et al. Membrane transporters in drug development. Nat Rev Drug Discov. *2010*, 9, 215–236.

Gillet, J.P. & Gottesman, M.M. Advances in themolecular detection of ABC transporters involved in multidrug resistance in cancer. Curr Pharm Biotechnol. *2011*, 12, 686–692.

Goldman, B. Multidrug resistance: can new drugs help chemotherapy score against cancer? J Natl Cancer Inst. *2003*, 95, 255–257.

Guns, E.S.; Denyssevych, T.; Dixon, R.; Bally, M.B.; Mayer, L. Drug interaction studies between paclitaxel (Taxol) and OC144-093 – a new modulator of MDR in cancer chemotherapy. Eur J Drug Metab Pharmacokinet. *2002*, 27, 119–126.

Heike, Y. et al. Monoclonal anti-P-glycoprotein antibody-dependent killing of multidrug-resistant tumor cells by human mononuclear cells. Jpn J Cancer Res. *1990*, 81, 1155–1161.

Higgins, C.F.; Gottesman, M.M. Is the multidrug transporter a flippase? Trends Biochem Sci. *1992*, 17, 18–21.

Hoffmeyer, S.; Burk, O.; von Richter, O.; Arnold, H.P.; Brockmoller, J.; Johne, A. et al. Functional polymorphisms of the human multidrug-resistance gene: Multiple sequence variations and correlation of one allele with P-glycoprotein expression and activity in vivo. Proc Natl Acad Sci U S A. *2000*, 97, 3473–3478.

Hollt, V.; Kouba, M.; Dietel, M.; Vogt, G. Stereoisomers of calcium antagonists which differ markedly in their potencies as calcium blockers are equally effective in modulating drug transport by P-glycoprotein. Biochem Pharmacol. *1992*, 43, 2601–2608.

Homolya, L.; Hollo, Z.; Germann, U.A.; Pastan, I.; Gottesman, M.M.; Sarkadi, B. Fluorescent cellular indicators are extruded by the multidrug resistance protein. J Biol Chem. *1993*, 268, 21493–21496.

Huang, Y. Pharmacogenetics/genomics of membrane transporters in cancer chemotherapy. Cancer Metastasis Rev. *2007*, 26, 183–201.

Kang, H. et al. Inhibition of MDR1 gene expression by chimeric HNA antisense oligonucleotides. Nucleic Acids Res. *2004*, 32, 4411–4419.

Kelly, R.J.; Draper, D.; Chen, C.C. et al. A pharmacodynamic study of docetaxel in combination with the P-glycoprotein antagonist tariquidar (XR9576) in patients with lung, ovarian, and cervical cancer. Clin Cancer Res. *2011*, 17, 569–580.

Kioka, N.; Tsubota, J.; Kakehi, Y.; Komano, T.; Gottesman, M. M.; Pastan, I. et al. P-glycoprotein gene (MDR1) cDNA from human adrenal: Normal P-glycoprotein carries Gly185 with an altered pattern of multidrug resistance. Biochem Biophys Res Commun. *1989*, 162, 224–231.

Lampidis, T.J. et al. Circumvention of P-GP MDR as a function of anthracycline lipophilicity and charge. Biochemistry *1997*, 36, 2679–2685.

Lehne, G. et al. The cyclosporin PSC 833 increases survival and delays engraftment of human multidrug-resistant leukemia cells in xenotransplanted NOD-SCID mice. Leukemia *2002*, 16, 2388–2394.

Lepper, E.R.; Nooter, K.; Verweij, J.; Acharya, M.R.; Figg, W.D.; Sparreboom, A. Mechanisms of resistance to anticancer drugs: The role of the polymorphic ABC transporters ABCB1 and ABCG2. Pharmacogenomics *2005*, 6, 115–138.

List, A.F.; Kopecky, K.J.; Willman, C.L.; Head, D.R.; Persons, D.L.; Slovak, M.L.; ... & Appelbaum, F. R. Benefit of cyclosporine modulation of drug resistance in patients with poor-risk acute myeloid leukemia: a Southwest Oncology Group study. Blood *2001*, 98(12), 3212–3220.

Loo, T.W. & Clarke, D.M. Blockage of drug resistance in vitro by disulfiram, a drug used to treat alcoholism. J Natl Cancer Inst. *2000*, 92, 898–902.

Loo, T.W. & Clarke, D.M. Do drug substrates enter the common drug-binding pocket of P-glycoprotein through "gates"? Biochem Biophys Res Commun. *2005*, 329, 419–422.

Mechetner, E. B. & Roninson, I. B. Efficient inhibition of P-glycoprotein-mediated multidrug resistance with a monoclonal antibody. Proc Natl Acad Sci U S A. *1992*, 89, 5824-8.

Minderman, H.; O'Loughlin, K.L.; Pendyala, L.; Baer, M.R. VX-710 (biricodar) increases drug retention and enhances chemosensitivity in resistant cells overexpressing P-glycoprotein, multidrug resistance protein, and breast cancer resistance protein. Clin Cancer Res. *2004*, 10, 1826–1834.

Mistry, P.; Stewart, A.J.; Dangerfield, W.; Okiji, S.; Liddle, C.; Bootle, D.; Plumb, J. A.; Templeton, D.; Charlton, P. In vitro and in vivo reversal of P-glycoprotein-mediated multidrug resistance by a novel potent modulator, XR9576. Cancer Res. *2001*, 61, 749–758.

Monks, A. et al. Feasibility of a high-flux anticancer drug screen using a diverse panel of cultured human tumor cell lines. J. Natl Cancer Inst. *1991*, 83, 757–766.

Morizono, K. et al. Lentiviral vector retargeting to P-glycoprotein on metastatic melanoma through intravenous injection. Nature Med. *2005*, 11, 346–352.

Ozols, R.F.; Cunnion, R.E.; Klecker, R.W. Jr. et al. Verapamil and adriamycin in the treatment of drug-resistant ovarian cancer patients. J Clin Oncol. *1987*, 5, 641–647.

Pauli-Magnus, C. & Kroetz, D.L. Functional implications of genetic polymorphisms in the multidrug resistance gene MDR1 (ABCB1). Pharm Res. *2004*, 21, 904–913.

Pawlak-Roblin, C. et al. Inhibition of multidrug resistance by immunisation with synthetic P-glycoprotein-derived peptides. Eur J Cancer *2004*, 40, 606–613.

Perego, P. et al. A novel 7-modified camptothecin analog overcomes breast cancer resistance protein-associated resistance in a mitoxantrone-selected colon carcinoma cell line. Cancer Res. *2001*, 61, 6034–6037.

Pichler, A.; Zelcer, N.; Prior, J.L.; Kuil, A.J.; Piwnica Worms, D. In vivo RNA interference-mediated ablation of MDR1 P-glycoprotein. Clin Cancer Res *2005*, 11, 4487–4494.

Pleban, K. & Ecker, G.F. Inhibitors of p-glycoprotein - lead identification and optimisation. Mini Rev Med Chem. *2005*, 5, 153–163.

Polgar, O. & Bates, S.E. ABC transporters in the balance: is there a role in multidrug resistance? Biochem Soc Trans. *2005*, 33, 241–245.

Puisset, F.; Chatelut, E.; Dalenc, F.; Busi, F.; Cresteil, T.; Azema, J. et al. Dexamethasone as a probe for docetaxel clearance. Cancer Chemother Pharm. *2004*, 54, 265–272.

Relling, M.V. Are the major effects of P-glycoprotein modulators due to altered pharmacokinetics of anticancer drugs? Ther Drug Monit. *1996*, 18, 350–356.

Robert, J. & Jarry, C. Multidrug resistance reversal agents. J Med Chem. *2003*, 46, 4805–4817.

Rund, D.; Azar, I.; Shperling, O. A mutation in the promoter of the multidrug resistance gene (MDR1) in human hematological malignancies may contribute to the pathogenesis of resistant disease. Adv Exper Med Biol. *1999*, 457, 71–75.

Saeki, T.; LJeda, K.; Tanigawara, Y.; Hori, R.; Komano, T. Human P-gjycoprotein transp orts cyclosporin A and FK506. J Biol Chem. *1993*, 268, 6077–6080.

Sandler, A.; Gordon, M.; De Alwis, D.P. et al. A Phase I trial of a potent P-glycoprotein inhibitor, zosuquidar trihydrochloride (LY335979), administered intravenously in combination with doxorubicin in patients with advanced malignancy. Clin Cancer Res. *2004*, 10, 3265–3272.

Sarkadi, B.; Homolya, L.; Szakacs, G.; Varadi, A. Human multidrug resistance ABCB and ABCG transporters: participation in chemoimmunity defense system. Physiol Rev *2006*, 86, 1179–1236.

Schinkel, A.H. P-glycoprocein, a gatekeeper in the blood—brain barrier. Adv Drug Delivery Rev. *1999*, 36, 179–194.

Schinkel, A.H., & Jonker, J.W. Mammalian drug efflux transporters of the ATP binding cassette (ABC) family: an overview. Adv Drug Deliv Rev. *2012*, 64, Supplement, 138–153.

Scotto, K.W. Transcriptional regulation of ABC drug transporters. Oncogene *2003*, 22, 7496–7511.

Seelig, A. & Gatlik-Landwojtowicz, E. Inhibitors of multidrug efflux transporters: their membrane and protein interactions. Mini Rev Med Chem. *2005*, 5, 135–151.

Siarheyeva, A.; Liu, R.; Sharom, F.J. Characterization of an asymmetric occluded state of P-glycoprotein with two bound nucleotides: implications for catalysis. J Biol Chem. *2010*, 285, 7575–7586.

Sikic, B.I. New approaches in cancer treatment. Ann Oncol, *1999*, 10(suppl 6), S149–S153.

Sharom, F.J. ABC multidrug transporters: structure, function and role in chemoresistance. Pharmacogenomics *2008*, 9, 105–127.

Sharom, F.J. & Siarheyeva, A. Functional assays for identification of compounds that interact with P-gp. In Multidrug Resistance: Biological and Pharmaceutical Advances in Antitumour Treatment (Colabufo, N.A., ed.), Research Signpost: Trivandrum 2008; pp. 261–290.

Sharom, F.J. et al. Interaction of the P-glycoprotein multidrug transporter (MDR1) with high affinity peptide chemosensitizers in isolated membranes, reconstituted systems, and intact cells. Biochem Pharmacol. *1999*, 58, 571–586.

Smith, A.J.; Timmermans-Hereijgers, J.L.P.M.; Roelofsen, B.; Wirtz, K.W.A.; van Blitterswijk, W.J.; Smit, J.J.M.; Schinkel, A.H.; Borst, P. The human MDR3 P-glycoprotein promotes translocation of phosphatidylcholine through the plasma membrane of fibroblasts from transgenic mice. FEBS Lett. *1994*, 354, 263–266.

Smith, A.J.; Mayer, U.; Schinkel, A.H.; Borst, P. Availability of PSC833, a substrate and inhibitor of P-glycoproteins, in various concentrations of serum. J Natl Cancer Inst. *1998*, 90, 1161–1166.

Staud, F.; Ceckova, M.; Micuda, S. and Pavek, P. Expression and function of p-glycoprotein in normal tissues: effect on pharmacokinetics. Methods Mol Biol. *2010*, 596, 199–222.

Stein, U.; Walther, W.; Shoemaker, R.H. Vincristine induction of mutant and wild-type human multidrug-resistance promoters is cell-type-specific and dose-dependent. J Cancer Res Clin Oncol. *1996*, 122, 275–282.

Steinbach, D. & Legrand, O. ABC transporters and drug resistance in leukemia: was P-gp nothing but the first head of the Hydra? Leukemia *2007*, 21, 1172–1176.

Stewart, A. et al. Phase I trial of XR9576 in healthy volunteers demonstrates modulation of P-glycoprotein in CD56+ lymphocytes after oral and intravenous administration. Clin Cancer Res. *2000*, 6, 4186–4191.

Synold, T.W.; Dussault, I.; Forman, B.M. The orphan nuclear receptor SXR coordinately regulates drug metabolism and efflux. Nature Med. *2001*, 7, 584–590.

Szakacs, G.; Paterson, J.K.; Ludwig, J.A.; Booth-Genthe, C.; Gottesman, M.M. Targeting multidrug resistance in cancer. Nat Rev Drug Discov. *2006*, 5, 219–234.

Tarasova, N.I.; Rice, W.G.; Michejda, C.J. Inhibition of G-protein-coupled receptor function by disruption of transmembrane domain interactions. J Biol Chem. *1999*, 274, 34911–34915.

Tarasova, N.I. et al. Transmembrane inhibitors of P-glycoprotein, an ABC transporter. J Med Chem. *2005*, 48, 3768–3775.

Thiebaut, F.; Tsuruo, T.; Hamada, H.; Gottesman, M.M.; Pastan, I.; Willingham, M.C. Cellular localization of the multidrug-resiscance gene product P-glycoprotein in normal human tissues. Proc Natl Acad Sci USA. *1987*, 84, 7735–7738.

Tiwari, A.K.; Sodani, K.; Dai, C.L. et al. Revisiting the ABCs of multidrug resistance in cancer chemotherapy. Curr Pharm Biotechnol. *2011*, 12, 570–594.

Trompier, D. et al. Verapamil and its derivative trigger apoptosis through glutathione extrusion by multidrug resistance protein MRP1. Cancer Res. *2004*, 64, 4950–4956.

Tsuruo, T.; Iida, H.; Tsukagoshi, S. et al. Overcoming of vincristine resistance in P388 leukemia in vivo and in vitro through enhanced cytotoxicity of vincristine and vinblastine by verapamil. Cancer Res. *1981*, 41, 1967–1972.

Turzanski, J.; Grundy, M.; Shang, S.; Russell, N.; Pallis, M. P-glycoprotein is implicated in the inhibition of ceramide-induced apoptosis in TF-1 acute myeloid leukemia cells by modulation of the glucosylceramide synthase pathway. Exp Hematol. *2005*, 33, 62–72.

van Helvoort, A.; Smith, A.J.; Sprong, H.; Fritsche, I.; Schinkel, A.H.; Borst, P.; van Meer, G. MDR1 P-glycoprotein is a lipid translocase of broad specificity, while MDR3 P-glycoprotein specifically translocates phosphatidylcholine. Cell *1996*, 87, 507–517.

van Zuylen, L.; Nooter, K.; Sparreboom, A.; Verweij, J. Development of multidrug-resistance convertors: sense or nonsense? Invest New Drugs *2000*, 18, 205–220.

Vail, D.M. et al. Pegylated liposomal doxorubicin: proof of principle using preclinical animal models and pharmacokinetic studies. Semin Oncol. *2004*, 31, 16–35.

Warr, J.R.; Bamford, A.; Quinn, D.M. The preferential induction of apoptosis in multidrug-resistant KB cells by 5-fluorouracil. Cancer Lett. *2002*, 175, 39–44.

Wu, X.; Gu, J.; Wu, T.T.; Swisher, S.G.; Liao, Z.; Correa, A.M. et al. Genetic variations in radiation and chemotherapy drug action pathways predict clinical outcomes in esophageal cancer. J Clin Oncol. *2006*, 24, 3789–3798.

Xenova Group Limited Tariquidar [online], http://www.xenova.co.uk/dc_xr9576.html (2006).

Xu, D.; Ye, D.; Fisher, M.; Juliano, R.L. Selective inhibition of P-glycoprotein expression in multidrug-resistant tumor cells by a designed transcriptional regulator. J Pharmacol Exp Ther. *2002*, 302, 963–971.

Xu, H.B.; Li, L.; Liu, G.Q. Reversal of p-glycoprotein-mediated multidrug resistance by guggulsterone in doxorubicin-resistance human myelogenous leukemia (K562/DOX) cells. Pharmazie *2009*, 64, 660–665.

Kolitz, J.E. et al. Dose escalation studies of cytarabine, daunorubicin, and etoposide with and without multidrug resistance modulation with PSC-833 in untreated adults with acute myeloid leukemia younger than 60 years: final induction results of Cancer and Leukemia Group B Study 9621. J Clin Oncol. *2004*, 22, 4290–4301.

Xu, D.; Ye, D.; Fisher, M.; Juliano, R.L. Selective inhibition of P-glycoprotein expression in multidrug-resistant tumor cells by a designed transcriptional regulator. J Pharmacol Exp Ther. *2002*, 302, 963–971.

Zhou, S.F. Structure, function and regulation of P−glycoprotein and its clinical relevance in drug disposition. Xenobiotica *2008*, 38, 802–832.

Zhou, S.; Lim, L.Y.; Chowbay, B. Herbal modulation of P-glycoprotein. Drug Metab Rev. *2004*, 36, 57–104.

CHAPTER 9

FORMULATION AND EVALUATION OF SELF-NANOEMULSIFYING DRUG DELIVERY SYSTEM (SNEDDS) FOR ORAL DELIVERY OF KETOCONAZOLE

POONAM VERMA, SANDEEP K. SINGH, KOSHY M. KYMONIL, and SHUBHINI A. SARAF

ABSTRACT

The present study deals with the development and characterization of self-nanoemulsifying drug delivery system (SNEDDS) to improve the oral bioavailability of water insoluble Biopharmaceutical Classification System (BCS) Class II drug ketoconazole (KTZ). The solubility of KTZ in various oils was determined to identify the oil phase of SNEDDS. Various surfactants and cosurfactants were screened for their ability to emulsify the selected oil. Pseudoternary phase diagrams were constructed to identify the efficient self-emulsification region. SNEDDS were further evaluated for their macroscopic characteristics, refractive index, percentage transmittance, emulsification time, drug content, droplet size determination, and in vitro drug release study in comparison with pure drug and a marketed tablet formulation. The optimized formulation was composed of KTZ (60 mg), Triacetin (0.3% wt/wt), Tween 80 (2.205% wt/wt), PEG 400, and n-butanol (0.675% w/w). The mean droplet size and emulsification time was found to be 177.49 nm and 10 seconds, respectively. The in vitro dissolution of KTZ from SNEDDS was found to be significantly higher (95.4±2.07) in comparison to the marketed tablet (64.8±1.36) and pure drug (52.5±1.65) in 0.1 N HCl as dissolution medium. The results indicate that SNEDDS of KTZ, owing to its nanosized has potential to enhance the absorption of drug due to its higher dissolution.

9.1 INTRODUCTION

Oral route is the easiest and most convenient way of noninvasive administration. Approximately 40% of new chemical entities exhibit a poor aqueous solubility and present a major challenge to modern drug delivery systems which leads to poor oral bioavailability, high intra- and intersubject variability, and lack of dose proportionality. These drugs are classified as class II drug by the Biopharmaceutical Classification System (BCS), drugs with low aqueous solubility and high permeability. Different formulation approaches like micronization, solid dispersion, and complexation with cyclodextrins have been utilized to resolve such problems. Indeed, in some selected cases, these approaches have been successful but they offer many other disadvantages. The main problem with micronization is chemical/ thermal stability; many drugs may degrade and lose bioactivity when they are micronized by conventional methods. For solid dispersion, the amount of carriers used is often large, and thus if the dose of active ingredient is high, the tablet or capsules formed will be large in volume and difficult to swallow. Moreover, the carriers used are usually expensive and the freeze-drying or spray-drying method requires particular facilities and processes, leading to a high production cost. Though a traditional solvent method can be adopted instead, it is difficult to deal with coprecipitates with a high viscosity. Complexation with cyclodextrins techniques is not applicable for drug substances which are not soluble in both aqueous and organic solvents. The realization that the oral bioavailability of poor water soluble drugs may be enhanced when coadministered with a meal rich in fat has lead to increasing recent interest in the formulation of poorly water soluble drugs in lipids (1).

In recent years, much attention has been focused on lipid-based formulations to improve the oral bioavailability of poorly water soluble drug compounds. Most of them increase surface area of the drugs to improve solubilization behavior as well as permeation. Lipids have been extensively studied as components of various oily liquids and dispersions that are designed to increase solubility and oral bioavailability of BCS class II and IV drugs. Potential effects of lipid-drug delivery systems on oral drug absorption are given in Figure 9.1. Some of the potential advantages of self-emulsifying lipid formulations include physicochemical stability, enhanced oral bioavailability enabling reduction in dose, consistent temporal profiles of drug absorption, selective targeting of drug toward specific absorption window in gastrointestinal tract (GIT), control of drug delivery profiles, ability to increase Cmax (maximum serum concentration), area under the curve (AUC), and reduced tmax, linear AUC-dose relationship, reduced variability including effect of food, protection of sensitive drug substances, high drug payloads, and flexibility of designing liquid or solid dosage forms (2).

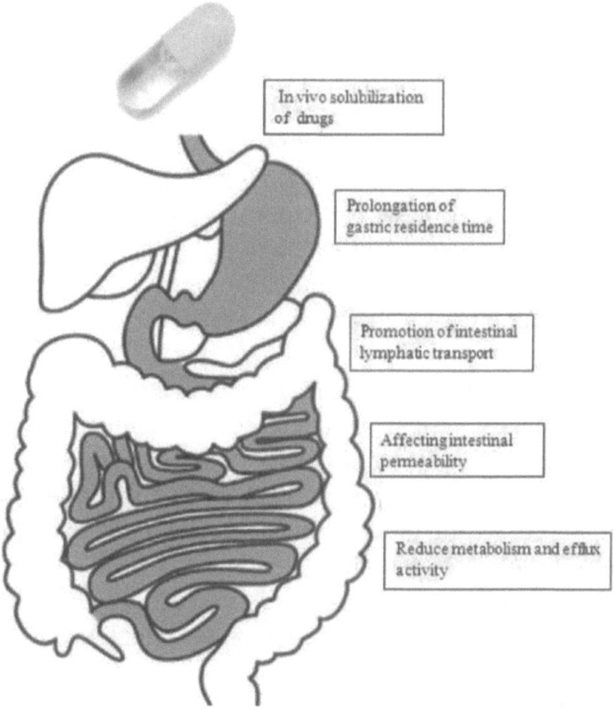

FIGURE 9.1 Potential effects of lipid-drug delivery systems on oral drug absorption.

9.1.1 SELF-NANOEMULSIFYING DRUG DELIVERY SYSTEM (SNEDDS)

SNEDDS are isotropic mixtures of oil, surfactants, and cosurfactants that form fine oil-in-water nanoemulsions upon mild agitation, followed by dilution with aqueous media, such as GI fluids. SNEDDS can be orally administered in soft or hard gelatin capsules due to their anhydrous nature enabling its administration as unit dosage form. They typically produce nanoemulsions with droplet sizes between 20 and 200 nm upon dilution.

When compared with emulsions, which are sensitive and metastable dispersed forms, SNEDDS are physically stable formulations that are easier to manufacture, and may offer an improvement in dissolution rates and extent of absorption, resulting in more reproducible blood time profiles due to the nanometer sized droplets present. Droplet size has been widely proposed as a key factor for the efficiency and fate of SNEDDS (3, 4).

SNEDDS are particularly useful when poorly water soluble compounds are to be predissolved in a suitable solvent and filled into capsules. The main benefit of this approach is that predissolving the compound overcomes the initial rate limiting step of particulate dissolution in the aqueous environment within the GI tract. However, a potential problem with this system is that the drug may precipitate out of solution when the formulation disperses in the GI tract, particularly if a hydrophilic solvent is used (e.g., polyethylene glycol). But alternatively, if the drug can be dissolved in a lipid vehicle there is less potential for precipitation on dilution in the GI tract, as partitioning kinetics will favor the drug remaining in the lipid droplets (5, 6).

Advantages of self-emulsifying drug delivery systems over conventional drug delivery system include ease of manufacture, rapid and extensive distribution of the drug during its passage through the GI tract, minimal irritation, large interfacial area for drug partitioning between oil and water, improved oral bioavailability, more consistent temporal profiles of drug absorption, selective drug targeting toward a specific absorption window in the GI tract, and protection of sensitive drugs from the hostile environment in the gut. Consequently, for lipophilic drug compounds that exhibit dissolution rate limited absorption, these systems may offer an improvement in the rate and extent of absorption and result in more reproducible blood time profiles.

Self-nanoemulsifying drug delivery systems, being liquid in nature, need to be delivered through either soft/hard gelatin or hydroxyl propyl methyl cellulose capsules. There are few issues associated with these systems when presented in capsules, such as incompatibility of components with the capsule shell in the long term, precipitation of drugs during fabrication and storage at low temperature and critical method of production, among others (7, 8).

9.1.2 APPLICATIONS

The various applications of SNEDDS in drug delivery are listed here under.

1. It acts as substitute for traditional oral formulations of lipophilic drugs.
2. It enhances the dissolution rate and hence, bioavailability of hydrophobic drugs.
3. It provides better consistent temporal profiles of drug absorption.
4. It helps in selective drug targeting toward a specific site in the GI tract.
5. It protects drug molecule from the hostile environment of GIT.

Ketoconazole (KTZ) belongs to the class of drugs called azole antifungals. KTZ is indicated for the treatment of candidiasis, chronic mucocutaneous candidiasis, blastomycosis, and paracoccidioidomycosis. KTZ is also indicated for the treatment of patients with severe recalcitrant cutaneous dermatophyte infections who are not responding to topical therapy or oral griseofulvin, or who are unable to take griseofulvin. The azole antifungal, ketoconazole, interferes with P450 enzyme activity and inhibits demethylization of 14-alpha-methylsterolsterols to ergosterol. Since ergosterol is essential to the fungal cell membrane, when it is depleted the fungal cells are destroyed. KTZ may also interfere with the conversion of lanosterol to cholesterol, affecting steroid hormone synthesis.

KTZ exhibit relatively low water solubility which may limit its effectiveness. Increasing the dissolution rate of this drug by incorporating it into a suitable vehicle can increase bioavailability, reduce side effects and variability, and improve effectiveness. Nanoemulsion-based formulations offer rapid dispersion and an enhanced drug absorption profile (7, 8). The objective of the present study was to develop self-nanoemulsified drug delivery system (SNEDDS) containing KTZ to achieve better dissolution rate which would further help in enhancing the oral bioavailability of poorly soluble KTZ compared with pure drug and marketed oral tablet.

9.2 EXPERIMENTAL METHODS

9.2.1 MATERIALS

Ketoconazole (KTZ) was received as a gift sample from Gufic Healthcare Ltd, Mumbai, India. Olive oil, coconut oil, isopropyl myristate, triacetin, liquid paraffin, castor oil, ethyl oleate, Tween 80, Tween 20, and Cremophor EL were purchased from S.d Fine–chem Limited, Mumbai, India. Water used was obtained from Milli-Q water purification system, Millipore, Synergy, Bangalore, India. All other chemicals and solvents used were of analytical grade.

9.2.2 METHODOLOGY

9.2.2.1 SELECTION OF OIL, SURFACTANT, AND COSURFACTANT

The various components for the SNEDDS formulation were selected by determining the solubility of KTZ in various oils (olive oil, ethyl oleate, isopropyl myristate, triacetin, coconut oil, and liquid paraffin), surfactants (Cremophor EL, Tween 80, and Tween 20), and cosurfactant (PEG 400 and propylene glycol). Surfactant and cosurfactant were further screened on the basis of their emulsification ability of the selected oil phase and percentage transparency of the mixture (9–12).

9.2.2.2 CONSTRUCTION OF PSEUDOTERNARY PHASE DIAGRAM

On the basis of solubility and emulsification study triacetin, Tween 80, and PEG 400 were selected as oil, surfactant, and cosurfactant, respectively for the formulation of SNEDDS. To determine the concentration of components of existing range of the SNEDDS, a pseudoternary phase diagram was constructed using water titration method at ambient temperature (25°C) with CHEMIX software. The surfactant and cosurfactant (Smix) were mixed in different weight ratios (1:1, 1:2, 1:3, 2:1, and 3:1). For each phase diagram, oil and specific Smix ratio was mixed thoroughly in different weight ratios from 1:9 to 9:1. Visual observation was carried out for transparent and easily flowable o/w nanoemulsions (13).

9.2.2.3 PREPARATION OF SNEDDS

A series of SNEDDS were prepared using triacetin as the oil phase, Tween 80 as surfactant and PEG 400 and n-butanol as the cosurfactants. Following the study and comparison of the constructed ternary phase diagrams, Smix ratio 3:1 was selected for drug incorporation. In all the formulations the quantity of KTZ (60 mg) was kept constant (Table 9.1). Accurately weighed KTZ was placed in beaker and oil, surfactant, and cosurfactant were added. The components were mixed by gentle stirring on a magnetic stirrer for 1 h and the resulting mixture was heated at 40°C, until the drug was completely dissolved. The homogenous mixtures were stored at room temperature and further subjected to thermodynamic stability studies (14–16).

TABLE 9.1 Composition of various SNEDDS formulations of KTZ

Ingredients	Formulations (% w/w) of Smix ratio 3:1								
	S1	S2	S3	S4	S5	S6	S7	S8	S9
Triacetin	0.3	0.6	0.9	1.2	1.5	1.8	2.1	2.4	2.7
Tween 80	2.025	1.8	1.575	1.35	1.125	0.9	0.675	0.45	0.225
PEG 400	0.3375	0.3	0.2625	0.225	0.1875	0.15	0.1125	0.075	0.0375
n-butanol	0.3375	0.3	0.2625	0.225	0.1875	0.15	0.1125	0.075	0.0375
Drug (mg)*	60	60	60	60	60	60	60	60	60

*Amount of drug in 3 g SNEDDS formulation.

9.3 CHARACTERIZATION OF SNEDDS

9.3.1 MACROSCOPIC EVALUATION

Macroscopic analysis of the selected formulations (S1, S2, S3, S4, S7, and S9) that were found to pass the thermodynamic stability study was carried out. Any change in color and transparency or phase separation occurring during normal storage condition (37°C) for three months was observed.

9.3.2 REFRACTIVE INDEX

The refractive index of the systems was measured using a refractometer and compared with refractive index of water (1.333).

9.3.3 PERCENTAGE TRANSMITTANCE

The percent transmittance of the formulations was measured at 450 nm spectrophotometrically (UV Pharmaspec 1700, Shimadzu, Japan) using distilled water as blank.

9.3.4 EMULSIFICATION TIME

One gram of each formulation was added to 200 mL of 0.1N HCl at 37°C with gentle agitation on a magnetic stirrer. The formulations were visually assessed for final appearance of the emulsion.

9.3.5 DRUG CONTENT

Weighed samples (1 g) were dissolved in dichloromethane (10 ml) and mixed by vortex stirring for 5 min. The solutions were filtered, suitably diluted, and the drug content was estimated spectrophotometrically at 249 nm (17).

9.3.6 MICROSCOPIC ANALYSIS

Transmission electron microscope (TEM) analysis was carried out to determine the surface morphology of the dispersed oil droplets. A drop of diluted SNEDDS was applied to a 300 mesh copper grid and was left for 1 min. After this the grid was kept inverted and a drop of phosphotungstic acid (PTA) (2% w/v) was applied to the grid for 10 s. Excess of PTA was removed by absorbing on a filter paper and the grid was analyzed using the JEM-2100F (JEOL, USA) operated at 200 kV operated with AMT image capture engine software.

9.3.7 DROPLET SIZE DETERMINATION

Fifty miligrams of the optimized SNEDDS formulation (S1) was diluted with triple distilled water to 100 mL in a flask, and gently mixed by hand shaking. The droplet size distribution of the diluted emulsion was determined by laser diffraction analysis using a particle size analyzer (Malvern Zetasizer, UK) (18).

9.3.8 IN VITRO DRUG RELEASE STUDY

In vitro drug release study of developed SNEDDS (S1) of KTZ, pure KTZ powder, and marketed KTZ tablet was carried out in USP XXIII apparatus I at 37 ± 0.5°C with a rotating speed of 100 rpm in 0.1 N HCl as the dissolution media. Then, 1 mL aliquots were removed at predetermined time intervals from the dissolution medium and replaced with fresh media. The amount of drug released was assayed using previously validated ultraviolet (UV) visible spectrophotometric methods (19).

9.3.9 DRUG RELEASE KINETICS

To study the release kinetics, data obtained from in vitro drug release studies was fitted into zero order, first order, Higuchi kinetics, and Korsemeyer –Peppas equation (20). The best fit model was confirmed by the value of correlation coefficient close to 1.

9.3.10 STABILITY STUDIES

The developed formulation (S1) was subjected to stability studies at 5 ± 1°C in refrigerator, 25 ± 2°C/60% ± 5 RH, and 40 ± 2°C /75 ± 5% RH in stability chamber for a period of 45 days. The samples were withdrawn after different time intervals and evaluated for physical appearance and percentage drug content. The degradation rate constant k and time required for degradation of 10% drug ($T_{10\%}$) values were determined at different temperature conditions as specified above (19, 21).

9.4 RESULTS AND DISCUSSION

9.4.1 FORMULATION DEVELOPMENT

9.4.1.1 SELECTION OF OIL

The various oils which were screened for drug partitioning were olive oil, isopropyl myristate, triacetin oil, ethyl oleate, liquid paraffin, and coconut oil. Triacetin oil was selected for SNEDDS formulation development as the drug KTZ showed maximum solubility in triacetin (Figure 9.2) as compared to other oils taken for the study. Thus, triacetin oil can have good proportion of soluble drug and may produce maximum sustained effect compared to other oils (22, 23).

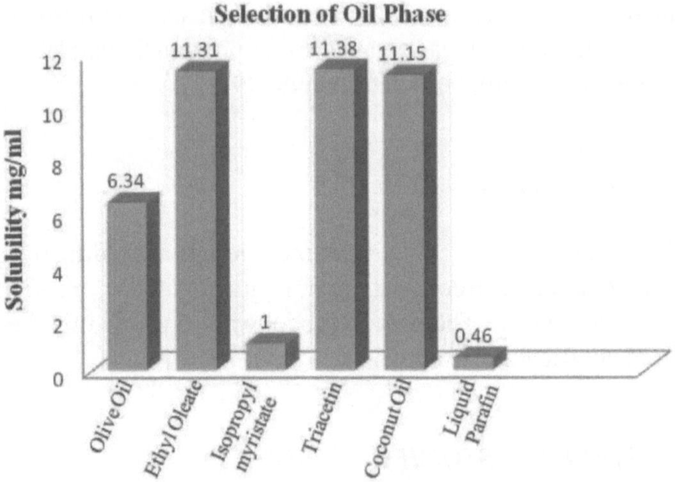

FIGURE 9.2 Selection of oil: equilibrium solubility of KTZ in various oils in milligram/milliliter.

9.4.2 SELECTION OF SURFACTANT

Nonionic surfactants are generally considered less toxic than ionic surfactants. They are usually accepted for oral ingestion. In this study, the three nonionic surfactants (Tween 80, Tween 20, and Cremophor EL) were screened. It has been reported that the well formulated SNEDDS is dispersed within seconds under gentle stirring conditions which ultimately depend on the emulsification ability of the surfactant. Results inferred that the oily phase triacetin exhibited the highest emulsification ability with Tween 80 (% transparency 94.2 and emulsification time 10 s) for the homogeneous emulsion formation. Also Tween 80 produced maximum solubility for the drug and therefore selected for the formulation of SNEDDS (Figure 9.3).

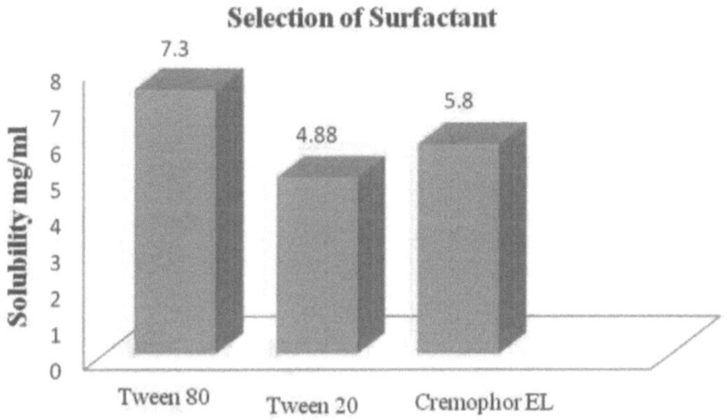

FIGURE 9.3 Selection of surfactant: equilibrium solubility of KTZ in various surfactants in milligram/milliliter.

9.4.3 SELECTION OF COSURFACTANT

Addition of the cosurfactant to the surfactant-containing formulation has been reported to improve dispersibility and drug absorption from the formulation. In view of the current investigation two cosurfactants such as PEG 400 and propylene glycol (PG) were screened. However, PEG 400 exhibited good emulsification showing maximum transmittance (93.6%) followed by PG (90.9%). Herein the solubility of drug in different cosurfactants judged the final selection. Results of the solubility study are shown in Figure 9.4 which inferred higher solubility in n-butanol and PEG 400. It is worthy to note that all the dispersions exhibited an instantaneous emulsion formation with only 15 flask inversions. This could content the importance of cosurfactant addition to the surfactant-containing dispersions.

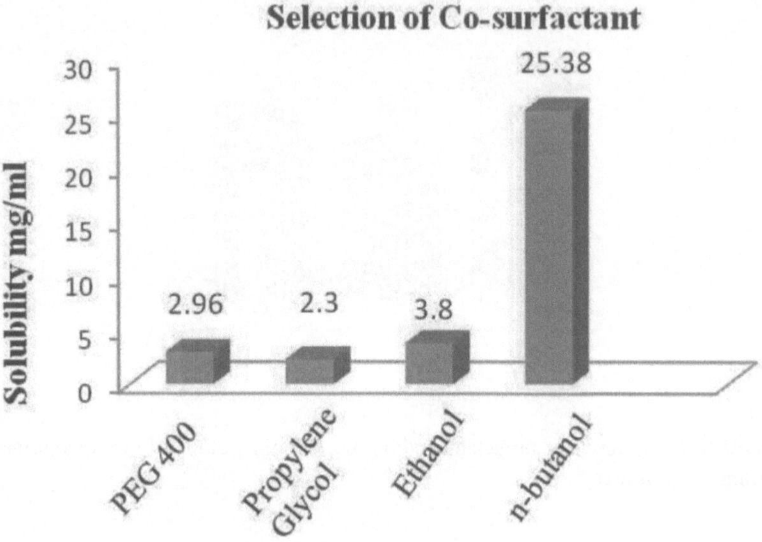

FIGURE 9.4 Selection of cosurfactant: equilibrium solubility of KTZ in various cosurfactants in milligram/milliliter.

9.4.4 CONSTRUCTION OF PSEUDOTERNARY PHASE DIAGRAM

A series of the SNEDDS were prepared and their self-emulsification properties were observed visually. Pseudoternary phase diagrams were constructed in the absence of KTZ to identify the self-emulsification regions and to optimize the concentration of oil, surfactant, and cosurfactant in the SNEDDS formulations (24). The different ratio of surfactant to cosurfactant was very effective for the stable and an efficient SNEDDS formation. The phase diagrams were constructed at surfactant/cosurfactant ratios of 1:1, 1:2, 1:3, 2:1, and 3:1(w/w). The maximum self-emulsification region was found to be at a ratio 3:1 (Figure 9.5).

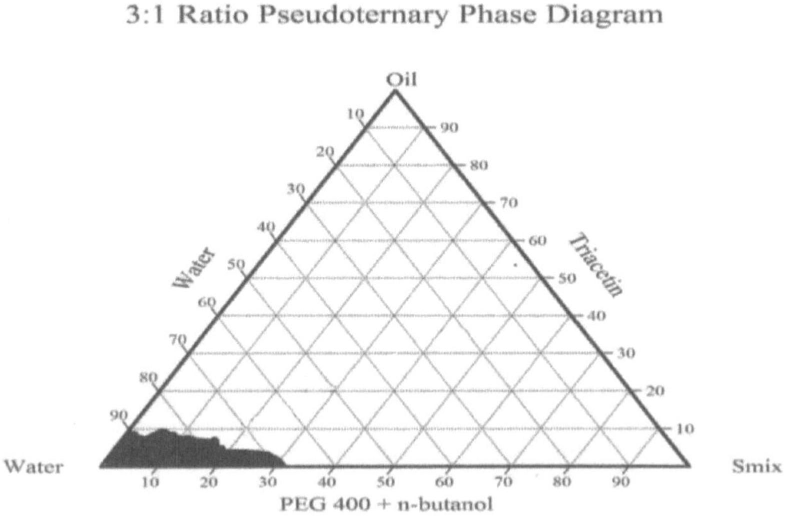

FIGURE 9.5 Pseudoternary phase diagram of systems containing Smix ratio 3:1 of surfactant/cosurfactant such as PEG 400 and n-butanol, triacetin as oil.

9.4.5 SELECTION OF OPTIMIZED BATCH OF SNEDDS

Thermodynamic stability study was performed involving heating cooling cycle, freeze thaw cycle, and centrifugation for the selection of final SNEDDS formulations. It was found that six formulations S1, S2, S3, S4, S7, and S9 passed the test and were selected for further characterization.

TABLE 9.2 Characterization of various optimized SNEDDS formulations of KTZ

Formulation code	Oil: Smix ratio	Smix ratio	Refractive index	Transmittance (%)	Emulsification time (s)	Drug content (%)
S1	**1:9**	**3:1**	**1.337**	**97.4%**	**10**	**96.9**
S2	2:8	3:1	1.341	90.1 %	13	95.55
S3	3:7	3:1	1.343	88.5 %	18	95.35
S4	4:6	3:1	1.345	94.2 %	20	96.7
S7	7:3	3:1	1.339	93.4 %	21	93.45
S9	9:1	3:1	1.344	78.4 %	16	94.8

9.5 PHYSICOCHEMICAL CHARACTERIZATION OF OPTIMIZED SNEDDS FORMULATIONS

9.5.1 MACROSCOPIC EVALUATION

Physical appearance (transparency and phase separation) of optimized formulations was studied. Formulations appeared uniform in color and transparency as well as there was no phase separation observed during normal storage (37 ± 2°C) condition under observation for three months.

9.5.2 REFRACTIVE INDEX

The mean values of the refractive index of drug-loaded formulations are found in the range of 1.337–1.344 (Table 9.2). When the refractive index values for formulations were compared with those of refractive index of water (1.333), it was found that there was no significant difference between the values. Therefore, it can be concluded that the SNEDDS formulations were not only thermodynamically stable but also chemically stable and remained isotropic; there were no interactions between self-nanoemulsion excipients and drug (25).

9.5.3 PERCENTAGE TRANSMITTANCE

The percentage transmittance of the six selected formulations was determined spectrophotometrically. Significant difference ($p<0.001$) was observed among the percentage transmittance of formulations (S1–S6). As the value closer to 100% is the formulation which is isotropic in nature therefore, optimized formulation S1 from Smix ratio of 3:1 gave maximum percentage transmittance (Table 9.2) as compared to other formulations. The droplet size of the emulsion is a crucial factor in self-emulsification performance, because it determines the rate and extent of drug release as well as absorption. Thus, the formulation has the capacity to undergo enhanced absorption and thus ability to have increased oral bioavailability.

9.5.4 DETERMINATION OF THE EMULSIFICATION TIME

The rate of emulsification was taken as an important index for the assessment of the efficiency of self-emulsification, that is, SNEDDS should disperse completely and quickly when subjected to dilution under mild agitation. All the formulations exhibited a rapid rate of emulsification ranging from 10 to 21 s (Table 9.2). It is obvious that rapid emulsification is correlated with lower content of oil and higher content of cosurfactant which result in lower viscosity of the system.

9.5.5 DRUG CONTENT DETERMINATION

Amongst all the six selected formulations, drug content was found to be the highest in S1, irrespective of less % wt/wt of oil and maximum amount Smix ratio, that is, 3:1 as the drug was more soluble in Smix than in the oil. Drug content was found to be in the range of 94.8–96.9% in all the formulations (Table 9.2).

9.5.6 MICROSCOPIC ANALYSIS

TEM images of S1, 24 h postdilution, in distilled water are shown in Figure 9.6. Spherical micelles were observed with no signs of coalescence even 24 h postdilution. The nanoemulsion droplets appeared dark on the light background. Furthermore, no signs of drug precipitation were observed inferring stability of the formed nanoemulsion.

9.5.7 DROPLET SIZE DETERMINATION

The droplet size of the emulsion is a critical factor in self-emulsification performance because it determines the rate and extent of drug release as well as absorption. All the developed SNEDDS formulations successfully attained a submicron droplet size ranging from 54.5 nm to 386.9 nm, while the mean droplet size of optimized S1 was found to be 177.49 nm (Figure 9.6).

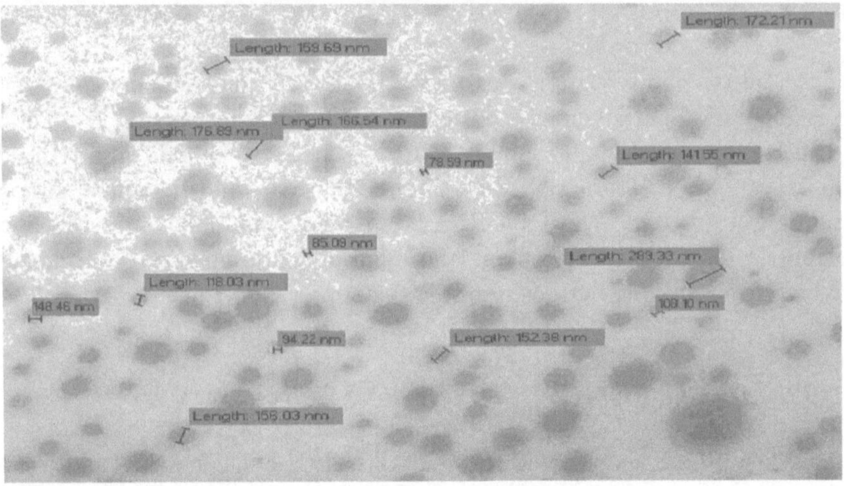

FIGURE 9.6 Transmission electron microscopic (TEM) positive image of the optimized SNEDDS formulation S1 showing size of some emulsion droplets.

9.5.8 *IN VITRO* DRUG RELEASE

Drug release from the SNEDDS formulation S1, was found to be significantly higher than that of the other SNEDDS formulations (Figure 9.7), pure KTZ powder, and marketed KTZ tablet as well (26). It could be attributed to the fact that the SNEDDS formulation resulted in spontaneous formation of a nanoemulsion with a small droplet size, which permitted a faster rate of drug release into the aqueous phase, much faster than that of marketed KTZ tablet. Thus, this greater availability of dissolved KTZ from the SNEDDS formulation could lead to higher absorption and higher oral bioavailability. The drug release from S1 formulation was found to be 95.4 ± 2.07% compared to 52.5 ± 1.65% and 64.8 ± 1.36%, respectively for pure drug and marketed tablet formulation.

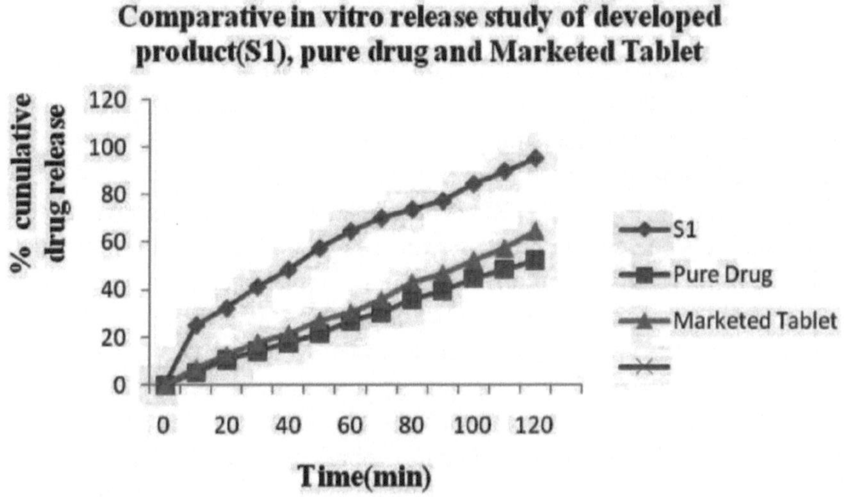

FIGURE 9.7 Comparative in vitro drug release profile of pure drug, marketed tablet, and optimized SNEDDS formulations (S1) in 0.1 N HCl.

9.5.9 DRUG RELEASE KINETICS

It was found that the in vitro drug release from S1 was best explained by Higuchi's equation (Table 9.3), as the plot showed the highest linearity (R^2=0.951), followed by zero order (R^2=0.923) and first order (R^2=0.922). The corresponding plot (log % cumulative drug release vs. log time) for the Korsmeyer-Peppas equation indicated good linearity (R^2=0.907). The release exponent "n" was found to be 0.047, which appears to indicate the quasi-fickian diffusion mechanism of drug release.

TABLE 9.3 Drug release kinetic study of optimized SNEDDS formulation S1

Formulation code	Zero order		First order		Higuchi matrix		Korsmeyer-Peppas	
S1	*K*	R^2	*K*	R^2	*K*	R^2	*n*	R^2
	8.065	0.923	0.093	0.922	7.012	0.951	0.047	0.907

9.5.10 STABILITY STUDIES

KTZ loaded formulation S1, was stored in amber color glass vials at 5 ± 1°C, 25 ± 2°C/60% RH ± 5%, and 40 ± 2°C for a period of 45 days and evaluated for physical appearance and drug content. However, S1 showed no significant change in the physical appearance at 5 ± 1°C, 25 ± 2°C/60% RH ± 5%, and 40 ± 2°C/75% RH. The percentage drug content was determined and log percentage drug content was plotted against time *T* which reflected almost linear relationship. Degradation rate constant *K* was calculated from the slope of straight line, between log of % drug content, and time interval. The time required for degradation of 10% drug was calculated as $T_{10\%}$.The *k* and $T_{10\%}$ values for S1 stored at 5 ± 1°C, 25 ± 2°C, and 40 ± 2°C were 2.46×10^{-4}, 5.87×10^{-4}, 9.27×10^{-4}, and 525, 210, and 117 days, respectively (Table 9.4). The $T_{10\%}$ obtained in case of formulation stored at 5 ± 1°C was found to be higher as compared with formulations stored at 25 ± 2°C and 40 ± 2°C. Thus, it was concluded that S1 is more stable at 5 ± 1°C and tends to degrade faster at higher temperature (27).

TABLE 9.4 Shelf-life of developed formulation (S1)

S. No.	Parameters	Storage conditions		
		5 ± 1°C	**25 ± 2°C/60% RH ± 5%**	**40 ± 2°C/75% RH ± 5% RH**
1	*K* (day^{-1})	2.46×10^{-4}	5.87×10^{-4}	9.27×10^{-4}
2	$t_{½}$ (days)	3,454	1,382	769
3	$T_{10\%}$ (days)	525	210	117

9.6 CONCLUSION

The composition of the developed SNEDDS formulation S1 was optimized as containing ketoconazole (2% wt/wt), triacetin (0.2% wt/wt), and PEG 400 + n-butanol (0.675% wt/wt). S1 showed greater rate and extent of dissolution than pure drug and conventional tablet. The study demonstrated the potential of SNEDDS for increasing the solubility and delivery of a hydrophobic drug, such as KTZ taken in the present study, by oral route.

ACKNOWLEDGMENT

The authors wish to thank Babu Banarasi Das National Institute of Technology and Management, Lucknow for providing infrastructure and facilities to carry out this work successfully.

KEYWORDS

- **Cosurfactant**
- **Ketoconazole**
- **n-Butanol**
- **Nanoemulsion**
- **SNEDDS**

REFERENCES

1. Patel, J.; Patel, A.; Raval, M.; Sheth, N. Formulation and development of self-nanoemulsifying drug delivery system of Irbesartan. J Adv Pharm Technol Res. 2011, 2(1), 9–16.
2. Patel, J.; Kevin, G.; Patel, A.; Raval, M. Design and development of self-nanoemulsifying drug delivery system for telmisartan for oral drug delivery. Int J Pharm Investig. 2011, 1(2), 112–118.
3. Thomas, N.; Holm, R.; Garmer, M.; Karlsson, J. J.; Müllertz, A.; Rades, T. Supersaturated self-nanoemulsifying drug delivery systems (super-SNEDDS) nhance the bioavailability of the poorly water-soluble drug simvastatin in dogs. AAPS J. 2013, 15(1), 219–227.
4. Rao, S. V. R.; Yajurvedi, K.; Shao, J., Self-nanoemulsifying drug delivery system (SNEDDS) for oral delivery of protein drugs: III. In vivo oral absorption study. Int J Pharm. 2008, 362 (1–2), 16–19.
5. Shahba, A-W.; Mohsin, K.; Alanazi, F. K. Novel self-nanoemulsifying drug delivery systems (SNEDDS) for oral delivery of cinnarizine: Design, optimization, and *in-vitro* assessment. AAPS Pharm Sci Tech. 2012, 13(3), 967–977.
6. Doha, H-J.; Junga, Y.; Balakrishnanb, P.; Choc, H-J.; Kimb, D-D. A novel lipid nanoemulsion system for improved permeation of granisetron. Colloids Surf B Biointerfaces. 2013, 101, 475–480.
7. Kuentz, M. Lipid-based formulations for oral delivery of lipophilic drugs. Drug Discov Today Technol. 2012, 9(2), e97–e104.
8. http://chemicalland21.com/lifescience/phar/KETOCONAZOLE.htm. Accessed Sept 10, 2011.

9. Date, A. A.; Nagarsenker, M. S., Design and evaluation of self-nanoemulsifying drug delivery systems (SNEDDS) for cefpodoxime proxetil. Int J Pharm. 2007, 329(1–2), 166–172.
10. Zhang, L.; Zhu, W.; Yang, C.; Guo, H.; Yu, A.; Ji, J.; Gao, Y.; Sun, M.; Zhai, G. A novel folate-modified self-microemulsifying drug delivery system of curcumin for colon targeting. Int J Nanomedicine. 2012, 7, 151–162.
11. Parmar, N.; Singla, N.; Amin, S.; Kohli, K., Study of cosurfactant effect on nanoemulsifying area and development of lercanidipine loaded (SNEDDS) self nanoemulsifying drug delivery system. Colloids Surf B Biointerfaces. 2011, 86(2), 327–338.
12. Balakumar, K.; Raghavan, C. V.; Selvan, N. T.; Prasad, R. H.; Abdu, S., Self nanoemulsifying drug delivery system (SNEDDS) of Rosuvastatin calcium: Design, formulation, bioavailability and pharmacokinetic evaluation. Colloids and Surfaces B: Biointerfaces. 2013, 112(0), 337–343.
13. Zhang, P.; Liu, Y.; Feng, N.; Xu, J. Preparation and evaluation of self-microemulsifying drug delivery system of oridonin. Int J Pharm. 2008, 355(1–2): 269–276.
14. Basalious, E. B.; Shawky, N.; Badr-Eldin, S. M. SNEDDS containing bioenhancers for improvement of dissolution and oral absorption of lacidipine. I: Development and optimization. Int J Pharm. 2010, 391(1–2), 203–211.
15. Zhang, J.; Peng, Q.; Shi, S.; Zhang, Q.; Sun, X.; Gong, T.; Zhang, Z. Preparation, characterization, and in vivo evaluation of a self-nanoemulsifying drug delivery system (SNEDDS) loaded with morin-phospholipid complex. Int J Nanomedicine. 2011, 6, 3405–3414.
16. Beg, S.; Swain, S.; Singh, H. P.; Patra, Ch. N.; Rao, M. B. Development, optimization, and characterization of solid self-nanoemulsifying drug delivery systems of valsartan using porous carriers. AAPS Pharm Sci Tech. 2012, 13(4), 1416–1427.
17. Nepal, P. R.; Han, H. K.; Choi, H. K., Preparation and in vitro–in vivo evaluation of Witepsol® H35 based self-nanoemulsifying drug delivery systems (SNEDDS) of coenzyme Q10. Eur J Pharm Sci. 2010, 39(4), 224–232.
18. Khan, F.; Islam, S.; Roni, M. A.; Jalil, R-U. Systematic development of self-emulsifying drug delivery systems of atorvastatin with improved bioavailability potential. Sci Pharm. 2012, 80(4), 1027–1043.
19. Dixit, R. P.; Nagarsenker, M. S. Self-nanoemulsifying granules of ezetimibe: Design, optimization and evaluation. Eur J Pharm Sci. 2008, 35 (3), 183–192.
20. Friedman, D.; Benita, S., A mathematical morel for drug release from 0/W emulsions: Application to controlled release mMrphine emulsions. Drug Dev Ind Pharm. 1987, 13(9–11), 2067–2085.
21. Pathan, I. B.; Setty, C. M. Nanoemulsion system for transdermal delivery of tamoxifen citrate: Design, characterization, effect of penetration enhancers and *in vivo* studies. Digest J Nanomater Biostruct. 2012, 7(4), 1373–1387.
22. Makhmalzadeh, B. S.; Torabi, S.; Azarpanah, A. Optimization of ibuprofen delivery through rat skin from traditional and novel nanoemulsion formulations. Iranian J Pharm Res. 2012, 11(1), 47–58.
23. Kumar, S.; Talegaonkar, S.; Negi, L. M.; Khan, Z. I. Design and development of ciclopirox topical nanoemulsion gel for the treatment of subungual onychomycosis. Ind J Pharm Edu Res. 2012, 46(4), 303–311.
24. Singh, B.; Singh, R.; Bandyopadhyay, S.; Kapil, R.; Garg, B. Optimized nanoemulsifying systems with enhanced bioavailability of carvedilol. Colloids Surf B Biointerfaces. 2013, 101(0), 465–474.

25. Shakeel, F.; Haq, N.; Alanazi, F. K.; Alsarra, I. A., Impact of various nonionic surfactants on self-nanoemulsification efficiency of two grades of Capryol (Capryol-90 and Capryol-PGMC). J Mol Liq. 2013, 182(0), 57–63.
26. Bandyopadhyay, S.; Katare, O. P.; Singh, B. Optimized self nano-emulsifying systems of ezetimibe with enhanced bioavailability potential using long chain and medium chain triglycerides. Colloids Surf B Biointerfaces. 2012, 100, 50–61.
27. Li, W.; Yi, S.; Wang, Z.; Chen, S.; Xin, S.; Xie, J.; Zhao, C., Self-nanoemulsifying drug delivery system of persimmon leaf extract: Optimization and bioavailability studies. Int J Pharm. 2011, 420(1), 161–171.

CHAPTER 10

RECENT ADVANCES IN THE APPLICATION OF BIOPOLYMER SCAFFOLDS FOR 3D CULTURE OF CELLS

MALA RAJENDRAN and ANANTHA S. SELVARAJ

ABSTRACT

Tissue engineering is a flourishing and promising biomedical engineering field which aims to develop viable substitutes to restore and maintain the function of damaged tissues. It is an interdisciplinary field employing amalgam of principles from biology and engineering. Cells, scaffolds, and signaling molecules are the basic tools and pillars of tissue engineering. Culturing of cells in three–dimensional (3D) microenvironments simulates normal cellular compartment and enhances adhesion, proliferation and differentiation of cells than in 2D. Scaffold functions as delivery vehicle, matrix for cell adhesion, and serves as a mechanical barrier against infiltrating surrounding tissues which hampers tissue repair and regeneration. Scaffolds can be synthetic or natural in origin. Different types of scaffold like porous, fibrous, customs, extracellular matrix, hydrogel, microspheres, and native tissue scaffolds are available and they influence the characteristics of cellular unit processes. A variety of techniques for fabrication of scaffolds were developed. There is no scaffold universally suitable for all cells and all applications. This review highlights the application of natural and synthetic biopolymers in 3D culture of cells, their fabrication techniques, and describes the biophysical, biochemical, and biomechanical properties of scaffold which influence cellular processes inside the scaffold.

10.1 INTRODUCTION

The term "tissue engineering" was coined by Fung from University of California at the National Science Foundation Bioengineering meeting at Washington in 1988. Tissue engineering is defined as the application of principles and methods of engineering and life sciences, to obtain a fundamental understanding of structural and functional relationships in normal and pathological mammalian tissues and the development of biological substitutes to restore, maintain, or improve tissue function (1). It also includes all efforts undertaken to perform biochemical functions with cells inside an artificial system. Organ failure can occur as a consequence of injury, disease, infection, autoimmune disorder, alcohol, or hormonal imbalance. Many types of treatment options are available to repair damaged organs like drug therapy, surgery, implanting internal medical devices, and transplantation. Even though transplantation restores the function of impaired organs and tissues, it encounters many challenges. Major one is the availability and compatibility of organs. Many thousands of people die because of shortage in the availability of organs. Infection by hepatitis C and human immunodeficiency virus (HIV) also complicates transplantation. Further, recipients generally must remain on expensive immunosuppressive drugs for the rest of their lives. Imbalance in immune surveillance due to immunosuppression can cause new tumor formation. For example, colon cancer occurs 20–30 years later in patients whose urine is diverted into the colon. Despite the expensiveness associated with transplantation, outcome studies have shown that survival rates after major organ transplants are poor. Implantable foreign body materials have produced dislodgment of the implant, infection, fracture, and migration over time.

Tissue engineering is a rapidly expanding field that solves the problem of organ and tissue shortage. It can offer a potentially viable solution to improve the structural integrity and function of damaged tissues. Generally in *in vitro* conditions, cells grow randomly into two dimensional (2D) layers of cells. But cells must grow in 3D to gain the anatomical shape of an organ. Tissue engineering attempts to replace damaged tissues using living cells housed in a scaffold that guides tissue development. Figure 10.1 depicts the general outline of tissue engineering process. Scaffolds function as a temporary 3D matrix providing highly porous microenvironment for cell adhesion and proliferation at the site of injury and serves as a carrier for growth factors (2). In 3D culture, 100% of cell surface are in contact with the extracellular matrix and adjacent cells. Scaffolds are fabricated from natural, synthetic, or hybrid biopolymers. It is degraded or integrates itself into the host tissue as new extracellular matrix resulting in the formation of a biologically functional tissue.

Promising application of tissue engineering includes repair or replacement of damaged skin, bones, cartilage, connective tissue, muscles, cornea, and blood vessels. Further, 3D culturing of cells is an excellent *in vitro* model to unravel and understand the secrets of growth, differentiation, metabolic processes, gene expression, and response to stimuli and in the investigation of all complex biological processes. Efficient and economic fabrication of scaffolds from biopolymers and the production of commercial quantities of viable cells without contamination or genetic changes are the two main challenges in the successful application of tissue engineering.

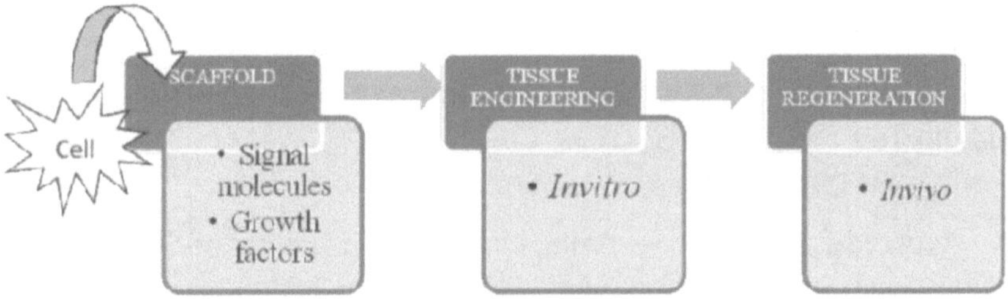

FIGURE 10.1 General tissue engineering process

10.2 HISTORY OF TISSUE ENGINEERING

Many principles of tissue engineering dates back to ancient times. The first verse of genesis mentioned in the *Bible* is the oldest sign of tissue engineering. It narrates that God turned one of Adam's ribs into Eve, the use of tissue to make a homologous human being of a different gender (Figure 10.2).

FIGURE 10.2 Eve being created from Adam's rib

Ayurvedic physician Sushruta, known as the "Father of Surgery" lived in India between 600 BC and 400 AD. The basic techniques used in modern plastic surgery today were outlined in Sushruta Samhita a medico-surgical compendium authored by Sushruta. In those days' ears or nose of criminals were amputated as a part of religious, social, or military punishment (3). Sushruta developed the forehead flap rhinoplasty technique that remains as a contemporary plastic surgical practice. Sushruta Samhita provided the first written record of free-graft Indian rhinoplasty (reconstructive nose surgery). It documented the use of autologous tissue to restore the aesthetic function of the mutilated nose and ears (4). Rotational flaps were taken from the forehead with the angular artery as a source of blood supply to reconstruct amputated noses as depicted in Figures 10.3–10.5 (5, 6).

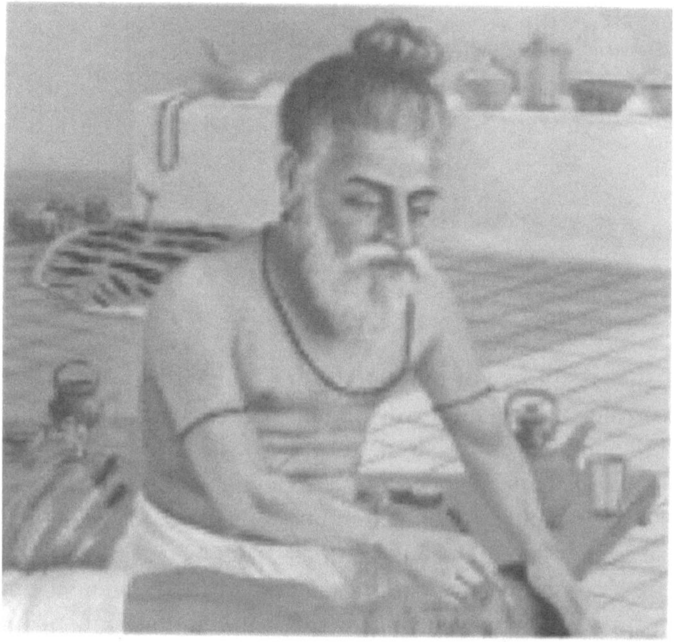

FIGURE 10.3 Skin grafting by Sushruta

FIGURE 10.4 Reconstruction of amputated ears by Sushruta

FIGURE 10.5 Indian rhinoplasty

First successful use of biomaterials in tissue engineering was the suturing of wounds with silk or linen/gum combination. It minimizes scarring and thus improves the strength of the healed wound. Archaeological records from ancient Egypt show that Egyptians used linen and animal sinew to close wounds. Linen is immunologically inert. Linen is effective in dealing with inflammatory conditions, reducing fever, and providing a healthy air exchange.Other natural materials used in ancient times were flax, hair, grass, cotton, silk, pig bristles, and animal gut. In Latin, flax means "being most useful". Flax thread appears to be the only natural material utilized for internal sutures in surgical practice (7).

Human dermal grafts were researched very earlier. First description of the use of porcine small intestinal submucosa for skin graft was documented by Arthur Bowen in 1936 for replacing damaged or lost skin parts. As natural biomaterials promote natural healing without eliciting any immune response, many techniques for decellularizing xenographic materials were developed in the 1980s. With the advance in the research of cell–scaffold interaction, the first *in vitro* generated full thickness skin graft was reported in 1981. Burke et al. (8) collaboratively investigated the application of collagen and silicone-based material as a template for the guided ingrowth of native skin and vessels from surrounding skin. Bell et al. (9) seeded autologous dermal fibroblasts and epidermal keratinocytes on collagen matrix and generated full thickness skin graft. This can be sutured or stapled in place during surgery without the risk of rejection. Wolter and Meyer in 1984 described the use of biosorbable scaffold as a carrier for cells (10).

Vacanti et al. (11) published the first tissue engineering paper as a study of cell transplantation using bioresorbable scaffold carriers. Researches were then centered on biomaterial scaffold fabrication and its influence on cell growth and proliferation. Publication of an article entitled "Tissue Engineering" in *Science journal* described critical areas for research within tissue engineering and its application in medicine and health care. Carticel, a chondrocyte expansion procedure was offered by Genzyme Biosurgery. It was approved by the Food and Drug Administration (FDA) and available in the market from 1990s. Apligraf, a collagen-based full thickness skin equivalent was available commercially to treat lower extremity diabetic and venous stasis ulcers. Vacanty et al. in 1997 created Vacanty mouse by subcutaneous transplantation of ear-shaped PGALA fiber mesh scaffold carrying chondrocytes at the back of athymic mouse (12).

10.3 BIOPOLYMERS AS SCAFFOLDS IN TISSUE ENGINEERING

Biomaterials are defined as those materials which are biocompatible and do not elicit any immune response (13). Broad spectrum of materials like polymers, ceramics, glasses, and metals can be designated as biomaterials. They are used to interact with biological system for the purpose of diagnosis or therapy. Polymers possess unique chemical and mechanical properties that can be tailored according to the tissue being engineered. So, they are widely used in tissue engineering applications. They facilitate the growth and differentiation of cells encapsulated or seeded into them (14).

10.3.1 CLASSIFICATION OF BIOPOLYMERS

Biopolymers are classified on the basis of their origin as natural, synthetic, and hybrid and are subdivided based on their chemical composition as given in Table 10.1 (15). Most materials commonly used in tissue engineering are adapted from surgical uses such as sutures, hemostatic agents, and wound dressings (16, 17).

TABLE 10.1 Classification of biopolymers

Natural	Synthetic	Hybrids
Proteins	**Organic**	PLAGA
Collagen	PLA	Bioactive glasses and hydroxy apatite
Gelatin	PGA	Biocomposite

Fibrinogen	PCL	Hybrids of natural polymers
Silk fibroin	**In organic**	
Elastin	Ceramic	
Polysaccharides	Hydroxy apatite	
Starch	Bioactive glasses	
Alginate		
Hyaluronic acid		
Chitin and chitosan		
Dextran		

Natural polymers are the attractive candidates for tissue engineering because of their biocompatibility and biodegradability. In spite of their potent advantages, widespread application of natural biomaterials is limited because of their vulnerability to viral infection, antigenicity, and heterogeneous material supply (18). Comparatively, synthetic polymers offer tremendous advantages with respect to homogeneity of the material, reproducibility, and flexibility in tailoring the polymer properties according to the tissue being engineered.

Biomaterial for repair should be carefully selected on the basis of their degradation rates compatible to the host tissue. Biodegradable polymers are those which are degraded *in vitro* or *in vivo* either into nontoxic normal metabolites of the body or into products that can be completely eliminated from the body. It should not be degraded before the tissue is repaired or reconstructed. Synthetic biodegradable polymers can be categorized into poly(ortho esters), poly(glycolic acid) (PGA), poly(lactic acid) (PLA), and poly(b-hydroxybutyrate) and copolymers with hydroxyvaleric acid, poly(ecaprolactone) (PCL), polyanhydrides, poly(trimethylene carbonate), and polyiminocarbonates (19, 20). Polymers should be processed to possess good thermomechanical properties. Copolymers of PLA and PGA (PLAGA) are used in sutures, bone pins, stents, drug delivery devices, and scaffolds for tissue engineering. The mechanical and degradation properties of these polymers can be tailored depending on the copolymer ratios. Application of polymers in the repair of nerve (21), liver (22), orthopedic suture (23, 24), and brain (25) were studied. Extensive investigations on the regeneration of bone by hydroxy apatite (26–32) and biocomposite (32–35) were also reported.

10.4 NATURAL BIOPOLYMERS

10.4.1 PROTEINS

Proteins are the major components of soft and hard tissues of human system. They are high molecular weight polymers composed of amino acids linked by peptide bonds. They are extensively studied for biomedical applications such as sutures, hemostatic agents, and scaffolds for tissue engineering and drug delivery.

10.4.2 COLLAGEN

Collagen is a group of closely related proteins having similar structural characteristics. At least 22 different types of collagen have been identified so far in the human body. Types I, II, III, and IV were studied extensively. Gelsea et al. (36) elaborated about various types of collagen and their molecular forms as represented in Table 10.2. Collagen is the most abundant protein being the major constituent of connective tissue such as bone, skin, ligament, cartilage, and tendon. Twenty-five percentage to thirty-five percentage of human body mass is formed of collagen. It also forms the structural framework for other tissues such as blood vessels. It is responsible for maintaining the biologic and structural integrity of the extracellular matrix. It is dynamic and undergoes constant remodeling for proper physiological functions (37). It is a triple helical polypeptide giving extraordinary tensile strength to the respective parts of body. Typically, they contain about 35% glycine, 11% alanine, and 21% proline and hydroxy proline. The amino acid sequence in collagen is generally a repeating tripeptide unit, Gly-Pro-Hydroxy proline. Tensile strength of collagen is more than that of a steel wire of same cross section.

TABLE 10.2 Types of collagen

S. No.	Type	Molecular formula	Tissue distribution	Polymerized form
1	I	$[\alpha 1(I)]2\alpha 2(I)$	Tendons skin, artery walls, cornea, the endomysium of myofibrils, fibrocartilage, and the bones and teeth	Fibers

TABLE 10.2 *(Continued)*

2	II	[α1(II)]3	Cartilage, intervertebral disks and vitreous humour of the eye	Fibers
3	III	[α1(III)]3	Arteries, intestine, and uterus	Fibers
4	IV	[α1(IV)]2α2(IV)	Basal lamina, eye lens, capillaries, and the glomeruli of nephron	Basement membrane
5	V	[α1(V)]2α2(V) and α1(V)α2(V)α3(V)	Most interstitial tissue	Fibers
6	VI	a1(VI), a2(VI), and a3(VI)	Most interstitial tissue, associated with type I	Basement membrane
7	VII	[α1(VII)]3	Forms anchoring fibrils in dermoepidermal junctions	Anchoring fibrils
8	VIII	[a1(VIII)]2a2(VIII)	Some endothelial cells	Anchoring fibrils
9	IX	α1(IX)α2(IX)α3(IX)	Cartilage, associated with type II and XI fibrils	FACIT
10	X	[a3(X)]3	Hypertrophic and mineralizing cartilage	Hexagonal network
11	XI	α1(XI)α2(XI)α3(XI)	Cartilage	Fibers
	XII	[α1(XII)]3	Perichondrium, ligaments, and tendon	FACIT
	XIII	[a1(XIII)]3	Epidermis, hair follicle, endomysium, intestine, chondrocytes, lungs, and liver	Transmembrane collagens
	XIV	[a1(XIV)]3	Dermis, tendon, vessel wall, placenta, lungs, and liver	FACIT
	XV	[a1(XV)]3	Fibroblasts, smooth muscle cells, kidney, and pancreas	Multiplexins
	XVI	[a1(XVI)]3	Fibroblasts, amnion, and keratinocytes	Multiplexins
	XVII	[a1(XVII)]3	Dermal–epidermal junctions	Transmembrane collagens
	XVIII	[a1(XVIII)]3	Lungs and liver	Multiplexins

TABLE 10.2 *(Continued)*

XIX	[a1(XIX)]3	Human rhabdomyosarcoma	FACIT
XX	[a1(XX)]3	Corneal epithelium, embryonic skin, sternal cartilage, and tendon	FACIT
XXI	[a1(XXI)]3	Blood vessel wall	FACIT

Advantageous feature of employing collagen in tissue engineering is that it can be fabricated into versatile forms like tubes, sponges, powders, fleeces, injectable solutions, and dispersions. The use of collagen as a suture material dates back a millennium, and one form of it, catgut, is still in use for surgery (38). Purified collagen taken from the submucosal layer of the small intestine of healthy ruminants are twisted together to form catgut suture (39). Abundant availability, versatile methods for isolation and purification, and biodegradability are the key advantages of using collagen as a biomaterial for tissue engineering (40). Collagen-based matrices also find wide applications in tissue engineering and have been extensively investigated in combination with growth factors (41). Due to its fibrous nature, collagen can withstand tensile loads and hence they are most suitable for bone tissue engineering.

In spite of the above mentioned advantages collagen also possesses a few limitations such as immune response and poor mechanical properties that restricts its wide application in clinical trials (42). Immunogenicity of collagen is dependent on the source from which collagen was isolated and purified and the technique is used to fabricate scaffold. Collagen from animal sources can transmit bovine spongiform encephalopathy. Other problems associated with the application of collagen are their high cost of production, heterogeneity, and difference in biodegradability (43).

Integra is a 3D porous matrix of collagen cross-linked with glycosaminoglycans. Apligraf and Orcel are FDA approved collagen-based skin equivalent. It consists of collagen layer seeded with fibroblasts and keratinocytes. Fabrication of collagen scaffold as a fiber was first reported with the use of polyethylene oxide (44–46). But fabrication into gels or sponges lacks mechanical strength. Strength can be improved by cross-linking with glutaraldehyde vapors, formaldehyde, and epoxy compounds. Type I collagen cross-linked with glutaraldehyde showed improved tensile strength (5). Main challenge in cross-linking is the cytotoxicity rendered by the cross-linker. In 2007 Barnes et al. (47) developed a technique for the cross-linking of electrospun collagen, without imparting cytotoxicity. They demonstrated the intact the nanofibrous structure, using 1-ethyl-3-(3-dimethylaminopropyl) carbodiimide hydrochloride (EDC) in ethanol. Mano et al. (48) also studied the morphology, growth, adhesion, motility, and osteogenic differentiation of human bone marrow-derived mesenchymal stem cells on EDC cross-linked electrospun collagen fibrous mats.

Collagen is used for tissue repair in two general ways: one is "top-down approach" where decellularized collagen is used as a scaffold. Next is "bottom-up approach" in which purified collagen can be fabricated into a scaffold of any 3D form microstructure like fibrous constructs, hydrogels, and foams (49). Gels are very low density 3D lattices of collagen nanofibrils (50, 51). When sedded with cells, they form stable integrin-mediated attachments with the fibrils (52). Aligned, fibrilar/cellular collagen hydrogel implants have been used successfully to guide peripheral nerve repair *in vivo* (53, 54). Collagen was used as as a nerve guide tube by Heijke et al. (55) for peripheral nerve regenration. Alovskaya et al. (56) reported the repair of spinal cord injury by collagen hydrogel in rat model. Fibrous scaffolds of collagen fabricated from native extracellular matrix are now commercially available from Koken Co, Ltd, Japan (57). Fibrillar collagen scaffolds with improved mechanical properties and matrix density has been developed by plastic compression (PC) from fibrillar collagen type I hydrogel. It produces sheets of dynamic shapes with the 40 μm thickness compressed sheets spiraled into 3D multilayered conduits (49, 51).

10.4.3 GELATIN

Gelatin is formed by the denaturation of collagen (58, 59). It consists of 19 amino acids. Denaturation eliminates all the limitations observed in collagen. It is highly biocompatible, nonantigenic, possess higher tensile moduli than collagen, and can be used as a delivery vehicle for growth factors (60, 61). One of the major limitation is it is soluble above 37°C and gels at room temperature. Gelatin is prepared in the form of hybrids by cross-linking with synthetic biopolymers. Many reports document the application of electrospun gelatin scaffolds were used for various applications such as wound healing (62, 63), nerve (64), dental (65), bone (66,

67), dermal tissue engineering applications (68, 69), and vascular grafts (70) cardiac tissue engineering applications.

10.4.4 FIBRINOGEN

Fibrinogen is a 340 kDa glycoprotein comprised of a pair of three polypeptide chains: 2Aα, 2Bβ, and 2γ. It is synthesized by the liver and circulates freely in the bloodstream. It is a naturally occurring key protein in blood clotting cascade and plays a vital role in wound healing (71, 72). The reaction of fibrinogen with thrombin produces fibrin. Interaction between oppositely charged molecules leads to the formation of fibrous clots. This functions as a transient matrix onto which tissues can rebuild and repair themselves (73, 74). Fibrinogen-based scaffolds have previously been developed in the form of fibrin gels and cables. These studies demonstrated that fibrinogen-based scaffolds were easily degradable, nonimunogenic, and promoted increased cell migration (75–77). The electrospinning of fibrinogen was first published by Wnek et al. (78) and are reported to be degradaed in a short time (79). Fibrin, is a temporary fibrous matrix that facilitates early repair. It is used as a "fibrin glue" by surgeons (80). It can be clinically used as a scaffold matrix for the delivery of keratinocytes (81, 82), mesenchymal stem cells transfected with growth factors, for spinal cord repair (83, 84), and repair of peripheral nerves (85). Fibrin has greater intrinsic strength than other scaffolds (86). Biomaterials containing mixtures of fibronectin/fibrinogen composite cables can be used as a scaffold for repair of peripheral nerve lesions (87, 88) and human dermal fibroblasts (89).

10.4.5 SILK FIBROIN

Silk fibroin (SF) a protein present in the natural silk fiber. It is present as two cores of fibroin covered with a layer of sericin (90). Sericin functions as an adhesive to maintain the structure of the cocoon. It is characterized by its high immunogenicity and should be removed before processing of fibroin for tissue engineering applications (91). It is made up of a heavy chain with molecular weight of 350 kDa and light chain of 25 kDa linked by disulfide bond (92). Extended β-sheet formation of silk fibroin provides it with extensive tensile strength (93). The unique mechanical property of silk fibroin is due to the shear alignment of the molecular chains and semicrystalline regions. Elasticity is provided by semicrystalline regions (94). Formic acid and water used in electrospinning enhanced the mechanical properties of the scaffolds (95). Electrospinning of silk fibroin was first reported and patented by Zarkoob et al. (96, 97). Soon after, Sukigara et al. (98, 99) reported the effects of the various electrospinning parameters on the morphology and fiber diameter of silk scaffold. SF concentration played a key role in producing uniform fibers. In 2002, Jin et al. (100) successfully electrospun SF from an aqueous solution by adding PEO to the silk solution in order to increase the viscosity (100). Rate of degradation of silk fibroin can be tailored by varying the parameters of electrospinning. It is vulnerabile to bacterial and enzymatic degradation. SF scaffolds have proven a feasible option for vascular grafts with their unique mechanical properties and flexibility. Porous tubular grafts can be fabricated from silk fibroin suitable for vascular applications (101). Tubular silk scaffolds fabricated out of formic acid resisted up to 575 mmHg in burst strength tests, which was more than four times the upper physiological pressure of 120 mmHg and twice that of pathological upper pressures of 180–220 mmHg. Aortic endothelial cell (EC) and arterial smooth muscle cell (SMC) growth and proliferation were reported to be promoted by electrospun aqueous SF scaffolds while withstanding vascular pulsating pressures (102, 103). Silk-based scaffolds meet all the ideal characteristics for a ligament and tendon tissue engineering (104, 105).

10.4.6 ELASTIN

Elastin is an important constituent of load-bearing tissues and extracellular matrix protein that provides elasticity to elastic human tissues like skin, lung and arteries, veins, ligaments, lung parenchyma, skin, and intestines (58, 106). It provides them with the ability to stretch and recoil and resume its original shape after contraction. For example, blood vessels stretch more than billion times in a human's life time. Its crucial role in arterial morphogenesis and in the maintenance of healthy tissue is well documented (107–110). It is insoluble and contains 36 domains with alternating hydrophobic and cross-linking characteristics (111, 112). It is composed of soluble 65 kDa protein called tropoelastin. Hydrophobic domain predominantly consists of glycine, proline valine, and leucine. Proline and hydroxyl proline residues occur as tandem repeats. Amino group of the basic amino acid lysine is converted to reactive aldehydes catalyzed by lysyl oxidase to facilitate the formation of interchain cross-links. Versatile nature of elastin is attributed to the wide range of Youngs modulus in different tissues from 100 to 600 kPa (113).

Loss of elastin due to burn injuries lead to scarring, wound contraction, and loss of skin extensibility (114). Renewal of elastin in adult tissues is poor as its expression is exclusive to early development (115). Therefore material which mimics the structure, properties, and function of elastin is in great demand for the regeneration of vascular tissue engineering (116). Mechanical properties of synthetic biomaterials used in vascular tissue engineering are incompatible with the natural arteries and is the sole reason for high failure rate in cardiovascular diseases (117). But elastin can restore the normal function of vasculature. It mediates the attachment and proliferation of endothelial cells from vascular origins (118, 119), dermal fibroblasts (114), and a chemoattractant for endothelial cells and monocytes (119). Repeated elastin-like sequences have been produced by recombinant means (120), and is used in engineering of vascular graft, skin (121), heart valves (122), and elastic cartilage (123). Decellularized tissues containing elastin is used for bladder augmentation (124), heart valves and vasculature for heart valve replacement and vascular grafts (125), and bovine pericardium to enhance bone formation (126). One limitation of decellularized tissue is the difficulty in obtaining highly purified preparations from intact tissue and a large batch-to-batch variation in the construct due to difference in the decellularization methodology applied (127). Various methods have been employed to prepare elastin including chemical extraction (128–135), oxalic acid solubilization (136), alkali solubilization (137), enzymatic solubilization (138, 139), and chemical synthesis (140–143). Modern biotechnological techniques like genetic engineering were also applied to prepare pure elastin (144–146), recombinant elastin and fibronectin (147), silk and elastin (148–149), and elastin and synthetic polymer (150–151). Limitations in the application of elastin gels documented include reduced proliferation of smooth muscle cells compared to tissue culture plates (TCPs) (152), incomplete 3D structure formation by electrospun elastin sheet (153), and *in vitro* scaffold contraction (154, 155).

10.5 POLYSACCHARIDES

Polysaccharides are high molecular weight homo or heteropolymers. They are available in abundance in nature. Reactive functional groups in polysaccharides make them easily amenable to modifications. They can be readily and easily fabricated into hydrogels which make them excellent candidates for tissue engineering.

10.5.1 STARCH

Starch is a natural homopolysaccharide composed of D-glucose units linked by α (1–4) and α (1–6) linkage (Figure 10.6). It is composed of two basic units amylose without branching and amylopectin with branch points. Natural starches contain 10–20% amylose and 80–90% amylopectin. Amylose molecules consist typically of 200–20,000 glucose units which form a helix as a result of the bond angles between the glucose units. Short side chains of about 30 glucose units are attached with 1α→6 linkages approximately every 20 to 30 glucose units along the chain.

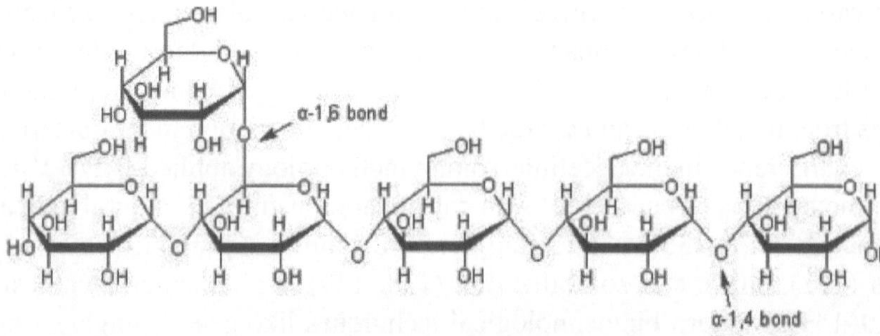

FIGURE 10.6 Structure of starch

Amylopectin molecules may contain up to two million glucose units. It is suitable for application in bone replacement (156, 157). In practice, starch is often used in combination with other biomaterials, such as with cellulose acetate (158), hydroxyapatite (HA) (159), poly(ethylene-vinylalcohol), and poly(lactic acid) to confer the scaffold with different properties suitable for different applications. *In vitro* study by Neves et al. (160) documented the proliferation of osteoblasts on starch scaffold.

10.5.2 ALGINIC ACID

Alginic acid is a linear heteropolysaccharide produced by brown algae. It is composed of D-mannuronic acid and L-guluronic acid (Figure 10.7). It readily forms gel in the presence of divalent ions like calcium. It is a suitable biopolymer for wound dressing (161). Alginates have been cross-linked with any divalent cation salts like calcium sulfate, calcium chloride, calcium carbonate, and barium chloride each with its characteristic gelation rate. Calcium sulfate kinetics results in the formation of nonuniform gel structures (162). Many researchers have studied the mechanical behavior of alginate under various conditions (163). Kuo et al. (164) have reported that slow gelling system resulted in the formation of uniform gels with high mechanical properties applicable for tissue engineering.

Genes et al. (165) showed that the cross-link density and substrate stiffness influenced the rate of cell attachment density. Concentration of calcium ions used for cross-linking influences the structural and functional integrity of scaffolds. This variation in structure and mechanical properties influences the adhesion of cells and the rate of proliferation (166–168), mechanical property of alginate gel, and on the loading capacity of the gel (166, 167). It also influences the structural and functional integrity of the constructs and their interaction with the surrounding hard and soft tissue of the host (168).

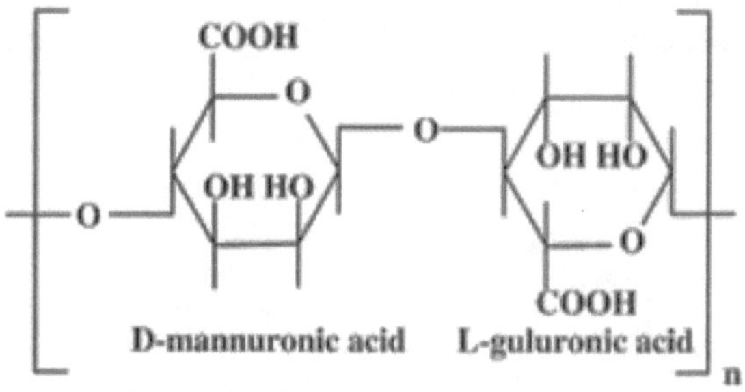

FIGURE 10.7 Structure of alginate

Alginates exhibit very low interaction with cells or proteins and hence they are suitable as matrices for anchorage-independent cells like chondrocytes (169). Spontaneous gelation, biocompatibility, and low cost of sodium alginate have made alginate as a suitable polymer for cultivation of chondrocytes (170). Many studies have reported the application of alginate for the culturing of chondrocytes, hepatocytes, and Schwann cells for nerve regeneration (171, 172). It is used in the repair of cartilage tissue (173, 174). The ability of rat bone marrow cells to proliferate and differentiate in alginate scaffold of differing composition and purity was studied by Wang et al. (175). It was recorded that high purity and high G-type alginate retained 27% of its initial strength after 12 days in culture (175). Tissue regeneration in the scaffold can be accelerated by stimulation with compressive loading (166).

Studies by Yan et al. (176) demonstrated the growth of human adipose derived adult stem cells in alginate gel. Their study reported the decrease in elastic modulus due to loss of calcium ion within 14 days. Mechanical properties of alginate reveal the rate of degradation of alginate. Simpson et al. (177) postulated that strength of the alginate gel network is provided by the number of alginate strands held together in the "egg-box" model.

10.5.3 HYALURONIC ACID

Hyaluronic acid is a naturally occurring linear anionic heteropolysaccharide consisting of D-glucuronic acid and 2-acetamido-2-deoxy-D-glucose monosaccharide units (Figure 10.8). It is soluble in water and forms very viscous solutions. It is an important component of articular cartilage and is widely distributed in the connective tissue, vitreous, and synovial fluids of mammals. Mesenchyme of developing embryos also contains hyaluronic acid. Rate of degradation of hyaluronic acid is determined by esterification (178).

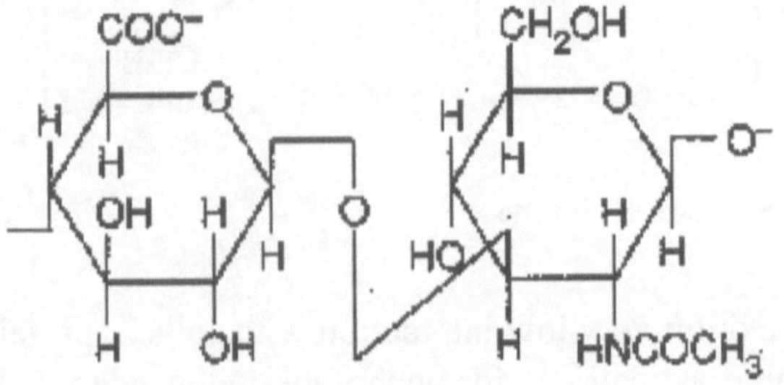

FIGURE 10.8 Structure of hyaluronic acid

10.5.4 CHITIN AND CHITOSAN

Chitin is the second most abundant natural polymer, next to cellulose. It is a semicrystalline linear polymer of (1→4) β-linked D-glucosamine residues with some randomly distributed *N*-acetyl glucosamine groups (Figure 10.9). It is present in the outer shell of crustaceans, insect exoskeletons, and fungal cell walls. Chitosan is completely soluble in aqueous solutions with pH lower than 5.0. It is degraded by lysozyme to nontoxic products (179). The rate of degradation of chitosan depends inversely on the degree of acetylation and crystallinity of the polymer. The feasibility of forming porous scaffolds may permit wide applications for this polymer in tissue engineering. This is particularly true for bone tissue engineering applications, as chitosan is known to support osteoblast proliferation and phenotype expression (180).

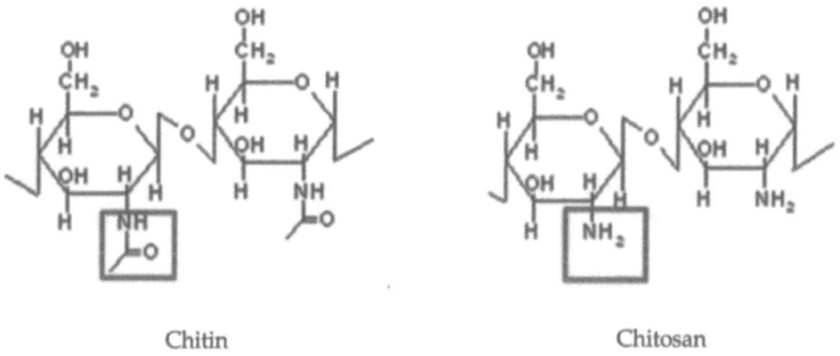

FIGURE 10.9 Structure of chitin and chitosan

10.5.5 DEXTRAN

Dextran is a polysaccharide similar to amylopectin, but the main chains are formed by 1α→6 glycosidic linkages and the side branches are attached by α(1→3) or α(1→4) linkages (Figure 10.10) with molecular weights greater than 1000 Dalton. On the basis of their structural features, dextrans can be classified into three classes. Class 1 dextrans contain the α(1→6)-linked D-glucopyranosyl backbone with small side chains of D-glucose branches with α(1→2), α(1→3), and α(1→4)-linkage. Isomaltose and isomaltotriose belong to class 1 dextran. Class 2 dextrans have a backbone of alternating α(1→3) and α(1→6)-linked D-glucopyranosyl units with α(1→3) branch. Class 3 dextrans have a backbone structure of consecutive α(1→3)-linked D-glucopyranosyl units with α(1→6)-linked branches. Properties of dextarn vary depending upon the microbial source used for the production. Biocompatible functionalized conjugates can be prepared by derivatization through the hydroxyl groups of dextran.

Lévesque et al. (181) fabricated interconnected microporous and macroporous gel- structure exploiting the advantage of the liquid–liquid immiscibility of poly(ethylene glycol) and methacrylated dextran during radical cross-linking of the methacrylated moieties. Zhou et al. (182) fabricated dextran-hyaluronate (Dex-HA)-based supermacroporous cryogel scaffolds for soft tissue engineering were prepared by free radical cryo-copolymerization of aqueous solutions containing the dextran methacrylate (Dex-MA) and hyaluronate methacrylate. They evaluated the porosity and permeability of dextran hydrogel for adipocyte cells. Influence of enzymatically cross-linked dextran and tyramine conjugate on the growth of chondrocytes was studied by Jin et al. (183).

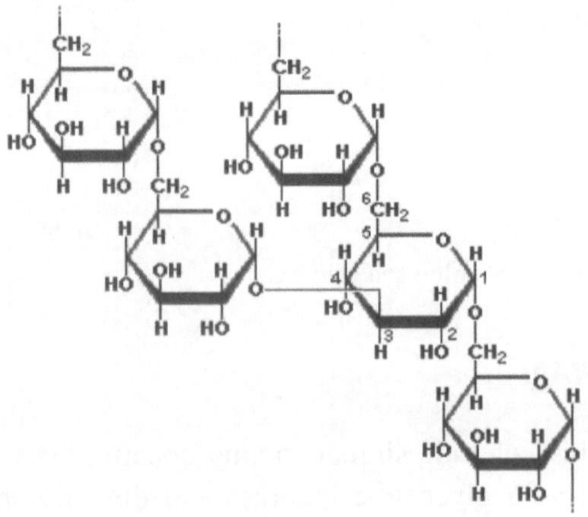

FIGURE 10.10 Structure of dextran

10.5.6 PLA

Polylactic acid or polylactide (PLA) is a biodegradable polymer of polyhydroxy ester (Figure 10.11). It is a thermoplastic aliphatic polyester derived from renewable resources, like corn starch, tapioca roots, chips, or starch or sugarcane It is approved by the FDA for medical application. PLA is one of the most preferred biomaterials for fabrication of scaffold in tissue engineering because of the ease in tailoring the physical and chemical properties according to the need. Matsusue et al. (184) studied the *in vitro* and *in vivo* degraradation of PLA when used for bone engineering. It can also be prepared as copolymer with PGA to form PLAGA which are used in surgical sutures and orthopedic fixation devices with extended success (185, 186).

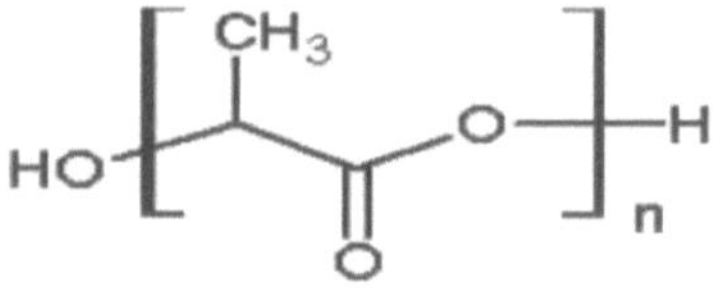

FIGURE 10.11 Structure of polylactic acid

10.5.7 POLYGLYCOLIC ACID (PGA)

PGA is a polymer of glycolic acid (Figure 10.12). It is a rigid thermoplastic material with high crystallinity (46–50%). Because of high crystallinity, it is not soluble in most organic solvents except highly fluorinated organic solvents such as hexafluoro isopropanol. Major application of PGA is in resorbable sutures. Chu et al. (187–189) have established a simple degradation mechanism via homogeneous erosion.

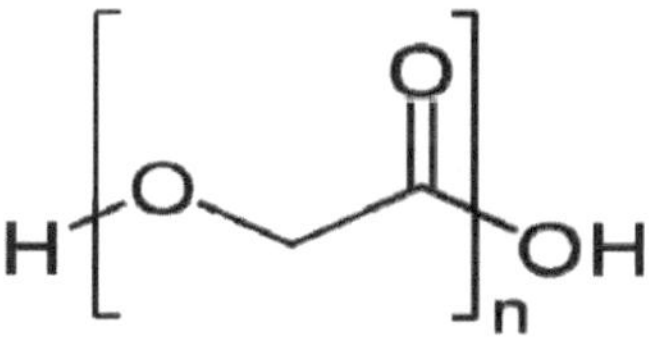

FIGURE 10.12 Structure of polyglycolic acid

PGA is degraded into glycolic acid which is a natural metabolite. PGA degradation occurs in two stages where initially it diffuses from matrix and then degraded in bulk leading to mechanical loss. Glycolic acid at high concentration can lower the surrounding pH causing tissue damage.

Finally it is metabolized to carbon dioxide and eliminated through respiratory system (190). Controversial to this metabolic fate of glycolic acid, Hollinger (191) has suggested that glycolic acid is converted into glyoxylate (by glycolate oxidase), which is then converted into glycine. Extrusion, injection, solvent casting, particular leaching, and compression moulding can be used to fabricate PGA into scaffold of various forms. The preferred method for preparing high molecular weight PGA is ring-opening polymerization of glycolide. Fabrication of PGA using melt polymerization was also reported (192, 193).

10.5.8 POLYCAPROLACTONE (PCL)

PCL is a semicrystalline linear aliphatic polyester (Figure 10.13). It is less widely used in tissue engineering due to its rigid and plasitic deformation charactereistics. Moreover, the degradation time is 2–3 years unsuitable for tissue engineering applications. To harness its biocompatibility for tissue engineering applications it must be blended with other polymers. It is approved by the FDA for medical and drug delivery applications. It is degraded via metabolites of tricarboxylic acid cycle intermediates. Gomes et al. (194) fabricated fiber mesh scaffolds from a blend of starch and PCL for the growth and proliferation of rat bone marrow stromal cells.

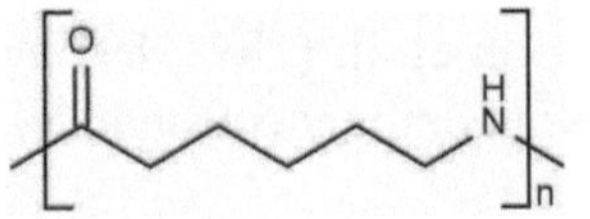

FIGURE 10.13 Structure of polycaprolactum

Properties of the scaffold depend on the nature of the solvent used and the substrate employed for casting. Zang et al. (195) fabricated PCL films by solvent casting and spin coating and evaluated the surface properties and biocompatibility for the growth of mouse fibroblast cells. Studies proved that finest pore size of the scaffold favored the adhesion of cells. PCL films have also been reported by spin cast on to glass coverslips by Ishaug-Riley et al. (196).

10.6 SYNTHETIC INORGANIC BIOMATERIALS

10.6.1 CERAMIC

The word "ceramic" in Greek means pottery "to burn". Ceramics can be classified based on the ability to interact with biological system into biologically inert or active which determines its medical application. Alumina, zirconia, and alumina–zirconia composites are biologically inert and are used in load-bearing applications. HA, bioactive glasses, and glass ceramics are biologically active and are used in nonload bearing applications, like fillers or as resorbable materials, in bone tissue engineering (tricalcium phosphate). For a ceramic to be used as an implant material, Youngs modulus should be compatible with the tissue. For example, the Young's modulus of cortical bone is approximately 20 GPa. Only pyrolytic carbon suits well for this as it has Young's modulus value approximately nearer to 30 GPa. Moduli of metals are much higher than those of bone. It is 210–250 GPa for Co–Cr–Mo alloy, 160–210 GPa for Ni–Cr dental alloy, 190–200 GPa for stainless steel, and 100–120 GPa for Ti alloy. So, they are unsuitable for bone engineering. New types of Ti alloys, like Ti–45Ni ("Nitinol"), have Young's modulus as low as 30 GPa. Most suitable biomaterial for elastocompatible implant is polymer–matrix composites only.

Bioactive bioceramics like hydroxy apatite and tricalcium phosphate are most suitable for bone engineering. Bone graft for tissue engineering should be able to support abundant bone formation (osteoconductive), able to induce bone formation (osteoinductive), able to form a continuous interface with surrounding bone tissue (osteointegrative), able to support angiogenesis, and be structurally and mechanically compatible with bone tissue (Shikinami and Okuno (197). For bone regeneration pore diameters of scaffold can be between 100 and 500 μm. Interconnected pore space is essential for perfusion of seeded cells into the scaffold matrix. TCP (tricalciumphosphate) resorbs at a very high rate, sometimes incompatible with bone regeneration rate. Two-phase calcium phosphate made from TCP and HA can be used as bone cements.

10.6.2 HYDROXY APATITE

HA is a naturally occurring mineral form of calcium apatite with the formula $Ca_5(PO_4)_3(OH)$. It is generally written as $Ca_{10}(PO_4)_6(OH)_2$ to denote that the crystal unit cell comprises two entities. It is the end member of the complex apatite group (28). It is widely used in clinical bone regeneration as they are highly stable and the ratio of calcium to phosphorus is close to natural bone (198). Bone tissue is considered as a composite of minerals and proteins.The minerals are mostly apatites such as hydroxyapatite (HA, Ca_{10} $(PO_4)_6(OH)_2$), fluorapatite, and carbonate-apatite according to Gineste et al. (199). Up to 50% of bone by weight is a modified form of HA (known as bone mineral).

It is well known for its biocompatibility, osseoconductivity, and osseoinductivity irrespective of its low degradation rate (200–202). It promotes osseointegration (30). HAs can be fabricated as gels, pastes, and solid blocks or even as porous matrices. Variety of techniques can be used to produce porous scaffold like gel casting (203, 204), gas foaming (205), 3D printing (206) slip casting (207, 208), fiber compacting (209), solid free form fabrication (210) freeze casting (208), and polymer casting (211, 212). Porous HA implants are used for local drug delivery in bone. Microcrystalline hydroxyapatite (MH) is marketed as a "bone-building" supplement with superior absorption in comparison to calcium. It is a second-generation calcium supplement derived from bovine bone (29). Teixeira et al. (210) prepared porous HA scaffolds by using polyurethane sponges as a template impregnated with ceramic slurry. Rapid prototyping holds an advantage that internal architecture can be tailored and fine-tuned and can be manufactured straight from a 3D data set in one step without using an additional mold (212). Leukers et al. (205), in 2005, fabricated high resolution porous scaffold by SD printing for the growth and proliferation of MC3T3-E1 murine fibroblasts. Interconnecting channels with mean diameters of about 500 μm were reported which facilitates enhanced cell adhesion and perfusion into the depth of porous scaffold. This pattern of growth will aid in osseointegration.

10.6.3 BIOACTIVE GLASSES

Bioactive glasses are a group of surface reactive glass–ceramic biomaterials. It was first developed by Hench and colleagues at the University of Florida. They are a unique range of dense, amorphous materials made of calcium, sodium, and phosphosilicate (CSPS). There are variations in the original composition which was approved by the FDA. This composition of bioglass termed as 45S5 is 46.1 mol% SiO_2, 26.9 mol% CaO, 24.4 mol% Na_2O, and 2.5 mol%

P_2O_5. Bioactivity represents the ability to form a mineralized HA layer on the surface of bone.It is capable of forming a strong chemical bond with the collagen of living tissues. Bonding of bone collagen to bioactive glass composition 45S5 was published in 1971. Further, 45S5 bioactive glass is the most bone bioactive material. In addition to being osteointegrative, the biocompatibility, osteoconductive, and osteoinductive nature of 45S5 bioactive glass have been well charatcterized (213, 214). But the disadvantages are its brittleness and lack of compressive and tensile strength (215). The underlying mechanisms that enable bioactive glasses to act as materials for bone repair occur in five stages. First is the ion exchange between glass (mostly Na^+) and external solution (hydronium$^+$). Second is the formation of Si-OH groups by hydrolysis of Si–O–Si bridges and the disruption of glass network. Once the disruption of glass network occurs, condensation of silanol group follows and forms a gel like surface layer. As a fourth stage, amorphous calcium phosphate layer is deposited on the gel layer by precipitation. As a fifth and final reaction, mineralization transforms the calcium phosphate layer into crystalline HA that mimics the mineral phase naturally contained with vertebrate bones.

10.6.4 HYBRID

10.6.4.1 PLAGA IN BONE ENGINEERING

PLAGA is a hybrid of PLA and PGA (Figure 10.14). Bone itself is a composite in nature with organic and inorganic phases, made up of bone-forming cells, bone-resorption cells, extracellular matrix, and inorganic bone mineral. Demand for bone engineering increases due to natural degeneration (216). Fabrication of scaffold with morphogenic proteins stimulated the differentiation of housed stem cells into osteoblasts. On degradation it was completely resorbed without the need for surgical intervention (217). Lu et al. (34) developed a microsphere-based, porous, PLAGA–BG composite scaffold by solvent casting to support the growth of human osteosarcoma cells. They reported that high porosity of microsphere decreased the compression strength of scaffold and also reported that addition of bioglass materials increased the hardness of the scaffold. Fabrication of polymer films by spin casting for the growth of human articular chondrocyte was investigated by Ishaug-Riley et al. (196). Surface properties of scaffold differ with the technique employed for fabrication. Heteropolymers like PLAGA possessed superior osteoconductive characteristics than the respective homopolymers (25). Astete and Sabliov (218) studied the synthesis and chracterization of PLAGA nanoparticles.

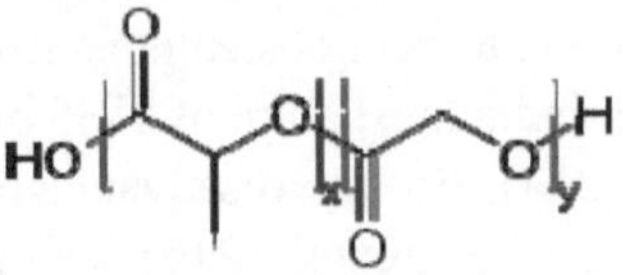

FIGURE 10.14 Structure of PLAGA (*x*=number of units of lactic acid and *y*=number of units of glycolic acid)

PLAGA scaffolds have been made from PLLA and PLGA by using chloroform as a binder. PLGA is one of the most common synthetic biopolymers used for fabricating 3D scaffold (219).

10.6.5 BIOACTIVE GLASSES AND HA

Bioactive glasses and HA composite are termed as active as they actively interact with the biological environment and can chemically integrate with surrounding bone tissue *in vivo*. They are excellent candidates for bone regeneration. Initially they form a calcium phosphate (Ca-P) layer, which is later modified by bone cells. Through this layer, the implant is chemically fixed to surrounding bone tissue (220). Ideally, a bone graft should be biocompatible. Ignjatovic et al. (186) investigated the role of HA and PLA in bone regeneration. Collagen, the major constituent of extracellular matrix, is arranged in different combinations and proportions in various tissues to provide varying tissue properties. In bone, it is aligned in parallel form along stress vectors. HA, the natural ceramic mineral component of the bone, is bonded to collagen (221). Glass and glass ceramic-based matrices have plenty of medical applications like dental implants, bone fixation devices, and implant coatings (222–223). Unique bioactive interaction of glass ceramics with bone is also documented (224). Inherently poor tensile strength and high compressive strength of bioactive glasses limit their widespread application in load-bearing situations (225).

10.6.6 PLAGA–BG BIOCOMPOSITE

PLAGA–BG is a biocomposite prepared by the combination of PLA, PGA, and bioglass by solvent casting. Preparation of a composite eliminates the disadvantage of each of the constituent components and possesses the advantage of all components. Lu et al. (226) produced PLAGA–BG composite disc by solvent casting and microspheres by water–oil–water emulsion. They combined 45S5 bioactive glass granules with polymeric matrix with the objective of improving the physical and chemical properties of the scaffold which will be mirrored in the osseoconductive and osseointegration potency of the scaffold. Glass substrate stiffens the polymeric matrix for the growth and proliferation of human osteosarcoma cells (SaOS-2). Biocomposite holds an additional advantage that acidity produced by the release of lactic acid and glycolic acid during degradation is neutralized by alkaline ions released by BG. Several studies were reported in the literature for the fabrication of polymer–glass composite. Fabrication of bioactive, resorbable, polylactide-glass microcarriers for culturing in the bioreactor was investigated and reported (227, 228).

10.6.7 HYBRIDS OF NATURAL BIOPOLYMER

Ma et al. (42) have studied the biodegradability of collagen scaffold fabricated with a cross-linking agent carbodiimide, 1-ethyl-3-(3-dimethylaminopropyl)-carbodiimide (EDAC) and N-hydroxysuccinimide (NHS) in the presence of glycine, glutamic acid, or lysine. Amino acids function as a cross-linking bridge between collagen molecular chains. They reported the delay in degradation of collagen scaffold by collagenase in the presence of lysine. Enhanced mechanical strength and seeding capacity and cellular interaction in collagen biocomposite with bioceramics was discussed (229–232). Collagraft is an FDA approved biocomposite composed of type I collagen and HA/tricalcium phosphate granules as a synthetic bone-graft substitute. Investigations were documented with composite of collagen and glycosaminoglycans and collagen and chondroitin sulfate for dermal regeneration. Collagen type II was fabricated to form a hybrid with chondroitin sulfate/dermatan sulfate/keratin sulfate for articular cartilage tissue engineering (233). Zhong et al. (234) fabricated collagen and glycoasamino glycan composite scaffold cross-linked by glutaraldehyde and studied its influence on proliferation of fibroblasts.

Thomas et al. (235) has prepared nanofibrous biocomposite scaffolds of type I collagen and nanohydroxyapatite (nanoHA) for bone tissue engineering applications. They reported the increase in surface roughness, fiber diameter, and the tensile strength of the scaffold in the presence of nanoHA. Collagen blended with chitosan nanofiber scaffold for the proliferation of EC and SMC nanofibers were also reported (236). Composite of collagen and silk fibroin was fabricated by Yeo et al. (237). He et al. (238) demonstrated that collagen type I blended PLLA-CL (poly (L-lactic acid)-co-poly(ε-caprolactone) improved the adhesion and proliferation of human coronary artery endothelial cells (HCAECs). A similar report was also documented by Kwon and Matsuda (239) for human umbilical vein endothelial cells. Lee et al. (240) developed PCL/collagen type I composite scaffolds which resist high degrees of pressurized flow over long durations for the growth of vascular cells. Polycaprolactone (PCL) is an inexpensive, bioresorbable polymer with excellent mechanical properties and slow degradation time (241). Schnell et al. (242) developed PCL/collagen scaffold for use as artificial nerve implants. Improved Schwann cell migration, neurite orientation, and formation of Schwann cells were recorded.

Electrospinning of gelatin with polyanaline was investigated by Li et al. (243). Fabrication of gelatin and polycaprolactum was studied by Heydarkhan-Hagvall et al. (244). Electrospinning of α-elastin and tropoelastin was reported by Li et al. (245). More recently, Nivison-Smith et al. (246). electrospun tropoelastin from HFP and cross-linked the scaffolds to form synthetic elastin microfibrous constructs.

The CS5 domain of fibronectin is important in promoting the attachment and spreading of endothelial cells over that of fibroblasts, vascular smooth muscle cells, and blood platelets (247). Hybrid artificial proteins consisting of elastin-like peptides interspersed with the CS5 domain of fibronectin were produced in *E. coli*. It promoted human umbilical vascular endothelial cell adhesion. These elastin–fibronectin polypeptides can be prepared with variable primary structure. Basic amino acid lysine is important in facilitating the cellular attachement. Glutaraldehyde is used to cross-link elastin-CS5 materials. Other cross-linkers including dissuccinimidyl suberate in DMF, hexamethylene diisocyanate in DMSO and bis(sulfosuccinimidyl) suberate were used (248, 249).

Recently, spider silk–elastin fusion proteins of a 51.2 kDa spider silk and a 94.2 kDa elastin-like protein (separated by a c-myc tag for detection) were prepared by a similar approach and were cytocompatible for human chondrocytes (250). Cappello et al. (148) synthesized polymers composed of fibroin-like crystalline blocks and elastin-like blocks in *E. coli*. Block composition and polymer length are of major importance for mechanical properties, biodegradability (which is primarily by enzymatic hydrolysis), and physiological properties of these proteins for drug-delivery and (251) gene delivery systems (252, 253). Synthetic elastin peptides can also be added to scaffolds, hydrogels or surface coatings, for example, elastin–laminin-like hybrid peptides to alginate dressings (254) and tropoelastin fragments coating for synthetic tubing (255).

Apligraf, a type I collagen gel with cultured human neonatal foreskin dermal fibroblasts and keratinocytes, little elastin was synthesised *in vitro* (256). Only few researchers have explored the use of elastin(-like) containing materials as skin substitutes. These include a decellularized porcine membrane of approximately 70% collagen and 30% insoluble elastin and minor amounts of glycosaminoglycans (257). Scaffolds of type I collagen coated with 3% α-elastin (258), hybrid peptides of elastin covalently linked to alginate dressings (186) and injected α-elastin (259). A hybrid of elastin–laminin peptides promoted attachment and proliferation of normal human dermal fibroblasts in culture. Recently, it was found that a proteolytic digest of elastin induced elastic fibre deposition in stimulated dermal fibroblasts injected in the skin of nude mice as well as in cultured human skin explants (260). Fibronectin and laminin were fabricated as a scaffold by conjugating with PEG (261). Zisch et al. (262) evaluated collagen–chitosan (COL-CS) scaffold supplemented with of aloe vera for their possible application in tissue engineering. Seliktar et al. (263) fabricated PEGylated fibrinogen hydrogel cross-linked in the presence of cells.

Another research effort by Miyamoto et al. (264) focused on the fabrication of alginate scaffold with nondecay type fast-setting calcium phosphate cement. Stitzel et al. (265) fabricated a hybrid made up of PLA and collagen type I fibers for vascular grafting. Collagen fibers were wound to mimic the extracellular matrix structure in an artery. PLA matrix was spun around this fiber to hold the cells.They reported the formation of confluent layers of human aortic smooth muscle cells in the luminal and external surfaces of the vascular graft. Layer-by-layer deposition of type I collagen, styrenated gelatin (ST-gelatin), and segmented polyurethane (SPU) was fabricated using electrospinning by Kidoki et al. (266). This hierarchically layered scaffold provided natural environment of arteries for artificial grafts (267).

10.7 TECHNIQUES FOR FABRICATION OF SCAFFOLD

Numerous techniques are available to fabricate polymers into various forms like films, beads, or foams with different degrees of porosity. The success of tissue engineering depends on the macro and microarchitecture of the scaffold used to harbor the cells. Type of scaffold, its thermo mechanical characteristics, biocompatibility, and ability to interact with adjacent cells for optimum cell-to-cell interaction determines the loading density, its ability to withstand shear stress and strain as the cells proliferate, and the differentiation of cells into appropriate tissue type. Scaffold should also serve as a vehicle for biochemical signaling molecules to program and guide cellular growth. But there are no universal characteristics applicable to all types of scaffolds. Characteristics of scaffold vary depending on the tissue being engineered and duration for which the scaffold is expected to act as a matrix.

Therefore versatile fabrication techniques are developed to meet dynamic requirements. There are many techniques/methods to fabricate scaffolds from biomaterials. Fabrication techniques can be broadly classified into conventional and solid freeform fabrication techniques (SSF). Conventional fabrication techniques are defined as the processes that build scaffolds with only bulk or porous structure without any channels. These are often relevant for hard tissue engineering like bone and cartilage tissues. Solid freeform fabrication produces precisely an accurately controlled hollow or tubular structures designed by computer-generated models. Many investigations were centered toward the SSF technique (268–271). It integrates cell seeding within the scaffold fabrication process, and thus avoids with poor cell infiltration (272).

Two types of strategies are utilized in developing scaffolds. In the first strategy, the scaffold has to provide support *in vivo*. In the second strategy, the scaffold only provides support *in vitro* until the cells are strong enough to support themselves *in vivo*. Biomimetic modeling and design can address some of the above concerns. Fabrication techniques reported are solvent casting, high-pressure gas foaming, lyophilization, phase separation, emulsion templating, and particle leaching (273).

10.7.1 SOLVENT CASTING

Solvent casting exploits the evaporative property of solvents in order to form scaffolds. Polymer solution will be deposited on the substrate and the scaffold will be formed after drying. Scaffolds can be fabricated by two ways either dipping the mold in polymer solution or by placing the polymer solution in the mould. This is a simple and easy method without any need for sophisticated equipments. But the disadvantage with this method is the use of toxic solvents which can denature proteinacious morphogens in the scaffold. Another disadvantage is the possibility of residual solvents retained in the scaffold. It can be solved by labor intensive and time-consuming vacuum drying process. Generally used solvents are 2,2,2-trifluoroethanol, methylene chloride, toluene, benzene, cyclohexane, acetone, and tetrahydrofuran (274).

10.7.2 PARTICULATE LEACHING

Leaching is employed to fabricate porousscaffolds (275, 276). Pores or channels are created using porogens. Porogens of desired size and shape are poured int molds. Polymer solublized in a suitable solvent is cast over the mold with porogens. On evaporation of the solvent, a composite of polymer/porogen will be formed. Excess unbound porogens are removed by washing with appropriate solvent. This method requires less volume of polymer and is easy to fabricate scaffolds with different microstructures by varying the size and shape of porogens (277). Acetone and sodium bicarbonate (278), toluene, hexane, cyclohexanone, 2-ethylhexanol, *p*-xylene, *n*-heptane (279) poly(ethylene glycol) (280), sodium hydrogen carbonate, and supercritical carbon dioxide (281) were also reported to be used as porogens.

10.7.3 PHASE SEPERATION

Phase separation technique is also known as immersion precipitation. Principle of the technique relays on the ability of the polymer solution to form a thin film on an inert support and the ability of the film to separate on contact with a nonsolvent. Polymer solution is cast into a mold to form a thin film. The mold with the thin film is immersed into a nonsolvent which results in the separation into different phases

10.7.4 FIBER MESH

The principle of this method is the deposition of polymer solution over a nonwoven mesh of another polymer followed by subsequent evaporation (282). This method provides good pore size and channels for nutrient transfer to cells (283). Lack of structural stability lmits the application of this technique.

10.7.5 FIBER BONDING

Fiber bonding technique develops scaffold with high mechanical strength (284). Cima et al. (285) first developed fiber bonding method. Synthetic polymer (PLLA) was dissolved in chloroform followed by the addition of nonwoven mesh of PGA added. Subsequently, the solvent was removed by evaporation (277). The scaffolds can also be fabricated by bonding with collagen matrix (286). Bondig was enhanced by sintering (287). This process yields the scaffolds of PGA fiber that is bonded together by heat treatment. PGA mesh provides the high porosity and surface area (288). Mechanical stability allows the penetration of cells (289). In fiber bonding, selective solvents used are toxic (290). This could interfere with the proliferation of cells. Principle of this method is similar to solvent casting but the polymer solution is casted on a mesh. PLLAGA constructs were made by this technique. This produces scaffold with large surface area that enhances seed-loading capacity. Interconnectivity between pores improves the perusion of cells and easy exchange of nutrients for the proliferation of cells PCL.

10.7.6 SPIN CASTING OR SPIN COATING

Principle of spin casting or spin coating is as solvent casting with an exception that the polymer solution is casted onto the glass or silicon wafer substrate held in vacuum under centrifugal field. Thickness of the film can be varied by varying the spinning time, speed, and solution characteristics. Only a narrow range of thickness can only be casted by this technique.

10.7.7 SUPERCRITICAL FLUID GASSING PROCESS

This technique is based on the principle of plasticizing polymers by gases under high pressure. This permits the incorporation of thermally labile proteins and drugs into scaffolds. It fabricates scaffold with pore size from 50 to 400 μm. But the limitation of this process is that it can process only polymers with a high amorphous fraction. This can be solved by using this technique along with particle leaching. Mooney and coworkers (282) were the first to describe the use of supercritical foaming for the preparation of macroporous scaffolds for tissue engineering applications. A recent study reported by Ginty et al. (291, 292) showed that mammalian cells can survive in a supercritical environment for up to 5 min. Myoblastic C2C12 cell line, 3T3 fibroblasts, chondrocytes, and hepatocytes were also tested which led to the development of a new injection system for the production of polymer/mammalian cell composites. In a single step, cells are loaded during scaffold processing.

10.7.8 FREEZE-DRYING

Freeze drying method is a thermally induced phase separation process in which the homogenous polymer solution is poured on to a mold and the temperature is decreased by freezing. As the processing is not automated it is subjected to variation and reproducibility is comparatively less resulting in the production of scaffold with wide variations in internal architecture (293).

10.7.9 MEMBRANE LAMINATION

Membrane lamination fabricates scaffold with precise anatomical shapes. It uses computer-assisted design to construct template of the implant shape. Membrane lamination is prepared by solvent casting and particle leaching and and assembles proteins layer-by-layer during the fabrication process. The membranes are soaked in solvent, and stacked up in 3D assemblies

(294). Limitation of this technique is comparatively less interconnectivity between pores of layers and only thin membranes can be used (295).

10.7.10 MELT MOLDING

Melt molding principle is based on the melting of polymers in the presence of porogens and casting into a desired mould. Porosity can be achieved by removing the porogens with water or suitable solvent. Thompson et al. (296, 297) used this principle first for the fabrication of PLAGA scaffold. This technique produces scaffold with homogeneously distributed pores. But high temperature required to melt limits the widespread use of this technique. Uniform distribution of HA fiber throughout the PLGA scaffolds could only be achieved by using the solvent casting technique to prepare the composite material of HA fiber, PLGA matrix, and gelatin or salt porogen, which are used in melt molding process (298).

10.7.11 HYDROCARBON TEMPLATING

This process is a combination of two distinct foam processes: (1) leaching of a fugitive phase with (2) polymer precipitation. It employs water insoluble hydrocarbon porogen for pore formation and simultaneous precipitation of polymer. Porogens can be removed by organic solvents according to Shastri et al. (299).

10.7.12 RAPID PROTOTYPING

Rapid prototyping (RP) exploits computer-aided design (CAD), computed tomography (CT), and magnetic resonance imaging (MRI) data for fabrication of tailor-made scaffolds with desirable properties. The digital information is then processed into a device specific cross-sectional format, expressing the model as series of layers. The file is then implemented on the SFF machine. It can fine-tune filament diameter, filament gap, and lay-down pattern which are correlated to the porosity, pore connectivity, and mechanical stability of the scaffolds (300). RP technology is launched in the market during the late 1980s with the introduction of the stereolithography system by 3D Systems Inc (301, 302). RP is also termed as "solid freeform fabrication". Over the past two decades, more than 20 SFF systems have been developed and commercialized, these include stereolithography (SLA), laminated object manufacturing (LOM), selective laser sintering (SLS), fused deposition modeling (FDM), and ink jet printing (IJP). Detailed information on the various SSF technologies is widely available in the literature (303–308). SFF allows the fabrication of objects with unique materials, combinations, and delicate geometries which could not be attained by traditional manufacturing methods. SFF is seeing increased use in biomedical engineering especially for tissue engineering scaffold fabrications as it can produce scaffolds with delicate internal architecture.

10.7.13 STEREO LITHOGRAPHY (SLA)

This technique involves fabrication of scaffold layer-by-layer by a computer-guided laser beam in accordance to the CAD cross-sectional data. It employs liquid photocurable monomer. Monomer is polymerized by guided UV laser beam. Once a layer is built it is sintered and lower to permit the layering of next layer till the model is completed. Chu et al. (209) have fabricated porous HA by SLA. Lin et al. (24) employed this method for engineering human trabecular bone.

10.7.14 FUSED DEPOSITION MODELING (FDM)

This technique exploits the principle of ink jet printers. Binder is extruded through a moving heated nozzle in a layered fashion on to a polymer. Binder dissolves the polymer and cross-links polymer particles. Layer-by-layer is assembled one above the other till the entire layers are fabricated. A variety of polymers can be fabricated into 3D scaffold using this FDM machine. PEGT/PBT block copolymer scaffold used for articular cartilage was produced by Woodfield et al. (308). Honeycomb-like architecture was produced with scaffolds PCL filaments by Hutmacher et al. (268). But the limitation is that it cannot be integrated with the addition of morphogens. The attractive feature of FDM is that it does not use toxic organic solvents Bredt et al. (309). This technique can produce scaffolds with pore sizes of 250–1,000 μm, with complete pore interconnectivity (310).

10.7.15 3D PRINTING (3DP)

In 3DP, binder is ejected from a jet head. It moves onto a polymer powder surface according to CAD cross-sectional data to form a layer. Chloroform can be used as a binder. It acts to swell, and bind adjacent particles once the solvent has evaporated (311). The limitation of this technique is the incomplete removal of solvent. Supercritical carbondioxide can be used to remove residual particles and solvent below the toxic level of 50 ppm (Koegler et al. 2002 (312)). The TheriForm 3DP-based fabrication process is a licensed technology developed at MIT. Zeltinger et al. (313) evaluated the influence of PLA porosity on fibroblasts, SMCs, and epithelial cells' growth.

10.7.16 3D DEPOSITION (3DD)

It is a technique of deposition using extrusion through nozzle or syringe. It is used to fabricate a variety of bioplolymers like thermoplastics, pastes, and hydrogels. Deposition occurs in a sterile environment avoiding contamination. So, this technique can use thermally sensitive biopolymers also. It also facilitates the incorporation of signaling molecules during fabrication. Ang et al. (314) fabricated chitosan–HA scaffolds using 3D plotting technique.

10.7.17 DESKTOP ROBOT-BASED RP (DRBRP)

Hoque et al. (315) have developed desktop robot-based RP (DRBRP) capable of extruding biopolymer for freeform construction of 3D tissue engineering scaffold. The DRBRP system was evaluated with PCL, PCL–PEG, and PCL–PEG–CL, and lay-down patterns. This technique extrudes a wide variety of biomaterials like hot melts, solutions, pastes, and dispersions of polymers as well as monomers and reactive oligomers. It completely utilizes polymer feed without wastage and without binders. Besides, this process is very appropriate to produce scaffolds for hard tissue engineering (e.g., bone). The DRBRP system consists of a computer-guided desktop robot metallic chamber heated by an electrical band heater, and a pneumatic dispenser. The dispenser consisted of an air filter, regulator, lubricator, a solenoid valve, and a nozzle. The thermoplastic polymers were melted in the stainless steel chamber and extruded out by means of compressed air pressure through a mini nozzle to build scaffold. Layer can be deposited with desired width and thickness. Software was made up of a slicing and dispensing program and allows generation of user friendly, geometrical data of 3D scaffolds. The scaffold was built in an additive manner: line-by-line to form a 2D layer and layer-by-layer to form the 3D structure. Once a layer was completed, the dispenser moved up vertically in the Z-direction for the next layer.

10.7.18 3D PLOTTING

This method uses compressed air to make a paste of plotting medium. The polymer is solidified on contact with the deposited layer. This permits wide array of physical form of polymer like solutions, slurries, and pastes along with growth factors simultaneously. Low viscosity in aqueous solutions can cause collapse of the 3D structure due to gravitational forces. So 3D plotting should be performed in a liquid medium whose viscosity is same as plotting material to compensate the gravitational forces. The temperature of the plotting medium was set to significantly higher by double jacket cartridge to delay gelation. This provides better interdiffusion. The ratio of gelation temperature and plotting temperature, nozzle type, densities of the deposited material and liquid medium determines the efficiency of 3D plotting.

10.7.19 ELECTROSPINNING

Electrospinning is a technology with the potential to fulfill the requirements of an ideal scaffold. This method employs electric field to create electrically charged jet of polymer droplets which form a mat of fibers upon evaporation of the solvent (316, 317). Briefly, electrospinning is accomplished by inducing a large electric potential (15–30 kilovolts DC) in a polymer solution and separating that polymer from an oppositely charged target. This charge separation creates a static electric field. As the field strength grows, the charge separation overcomes the surface tension of the solution and a thin jet of entangled polymer chains is ejected from the polymer reservoir. As this jet travels toward the target, instabilities within the charged jet define its orientation in space (condition previously described as whipping). By the time the jet reaches the target, the solvent has evaporated and a dry fiber is collected in the form of a nonwoven scaffold. Electro spinning produces nanofibrous scaffolds with a variety of polymers. This technique is suitable for a variety of polymers (318–323). It can produce the nanofibers with special orientation, high aspect ratio, high surface area, and having control over pore geometry (324–326). It offers an advantage to spin a blend of natural and synthetic polymers. It does not use toxic chemicals like other methods and are biologically safe. But there are limitations like low rigidity, uncontrolled pore size, and shape.

KEYWORDS

- **3D culture**
- **Biomaterials**
- **Biopolymers**
- **Collagen**
- **Electrospinning**
- **Scaffold**
- **Tissue engineering**

REFERENCES

1. Shalak, R.; Fox, C.F.; Liss, A.R. Preface. In: Tissue engineering; Shalak, R.; Fox, C. F.; Liss, A.R. Ed. Manhattan Publisher, New York, 1998; pp 26–29.
2. Langer, R.; Vacanti, J.P. Tissue engineering. Science. *1993*, 260(5110), 920–926.
3. Bhishagratna, K.K. Chapter VII. Surgical appliances, their uses and construction. (Yantra-Vidhimadhyayam).In: The Sushruta Samhita: An English translation based on original texts. *2006*, 978–981.
4. Eisenberg, I. A history of rhinoplasty. S. Afr. Med. J. 62. *1982,* 286–292.
5. Zimmerman, L.M.; Veith, I. Great ideas in the history of surgery. Norman Publishers: New York, 1993, p 587.
6. Sperati, G. Amputation of the nose throughout history. Acta Otorhinolaryngol. Ital. *2009*, 29(1), 44–50.
7. Madsen, E.T. An experimental and clinical evaluation of surgical suture materials—III. Surg. Gynecol. Obstet., *1958*, 106, 216–224.
8. Burke, J.F.; Yannas, I.V.; Quinby, W. C.; Jr.; Bondoc, C.C.; Jung, W.K. Successful use of a physiologically acceptable artificial skin in the treatment of extensive burn injury. Ann. Surg. *1981*, 194, 413–428.

9. Bell, E.; Ehrlich, H.P.; Buttle, D.J.; Nakatsuji, T. Living tissue formed *in vitro* and accepted as skinequivalent tissue of full thickness. Science, *1981*, 211, 1052–1054.
10. Wolter, J.R.; Meyer, R.F. Sessile macrophages forming clear endothelium-like membrane on inside of successful keratoprosthesis. Trans. Am. Ophthalmol. Soc. *1984*, 82, 187–202.
11. Vacanti, J.P.; Morse, M.A.; Saltzman, W.M.; Domb, A.J.; Perez-Atayde, A.; Langer, R. Selective cell transplantation using bioabsorbable artificial polymers as matrices. J. Pediatr. Surg. *1988*, 23, 3–9.
12. Vacanti, J.P.; Cao, Y,; Paige, K.T.; Upton, J.; Vacanti, C.A. Transplantation of chondrocytes utilizing a polymer-cell construct to produce tissue-engineered cartilage in the shape of a human ear. Plast Reconstr. Surg. *1997*, 100, 297–302.
13. Williams, D.F. The Williams dictionary of biomaterials. Liverpool University Press, Liverpool, 1999.
14. Sipe, J.D. Tissue engineering and reparative medicine. Ann. NY Acad. Sci. *2002*, 961, 1–9.
15. Griffith, L.G. Emerging design principles in biomaterials and scaffolds for tissue engineering. Ann. NY Acad. Sci. *2002,* 961, 83–95.
16. Harikumar, S.; B. Ramesh, U.; Rajesh Kumar, G.; Ruthra Moorthy. Biomaterials for tissue engineering applications – a review. Int. J. Med. Pharm. Sci. *2013*, 3(2), 21–24.
17. Linda G. Griffith. Biomaterials. In: WTEC Panel Report on Tissue Engineering Research. Larry V. McIntire. Ed. Academic Press, Amsterdam; Boston, 2003; pp 9–22.
18. Shin, H.; Jo, S.; Mikos, A.J. Biomimetic materials for tissue engineering. Biomaterials. *2003,* 24, 4353–4364.
19. Engelberg, I.; Kohn, J. Physico-mechanical properties of degradable polymers used in medical applications: a comparative study. Biomaterials. *1991,* 12, 292–304.
20. Middleton, J.C,; Tipton, A.J. Synthetic biodegradable polymers as orthopedic devices. Biomaterials. 2000, 21(23), 2335.
21. Scanga, V.I.; Alex, G.; Nasser, N.; Shoichet, M.; Morshead, C.M. Biomaterials for neural tissue engineering – chitosan support the survival, migration and differentiation of adult derived neural stem and progenitor cells. Can J.Chem. *2009*, 88, 277–287.
22. Ingber, D.E.; Mow, V.C.; Butler, D.; Niklason, L.; Huard, J.; Moa, J.; Yannas, I.; Kaplan, D.; Novakovic, G.V. Tissue engineering and developmental biology: going biomimetic. J. Alter. Complement. Med. *2006*, 12, 3265–3283.
23. Leenslag, J.W.; Pennings, A.J.; Bos, R.R.; Rozema, F.R.; Boering, G. Resorbable materials of poly(L-lactide). VII. In vivo and invitro degradation. Biomaterials. *1987*, 8, 311–314.
24. Lin, F.H.; Chen, T.M.; Lin, C.P., Lee, C.J. The merit of sintered PDLLA/TCP composites in management of bone fracture internal fixation. Artif. Organs. *1999*, 23, 186–194.
25. Pettikiriarachchi, J.T.S.; Clare, L.P.; Molly, S.S.; John, S.F.; David, R.N. Biomaterial for brain tissue engineering. Aust. J. Chem. *2010,* 63, 1143–1154.
26. Ouyang, H.W.; Goh, J.C.H.; Mo, X.M.; Teoh, S.H.; Lee, E.H. Characterisation of anterior cruciate ligment cells and bone marrow stromal cells on various biodegradable polymeric films. Mater. Sci. Eng. *2002,* 20, 63–69.
27. Zhang, Y.; Huang, Z.; Xu, X. Preparation of core-shell structured PCL-r-gelatin bicomponent nanofibers by coaxial electrospinning. Chem. Mat. *2004,* 16, 3406–3409.

28. Junqueira, L.C.; José Carneiro, F.; Janet, L.; Harriet, B.; Peter, J. Ed. Basic Histology, Text & Atlas (10th ed.). McGraw-Hill Companies, 2003, p. 144.
29. Kundu, B.; Lemos, A.; Soundrapandian, C.; Sen, P.S.; Datta, S.; Ferreira, J.M.F.; Basu, D. Development of porous HAp and β-TCP scaffolds by starch consolidation with foaming method and drug-chitosan bilayered scaffold based drug delivery system. J. Mater. Sci. Mater. Med. *2010,* 21(11), 2955–2969.
30. Straub, D.A. Calcium supplementation in clinical practice: a review of forms, doses, and indications. NCP- Nutr. Clin. Prac. *2007,* 22(3), 286–296.
31. Husing, B.; Buhrlen, B.; Gaisser, S. Human tissue engineering products – today's markets and future prospects. Fraunhofer institute for systems and innovation research press, 2003.
32. Calvert, J.W.; Marra, K.G.; Cook, L.; Kumta, P.N.; DiMilla, P.A.; Weiss, L.E. Characterization of osteoblast-like behaviour of cultured bone marrow stromal cells on various polymer surfaces. J. Biomed. Mater. Res. *2000,* 52, 279–284.
33. Hench, L.L. Bioceramics: from concept to clinic. J. Am. Ceram. Soc. *1991*, 74 (7), 1487–1510.
34. Lu, H.H.; El-Amin, S.; Laurencin, C.T. Book of Abstracts, Sixth World Biomaterials Congress Transactions, May, 15–20, 2000; Kamuela (Big Island), Hawaii, USA, 3-D Porous polymerbioactive glass composite promotes collagen synthesis and mineralization of human osteoblast-like cells. 972–972.
35. Qiu, Q.Q.; Ducheyne, P.; Ayyaswamy, P.S. New bioactive, degradable composite microspheres as tissue engineering substrates. J. Biomed. Mater. Res. *2000,* 52, 66–76.
36. Gelsea, K.; Po¨schlb, E.; Aigner, T. Collagens—structure, function, and biosynthesis. Adv. Drug Deliv. Rev. *2003*, 55, 1531–1154.
37. Farach-Carson, M.C.; Wagner, R.C.; Kiick, K.L. Extracellular matrix: structure, function, and applications to tissue engineering. In Tissue Engineering; Fisher, J.P., Mikos, A.G., Bronzino, J.D., Eds.; CRC Press: Boca Raton, FL, 2007; pp. 3-1-3-22.
38. Othman, M.O.; Quassem, W.; Shahalam, A.P. The mechanical properties of catgut in holding and bonding fractured bones. Med. Eng. Phys. *1996*, 18, 584.
39. Wray; David, B. Sutures and suture materials: absorbable sutures: surgical gut. In Remington: The Science and Practice of Pharmacy, 21st Ed. Lippincott Williams & Wilkins Publishers, Baltimore, MA; Philadelphia, PA, 2006.
40. Sell, S.A.; McClure, M.J.; Garg, K.; Wolfe, P.S.; Bowlin, G.L. Electrospinning of collagen/biopolymers for regenerative medicine and cardiovascular tissue engineering. Adv. Drug Deliv. Rev. *2009*, *61*, 1007–1019.
41. Royce, P.M, Kato, T.; Ohsaki, K.; Miura, A. The enhancement of cellular infiltration and vascularisation of a collagenous dermal implant in the rat by platelet-derived growth factor BB. J. Dermatol. Sci. *1995*, 10, 42.
42. Ma, L.; Gao,C.; Mao, Z.; Zhou, J.; She, J. Enhanced biological stability of collagen porous scaffolds by using amino acids as novel cross-linking bridges. Biomaterials. *2004,* 25, 2997–3004.
43. Lynn, A.K.; Yannas, I.V.; Bonfield, W. Antigenicity nd immunogenicity of collagen. J. Biomed. Mater. Res. B- Appl. Biomater. *2004*, 71B, 343–354.
44. Huang, L.; Nagapudi, K.; Apkarian, R.P.; Chaikof, E.L. Engineered collagen-PEO nanofibers and fabrics. J. Biomater. Sci. Polym. Ed. *2001*, *12*, 979–993.

45. Matthews, J.A.; Boland, E.D.; Wnek, G.E.; Simpson, D.G.; Bowlin, G.L. Electrospinning of collagen type II: a feasibility study. J. Bioact. Compat. Polym. *2003*, 18, 125–134.
46. Barnes, C.P.; Sell, S.A.; Knapp, D.C.; Walpoth, B.H.; Brand, D.D.; Bowlin, G.L. Preliminary investigation of electrospun collagen and polydioxanone for vascular tissue engineering applications. Int. J. Electrospun Nanofiber. Appl. *2007*, 1, 73–87.
47. Barnes, C.P.; Pemble, C.W.; Brand, D.D.; Simpson, D.G.; Bowlin, G.L. Cross-linking electrospun type II collagen tissue engineering scaffolds with carbodiimide in ethanol. Tissue Eng. *2007*, *13*, 1593–1605.
48. Mano, J.F.; Silva, G.A.; Azevedo, H.S.; Malafaya, P.B.; Sousa, R.A.; Silva, S.S.; Boesel, L.F.; Oliveira, J.; Santos, T.C.; Marques, A.P.; Neves, N.M.; Reis, R.L. Natural origin biodegradable systems in tissue engineering and regenerative medicine: present status and some moving trends. J. Royal Soc. Interf. *2007,* 4, 999–1030.
49. Brown RA, Blunn GW, Ejim OS. Preparation of oriented fibrous mats from fibronectin: composition and stability. Biomaterials. *1994*, 15, 457–464.
50. Cheema, U.; Nazhat, S.W.; Alp, B.; Foroughi, F.; Anadagoda, N.; Mudera, V., Ra, B. Fabricating tissues. Biotech. Bioproc. Eng. *2007*, 12(1), 9–14.
51. Brown, R.A.; Wiseman, M.; Chuo C-B, Cheema U, Nazhat SN. Ultrarapid engineering of biomimetic materials and tissues: fabrication of nano- and microstructures by plastic compression. Adv. Func. Mat. *2005,* 15, 1762–1770.
52. Phillips, J.B; Bunting, S.C.J.; Hall, S.M.; Brown, R.A. Neural tissue engineering: a self- organizing collagen guidance conduit. Tissue Eng. *2005,* 11, 1612–1618.
53. Eastwood, M.; Mudera, V.C.; McGrouther,; D.A.; Brown, R.A. Effect of precise mechanical loading on fibroblast populated collagen lattices: morphological changes. Cell Mot. Cytosk. *1998*, 40, 13–21.
54. Mudera, V.C; Al, E. Molecular responses of human dermal fibroblast to dual cues: Contact guidance and mechanical load. Cell. Mot. Cytosk. *2000*, 45, 1–9.
55. Heijke, G.; Klopper, P.; Van Doorn, I.; Baljet, B.; Processed porcine collagen tubulization versus conventional suturing in peripheral nerve reconstruction: an experimental study in Rabbits. Microsurg. *2001*, 21, 84–95.
56. Liu, T.; Houle, J.D.; Xu, J.; Chan, B.P.; Chew, S.Y. Nanofibrous collagen nerve conduits for spinal cord repair. Tissue Eng. Part A. *2012*, 18(9–10), 1057–1066.
57. Yoshii, S.; Oka, M. Peripheral nerve regeneration along collagen filaments. Brain. Res. *2001,* 888, 158–162.
58. Barnes, C.P.; Sell, S.A.; Boland, E.D.; Simpson, D.G.; Bowlin, G.L. Nanofiber technology: designing the next generation of tissue engineering scaffolds. Adv. Drug Deliv. Rev. *2007*, *59*, 1413–1433.
59. Yoon, D.M.; Fisher, J.P. Polymeric scaffolds for tissue engineering applications. In Tissue Engineering; Fisher, J.P., Mikos, A.G., Bronzino, J.D., Eds.; Taylor and Francis Group, LLC: Boca Raton, FL, 2007, p. 18.
60. Hanson, S. Blood coagulation and blood-materials interactions. In BioMaterials Science: An Introduction to Materials in Medicine; Elsevier Academic Press: San Diego, CA, 2004; pp. 332–338.
61. Zhang, S.; Huang, Y.; Yang, X.; Mei, F.; Ma, Q.; Chen, G.; Ryu, S.; Deng, X. Gelatin nanofibrous membrane fabricated by electrospinning of aqueous gelatin solution for guided tissue regeneration. J. Biomed. Mater. Res. *2009*, 90(3), 671–679.

62. Chong, E.J.; Phan, T.T.; Lim, I.J.; Zhang, Y.Z.; Bay, B.H.; Ramakrishna, S.; Lim, C.T. Evaluation of electrospun PCL/gelatin nanofibrous scaffold for wound healing and layered dermal reconstitution. Acta Biomater. *2007*, *3*, 321–330.
63. Kim, S.E.; Heo, D.N.; Lee, J.B.; Kim, J.R.; Park, S.H.; Jeon, S.H.; Kwon, I.K. Electrospun gelatin/polyurethane blended nanofibers for wound healing. Biomed. Mater. *2009*, *4*, 1–11.
64. Ghasemi-Mobarakeh, L.; Prabhakaran, M.P.; Morshed, M.; Nasr-Esfahani, M.H.; Ramakrishna, S. Electrical Stimulation of nerve cells using conductive nanofibrous scaffolds for nerve tissue engineering. Tissue Eng. A *2009*, *15*, 3605–3619.
65. Ohkawa, K.; Hayashi, S.; Kameyama, N.; Yamamoto, H.; Yamaguchi, M.; Kimoto, S.; Kurata, S.; Shinji, H. Synthesis of collagen-like sequential polypeptides containing O-Phospho-L-hydroxyproline and preparation of electrospun composite fibers for possible dental application. Macromol. BioSci. *2009*, *9*, 79–92.
66. Francis, L.; Venugopal, J.; Prabhakaran, M.P.; Thavasi, V.; Marsano, E.; Ramakrishna, S. Simultaneous electrospin-electrosprayed biocomposite nanofibrous scaffolds for bone tissue regeneration. Acta Biomater. *2010*, *6*, 4100–4109.
67. Sisson, K.; Zhang, C.; Farach-Carson, M.C.; Chase, D.B.; Rabolt, J.F. Fiber diameters control osteoblastic cell migration and differentiation in electrospun gelatin. J. Biomed. Mater. Res. *A 2010*, *94A*, 1312–1320.
68. Dhandayuthapani, B.; Krishnan, U.M.; Sethuraman, S. Fabrication and characterization of chitosan-gelatin blend nanofibers for skin tissue engineering. J. Biomed. Mater. Res. B Appl. BioMater. *2010*, *94B*, 264–272.
69. Powell, H.M.; Boyce, S.T. Fiber density of electrospun gelatin scaffolds regulates morphogenesis of dermal-epidermal skin substitutes. J. Biomed. Mater. Res. *2008*, *84A*, 1078–1086.
70. Wang, S.; Zhang, Y.; Wang, H.; Yin, G.; Dong, Z. Fabrication and properties of the electrospun polylactide/silk fibroin-gelatin composite tubular scaffold. Biomacromol. *2009*, *10*, 2240–2244.
71. Clark, R.A. Fibrin is a many splendored thing. J. Invest. Dermatol. *2003*, *121*, 21–22.
72. Francis, C.W. Disorganized wound healing in fibrinogen-deficient mice. Blood, *2001*, 97(12), 3681.
73. Rybarczyk, B.J.; Lawrence, S.O.; Simpson-Haidaris, P.J. Matrix-fibrinogen enhances wound closure by increasing both cell proliferation and migration. Blood. *2003*, *102*, 4035–4043.
74. Drew, A.F.; Liu, H.; Davidson, J.M.; Daugherty, C.C.; Degen, J.L. Wound-healing defects in mice lacking fibrinogen. Blood. *2001*, *97*, 3691–3698.
75. Mosesson, M.W.; Siebenlist, K.R.; Meh, D.A. The structure and biological features of fibrinogen and fibrin. Ann. NY. Acad. Sci. *2001*, *936*, 11–30.
76. Kollman, J.M.; Pandi, L.; Sawaya, M.R.; Riley, M.; Doolittle, R.F. Crystal structure of human fibrinogen. Biochem. *2009*, *48*, 3877–3886.
77. Kim, J.; Song, H.; Park, I.; Carlisle, C.R.; Bonin, K.; Guthold, M. Denaturing of single electrospun fibrinogen fibers studied by deep ultraviolet fluorescence microscopy. Microsc. Res. Tech. *2010*, 74(3), 219–224.
78. Wnek, G.E.; Carr, M.; Simpson, D.G.; Bowlin, G.L. Electrospinning of nanofiber fibrinogen structures. Nano Lett. *2003*, 3, 213–216.

79. McManus, M.C.; Boland, E.D.; Simpson, D.G.; Barnes, C.P.; Bowlin, G.L. Electrospun fibrinogen: feasibility as a tissue engineering scaffold in a rat cell culture model. J. Biomed. Mater. Res. A *2007*, 81, 299–309.
80. Currie, L.J.; Sharpe, J.R.; Martin, R. The use of fibrin glue in skin grafts and tissueengineered skin replacements: a review. Plast Reconstr. Surg. *2001*, 108, 1713–1726.
81. Horch, R.E.; Bannasch, H.; Stark, G.B. Transplantation of cultured autologous keratinocytes in fibrin sealant biomatrix to resurface chronic wounds. Transplant Proc. *2001,* 33, 642–644.
82. Grant, I.; Warwick, K.; Marshall, J.; Green, C.; Martin, R. The co-application of sprayed cultured autologous keratinocytes and autologous fibrin sealant in a porcine wound model. Br. J. Plast. Surg. *2002*, 55, 219–227.
83. Shimada, Y.; Hongo, M.; Miyakoshi, N.; Sugawara, T.; Kasukawa, Y.; Ando, S.; Ishikawa, Y.; Itoi, E.; Dural substitute with polyglycolic acid mesh and fibrin glue for dural repair: technical note and preliminary results. J. Orthop. Sci. *2006,* 11, 454–458.
84. Iannotti, C.; Ping Zhang, Y.; Shields, L.B.E.; Han, Y.; Burke, D.A.; Xu, X.M, Shields, C.B. Dural repair reduces connective tissue scar invasion and cystic cavity formation after acute spinal cord laceration injury in adult rats. J. Neurotr. *2006*, 23, 853–865.
85. Lee, Y.S.; Hsiao, I.; Lin, V.W. Peripheral nerve grafts and a fgf restore partial hindlimb function in adult paraplegic rats. J. Neurotr. *2002*, 19, 1203–1216.
86. Frenkel, S.R.; Di Cesare, P.E. Scaffolds for articular cartilage repair. Ann. Biomed. Eng. *2004*, 32, 26–34.
87. Underwood, S.; Afoke, A.; Brown, R.A.; MacLeod, A.J.; Dunnill, P. The physical properties of a fibrillar fibronectin-fibrinogen material with potential use in tissue engineering. Biopro. Eng. *1999*, 20, 239–248.
88. Ahmed, Z.; Underwood, S.; Brown, R.A. Low concentration of fibrinogen increase cell migration speed on fibronectin/fibrinogen composite cables. Cell Mot. Cytosk. *2000*, 46(1):6–16.
89. Gailit, J.; Clarke, C.; Newman, D.; Tonnesen, M.G.; Mosesson, M.W.; Clark, R.A.F. Human fibroblasts bind directly to fibrinogen at RGD sites through integrin alpha v beta 3. Exp. Cell Res. *1997*, 232, 118–126.
90. Perez-Rigueiro, J.; Elices, M.; Llorca, J.; Viney, C. Tensile properties of silkworm silk obtained by forced silking. J. Appl. Polym. Sci. *2001*, 82, 1928–1935.
91. Liu, H.; Ge, Z.; Wang, Y.; Toh, S.L.; Sutthikhum, V.; Goh, J.C.H. Modification of sericin-free silk fibers for ligament tissue engineering applications. J. Biomed. Mater. Res. B. *2007*, 82B, 129–138.
92. Nagarkar, S.; Patil, A.; Lele, A.; Bhat, S.; Bellare, J.; Mashelkar, R.A. Some mechanistic insights into the gelation of regenerated silk fibroin sol. Ind. Eng. Chem. Res. *2009*, 48, 8014–8023.
93. Zhang, Q.; Yan, S.; Li, M. Silk fibroin based porous materials. Mat. *2009,* 2, 2276–2295.
94. Altman, G.H.; Diaz, F.; Jakuba, C.; Calabro, T.; Horan, R.L.; Chen, J.; Lu, H.; Richmond, J.; Kaplan, D.L. Silk-based biomaterials. Biomat. *2003*, 24, 401–416.
95. Jeong, L.; Lee, K.Y.; Park, W.H. Effect of solvent on the characteristics of electrospun regenerated silk fibroin nanofibers. Key Eng. Mater. *2007*, 342, 813–816.
96. Zarkoob, S. Structure and morphology of regenerated silk nano-fibers produced by electrospinning. Ph.D. Dissertation, The University of Akron: Akron, OH, USA, 1998.

97. Zarkoob, S.; Reneker, D.H.; Ertley, D.; Eby, R.K.; Hudson, S.D. Synthetically spun silk nanofibers and a process for making the same. U.S. Patent, 6110590, August 2000.
98. Sukigara, S.; Gandhi, M.; Ayutsede, J.; Micklus, M.; Ko, F. Regeneration of *Bombyx mori* silk by electrospinning—Part 1: Processing parameters and geometric properties. Polymers. *2003*, 44, 5721–5727.
99. Sukigara, S.; Gandhi, M.; Ayutesede, J.; Micklus, M.; Ko, F. Regeneration of *Bombyx mori* silk by electrospinning. Part 2. Process optimization and empirical modeling using response surface methodology. Polymers. *2004,* 45, 3701–3370.
100. Jin, H.J.; Fridrikh, S.V.; Rutledge, G.C.; Kaplan, D.L. Electrospinning *Bombyx mori* silk with poly(ethylene oxide). Biomacromol. *2002*, 3, 1233–1239.
101. Marelli, B.; Alessandrino, A.; Fare, S.; Freddi, G.; Mantovani, D. Compliant electrospun silk fibroin tubes for small vessel bypass grafting. Acta Biomater. *2010*, 6, 403–408.
102. Soffer, L.; Wang, X.; Zhang, X.; Kluge, J.; Dorfmann, L.; Kaplan, D.L.; Leisk, G. Silk-based electrospun tubular scaffolds for tissue-engineered vascular grafts. J. BioMater. Sci. Polym. *2008*, 19, 653–664.
103. Zhang, X.; Baughman, C.B.; Kaplan, D.L. *In vitro* evaluation of electrospun silk fibroin scaffolds for vascular cell growth. Biomaterials. *2008*, 29, 2217–2227.
104. Seo, Y.K.; Choi, G.M.; Kwon, S.Y.; Lee, H.S.; Park, Y.S.; Song, K.Y.; Kim, Y.J.; Park, J.K. The biocompatibility of silk scaffold for tissue engineered ligaments. Key Eng. Mater. *2007*, 342, 73–76.
105. Liu, H.; Fan, H.; Wang, Y.; Toh, S.L.; Goh, J.C.H. The interaction between a combined knitted silk scaffold and microporous silk sponge with human mesenchymal stem cells for ligament tissue engineering. Biomat. *2008*, 29, 662–674.
106. Kielty, C.M.; Sherratt, M.J.; Shuttleworth, C.A. Elastic fibres. J. Cell. Sci. *2002,* 115, 2817–2828.
107. Li, D.Y., Brooke, B.; Davis, E.C.; Mecham, R.P.; Sorensen, L.K.; Boak, B.B; Eichwald, E.; Keating, M.T. Elastin is an essential determinant of arterial morphogenesis. *Nature, 1998,* 393, 276–280.
108. Almine, J.F.; Bax, D.V.; Mithieux, S.M.; Nivison-Smith, L.; Rnjak, J.; Waterhouse, A.; Wise, S.G.; Weiss, A.S. Elastin-based materials. Chem. Soc. Rev. *2010*, 39, 3371–3379.
109. Daamen, W.F.; Veerkamp, J.H.; van Hest, J.C.; van Kuppevelt, T.H. Elastin as a biomaterial for tissue engineering. Biomat. *2007*, 28, 4378–4398.
110. Faury, G. Function-structure relationship of elastic arteries in evolution: from microfi brils to elastin and elastic fi bres. Pathol. Biol (Paris) *2001*, 49, 310–325.
111. Mithieux, S.M.; Rasko, J.E.; Weiss, A.S. Synthetic elastin hydrogels derived from massive elastic assemblies of self-organized human protein monomers. Biomat. *2004*, 25, 4921–4927.
112. Rodgers, U.R.; Weiss, A.S. Cellular interactions with elastin. Pathol. Biol (Paris). *2005,* 53(7), 390–398.
113. Zou, Y.; Zhang, Y. An experimental and theoretical study on the anisotropy of elastin network. Ann. Biomed. Eng. *2009*. 37(8), 1572–1583.
114. Rnjak, J.; Wise, S.G, Mithieux, S.M.; Weiss, A.S. Severe burn injuries and the role of elastin in the design of dermal substitutes. Tissue Eng. Part B Rev. *2011*, 17(2), 81–91.

115. Mecham, R.P. Elastin synthesis and fiber assembly. Ann. NY. Acad. Sci. *1991,* 624, 137–146.
116. Lloyd-Jones, D.; Adams, R.J.; Brown, T.M.; Carnethon, M.; Dai, S.; De Simone,G. Heart disease and stroke statistics--2010 update: a report from the American Heart Association. Circul. *2010,* 121(7), e46–e215.
117. Chlupac, J.; Filova, E.; Bacakova, L. Blood vessel replacement: 50 years of development and tissue engineering paradigms in vascular surgery. Physiol. Res. *2009,* 58(Suppl 2), S119– S139.
118. Williamson, M.R.; Shuttleworth, C.A.; Canfield, A.E.; Black, R.A.; Kielty, C.M. The role of endothelial cell attachment to elastic fibre molecules in the enhancement of monolayer formation and retention, and the inhibition of smooth muscle cell recruitment. Biomat. *2007,* 28(35), 5307–5318.
119. Wilson, B.D.; Gibson, C.C.; Sorensen, L.K.; Guilhermier, M.Y.; Clinger, M.; Kelley, L.L, Shiu, Y.T.; Li, D.Y. Novel approach for endothelializing vascular devices: understanding and exploiting elastin-endothelial interactions. Ann. Biomed. Eng. *2010,* 39(1), 337–46.
120. Huang, L.; McMillan, R.A.; Apkarian, R.P.; Pourdeyhimi, B.; Conticello, V.P.; Chaikof, E.L. Generation of synthetic elastin-mimetic small diameter fibers and fiber networks. Macromol. *2000,* 33, 2989–2997.
121. Lamme, E.N.; van Leeuwen, R.T.; Jonker, A.; van Marle, J.; Middelkoop, E. Living skin substitutes: Survival and function of fibroblasts seeded in a dermal substitute in experimental wounds. J. Invest. Dermatol. *1998*, 111, 989–995.
122. Neuenschwander, S.; Hoerstrup, S.P. Heart valve tissue engineering. Transplant. Immunol. *2004,* 12, 359–336.
123. Xu, J.W.; Johnson, T.S.; Motarjem, P.M.; Peretti, G.M.; Randolph, M.A.; Yaremchuk, M.J. Tissue-engineered flexible ear-shaped cartilage. Plast. Reconstr. Sur. *2005*, 115, 1633–1641.
124. Brown, A.L.; Farhat, W.; Merguerian, P.A.; Wilson, G.J.; Khoury, A.E; Woodhouse, K.A. 22 week assessment of bladder acellular matrix as a bladder augmentation material in a porcine model. Biomat. *2002*, 23, 2179–2190.
125. Schoen, F.J.; Levy, R.J. Tissue heart valves: current challenges and future research perspectives. J. Biomed. Mater. Res. *1999*, 47, 439–465.
126. Rosa, F.P, Lia, R.C.; de Souza, K.O.; Goissis, G.; Marcantonio, E. Tissue response to polyanionic collagen: elastin matrices implanted in rat calvaria. Biomat. *2003*, 24, 207–212.
127. Grauss, R.W.; Hazekamp, M.G.; Oppenhuizen, F.; van Munsteren, C.J.; Gittenberger-de Groot, A.C, DeRuiter, M.C. Histological evaluation of decellularised porcine aortic valves: matrix changes due to different decellularisation methods. Eur. J. Cardiotho.Surg. *2005*, 27, 566–571.
128. Hinds. M.T.; Courtman, D.W.; Goodell, T.; Kwong, M.; Brant-Zawadzki, H.; Burke, A.; Fox, B.A.; Gregory, K.W. Biocompatibility of a xenogenic elastin-based biomaterial in a murine implantation model: therole of aluminum chloride pretreatment. J. Biomed. Mater. Res. *2004*, 69A, 55–64.
129. Bader, A.; Schilling, T.; Teebken, O.E.; Brandes, G.; Herden, T.; Steinhoff, G.; Haverich, A. Tissue engineering of heart valves--human endothelial cell seeding of detergent acellularized porcine valves. Eur. J. Cardiothorac Surg. *1998*, 14, 279–284.

130. Lansing, A.I.; Rosenthal, T.B.; Alex, M.; Dempsey, W. The structure and chemical characterization of elastic fibers as revealed by elastase and by electron microscopy. Anat. Rec. *1952*, 114, 555–575.
131. John, R.; Thomas, J. Chemical compositions of elastins isolated from aortas and pulmonary tissues of humans of different ages. Biochem. *1972*, 127, 261–269.
132. Starcher, B.C.; Galione, M.J. Purification and comparison of elastins from different animal species. Anal Biochem. *1976*, 74, 4417–4447.
133. Daamen, W.F.; Hafmans, T.; Veerkamp, J.H, Van Kuppevelt, T.H. Isolation of intact elastin fi bers devoid of microfi brils. Tissue Eng. *2005*, 11, 1168–1176.
134. Sandberg, L.B.; Wolt, T.B. Production and isolation of soluble elastin from copper-defi cient swine. Methods Enzymol. *1982*, 82 Pt A, 657–665.
135. Rucker, R.B. Isolation of soluble elastin from copper deficient chick aorta. Methods Enzymol. *1982*, 82 Pt A, 650–657.
136. Partridge, S.M.; Davis, H.F.; Adair, G.S. The chemistry of connective tissues. 2 – Soluble proteins derived from partial hydrolysis of elastin. Biochem. *1955*, 61, 11–21.
137. Jacob, M.P.; Hornebeck, W. Isolation and characterisation of insoluble and kappa-elastins. In: Robert L, Moczar M, Moczar E, eds. Methods of Connective Tissue Research. Basel: Karger, *1985*, 92–123.
138. Hattori, M.; Yamaji-Tsukamoto, K.; Kumagai, H.; Feng, Y.; Takahashi, K. Antioxidative activity of soluble elastin peptides. J. Agric. Food. Chem. *1998*, 46, 2167–2170.
139. Wei, S.M.; Katona, E.; Fachet, J.; Fülöp Jr, T.; Robert, L.; Jacob, M.P. Epitope specifi city of monoclonal and polyclonal antibodies to human elastin. Int. Arch. Allergy. Immunol. *1998*, 115, 33.
140. Lee, J.; Macosko, C.W.; Urry, D.W. Elastomeric polypentapeptides cross-linked into matrixes and fi bers. Biomacromol. *2001*, 2, 170–179.
141. Martino, M.; Tamburro, A.M. Chemical synthesis of cross-linked poly(KGGVG), an elastin-like biopolymer. Biopol. *2001*, 59, 29–37.
142. McMillan, R.A.; Lee, T.A.T.; Conticello, V.P. Rapid assembly of synthetic genes encoding protein polymers. Macromol. *1999*, 32, 3643–3648.
143. Nagapudi, K.; Brinkman, W.T.; Thomas, B.S.; Park, J.O.; Srinivasarao, M.; Wright, E.; Conticello, V.P.; Chaikof, E.L. Viscoelastic and mechanical behavior of recombinant protein elastomers. Biomat. *2005*, 26, 4695–4706.
144. Heilshorn, S.C.; Liu, J.C.; Tirrell, D.A. Cell-binding domain context affects cell behavior on engineered proteins. Biomacromol. *2005*, 6, 318-323.
145. Welsh, E.R.; Tirrell, D.A. Engineering the extracellular matrix: a novel approach to polymeric biomaterials. I. Control of the physical properties of artifi cial protein matrices designed to support adhesion of vascular endothelial cells. Biomacromol. *2000*, 1, 23–30.
146. Ritz-Timme, S.; Laumeier, I.; Collins, M.J. Aspartic acid racemization: evidence for marked longevityof elastin in human skin. Br. J. Dermatol. *2003*, 149, 951–959.
147. Heilshorn, S.C.; DiZio, K.A.; Welsh, E.R.; Tirrell, D.A. Endothelial cell adhesion to the fibronectin CS5 domain in artifi cial extracellular matrix proteins. Biomat. *2003*, 24, 4245–4252.
148. Cappello, J.; Crissman, J.; Dorman, M.; Mikolajczak, M.; Textor, G.; Marquet, M.; Ferrari, F. Genetic engineering of structural protein polymers. Biotechnol. Prog. *1990*, 6, 198–202.

149. Scheller, J.; Henggeler, D.; Viviani, A.; Conrad, U. Purification of spider silk-elastin from transgenic plants and application for human chondrocyte proliferation. Transgenic. Res. *2004,* 13, 51–57.
150. Gobin, A.S.; West, J.L. Val-ala-pro-gly, an elastin-derived non-integrin ligand: smooth muscle cell adhesion and specificity. J. Biomed. Mater. Res. *2003*, 67A, 255–259.
151. Kakisis, J.D.; Liapis, C.D.; Breuer, C.; Sumpio, B.E. Artifi cial blood vessel: the Holy Grail of peripheral vascular surgery. J. Vasc. Surg. *2005*, 41, 349–354.
152. Lamprou, D.; Zhdan, P.; Labeed, F.; Lekakou, C. Gelatine and gelatine/elastin nanocomposites for vascular grafts: processing and characterisation. Journal of Biomaterials Applications. *2011*, 26(2), 209–226.
153. Li. M.; Mondrinos, M.J.; Gandhi, M.R.; Ko, F.K.; Weiss, A.S.; Lelkes, P.I. Electrospun protein fibers as matrices for tissue engineering. Biomat. *2005*, 26(30), 5999–6008.
154. Stitzel, J.; Liu, J.; Komura, M.; Berry, J.; Soker, S. Controlled fabrication of a biological vascular substitute. Biomat. *2006,* 27(7), 1088–1094.
155. Lee, S.J.; Yoo, J.J.; Lim, G.L.; Atala, A.; Stitzel, J. In vitro evaluation of electrospun nanofiber scaffolds for vascular graft application. J. Biomed. Mat. Res. *2007,* A 83(4), 999–1008.
156. Lam, C.X.F.; Mo, X.M.; Teoh, S.H.; Hutmacher, D.W. Scaffold development using 3D printing with a starch-based polymer, Mat. Sci. Eng. C. *2002*, 20, 49–56.
157. Levy, I.; Paldi, T.; Shoseyov, O.; Engineering a bifunctional starch-cellulose crossbridge protein. Biomat. *2004*, 25(10), 1841–1849.
158. Salgado, A.J.; Gomes, M.E.; Chou, A.; Coutinho, O.P.; Reis, R.L.; Hutmacher, D.W. Preliminary study on the adhesion and proliferation of human osteoblasts on starch-based scaffolds. Mat. Sci.Eng. C. *2002*, 20(1–2), 27–33.
159. Marques, A.P.; Reis, R.L. Hydroxyapatite reinforcement of different starch-based polymers affects osteoblast-like cells adhesion/spreading and proliferation. Mater. Sci. Eng. C, *2005*, 25(2), 215–229.
160. Neves, N.M.; Kouyumdzhiev, A.; Reis, R.L. The morphology, mechanical properties and ageing behavior of porous injection molded starch-based blends for tissue engineering scaffolding. Mater. Sci. Eng. C, *2005*, 25(2), 195.
161. Vasquez, Zimmermann, U. A highly sensitive cell assay for validation of purification regimes of alginates. Biomaterials. *2003*, 24, 4161–4172.
162. Skjak-Braek, G.H.; Smidsrod, O. Inhomogeneous polysaccharide ionic gels. Carb. Polym. *1989*, 1, 31–54.
163. Svensson, A.; Nicklasson, E.; Harrah, T.; Panilaitis, B.; Kaplan, G.L.; Brittberg, M.; Gatenholm, P. Bacterial cellulose as a potential scaffold for tissue engineering of cartilage. Biomat. *2005*, 26, 419–431.
164. Kuo, C.K.; Ma, P.X. Ionically crosslinked alginate hydrogels as scaffolds for tissue engineering: Part 1. Structure, gelation rate and mechanical properties. Biomaterials. *2001*, 22, 511–512.
165. Genes, N.G.; Rowley, J.A.; Mooney, D.J.; Bonassar, L.J. Effect of substrate mechanics on chondrocyte adhesion to modified alginate surfaces. Arch. Biochem. Biophy. *2004.* 422, 161–167.
166. Freyman, T.M.; Yannas, I.V.; Gibson, L.J. Cellular materials as porous scaffolds for tissue engineering. Prog. Mat. Sci. *2001*, 46, 273–282.

167. LeRoux, M.A.; Guilak, F.; Setton, L.A. Compressive and shear properties of alginate gel: effects of sodium ions and alginate concentration. J. Biomed. Mater. Res. *1999,* 47, 46–53.
168. Babensee, J.E.; Cornelius, R.M.; Brash, J.L.; Sefton, M.V. Immunoblot analysis of proteins associated with HEMA-MMA microcapsules: Human serum proteins *in vitro* and rat proteins following implantation. Biomat. *1998*, 19, 839–849.
169. Awad, H.A.; Erickson, G.R.; Guilak, F. Material selection for engineering cartilage. In: Lewandrowski K-U et al. (eds) Tissue engineering and biodegradable equivalents: scientific and clinical applications. Marcel Dekker, New York, 2002, p 195.
170. Drury, J.L.; Mooney, D.J. Hydrogels for tissue engineering: scaffold design variables and applications. Biomaterials. *2003,* 24, 4337–4351.
171. Stevens, M.M.; Qanadilo, H.F.; Langer, R.; Shastri, V.P. A rapid-curing alginate gel system: utility in periosteum-derived cartilage tissue engineering. Biomat. *2004*, 25, 887–894.
172. Dausse, Y.; Grossin,L.; Miralles,G.; Pelletier, S.; Mainard, D.; Hubert, P.; Baptiste, D.; Gillet, P.; Dellacherie, E.; Netter, P.; Payan, E. Cartilage repair using new polysaccharidic biomaterials: macroscopic, histological andbiochemical approaches in a rat model of cartilage defect. Osteoarthritis and Cartilage. *2003*, 11, 16–28.
173. Mauck, R.L.; Wang, C.C.B.; Oswald, E.S.; Ateshian, G.A.; Hung, C.T. The role of cell seeding density and nutrient supply for articular cartilage tissue engineering with deformational loading. Osteoarthritis and Cartilage. *2003*, 11, 879–890.
174. Draget, O.K.; Smidsrod, O. Homogeneous alginate gels: a technical approach. Carb. Polym. *1991*, 14, 159–178.
175. Wang, L.; Shelton, R.M.; Cooper, P.R.; Lawson, M.; Triffitt, J.T.; Barralet, J.E. Evaluation of sodium alginate for bone marrow cell tissue engineering. Biomat. *2003*, 24, 3475–3481.
176. Yan,Y.N.; Xiong, Z.; Hu, Y.Y.; Wang, S.G.; Zhang, R.J.; Zhang, C. Layered manufacturing of tissue engineering scaffolds via multi-nozzle deposition. Mat. Lett. *2003*, 57, 2623–2628.
177. Simpson, N.E.; Stabler, C.L.; Simpson, C.P.; Sambanis, A.; Constantinidis, L. The role of the CaCl2-guluronic acid interaction on alginate encapsulated beta TC3 cells. Biomat. *2004*, 25, 2603–2610.
178. Collins, M.M.; Birkinshaw, C.; Hyaluronic acid based scaffolds for tissue engineering—a review. Carb. Poly. *2013*, 92(2), 1262–1279
179. Khor, E.; Lim, L.Y. Implantable applications of chitin and chitosan. Biomat. *2003*, 24, 2339–2349.
180. Lahiji, A.; Sohrabi, A.; Hungerford, D.S.; Frondoza, C.G. Chitosan supports the expression of extracellular matrix proteins in human osteoblasts and chondrocytes. J. Biomed. Mater. Res. *2000*, 51, 586–595.
181. Levesque, S.G.; Lin, R.M.; Shoichet. Macroporous interconnected dextran scaffolds of controlled porosity for tissue engineering application.Biomat. *2005*, 26(35), 7436–46.
182. Zhou, D.; Shen,S.; Yun, J.; Yao, K.; Lin, D. Cryo copolymerization preparation of dextran hyaluronate based super macroporous cryogel scaffolds for tissue engineering applications. Front. Chem. Sci. Engg. *2012*, 6,(3), 339–347.

183. Jin, R.; Moreira, Teixeira, L.S, Dijkstra, P.J.; Zhong, Z.; van Blitterswijk, C.A.; Karperien, M.; Feijen, J. Enzymatically crosslinked dextran-tyramine hydrogels as injectable scaffolds for cartilage tissue engineering. Tissue. Eng. Part A. *2010*, 16(8), 2429–2440.
184. Matsusue, Y.; Yamamuro, T.; Oka, M.; Shikinami, Y.; Hyon, S.H.; Ikada, Y. In vitro and in vivo studies on bioabsorbable ultrahigh-strength poly(L-lactide) rods. J. Biomed. Mater. Res. *1992*, 26, 1553–1567.
185. Shikinami, Y.; Okuno, M. Bioresorbable devices made of forged composites of hydroxyapatite (HA) particles and poly-Llactide(PLLA): Part I. Basic characteristics. Biomat. *1999*, 20, 859–877.
186. Ignjatovic, N.; Tomic, S.; Dakic, M.; Miljkovic, M.; Plavsic, M.; Uskokovic,D. Synthesis and properties of hydroxyapatite/poly-Llactide composite biomaterials. Biomat. *1999*, 20, 809–816.
187. Chu, C.C. An in-vitro study of the effect of buffer on the degradation of poly(glycolic acid) sutures. J Biomed Mater Res. *1981*, 15, 19–27.
188. Chu, C.C. The in-vitro degradation of poly(glycolic acid) sutures- effect of pH. J. Biomed. Mater. Res. *1981,* 15, 795–804.
189. Chu, C.C. Hydrolytic degradation of polyglycolic acid: tensile strength and crystallinity study. J Appl Polym Sci. *1981*, 26, 1727–1734.
190. Gilding, D.K. (1981) In: Biodegradable Polymers. Biocompatibility of Clinical Implant Materials Vol II. Williams DF, ed. CRC Press, Boca Raton, FL. Vol 2, pp 209–232.
191. Hollinger, J.O. Preliminary report on osteogenic potential of a biodegradable copolymer of polylactide (PLA) and polyglycolide (PGA). J. Biomed Mater Res. *1983*, 17, 71–82.
192. Hollinger J.O.; Jamiolkoski, D.D.; Shalaby, S.W. Bone repair and a unique class of bidegradable polymers: The polyesters. In: Biomedical Applications of Synthetic Biodegradable Polymers. Hollinger, J.O. Ed. CRC Press, Boca Raton, FL, 1997, pp. 197–222.
193. Sawhney, A.S.; Drumheller, P.D. Polymer synthesis. In: Frontiers in Tissue Engineering. Patrick CW, Mikos, AG, McIntire, LV, eds. Pergamon, Newyork, 1998, pp. 83–106.
194. Gomes, M.E.; Bossano, C.M.; Johnston, C.M.; Reis, R.L.; Mikos, A.G. In vitrolocalization of bone growth factors in constructs of biodegradable scaffolds seeded with marrow stromal cells and cultured in a flow perfusion bioreactor. Tissue. Eng. *2006*, 12(1), 177–188.
195. Zhang, Y.Z.; Venugopal, J.; Huang, Z.M.; Lim, C.T.; Ramakrishna, S. Characterization of the surface biocompatibility of the electrospun PCL-collagen nanofibers using fibroblasts. Biomacromol. *2005*, 6(5), 2583–2589.
196. Ishaug-Riley, S.L.; Okun, L.E.; Prado, G.; Applegate, M.A.; Ratcliffe, A.; Human articular chondrocyte adhesion and proliferation on synthetic biodegradable polymer films. Biomat. *1999,* 20, 2245–2256.
197. Tadic, D.; Epple, M. A thorough physicochemical characterization of 14 calcium phosphate-based bone substitution materials in comparison to natural bone. Biomat. *2003*, 25, , 987.

198. Gineste, L.; Gineste, M.; Ranz, X.; Ellefterion, A.; Guilhem, A.; Rouquet, N.; Frayssinet, P. Degradation of hydroxyapatite, fluoroapatite, and fluorohydroxyapatite coatings of dental implants in dogs. J. Biomed. Mater. Res. *1999,* 48, 224–234.
199. Wozney, J.M.; Rosen, V. Bone morphogenetic protein and bone morphogenetic protein gene family in bone formation and repair. Clin. Orthop. Rel. Res. *1998,* 346, 26–37.
200. Myung, C.C.; Ching-Chang, K.; William, H.D. Preparation of hydroxyapatite-gelatin nanocomposite. J. Biomat. *2003,* 24, 2853–2862.
201. Ramay, H.R.; Zhang, M. Preparation of porous hydroxyapatite scaffolds by combination of the gel-casting and polymer sponge methods.Biomat. *2003*, 24(19), 3293–3302.
202. Sepulveda, P.; Ortega, F.S.; Innocentini, M.D.M.; Pandolfelli, V.C. Properties of highly porous hydroxy apatite obtained by gel casting foams. J. Am. Ceram. Soc. *2000*, 83(12), 3021–3024.
203. Woodard, J.R.; Hilldore, A.J.; Lan, S.K.; Park, C.J.; Morgan, A.W.; Eurell, J.A. Clark, S.G.; Wheeler, M.B.; Jamison, R.D.; Wagoner Johnson, A.J. The mechanical properties and osteoconductivity of hydroxyapatite bone scaffolds with multi-scale porosity. Biomat. *2007*, 28(1), 45–54.
204. Kim, S.S.; Ahn, K.M.; Park, M.S.; Lee, J.H.; Choi, C.Y.; Kim, B.S. A poly(lactide-co-glycolide)/hydroxyapatite composite scaffold with enhanced osteoconductivity. J. Biomed. Mat. Res A. *2007*, 80(1), 206–215.
205. Leukers, B.; Lkan, H.; Irsen, S.H.; Milz, S.; Tille, C.; Schieker, M.; Seitz, H. Hydroxyapatite scaffolds for bone tissue engineering made by 3D printing. J. Mat. Sci. Mat. Med. *2005*, 16, 1121–1124.
206. Cyster,L.A.; Grant,D.M.; Howdle, S.M.; Rose,F.R.; Irvine, D.J.; Freeman, D. Scotchford, C.A.; Shakesheff, K. M. The influence of dispersant concentration on the pore morphology of hydroxyapatite ceramics for bone tissue engineering.Biomaterials. *2005*, 26(7), 697–702.
207. Fu, Q.; Rahaman, M.N.; Bal, B.S.; Huang, W.; Day, D.E. Preparation and bioactive characteristics of a porous 13-93 glass, and fabrication into the articulating surface of a proximal tibia. J. Biomed. Mat.Res. *2007*, A 82(1), 222–229.
208. Chang, B.S.; Lee, C.K.; Hong, K.S.; Youn, H.J.; Ryu, H.S.; Chung, S.S.; Park, K.W. Osteoconduction at porous hydroxyapatite with various pore configurations. Biomat. *2000*, 21(12), 1291–1298.
209. Chu, T.M.; Orton, D.G.; Hollister, S.J.; Feinberg, S.E.; Halloran, J.W. Mechanical and in vivo performance of hydroxyapatite implants with controlled architectures. Biomat. *2002,* 23(5), 1283–1293.
210. Teixeira, S.; Ferraz, M.P.; Monteiro, F.J. Physical characterization of hydroxyapatite porous scaffolds for tissue engineering. J. Mat. Sci.Mat. Med. *2008*, 19(2), 855–859.
211. Queiroz, A.C.; Teixeira, S.; Santos, J.D.; Monteiro, F.J. Production of porous hydroxyapatite with potential for controlled drug. Adv.Mat. Forum II. *2004,* 455–456, 358–360.
212. Yang, S.; Leong, K.F.; Du, Z.H., Chua, C.H. The design of scaffold for use in tissue engineering. Part II. Rapid prototype. Thech. Tissue Engg. *2002*, 8(1), 1–11.

213. Oonishi, H.; Kushitani, S.; Yasukawa, E.; Iwaki, H.; Hench, L.L.; Wilson, J.; et al. Particulate bioglass compared with hydroxyapatite as a bone graft substitute. Clin Orthop. *1997*, 334, 316–325.
214. Davies, J.E.; Baldan, N.; Scanning electron microscopy of the bone-bioactive implant interface. J. Biomed Mater. Res. *1997*, 36, 429–440.
215. Yaszemski, M.J.; Payne, R.G.; Hayes, W.C.; Langer, R.; Mikos, A.G. Evolution of bone transplantation: molecular, cellular and tissue strategies to engineer human bone. [Review] Biomat. *1996*, 17, 175–185.
216. Mrunal, S.C. Tissue engineering: challenges and opportunities. Biomat. *2000,* 53, 617–620.
217. Agrawal, C.M.; Ray, R.B. Biodegradable polymeric scaffolds for musculoskeletal tissue engineering. J. Biomed. Mater. Res. *2001*, 55, 141–50.
218. Astete, C.E.; Sabliov, C.M. Synthesis and characterization of PLGA nanoparticles. J. Biomat. Sci. Polym. Ed. *2006*, 17(3), 247–289.
219. Seal, B.L.; Otero, T.C.; Panitch, A. Polymeric biomaterials for tissue and organ regeneration. Mater Sci Eng: R: Rep. *2001*, 34, 147–230.
220. El-Ghannam, A.; Ducheyne, P.; Shapiro, I.M.; Bioactive material template for in vitro synthesis of bone. J Biomed. Mat. Res. *1995*, 29, 359–370.
221. Hench, L.L.; Splinter, R.J.; Greenlee, T.K.; Allen, W.C. Bonding mechanisms at the interface of ceramic prosthetic materials. J. Biomed. Mat. Res. *1971*, 2, 117–141.
222. Bond, G.; Piotrowski, G.; Hench, L.L.; Allen, W.C.; Miller, G.J. Mechanical studies of the bone-bioglass. Interf. J. Biomed. Mater. Res. Symp. *1975,* 9(4), 47–61.
223. Hench, L.L.; Paschall, H.A.; Allen, W.C.; Piotrowski, G. Interfacial behavior of ceramics implants. Natl. Bureau Stand. Spl. Publ. *1975,* 415, 19–35.
224. Shoulders, M.D.; Raines, R.T.; Collagen structure and stability, Ann. Rev. Biochemistry. *2009*, 78, 929–958.
225. Ratner, B.D.; Hoffman, A.S.; Schoen, F.J.; Lemons, J.E. eds, Biomaterials Science, 3rd Edition, Academic Press, Elsivier, Oxford, U.K.; 2013, pp 198.
226. Lu, H.H.; El-Amin, S.F.; Scott, K.D.; Laurenci, C.T. Three-dimensional, bioactive, biodegradable, polymer–bioactive glass composite scaffolds with improved mechanical properties support collagen synthesis and mineralization of human osteoblast-like cells in vitro. J. Biomed. Mater. Res. *2003*, 64A, 465–474.
227. Niederauer, G.G.; Cullen, L.C.; Kieswetter, K.; Leatherbury, N.C.; Walter, M.A.; Greenspan, D.C. Development of polylactideco- glycolide/bioglass composites. Biomater. Proc. *1997*, 23, 162–162.
228. Marcolongo, M.; Ducheyne, P.; Garino, J.; Schepers, E. Bioactive glass fiber/polymeric composites bond to bone tissue. J. Biomed. Mater. Res. *1998*, 39, 161–170.
229. Ushida, G.P., Tateishi, T. Development of biodegradable porous scaffolds for tissue engineering. Mater. Sci. Eng. *2001*, 17, 63–69.
230. Kikuchi, M.; Ikoma, T.; Itoh, S.; Matsumoto, H.N.; Koyama, Y.; Takakuda, K.; Shinomiya, K.; Tanaka, J. Biomimetic synthesis of bone-like nanocomposites using the self-organization mechanism of hydroxyapatite and collagen. Comp. Sci. Technol. *2004*, 64, 819–825.
231. Rodrigues, C.V.M.; Serricella, P.; Linhares, A.B.R.; Guerdes, R.M.; Borojevic, R.; Rossi, M.A.; Duarte, M.E.L.; Farina, M.; Characterization of a bovine collagen-hydroxyapatite composite scaffold for bone tissue engineering, Biomaterials. *2003*, 24, 4987–4997.

232. Sukhodub, L.F.; Moseke, C.; Sukhodub, L.B.; Sulkio-Cleff, B.; Maleev, V.Y.; Semenov, M.A.; Bereznyak, E.G.; Bolbukh, T.V. Collagen-hydroxyapatite-water interactions investigated by, X.R.D., piezogravimetry, infrared and Raman spectroscopy. J. Mol. Struct. *2004*, 704(1–3), 53–58.
233. Jithendra, P.; Rajam, A.M.;, Kalaivani, T.; Mandal, A.B.; Rose, C. Preparation and characterization of aloe vera blended collagen-chitosan composite scaffold for tissue engineering applications. ACS Appl. Mater. Interfaces. *2013*, 5(15), 7291–7298.
234. Zhong, S.; Teo, W.E.; Zhu, X.; Beuermn, R.W.; Ramakrishna, S.; Yung L.Y.L. An aligned nanofibrous collagen scaffold by electrospinning and its effects on in vitro fibroblast culture. J. Biomed. Mat. Res A. *2006,* 79(3), 456–463.
235. Thomas, V.; Dean, D.R.; Jose, M.V.; Mathew, B.; Chowdhury, S.; Vohra, Y.K. Nanostructured biocomposite scaffolds based on collagen coelectrospun with nanohydroxyapatite. Biomacromol. *2007*, 8, 631–637.
236. Chen, Z.G.; Wang, P.W.; Wei, B.; Mo, X.M.; Cui, F.Z. Electrospun collagen-chitosan nanofiber: A biomimetic extracellular matrix for endothelial cell and smooth muscle cell. Acta Biomater. *2009*, 6, 372–382.
237. Yeo, I.S.; Oh, J.E.; Jeong, L.; Lee, T.S.; Lee, S.J.; Park, W.H.; Min, B.M. Collagen-based biomimetic nanofibrous scaffolds: preparation and characterization of collagen/silk fibroin bicomponent nanofibrous structures. Biomacromol. *2008*, 9, 1106–1116.
238. He, W.; Yong, T.; Teo, W.E.; Ma, Z.; Ramakrishna, S. Fabrication and endothelialization of collagen-blended biodegradable polymer nanofibers: potential vascular graft for blood vessel tissue engineering. Tissue Eng. *2005*, 11, 1574–1588.
239. Kwon, I.K.; Matsuda, T. Co-electrospun nanofiber fabrics of poly(L-lactide-co-epsilon-caprolactone) with type I collagen or heparin. Biomacromol. *2005*, 6, 2096–2105.
240. Lee, S.J.; Liu, J.; Oh, S.H.; Soker, S.; Atala, A.; Yoo, J.J. Development of a composite vascular scaffolding system that withstands physiological vascular conditions. Biomat. *2008*, 29, 2891–2898.
241. Tillman, B.W.; Yazdani, S.K.; Lee, S.J.; Geary, R.L.; Atala, A.; Yoo, J.J. The *in vivo* stability of electrospun polycaprolactone-collagen scaffolds in vascular reconstruction. Biomat. *2009*, 30, 583–588.
242. Schnell, E.; Klinkhammer, K.; Balzer, S.; Brook, G.; Klee, D.; Dalton, P.; Mey, J. Guidance of glial cell migration and axonal growth on electrospun nanofibers of poly-epsilon-caprolactone and a collagen/poly-epsilon-caprolactone blend. Biomat. *2007*, 28, 3012–3025.
243. Li, M.; Guo, Y.; Wei, Y.; MacDiarmid, A.G.; Lelkes, P.I. Electrospinning polyaniline-contained gelatin nanofibers for tissue engineering applications. Biomat. *2006*, 27, 2705–2715.
244. Heydarkhan-Hagvall, S.; Schenke-Layland, K.; Dhanasopon, A.P.; Rofail, F.; Smith, H.; Wu, B.M.; Shemin, R.; Beygui, R.E.; MacLellan, W.R. Three-dimensional electrospun ECM-based hybrid scaffolds for cardiovascular tissue engineering. Biomat. *2008*, 29, 2907–2914.
245. Li, M.; Mondrinos, M.J.; Gandhi, M.R.; Ko, F.K.; Weiss, A.S.; Lelkes, P.I. Electrospun protein fibers as matrices for tissue engineering. Biomat. *2005*, 26, 5999–6008.

246. Nivison-Smith, L.; Rnjak, J.; Weiss, A.S. Synthetic human elastin microfibers: stable cross-linked tropoelastin and cell interactive constructs for tissue engineering applications. Acta Biomater. *2010*, 6, 354–359.
247. Massia, S.P.; Hubbell, J.A. Vascular endothelial cell adhesion and spreading promoted by the peptide REDV of the IIICS region of plasma fibronectin is mediated by integrin alpha4 beta1. J. Biol. Chem. *1992*, 267, 14019–14026.
248. Zio, K.D.; Tirrell, D.A. Mechanical properties of artifi cial protein matrices engineered for control of cell and tissue behaviour. Macromol. *2003*, 36, 1553–1558.
249. Nowatzki, P.J.; Tirrell, D.A. Physical properties of artifi cial extracellular matrix protein films prepared by isocyanate crosslinking. Biomat. *2004*, 25, 1261–1267.
250. Scheller, J.; Guhrs, K.H.; Grosse, F.; Conrad, U.; Production of spider silk proteins in tobacco and potato. Nat. Biotechnol. *2001*, 19, 573–577.
251. Nagarsekar, A.; Crissman, J.; Crissman, M,; Ferrari, F.; Cappello, J.; Ghandehari, H. Genetic engineering of stimuli-sensitive silkelastin-like protein block copolymers. Biomacromol. *2003*, 4, 602–607.
252. Cappello, J.; Crissman, J.W.; Crissman, M.; Ferrari, F.A.; Textor, G.; Wallis, O.; Whitledge, J.R.; Zhou, X.; Burman, D.; Aukerman, L.; Stedronsky, E.R. In-situ self-assembling protein polymer gel systems for administration, delivery, and release of drugs. J. Cont. Rel. *1998*, 53, 105–117.
253. Megeed, Z.; Haider, M.; Li, D.; O'Malley, B.W.; Jr.; Cappello, J.; Ghandehari, H. In vitro and in vivo evaluation of recombinant silk-elastinlike hydrogels for cancer gene therapy. J. Cont. Rel. *2004*, 94, 433–445.
254. Haider, M.; Leung, V.; Ferrari, F.; Crissman, J.; Powell, J.; Cappello, J.; Ghandehari, H. Molecular engineering of silk-elastinlike polymers for matrix-mediated gene delivery: biosynthesis and characterization. Mol. Pharm. *2005*, 2, 139–150.
255. Hashimoto, T.; Suzuki, Y.; Tanihara, M.; Kakimaru, Y.; Suzuki, K. Development of alginate wound dressings linked with hybrid peptides derived from laminin and elastin. Biomat. *2004*, 25, 1407–1414.
256. Woodhouse, K.A.; Klement. P.; Chen, V.; Gorbet, M.B.; Keeley, F.W.; Stahl, R.; Fromstein, J.D.; Bellingham, C.M. Investigation of recombinant human elastin polypeptides as non-thrombogenic coatings. Biomat. *2004*, 25, 4543–4553.
257. Casasco, M.; Casasco, A.; Icaro, C.A.; Farina, A.; Calligaro, A. Differential distribution of elastic tissue in human natural skin and tissue-engineered skin. J. Mol. Histol. *2004*, 35, 421–428.
258. Hafemann, B.; Ensslen, S.; Erdmann, C.; Niedballa, R.; Zuhlke, A.; Ghofrani, K.; Kirkpatrick, C.J. Use of a collagen/elastin-membrane for the tissue engineering of dermis. Burns. *1999*, 25, 373–384.
259. De Vries, H.J.; Zeegelaar, J.E.; Middelkoop, E.; Gijsbers, G.; van Marle, J.; Wildevuur, C.H.; Westerhof, W. Reduced wound contraction and scar formation in punch biopsy wounds. Native collagen dermal substitutes. A clinical study. Br. J Dermatol. *1995*, 132, 690–697.
260. Hinek, A.; Wang, Y.; Liu, K.; Mitts, T.F.; Jimenez, F. Proteolytic digest derived from bovine Ligamentum Nuchae stimulates deposition of new elastin-enriched matrix in cultures and transplants of human dermal fi broblasts. J. Dermatol. Sci. *2005*, 39(3), 155–166.

261. Heydarkhan-Hagvall, S.; Schenke-Layland, K.; Dhanasopon, A.P.; Rofail, F.; Smith, H.; Wu, B.M.; Shemin, R.; Beygui, R.E.; MacLellan, W.R. Three-dimensional electrospun ECM-based hybrid scaffolds for cardiovascular tissue engineering. Biomat. *2008*, 29, 2907–2914.
262. Zisch, A.H.; Lutolf Ehrbar, M.; Raeber, G.P.; Rizzi, S.C.; Davies, N.; Schmökel, H.; Bezuidenhout, D.; Djonov, V.; Zilla, P.; Hubbell, J.A. Cell-demanded release of VEGF from synthetic, biointeractive cell-ingrowth matrices for vascularized tissue growth. J. Fed. Am. Soc. Exp. Biol. *2003*, 17(13), 2260–2262.
263. Seliktar, D.; Zisch, A.H.; Lutolf, M.P.; Wrana, J.L, Hubbell, J.A. MMP- 2 sensitive, VEGF-bearing bioactive hydrogels for promotion of vascular healing. J. Biomed. Mater. Res. *2004*, 68A(4), 704–716.
264. Miyamoto, K.; An, H.S.; Sah, R.L.; Akeda, K.; Okuma, M.; Otten, L.; Thonar, E. J.M.A.; Masuda, K. Exposure to pulsed low intensity ultrasound stimulates extracellular matrix metabolism of bovine intervertebral disc cells cultured in alginate beads. Spine. *2005*, 30(21), 2398–2405.
265. Stitzel, J.D.; Pawlowski, K.J.; Wnek, G.E.; Simpson, D.G.; Bowlin, G.L. Arterial smooth muscle cell proliferation on a novel biomimicking, biodegradable vascular graft scaffold. J. Biomater. Appl. *2001*, 16, 22–33.
266. Kidoaki, S.; Kwon, I.K.; Matsuda, T. Mesoscopic spatial designs of nano- and microfiber meshes for tissue-engineering matrix and scaffold based on newly devised multilayering and mixing electrospinning techniques. Biomat. *2005*, 26, 37–46.
267. Boland, E.D.; Coleman, B.D.; Barnes, C.P.; Simpson, D.G.; Wnek, G.E.; Bowlin, G.L. Electrospinning polydioxanone for biomedical applications. Acta Biomater. *2005*, 1, 115–123.
268. Hutmacher, D.W.; Sittinger, M.; and Risbud, M.V. Scaffoldbased tissue engineering: rationale for computer-aided design and solid free-form fabrication systems. Trends Biotechnol. *2004*, 22(7), 354–362.
269. Lin, L.; Ju, S.; Cen, L.; Zhang, H.; Hu, Q. Fabrication of porous, TCP scaffolds by combination of rapid prototyping and freeze drying technology. Yi Peng, Xiaohong Weng (Eds.), 2008, pp. 88–91. APCMBE 2008, IFMBE Proceedings 19, Springer-Verlag, Berlin Heidelberg.
270. Taboas, J.M.; Maddox, R.D.; Krebsbach, P.H.; Hollister, S.J. Indirect solid free form fabrication of local and global porous, biomimetic and composite 3D polymerceramic scaffolds. Biomaterials. *2003*, 24(1), 181–194.
271. Xiong, Z.; Yan, Y.; Wang, S.G.; Zhang, R.J.; Zhang, C. Fabrication of porous scaffolds for bone tissue engineering via low temperature deposition. Scripta Mat. *2002,* 46: 771–776.
272. Salusbury, I. Bone in contention. Mat. World. *2005*, 13(4): 25–27.
273. Tang, T.G.; Black, R.A.; Curran, J.M.; Hunt, J.A.; Rhodes, N.P.; Williams. D.F. Surface properties and biocompatibility of solvent-cast poly[e-caprolactone] films. Biomaterials. *2004*, 25, 4741–4748.
274. Huang, Y. C.; Mooney, D. J. Gas foaming to fabricate polymer scaffolds in tissue engineering. In*:* Scaffoldings in tissue engineering, Ma X P., Elisseeff J., (Ed.), 2005. Pp. 159, Taylor and Francis Group. CRC Press, 206, 61–72.
275. Ma, P. X.; Langer, R. Fabrication of biodegradable polymer foams for cell transplantation and tissue engineering. In Tissue Engineering Methods and Protocols, Morgan, J., and Yarmush, M. (eds.) 1999, Humana Press, Totowa, NJ, 47.

276. Lu, L.; Peter, S.; Lyman, M.; Lai, H.; Leite, S.; Tamada, J.; Uyama, S.; Vacanti, J.; Langer, R.; Mikos, A. *In vitro* and *in vivo* degradation of porous poly (DL-lactic-coglycolic acid) foams. Biomaterials. *2000*, 21, 1595–1605.
277. Lippiello, L. Glucosamine and chondroitin sulfate: biological response modifiers of chondrocytes under simulated conditions of joint stress. Osteoarthr Cartil. *2003*, 11, 335–342.
278. Jabbari, E.; Wang, S.; Lu, L.; Gruetzmacher, J.A.; Ameenuddin, S.; Hefferan, T.E.; Currier, B.L.; Windebank, A.J.; Yaszemski, M.J. Synthesis, material properties and biocompatibility of a novel self-crosslinkable poly(caprolactone fumarate) as an injectable tissue engineering scaffold. Biomacromol. *2005*, 6(5), 2503–2511.
279. Ortiz-Palacios, J.; Cardoso, J.; Manero, O. Production of macroporous resins for heavy-metal removal. I. Non functionalized polymers. J. Appl. Polym. Sci. *2008*, 107, 2203–2210.
280. Guettaf, H.; Iayadene, F.; Bencheikh, Z.; Rabia, I. Structure and properties of styerene-divinylbenzene-methylmethacrylateterpolymers-ii- effect of methylacrylate at different n-heptane 2-ethyl-1-hexanol diluent composition. Eur. Polym. J. *1998*, 34, 241–246.
281. Zou, H.; Wu, S.; Shen, J. Polymer/silica nanocomposites: preparation, characterization, properties and applications. Chem. Rev. *2008*, 108, 3893–3957.
282. Ikada, Y. Scope of tissue engineering In: Tissue engineering: fundamental andapplications, Ikada Y. (Ed.). Academic Press, Amsterdam; Boston, 2006; p 29.
283. Chen, G.; Ushida, T.; Tateishi, T. Development of biodegradable porous scaffolds for tissue engineering. J. Mater. Sci. Eng. C. *2002*, 17, 63–69.
284. Mikos, A.G.; Bao, Y.; Cima, L.; Ingber,D.; Vacanti, J.P.; Langer, R. Preparation of poly (glycolic acid) bonded fiber structures for cell attachment and transplantation. J. Biomed. Mater. Res. *1993*, 27, 183.
285. Cima, L.G.; Vacanti, J.P.; Vacanti, C.; Ingber, D.; Mooney, D.; Langer, R. Tissue engineering by cell transplantation using degradable polymer substrates. J. Biomech. Eng.*1991,* 113, 143–151.
286. Chen, G.Q.; Wu, Q. The application of polyhydroxyalkanoates as tissue engineering materials. Biomat. *2005*, 26, 6565–6578.
287. Sachlos, E.; Reis, N.; Ainsley, C.; Derby, B.; Czernuszka, J.T.; Novel collagen scaffolds with predefined internal morphology made by, SFF. Biomat. *2003*, 24, 1487–1497.
288. Moone, D.J.; Baldwin, D.F.; Suh, N.P.; Vacanty, J.P.; Langer, R. Novel approach to fabricate porous sponges of poly(D,L-lacto-co-glycolicacid) without the use of organic solvents. Biomat. *1996*, 17, 1417–1422.
289. Moroni, L.; Hamann, D.; Paoluzzi, L.; Pieper, J.; de Wijn J. R.; van Blitterswijk C. A. Regenerating articular tissue by converging technologies. PLoS One. 3(8) e3032.
290. Miko, A.G.; Temenoff, J.S.; Formation of highly porous biodegradable scaffolds for tissue engineering. EJB Electronic J. Biotechnol. *2000*, 13(2): 114–119.
291. Ginty, P.J., Howard, D., Rose, F.R.A.J., Whitaker, M.J., Barry, J.J.A., Tighe, P. et al. Mammalian cell survival and processing in supercritical CO_2. PNAS, *2006*, 103(19), 7426–7431.

292. Ginty, P.J.; Howard, D.; Upton, C.E.; Barry, J.J.A.; Rose, F.R.A.J.; Shakesheff, K.M. A supercritical CO_2 injection system for the production of polymer/mammalian cell composites. J. Supercrit. Fluids. *2008*, 43(3), 535–541.
293. Mandal, B. B.; Kundu, S. C. Osteogenic and adipogenic differentiation of rat bone marrow cells on non-mulberry and mulberry silk gland fibroin 3D scaffolds. Biomaterials. *2009*, 30, 5019–5030.
294. Maquet, V.; Jerome, R. Design of macroporous biodegradable polymer scaffolds for cell transplantations, Mat. Sci Forum, *1997*, 250, 15–24.
295. Hutmacher, D.W. Scaffold design and fabrication technologies for engineering tissues- state of the art and future perspectives. J. Biomat. Sci. Polym. Ed .*2001,* 12, 107–124.
296. Thompson, R.C.; Wake, M.C.; Yaszemski; Mikos, A.G. Biodegradable polymer scaffolds to regenerate organs. Adv. Polym. Sci. *1995,* 122, 245–274.
297. Thompson, R.C.; Yaszembksi, M.J.; Powers, J.M.; Harrigan, T.P.; Mikos, A.G. Poly (a-hydroxy ester)/short fiber hydroxyapatite composite foams for orthopaedic applications. In: Polymers in Medicine and Pharmacy, Vol 394. Mikos AG, Leong KW, Yaszemski MJ, (Ed.). 1995 .pp. 25–30, Materials Research Society Symposium Proceedings, Pittsburgh USA.
298. Hou, Q.P.; Grijpma, D.W.; Feijen, J. Porous polymeric structures for tissue engineering prepared by a coagulation, compression moulding and salt leaching technique. Biomat. *2003*, 24, 1937–1947.
299. Shastri, V.P.; Martin, I.; Langer, R. Macroporous polymers foams by hydrocarbon templating. Proc. Natl. Acad.Sci. U.S.A. *2000*, 97, 1970,
300. Das, S.; Hollister, S.J.; Flanagan, C.; Adewunmi, A.; Bark, K.;Chen, C.; Ramaswamy, K.; Rose, D. Widjaja, E. Freeform fabrication of Nylon-6 tissue engineering scaffolds. Rapid Prototyping J.*2003*, 9(1), 43–49.
301. Radisic, M.; Park. H.; Chen, F.; Salazar-Lazzaro, J.E.; Wang, Y.D.; Dennis, R.; Langer, R.; Freed, L.E.; Vunjak- Novakovic, G. Biomimetic approach to cardiac tissue engineering: oxygen carriers and channeled scaffolds. Tissue Eng. *2006*, 12, 2077–2091.
302. Sundback, C.A.; Shyu, J.Y.; Wang, Y.D.; Faquin, R.S.; Langer, J.P.; Vacanti, T.A. Biocompatibility analysis of poly(glycerol sebacate) as a nerve guide material. Biomat. *2005*, 26, 5454–5464.
303. Fang, Z.; Starly, B.; Sun, W.; Computer-aided characterization for effective mechanical properties of porous tissue scaffolds. Computer-Aided Design. *2005*, 37(1), 65–72.
304. Hutmacher, D.W.; Sittinger, M.; Risbud, M.V. Scaffoldbased tissue engineering: rationale for computer-aided design and solid free-form fabrication systems. Trends in Biotechnol. *2004*, 22(7), 354–362.
305. Khalil, S.; Nam, J.; Sun, W.; Multi-nozzle deposition for construction of 3D biopolymer tissue scaffolds. Rapid Prototyping J. *2005*, 11(1), 9–17.
306. Sun,W.; Lal, P. Recent development on computer aided tissue engineering—a review. Comput. Methods Programs Biomed. *2002*, 67(2), 85–103,
307. Wilson, C.E.; de Bruijn, J.D.; Van Blitterswijk, C.A.; Verbout, A.J.; Dhert, W.J.A.; Design and fabrication of standardized hydroxyapatite scaffolds with a defined mac-

ro-architecture by rapid prototyping for bone-tissue-engineering research. J. Biomed. Mater. Res. *2004*, 68A, 123–132.
308. Woodfield, T. B.; Guggenheim, M.; von Rechenberg, B.; Riesle, J.; van Blitterswijk, C. A.; Wedler, V. Rapid prototyping of anatomically shaped, tissue engineeredimplants for restoring congruent articulating surfaces in small joints. Cell. Prolif. *2009,* (4), 485–497.
309. Bredt, J.F.; Sach, E.; Brancazio, D.; Cima, M.; Curodeau, A.; Fan, T. Three dimensional printing system. US Patent, *1998*, 5807437.
310. Lee, C.T.; Kung, P.H.; Lee, Y.D., Preparation of poly(vinyl alcohol)- chondroitin sulfate hydrogel as matrices in tissue engineering. Carb. Polym. *2005*, 61, 348–54.
311. Sachlos, E.; Czernuszka, J.T. Making tissue engineering scaffold work: review on the application of SFF technology to the production of tissue engineering scaffolds. Eur. Cells. Mat. *2003*, 5, 29–40.
312. Koegler, W.S.; Patrick, C.; Cima, M.J.; Griffith, L.G.; Carbon dioxide extraction of residual chloroform from biodegradable polymers, Journal of Biomedical Materials Research. *2002*, 63, 567–576.
313. Zeltinger, J.; Sherwood, J.K.; Graham, D.A.; Mueller, R.; Griffith, L.G. Effect of pore size and void fraction on cellular adhesion, proliferation, and matrix depostion. Tissue Eng.. *2001,* 7(5), 557–572.
314. Ang, T.H.; Sultana, F.S.A.; Hutmacher, D.W.; Wong, Y.S.; Fuh, J.Y.H.; Mo, X.M.; Loh, H.T.; Burdet, E.; Teoh, S.H.; Fabrication of 3D chitosan-hydroxyapatite scaffolds using a robotic dispensing system. Mat. Sci. Engg. C, *2002*, 20, 35–42.
315. Hoque, M.E.; Hutmacher, D.W.; Feng, W., Li, S.; Huang M.H.; Vert, M.; Wong, YS.; Fabrication using a rapid prototyping system and in vitro characterization of PEG-PCL-PLA scaffolds for tissue engineering. J. Biomat. Sci.Polym. Edn. *2005*, 16(12), 1595–1610.
316. Doshi, J.; Reneker, D. H.; Electrospinning process and application of electrospunfibers. J. Electrostat. *1995,* 35, 151–160.
317. Reneker, D. H.; Chun, I. Nanometer diameter fibres of polymer, produced by electrospinning Nanotechnol. *1996*, 7, 216–223.
318. Matthews, J.A.; Wnek, G.E.; Simpson, D.G.; Bowlin G.L. Electrospinning of collagen nanofibers. Biomacromol. *2002,* 3, 232–238.
319. Jin, H.J.; Chen, J.; Karageorgiou, V.; Altman, G.H.; Kaplan, D.L. Human bone marrow stromal cell responses to electrospun silk fibroin mats. Biomat. *2004*, 25, 1039–1047.
320. Ohkawa, K.; Cha, D.; Kim, H.; Nishida, A.; Yamamoto, H. Electrospinning of Chitosan. Macromol. Rapid Comm. *2004*, 25, 1600–1605.
321. Zarkoob, S.; Eby, R.K.; Reneker, D.H.; Hudson, S.D.; Ertley D.; Adams, W.W. Structure and morphology of electrospun silk nanofibers. Polymer. *2004,* 45, 3973–3977.
322. Baker, B.M.; Gee. A.O.; Metter, R.B.; Nathan, A.S.; Marklein, R.A.; Burdick, J.A.; Mauk, R.L.The potential to improve cell infiltration in composite fiber-aligned electrospun scaffolds by the selective removal of sacrificial fibers. Biomat. *2008*, 29, 2348–2358.
323. Ma, P. X. Biomimetic materials for tissue engineering. Adv. Drug Deliv. Rev. *2008*, 60, 184–98.

324. Li, W.J.; Tuan R.S. Fabrication and application of nanofibrous scaffolds in tissue engineering. Curr. Protoc. Cell. Biol. *2009,* 25 Unit 25.2.
325. Liang, D.; Hsiao, B.S.; Chu, B. Functional electrospun nanofibrous scaffolds for biomedical applications. Adv. Drug Deliv. Rev. *2007,* 59, 1392–1412.
326. Leong, M.F.; Rasheed, M.Z.; Lim, T.C.; Chian, K.S. Invitro cell infiltration and in-vivo cell in filtration and vascularization in fibrous highly porous poly (D,L-Lacticacid) scaffold fabrication by electrospining technique. J. Biomed. Res. *2008*, A91, 231–240.

CHAPTER 11

ELECTROSPUN MATRICES FOR BIOMEDICAL APPLICATIONS: RECENT ADVANCES

DEEPA P. MOHANAN, ROBIN AUGUSTINE,
NANDAKUMAR KALARIKKAL, RADHAKRISHNAN E. K., and
SABU THOMAS

ABSTRACT

Electrospinning is a unique, versatile, and facile approach for engineering nanofibrous scaffolds with characteristic micro to nanoscale topography and high porosity similar to the natural extracellular matrix (ECM). These nanofibrous matrices influence cellular activities both *in vitro* and *in vivo*. Biodegradable polymers used for the electrospinning can be of synthetic (PCL, PVA, PLA, PLGA, PVP, etc.) or natural (alginate, chitosan, collagen, elastin, fibrinogen, gelatin, keratin, etc.) origin. High surface to volume ratio and porous nature of the electrospun scaffolds help in the attachment of cells and incorporation of proteins, DNA, RNA, and drugs ranging from antibiotics to anticancer agents. Thus, they find promising applications in drug delivery, tissue engineering, wound dressing, and in other areas. In this chapter, electrospinning method for the generation of nanofibrous scaffolds and its important biomedical applications are reviewed.

11.1 INTRODUCTION

During the past few decades, suggestive advances have been made in the development of biodegradable polymeric biomaterials for biomedical applications. An ideal biomedical scaffold should induce high cellular activities, have a large surface area to provide good attachment and proliferation of cells, tunable physicochemical property, and should be biocompatible (Kretlow and Mikos, 2008). For biomedical applications, an adequate rate of *in vivo* degradation or stability of polymers is also very important (Auras et al., 2010). The degradation involves cleavage of hydrolytically or enzymatically sensitive bond in the polymer leading to polymer bulk degradation and the rate of which is depend on parameters like polymer molecular weight and polydispersity (Zhang et al., 2010). Other parameters like ionic strength, pH, temperature, and buffer capacity of the medium in which degradation occurs also affect the degradation kinetics (Zhou et al., 2010).

Electrospinning technique is used for the generation of nanofibers of biodegradable polymers with ultrafine macro to nano sizes offers considerable advantages and the method was first patented by Antonio Formhals in 1934 (Formhals, 1934). Temporary biomimetic extracellular matrix (ECM) formed by the polymeric electrospun nanofibrous scaffolds support cell growth and enhances neotissue regeneration (Szentivanyi et al., 2011). In addition to its application in tissue engineering (TE), these scaffolds can have potential applications in drug and gene delivery, enzyme immobilization, wound dressing (WD), and biomedical imaging (Chong et al., 2007). Most commonly used electrospinnable biodegradable synthetic polymers include polylactic acids (PLA), polyglycolic acids (PGA) and their copolymers (PLGA), polycaprolactone (PCL), polyvinyl alcohol (PVA), polyethylene glycol (PEG), etc. (Kim and Kim, 2012; Li et al., 2012a; Ligia et al., 2013). Among electrospinnable natural polymers, collagen, chitosan (CS), cellulose, hyaluronic acid (HA), alginate, elastin, and keratin are the most popular for their biomedical applications (Kaina et al., 2012; Kovacina et al., 2012; Maurice and Colin, 2013). In this chapter, the nonwoven mats developed by electrospinning method and its biomedical applications in TE, WD, and drug and gene delivery are reviewed.

11.1.1 ELECTROSPINNING

Electrospinning technique is most remarkable for its technical simplicity and easy adaptability. An electrospinning system consists of a high-voltage power supply, a syringe pump, and a grounded collecting plate. The typical setup of electrospinning apparatus is shown in Figure 11.1. The principle of electrospinning process is that strong mutual electrical repulsive forces overcome weaker forces of surface tension in the charged polymer liquid. Under the influence of applied high electrical field the polymer jet rapidly travels to the grounded collector and deposite as nonwoven mats. Several polymers such as natural, synthetic, their blends, and several composites have been electrospun into three-dimensional (3-D) structures. Scanning electrospun micrographic (SEM) image of pure electrospun PCL scaffold has been shown in Figure 11.2. Developing biodegradable, biofunctional, and biocompatible nanostructured materials are of great global interest because of its wide applications in therapeutic agents delivery, TE, and regenerative medicine. Different types of polymers used in electrospinning and their applications are listed in Tables 11.1, 11.2, 11.3, and 11.4.

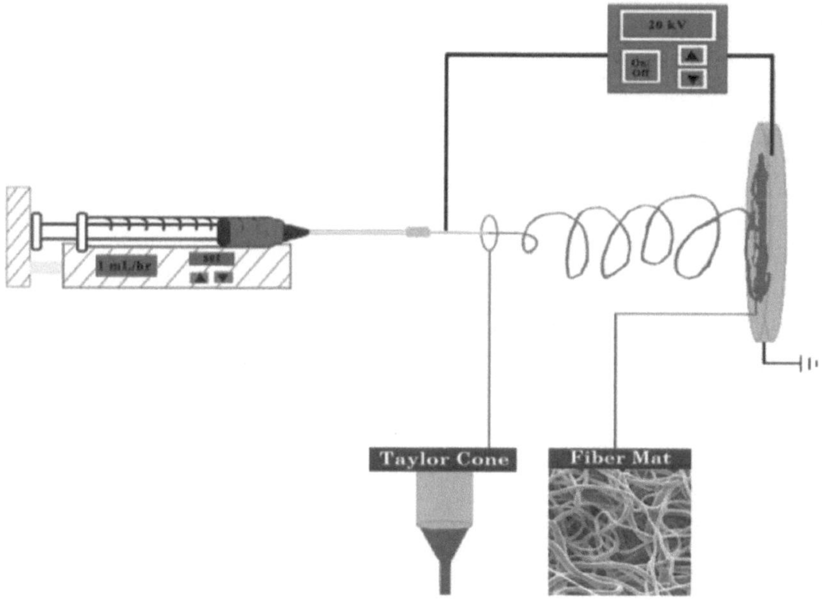

FIGURE 11.1 Schematic diagram of electrospinning setup

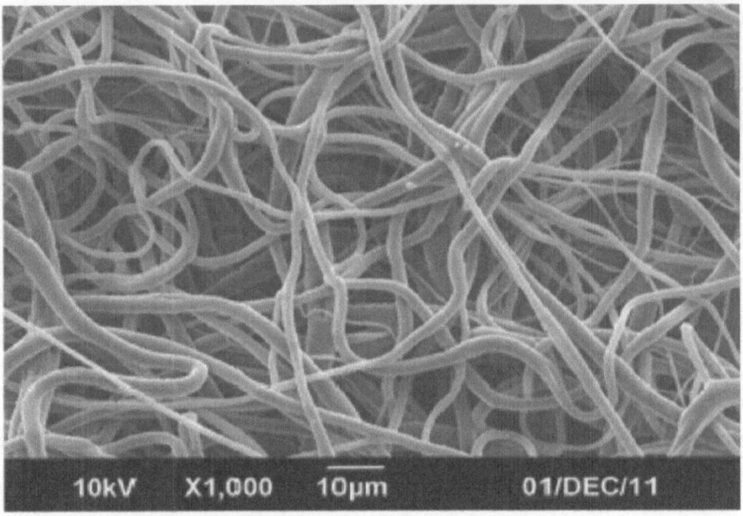

FIGURE 11.2 SEM image of electrospun PCL membrane

TABLE 11.1 Natural polymers used in electrospinning and their applications

Electrospun scaffold material	Proposed biomedical applications	References
N-methylene phosphonic chitosan (NMPC)	Bone grafting applications	Datta et al. (2012)
Elastin	small-diameter vascular graft	McKenna et al. (2012)
Silk fibroin (SF)	Biosensing, wound dressing, controlled drug delivery, and tissue engineering	Cervantes et al. (2012),
Chitosan (CS)	Drug-delivery biomaterial	Fan et al. (2013)
Cellulose acetate (CA)	Wound healing application	Norowski et al. (2012)
Keratin	Wound dressing	Suwantong et al. (2007) Xin et al. (2013) Yuan et al. (2009)

TABLE 11.2 Synthetic polymers used in electrospinning and their applications

Electrospun scaffold material	Proposed biomedical applications	References
Polycaprolactone (PCL)	Vascular graft,	Wu et al. (2010)
	Bone tissue regeneration,	Lee et al. (2012a, 2012b), Lin et al. (2012a, 2012b)
	Tissue engineering repair,	Shao et al. (2012), Kim and Kim (2012)
	Drug-delivery vehicle against pathogenic microorganisms,	Ruckh et al. (2012)
	neural grafts, and	Wang et al. (2012b)
	wound healing applications.	Merrell et al. (2009)
Poly lactide-co-glycolide (PLGA)	Tissue engineering application,	Gang et al. (2012)
	sustained drug delivery, and	
	effective bone regeneration	Lee et al. (2011)
	Cell delivery in transplantation therapy	Shin et al. (2011)
Poly *L*-lactide-*co*-ε- caprolactone (PLCL)	Retinal cell transplant therapies and	
polylactic acids (PLA)		Kador et al. (2013)
	Transdermal drug-delivery system	Toncheva et al. (2012)
polyninyl alcohol (PVA)	Drug-delivery scaffold and	Taepaiboon et al. (2006)
	wound dressing, tissue engineering,	
	implantable glucose biosensors,	
Polyurethane (PU)	vascular substitutes and	Wang et al. (2013)
		Bergmeister et al. (2013)
	wound healing application.	Guo et al. (2012), Macocinschia et al. (2013)
Polyvinyl pyrrolidone (PVP)	Wound healing agent as well as drug-delivery system.	Dai et al. (2012)

TABLE 11.3 Composites used in electrospinning and their applications

Electrospun scaffold material	Proposed biomedical applications	References
PCL/nanoclay	Superior scaffolding material for bone tissue engineering	Nitya et al. (2012)
Nanohydroxyapatite (nHA)/PLGA	Drug delivery in cancer therapy and	Zheng et al. (2013a)
	Drug delivery as well as TE application	Zheng et al. (2013b)
	Respirators and antibacterial coatings	Wu et al. (2009)
Hydroxyapatite (HA)/Ag/AgBr/TiO2	Supportive matrices for osteoregeneration	Phipps et al. (2012)
PCL/CG I/HA	Bone tissue engineering material	Fu et al. (2012)
poly(ε-caprolactone)-poly(ethylene glycol)-poly(εcaprolactone (PCEC)/n-HA	Bone regeneration	Li et al. (2012a)
nHA/PCL	Bone tissue engineering application	Jaiswal et al. (2013)
poly-L-lactic acid (PLLA)/gelatin/HA	Tissue engineering applications	Sheikh et al. (2010)
PVA/HA	Tissue engineering applications	Li et al. (2012a)
PCL/SF/HA		

TABLE 11.4 Blends used in electrospinning and their applications

Electrospun scaffold material	Proposed biomedical applications	References
CG/hyaluronic acid	Substitute for bone tissue engineering	Fischer et al. (2012)
Elastin/Collagen (CG)	Dermal tissue engineering	Kovacina et al. (2012)
CG/PCL	Tubular tissue-engineered vascular graft	Zhang et al. (2012)
PCL/gelatin	Tissue engineering applications	Lee et al. (2012c)
PELCL/PELGA/CS		Zhang et al. (2013)

PLCL/CG	Substitute for small-diameter vascular regeneration	Jang et al. (2012)
PLLA/CS	Vascular substitute material for cardiovascular applications	Cui et al. (2012)
PCL/PLGA	Tissue engineering application	Franco et al. (2011)
Sodium alginate (SA)/PEO	Wound dressings/tissue regeneration	Ma et al. (2012)
PLLA/CG	Tissue engineering application	Ghaedi et al. (2012)
Xylan/PVA	Liver tissue engineering	Krishnan et al. (2012)
PCL/gealatin	Template for skin tissue regeneration	Chong et al. (2007)
PLA/CS	Cost-efficient alternative for wound dressing	Li et al. (2012b)
CS/PVA	Wound healing dressing	Vimala et al. (2011)
PLGA/CG	Antibacterial wound dressing material	Chen et al. (2012)
CECS/PVA/SF	Wound healing application	Zhou et al. (2013)
Carboxymethyl chitosan/PEO	Wound dressing for skin regeneration	Fouda et al. (2013)
CS/PVA	Protective antimicrobial wound dressings	Park et al. (2013)
PEG/PLA	Enhanced and continuous antibacterial applications	Xu et al. (2010)
PCL/poly(ethylene-co-vinyl acetate (PEVA)/PCL	Wound dressing material for diabetic foot ulcer treatment	Alhausein et al. (2012)
PU/gelatin	Wound healing application	Wang et al. (2011a)
PCL/CS		Zhao et al. (2013)
CG/CS		Sundar and Sangeetha (2012)
CS/alginate		

11.1.2 MEDICAL APPLICATIONS OF ELECECTROSPUN SCAFFOLDS

Biodegradable polymers such as PLGA, PCL, PVA, PLA, CS, silk fibroin (SF), and alginate are some of the materials which have been electrospun for biomedical applications by a number of research groups (Jao et al., 2012; Khalil et al., 2013; Mahoney et al., 2012; Sakai et al., 2008).

Recently, a huge number of composite scaffolds such as PCL/PLGA, nHA/PCL, Xylan/PVA, PEGMA/PU, PVA/ Hap, PCL-PEG-PCL, PCL/collagen/HA, PCL/SF/HA, SA/PEO, Elastin-collagen, CG/PNIPAA/CS, AMX/n-HA/PLGA, PLA/CS, PHBV/m-keratin, PLGA/collagen, PEG-PLA, etc., have been developed for biomedical applications through electrospinning technology (Chen et al., 2012; Franco et al., 2011; Fu et al., 2012; Jiang et al., 2009; Kovacina et al., 2012; Krishnan et al., 2012; Li et al., 2012c; Ma et al., 2012; Phipps et al., 2012; Sundar and Sangeetha, 2012; Xu et al., 2010; Zheng et al., 2013a).

11.1.3 COMMONLY USED NATURAL POLYMERS

Chitosan (CS): CS can be obtained by deacetylation of chitin, isolated from fungal cell wall, shells of crustaceans, insects, and other sources. As a natural polymer, CS exhibits desirable properties such as biocompatibility, biodegradability, antimicrobial activity, and nontoxicity. Due to these properties, CS is widely used in biomedical applications. During the past two decades, electrospinning has been found to be a new method to produce CS nanofibers. Highly viscous CS can easily dissolve in solvents like tetrahydrofuran (THF), aqueous acetic acid solution (aq AA), dimethyl formamide (DMF), and chloroform. Cross-linking with glutaraldehyde (GA) makes it a mechanically stable form. (Cui et al., 2012; Datta et al., 2012; Du and Hsieh, 2007; Franco et al., 2011; Jayakumar et al., 2010; Vrieze et al., 2007; Zhang et al., 2013).

Collagen: Collagen is one of the chief structural proteins of bones, ligaments, tendons, blood vessels, skin, and cartilages. It constitutes approximately 30% of all vertebrate body protein (Hsu et al., 2010). It has known for low antigenic, high mechanical, biodegradable, and good biocompatible properties. All these properties make collagen an important protein for medical applications. Mechanical property can be improved by cross-linking with GA. Different concentrations of collagen can be dissolved in 1,1,1,3,3,3 hexafluoro-2-propanol (HFP) solvent (Fischer et al., 2012; Kovacina et al., 2012; Phipps et al., 2012; Matthews et al., 2002).

Gelatin: Gelatin is a natural polymer containing peptides and proteins which are derived from collagen. It is commonly used for pharmaceutical and medical applications because of its biodegradability and biocompatibility in physiological environments. GA facilitates the cross-linking of gelatin molecules. This is well dissolved in solvents like 2,2,2-trifluoroethanol (TFE), dimethyl sulfoxide (DMSO), and glacial acetic acid (AA) (Wang et al., 2011a; Zhang et al., 2006).

Alginate: Alginate is a naturally derived water-soluble polysaccharide extracted from brown seaweed. Due to its biodegradability, biocompatibility, nonimmunogenicity, and bioresorption properties, alginate is having various medical applications. Alginate forms a mechanically stable gel when cross-linked with divalent calcium ions or poly(ethylene glycol)-diamines (PEG). The viscous form of alginate is suitable for electrospinning technique (Chang et al., 2012; Eiselt et al., 1999;).

Hyaluronic acid (HA): HA is a heavily hydrated glycosaminoglycan (GAG) polymer which is widely distributed in ECM. Due to its viscoelastic property, good biocompatibility, and biodegradability, it is a good candidate for medical field (Fischer et al., 2012).

Silk fibroin (SF): SF is a naturally occurring protein in variety of insects and spiders. The water-soluble and absorption nature with excellent biocompatibility and functionality make its application in WDs, TE scaffolds, artificial ligaments, and nerve conduits. Synthetic polymer like polyethylene oxide (PEO) when cross-linked with SF was found to have excellent mechanical stability (Ghosh et al., 2009).

Keratin: Keratin is a fibrous protein present in hair, wool, feathers, nails, and horns of animals. Extracted keratin proteins have an intrinsic ability to self-assemble and polymerize into fibrous and porous films, gels, and scaffolds. The poor mechanical properties of keratin prevent its practical applications in biomedical fields. But cross-linking it with GA or some other polymers like PEO, SF, etc. can significantly improve the mechanical strength. This is dissolved in solvents like HFP, trifluoroethanol (TFE), etc. (Aluigi et al., 2007; Yuan et al., 2009).

Elastin: Elastin is an ECM protein providing elasticity to tissues and organs. This is commonly present in skin, lung alveoli, blood vessels, heart, mucous membranes, gut lining, and elastic cartilage. Nowadays it is widely used in skin and vascular TE because of its ability to promote cell adhesion and proliferation. Elastin dissolves in hexafluro propanol solvent (Kovacina et al., 2012; McKenna et al., 2012).

11.1.4 COMMONLY USED SYNTHETIC POLYMERS

Polycaprolactone (PCL): PCL is a biodegradable semicrystalline aliphatic polyester with a low melting point of around 60°C and a glass transition temperature of about −60°C. PCL is prepared by ring opening polymerization of ε-caprolactone using catalyst such as stannous octoate. Lack of toxicity, high mechanical property, good biocompatibility, and relatively lower cost makes it an ideal scaffold material for biomedical applications (Franco et al., 2011; Li et al., 2012a).

Poly(lactide-co-glycolide) (PLGA): PLGA is the copolymer of polylactides and polyglycolides. In nanomedicine, PLGA electrospun scaffold is widely used to deliver various drugs or biomolecules to the target site. This has good hydrophilic, biocompatible, biodegradable, nonimmunogenic, and nontoxic characters. So this polymer has very important place in biomedical field (Chen et al., 2012; Franco et al., 2011; Zheng et al., 2013b).

Polyurethane (PU): PU is a polymer containing carbamate (urethane) linkage. Its microphase-separated structures provide excellent mechanical strength. PU is suitable in the field of biomedical materials due to the very good mechanical and elastic properties (Blit et al., 2012; Wang et al., 2012a).

Polyvinyl alcohol (PVA): PVA is a relatively cheap synthetic polymer possessing biodegradable, biocompatible, and nontoxic properties. This is widely used to produce spongy matrices for tissue regeneration and other medical applications (Vimala et al., 2011).

11.2 APPLICATIONS OF ELECTROSPUN SCAFFOLDS

11.2.1 TISSUE ENGINEERING (TE) APPLICATIONS

TE is an interdisciplinary field that applies the principles of engineering and life sciences for the development of functional substitutes for damaged system. Both biodegradable natural and synthetic polymers are used as scaffolding structures for regenerative medicine. Polymers are often reinforced with other functional materials and used as composites for different TE applications. An appropriate selection of biomaterial scaffold is one of the key factors for successful TE. The biomedical scaffolds used should be safe, nonallergic, nonpyrogenic, nontoxic, noncarcinogenic, immunologically compatible, promote ECM formation, and repair damaged tissue. ECM provides anchorage to the cells, forms a supportive meshwork around cells, and separates different tissues. Due to the resemblance to ECM structure, the electrospun scaffolds are excellent candidates for TE applications.

The scaffolds with potential TE applications include elastin/collagen, porous silks, PCL matrix, poly(lactide-co-glycolide) (PLGA)-based meshes, PVA/Hap, alginate/chitosan, PLA/CS, PLGA/collagen, PCL/PLGA, and PEGMA/PU. Some of the important TE applications are as follows.

11.2.1.1 BONE TISSUE ENGINEERING APPLICATIONS

Nowadays many people are suffering from severe bone diseases. Autografts and allografts are the commonly used bone substitutes. But they have some negative impacts such as immune response, disease transmission, and donor site morbidity. Bone tissue engineering is an alternative way to generate bone by utilizing a combination of biomaterials and cells. The biomimetic scaffold provides temporary physical support for neotissue formation and also preferably delivers bioactive agents. Here, the scaffold used generates a synthetic osteogenic microenvironment to facilitate the natural ossification processes.

Many recent studies show the promising potential of electrospinning technique in bone tissue engineering application. Electrospun PCL/fucoidan mats was identified as a potential biomaterial for bone tissue regeneration (Lee et al., 2012b). Fucoidan in the new biocomposite had an inherent anticoagulant, antiviral, and immunomodulatory activity. The studies also showed electrospinability, mechanical, and hydrophilic properties of various weight fractions (1, 2, 3, and 10 wt. %) of fucoidan in PCL and also the cellular behavior on the new mat as a bone tissue regeneration scaffold in addition to other properties. The construction of "MSC-homing device" by using peptide sequence (E7) covalently conjugated onto PCL electrospun mesh is very interesting. As the E7 had a high specific affinity toward bone marrow-derived mesenchymal stem cells (MSCs), the product was very efficient for recruiting MSCs of both in vitro and in vivo and thereby it offers a promising way to improve MSC-based TE repair (Shao et al., 2012). When the pore size of polycaprolactone/collagen I/hydroxyapatite (PCL/col/HA) scaffold was enhanced by incorporating PEO fibers, it promoted human MSC attachments indicating its use as supportive matrices for osteoregeneration (Phipps et al., 2012).

Electrospinning technique was also used for the development of N-methylene phosphonic chitosan (NMPC)-based scaffold fabrication (Datta et al., 2012). The NMPC used is known to provide nucleation sites for biomineralization and hence, the presence of a large number of nucleation sites on matrix helped the increased mineralization process. This makes its application as osteoconductive and osteoinductive matrixes for bone grafting applications.

Mechanical properties of biopolymer matrices are reported to be improved by the incorporation of nanohydroxyapatite (nHA). An example for this is an electrospun mat of PCEC/nHA developed from poly(ε-caprolactone)-poly(ethylene glycol)-poly(ε-caprolactone) (PCL-PEG-PCL, PCEC) with nHA which was shown to have potential use as bone tissue engineering material (Fu et al., 2012).

Results of culturing of MG63 osteoblast-like cells on electrospun MBG/PCL scaffold showed excellent attachment, proliferation, and growth of the cells on to the scaffold. This makes its use as bone graft for bone regeneration (Lin et al., 2012a). When osteoblasts and osteoclast-like cells were cultured on biphasic hydroxyapatite/**β**-tricalcium phosphate (β-TCP), nanobioceramic scaffold, cell migration, and proliferation in the scaffold were observed (Wepener et al., 2012). This shows biocompatible nature of electrospun scaffolds, its effect on osteoblasts or osteoclast-like cells in vitro, and thereby its use as synthetic bioactive bone substitute. Promising potential of A187-HA/PCL electrospun mat to be used as bone regeneration scaffolds is also reported (Li et al., 2012a).

11.2.1.2 BLOOD VESSEL TISSUE ENGINEERING APPLICATIONS

Peripheral vascular pathology and coronary artery diseases are major vascular compliances causing the mortality in huge number of patients. Autografts or allografts are now widely used to replace blood vessels. Very recently, electrospun scaffolds from many natural and synthetic polymers were found to have promising application in vascular graft replacement due to their ease and biocompatibility. The features of vascular grafts to avoid vascular thrombogenicity and to promote good tissue regeneration are very important. A tubular scaffold can be constructed by allowing the deposition of nanofibers on to the external surface of the rotating mandrel.

Xiang et al. (2011) successfully constructed an electrospun tubular scaffold for TE blood vessels by using polymer blends of recombinant spider silk protein (pNSR32), PCL, and gelatin. The obtained scaffold was with a length of 3 cm, wall thickness of 0.3 mm, and an inner diameter of 3 mm.

Very recently Zhang et al. (2013) have developed a double-layered electrospun membrane with the combination of poly(ethylene glycol)-b-poly(L-lactide-co-caprolactone) (PELCL), methoxy poly(ethylene glycol)-b-poly(L-lactide-co-glycolide) (PELGA), and CS hydrogel. The study was conducted to encapsulate vascular endothelial growth factor (VEGF) and platelet-derived growth factor-bb (PDGF) into the scaffold for improved blood vessel regeneration.

3-D PCL nanofibrous tubular scaffold developed with controllable nanofiber orientations by electrospinning was shown to have significant vascular graft applications (Wu et al., 2010). In a recent study, Wang et al. (2012a) fabricated poly (ethylene glycol) methacrylate (PEGMA)/PU electrospun scaffold for TE application such as blood vessel grafts.

Bilayer electrospinning technology to fabricate a tubular scaffold composed of a PU fibrous outer layer and a gelatin-heparin fibrous inner layer was developed by Wang et al. (2011a). This fibrous scaffold is of promising potential that can be used as artificial blood vessels because of its elastic property, mechanical property, and hemocompatibility. Since elastin is an essential ECM helping in the formation of vascular tissues, studies were also conducted to develop an electrospun scaffold with PU and elastin-like polypeptide-4 (ELP4) (Blit et al., 2012). Results of the study suggests the application of ELP4 modified degradable PUs as candidate scaffolds for the fabrication of contractile tissue-engineered SMC-rich vascular medial layers. Recombinant human tropoelastin (rTE), the monomer unit of elastin, was also reported to have been used in the production of small-diameter vascular graft (McKenna et al., 2012).

By using a combination of bioactive polypeptide RGD (arginine–glycine–aspartate) and spider silk protein, a biocomponent named as pNSR32 was developed (Zhao et al., 2013). Small-diameter scaffold was then prepared from pNSR32 by incorporating it to CS and PCL. Since the scaffold displayed outstanding performance in the Sprague-Dawley (SD) rat abdominal aortic transplant surgery, it can be used as vascular TE support.

11.2.1.3 SKIN TISSUE ENGINEERING APPLICATIONS

Electrospun skin substitutes promote fibroblast adhesion, proliferation, and growth, which enhance and accelerate epidermal and dermal skin regeneration. Recently human dermal fibroblasts were cultured well on electrospun elastin/collagen composite scaffold for dermal TE (Kovacina et al., 2012). Mouse fibroblast implantation study showed excellent fibroblast infiltration, collagen deposition, and new capillary formation. Template for skin tissue regeneration developed from nanofibrous Xylan/PVA scaffold exhibited good fibroblast adhesion, proliferation, and improved cell matrix interaction (Krishnan et al., 2012).

11.2.1.4 LIVER TISSUE ENGINEERING APPLICATIONS

Ghaedi et al. (2012) developed a new electrospun PLLA scaffold, which facilitated the hepatic differentiation of human bone marrow-derived mesenchymal stem cells (hMSCs). This material was found to be very suitable for liver TE applications. Chua et al. (2007) produced a novel poly (e-caprolactone-co-ethyl ethylene phosphate) polymer (PCLEEP) scaffold by electrospinning method. This scaffold showed increased attachment and growth of hepatocytes.

11.2.1.5 NERVE TISSUE ENGINEERING APPLICATIONS

Damaged nerve tissue repair and regeneration is one of the major compliances in medical science. Electrospun scaffolds from various biodegradable polymers have been widely used for neural graft replacement due to their excellent biocompatibility. Nowadays several reports have already appeared demonstrating outstanding performances of these electrospun scaffolds. A recent report by Wang et al. (2012b) highlights the importance of electrospinning method for neural grafts. They developed a nanofibrous PCL scaffold and biofunctionalized with immobilized glial cell-derived neurotrophic factor (iGDNF). They identified the ability of the newly formed scaffold to enhance the neural graft integration.

Mobarakeh et al. (2008) developed PCL/gelatin scaffold which had proved to be a promising biomaterial for nerve tissue regeneration. Recently, Prabhakaran et al. (2009) demonstrated that PLCL/collagen nanofibrous scaffolds were capable of differentiating MSCs from neuronal cells. The obtained results proposed the scaffolds' potential applications toward neuroinjuries and neurodegenerative diseases.

11.2.1.6 CARTILAGE TISSUE ENGINEERING APPLICATIONS

An electrospun cartilage implant supports the adhesion, proliferation, and growth of the chondrocyte cells. In a recent study, Xue et al. (2013) have fabricated an ear-shaped cartilage with electrospun PCL/gelatin nanofibrous scaffold. The seeded chondrocytes in the material showed excellent growth.

11.2.2 DRUG-DELIVERY APPLICATIONS

Polymeric nanofiber matrices have been used as carriers to deliver therapeutic agents. The large surface area and high porosity of nanofiber-based systems could facilitate drug diffusion and also reduce drug dosage and nontarget site toxicities.

Sundar and Sangeetha (2012) fabricated collagen/poly(N-isopropyl acrylamide)/chitosan (CG/PNIPAA/CS) complex with 5-fluorouracil, an anticancer drug by the method of electrospinning. From the results it was observed that the CS concentration played a vital role in the characteristics of the scaffold and increased CS concentration lead to the slower rate of drug release and disturbances in the fiber morphology. The excellent antithrombogenic effect of PNIPAA makes it a good blood-contacting biomaterial. The model prepared thus can be used in the field of cancer therapy as a drug-delivery agent in postsurgical treatment of cancer and as blood-contacting biomaterial.

Electrospinning was also used to develop antibiotic rifampicin (RIF)-loaded PCL scaffolds (Ruckh et al., 2012). The antibiotic-release kinetics and bactericidal efficacy of RIF against the bacterial strains *Pseudomonas aeruginosa* and *Staphylococcus epidermidis* were observed. Very clear differences between bacterial growth on RIF-free scaffold and PCL-RIF scaffolds indicate its promising applications. RIF-containing scaffold was also shown to have inhibitory effect on biofilm formation.

Studies using electrospun scaffold with poly(a-Lalanine) (PLLA) and chlorhexidine (CHX)-gluconate showed that biodegradable drug-delivery membranes can steadily and continuously inhibit the growth of bacteria which indicate its potential drug-delivery applications (Lin et al., 2011). Aspirin is an antiplatelet drug commonly used against vascular diseases. Studies using aspirin-loaded PCL fibrous scaffold generated by electrospinning showed a sustained release of the drug (Del Gaudio et al., 2012).

Antibiotic amoxicillin containing AMX/n-HA/PLGA electrospun mat was shown to have excellent activity against *Staphylococcus aureus* (Zheng et al., 2013b). Four model drugs, sodium salicylate, diclofenac sodium (DS), naproxen (NAP), and indomethacin (IND) were successfully incorporated on to the electrospun PVA scaffold by Taepaiboon et al. (2006). This drug-loaded electrospun PVA mats exhibited much better release characteristics of the model drugs than drug-loaded cast films. These scaffolds can be used as carriers of drugs for a transdermal drug-delivery system.

11.2.3 WOUND DRESSING APPLICATIONS

WDs act as a physical barrier to protect the wound from the invasion of exogenous microorganisms and thereby they facilitate wound healing and reduce scarring. Desirable properties of WDs include ease of application and removal, nontoxic and nonallergic character, nonsensitizing to both medical staff and patients, provide thermal insulation, allow fluid and gaseous exchanges, absorb excess exudates without leakage, facilitate some actions like removal of foreign materials, dead tissues, and encourage rapid healing. Additionally, mechanical properties are also very important because these materials require certain flexibility, tensile strength, bending, sustainability, and elastic properties (Zahedi et al., 2010).

Nanofibrous electrospun scaffold is of having WD application because of its unique properties like porous structure and increased surface area. The modern dressing materials made of electrospun biopolymers contain various bioactive components such as antimicrobial and anti-inflammatory agents that are favorable to the healing of wounds. Both the natural and synthetic polymers and their composites can be used as electrospun WDs.

Dai et al. (2012) successfully fabricated electrospun polyvinyl pyrrolidone (PVP)/emodin nanofibrous scaffold. Emodin is a widely used drug against suppurative dermatitis, gonorrhea, arthralgia, and also it has the ability to accelerate wound healing. In vitro drug dissolution test of the scaffold showed interesting results indicating the use of electrospun composite as a topical wound healing agent as well as drug-delivery system.

Toncheva et al. (2012) have developed a PLA fibrous material containing DS and lidocaine hydrochloride (LHC) by dual spinneret electrospinning of two separate solutions each containing one drug only [(PLA/DS and PLA/LHC)] and compared this with materials obtained by electrospinning of a common solution (PLA/DS/LHC) using a single spinneret. Interestingly, the scaffold from dual spinneret approach exhibited an excellent antimicrobial activity against *Staphylococcus aureus* and was also biocompatible. This indicates the wound healing applications of the novel material generated.

Curcumin is a natural yellow-orange polyphenol compound. It has an antitumor, antioxidant, antimicrobial, and anti-inflammatory property. Suwantong et al. (2007) developed a new curcumin-loaded cellulose acetate (CA) electrospun nonwoven mat for WD application. The antibacterial and antifungal property of CS-PVA silver (CS/PVA,Ag np) nanocomposite films formed by electrospinning showed significant effects against various microorganisms like *Escherichia coli*, pseudomonas, staphylococcus, micrococcus, *Candida albicans*, and *Pseudomonas aeruginosa*. This shows the application of novel antimicrobial film in WD (Vimala et al., 2011).

Medicated biodegradable electrospun PEG-PLA nanofibrous membranes containing tetracycline hydrochloride (TCH) was found to have enhanced antibacterial effectiveness due to sustained release of TCH content from electrospun meshes (Xu et al., 2010). Tetracycline (Tet) HCl-loaded electrospun PCL/PEVA/PCL scaffold fabricated by Alhausein et al. (2012) was shown to have WD application. For repairing infected wounds, biodegradable sandwich-structured drug-eluting scaffold containing sandwich-structure formed by nanofibrous PLGA/collagen as the surface layer and PLGA/drugs (vancomycin, gentamicin, and lidocaine) as the core was found to be very effective as wound healing accelerators in the early stages of healing (Chen et al., 2012). Coaxial electrospinning produces micro to nanosized fibers comprising an inner-core material and an outer-shell material. This is generally used to fabricate scaffolds with tunable mechanical strength and stiffness without changing morphology or surface chemistry. Schematic diagram of coaxial spinneret for the production of core-sheath nanofiber are shown in Figure 11.3.

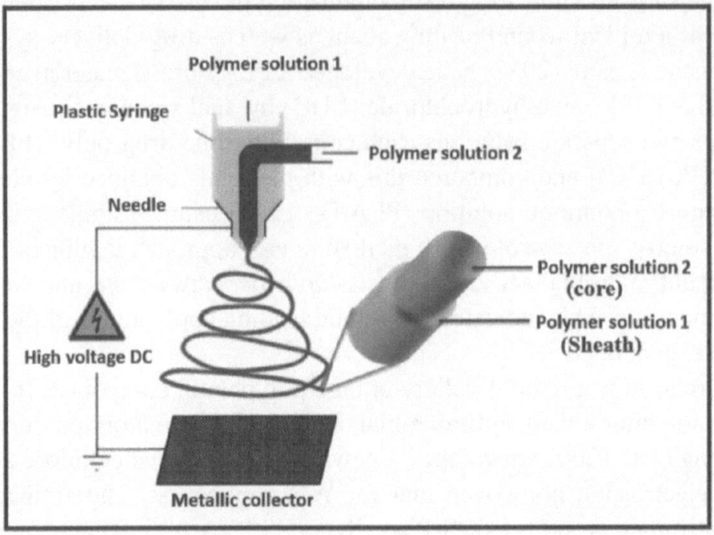

FIGURE 11.3 Schematic diagram of coaxial spinneret for production of core-sheath nanofibrous membranes.

Fouda et al. (2013) fabricated a carboxymethyl chitosan (CMCTS)/PEO/silver nanoparticle (AgNPs) scaffold for medical purposes. Infiltration of CMCTS in this scaffold acts as a protecting agent for silver ion and also as a reducing agent. At the same time, AgNPs were incorporated into the scaffold to explore its inherent antimicrobial property. The scaffold showed excellent antimicrobial activity against all tested microbes like *Staphylococcus aureus* ATCC 25923, *Pseudomonas aeruginosa* ATCC 27853, *Escherichia coli* ATCC 25922, and *Candida albicans* ATCC 10231. Thus this scaffold developed can be used as protective antimicrobial WD materials.

Using layer-by-layer (LBL) technique, Xin et al. (2013) deposited CS and pectin/organic rectorite (OREC) onto the surface of CA electrospun mat. Antibacterial assay studies of this novel scaffold also showed its use as ideal candidate for WDs.

11.2.4 DENTAL APPLICATIONS

TE technology facilitates the dentin regeneration of impaired dentin-pulp tissues. There are various examples of using biodegradable polymers, for dental therapy by electrospinning process. In a dental pulp capping therapy, Lee et al. (2012c) have developed a fibrous electrospun PCL meshes with mineral trioxide aggregates (MTA). Experimental results on the premolars of beagle dogs showed the formation of tubular dentin-like matrices and columnar-polarized odontoblast-like cells beneath the dentin bridge of the material. In another study, human dental pulp stem cells (DP-SCs) were grown on the nanofibrous PLLA (NF PLLA) scaffolds (Wang et al., 2011b) and the result shows promising potential of the scaffold in dental TE field. Recently, Norowski et al. (2012) have developed an anti-biotic minocycline loaded nanofibrous CS mat. In vitro study showed the growth inhibition of *Pseudomonas gingivalis*. This also shows promising capability in periodontal therapy.

11.3 OTHER IMPORTANT MEDICAL APPLICATIONS

CS/alginate electrospun mat was found to provide antiadhesion barrier to prevent peritoneal adhesions (Chang et al., 2012). Park et al. (2013) fabricated a nanofibrous CS/PVA membrane by electrospinning method. They immobilized hen egg-white lysozyme on to the scaffold with the help of cross-linked enzyme aggregates (CLEAs). This scaffold was identified as a promising material for enhanced and continuous antibacterial applications.

Wu et al. (2009) fabricated an excellent electrospun multiaction antibacterial nanofibrous membrane containing four active components such as Ag, AgBr, TiO_2, and hydroxyapatite. Each component has different functions and thereby the membrane exhibited broad antibacterial activity. This makes its applications in respirators, air conditioning filters, antibacterial coatings, etc.

11.4 CONCLUSION

Polymer electrospinning is a unique, scalable, versatile, and cost-effective technology by which continuous polymer nanofibers can be produced. The main advantages of this technique are the production of highly porous matrices with large specific surface areas, excellent mechanical properties, ease of functionalization for different purposes, and ease of process. All these properties provide wide range of opportunities for their use in various biomedical applications. The important medical applications range from TE, drug delivery, WD to artificial bioimplants. Large numbers of polymers have been electrospun by conventional or modified electrospinning methods like multiple needle electrospinning, coaxial electrospinning, needleless electrospinning, etc. This modified electrospinning techniques are able to produce fibers like continuous nanofibrous yarns, LBL-stacked films, hollow structures, core sheath, or as uniaxially aligned films. A dual porous electrospun scaffolds can improve the properties of the scaffold. They contain both nanosized and microsized pores. Nutrients and metabolic wastes may pass through the nanosized porosity. Cell growth and blood vessel invasion occurred through microsized porosity. There is no doubt that electrospun nanofibrous materials are going to take major position in future for various biomedical applications.

KEYWORDS

- **Biomedical applications**
- **Drug delivery**
- **Electrospinning**
- **Nanofibrous scaffolds**
- **Polymers**
- **Tissue engineering**
- **Wound dressing**

REFERENCES

Alhausein, N., Blagbrough, I.S. and Bank, P.A.D. 2012. Electrospun matrices for localised controlled drug delivery: release of tetracycline hydrochloride from layers of polycaprolactone and poly(ethylene-co-vinyl acetate), *Drug Delivery and Translational Research* 2: 477–488.

Aluigi, A., Varesano, A., Montarsolo, A., Vineis, C., Ferrero, F., Mazzuchetti, G. and Tonin, C. 2007. Electrospinning of keratin/Poly(ethylene oxide) blend nanofibers, *Journal of Applied Polymer Science* 104: 863–870.

Auras, R., Lim, L.T., Selke S.E.M. and Tsuji, H. 2010. Poly(lacitic Acid): Synthesis, structures, properties, processing, and application, *John Wiley & Sons, Inc.* http://onlinelibrary.wiley.com/book/10.1002/9780470649848

Bergmeister, H., Schreiber, C., Grasl, C., Walter, I., Plasenzotti, R., Stoiber, M., Bernhard, D. and Schima, H. 2013. Healing characteristics of electrospun polyurethane grafts with various porosities, *Acta Biomaterialia. 9*(4), 6032–6040

Blit, P.H., Battiston, K.G., Yang, M., Santerre, J.P. and Woodhouse, K.A. 2012. Electrospun elastin-like polypeptide enriched polyurethanes and their interactions with vascular smooth muscle cells, *Acta Biomaterialia* 8: 2493–2503.

Cervantes, S.A., Roca, M.I., Martinez, J.G., Olmo, L.M., Cenis, J.L., Moraleda, J.M. and Otero, T.F. 2012. Fabrication of conductive electrospun silk fibroin scaffolds by coating with polypyrrole for biomedical applications, *Bioelectrochemistry* 85: 36–43.

Chang, J.J., Lee, Y.H. Wu, M.H., Yanga, M.C. and Chien, C.T. 2012. Electrospun anti-adhesion barrier made of chitosan alginate for reducing peritoneal adhesions, *Carbohydrate Polymers* 88: 1304–1312.

Chen, D.W., Liao, J.Y., Liu, S.J. and Chan, E.C. 2012. Novel biodegradable sandwich-structured nanofibrous drug-eluting membranes for repair of infected wounds: an in vitro and in vivo study, *International Journal of Nanomedicine* 7: 763–771.

Chong, E.J., Phan, T.T., Lim, I.J., Zhang, Y.Z., Bay, B.H., Ramankrishna, S. and Lim, C.T. 2007. Evaluation of electrospun PCL/gelatin nanofibrous scaffold for wound healing and layered dermal reconstitution, *Acta Biomatererialia* 3: 321–330.

Chua, K.N., Tang, Y.N., Quek, C.H., Ramakrishna, S., Leong, K.W. and Mao, H.Q. 2007. A dual-functional fibrous scaffold enhances P450 activity of cultured primary rat hepatocytes, *Acta Biomaterialia* 3: 643–650.

Cui, W., Cheng, L., Li, H., Zhou, Y., Zhang, Y. and Chang, J. 2012. Preparation of hydrophilic poly(L-lactide) electrospun fibrous scaffolds modified with chitosan for enhanced cell biocompatibility, *Polymer* 53: 2298–2305.

Dai, X.Y., Nie, W., Wang, Y.C., Shen, Y., Li, Y. and Gan, S.J. 2012. Electrospun emodin polyvinylpyrrolidone blended nanofibrous membrane: a novel medicated biomaterial for drug delivery and accelerated wound healing, *Journal of Materials Science: Materials in Medicine* 23: 2709–2716.

Datta, P., Dhara, S. and Chatterjee, J. 2012. Hydrogels and electrospun nanofibrous scaffolds of N-methylene phosphonic chitosan as bioinspired osteoconductive materials for bone grafting, *Carbohydrate Polymers* 87: 1354–1362.

Del Gaudio, C.D., Ercolani, E., Galloni, P., Santilli, F., Baiguera, S., Polizzi, L. and Bianco, A. 2012. Aspirin-loaded electrospun poly(e-caprolactone) tubular scaffolds: potential small-diameter vascular grafts for thrombosis prevention, *Journal of Materials Science: Materials in Medicine. 24*(2): 523–532.

Du, J. and Hsieh, Y. 2007. PEGylation of chitosan for improved solubility and fiber formation via electrospinning, *Cellulose* 14: 543–552.

Eiselt, P., Lee, K.Y and Mooney, D.J. 1999. Rigidity of two-component hydrogels prepared from alginate and poly(ethylene glycol)-diamines, *Macromolecules* 32: 5561–5566.

Fan, S., Zhang, Y., Shao, H. and Hu, X. 2013. Electrospun regenerated silk fibroin mats with enhanced mechanical properties, *International Journal of Biological Macromolecules* 56: 83–88.

Fischer, R.L., McCoy, M.G. and Grant, S.A. 2012. Electrospinning collagen and hyaluronic acid nanofiber meshes, *Journal of Materials Science: Materials in Medicine* 23: 1645–1654.

Formhals, A. 1934. US patent, 1975504.

Fouda, M.M.G., Aassar, M.R.E. and Deyab, S.S.A. 2013. Antimicrobial activity of carboxymethyl chitosan/polyethylene oxide nanofibers embedded silver nanoparticles, *Carbohydrate Polymers* 92: 1012–1017.

Franco, R.A., Nguyen, T.H. and Lee, B.T. 2011. Preparation and characterization of electrospun PCL/PLGA membranes and chitosan/gelatin hydrogels for skin bioengineering applications, *Journal of Materials Science: Materials in Medicine* 22: 2207–2218.

Fu, S.Z., Ni, P.Y., Wang, B.Y., Chu, B.Y., Peng, J.R., Zheng, L., Zhao, X., Luo, F., Wei, Y.Q. and Qian, Z.Y. 2012. In vivo biocompatibility and osteogenesis of electrospun poly(ε-caprolactone)epoly(ethylene glycol)epoly(ε-caprolactone)/nano-hydroxyapatite composite scaffold, *Biomaterials* 33: 8363–8371.

Gang, E.H., Ki, C.S., Kim, J.W., Lee, J., Cha B.G., Lee, K.H and Park, Y.H. 2012. Highly porous three-dimensional poly(lactide-co-glycolide) (PLGA) microfibrous scaffold prepared by electrospinning method: a comparison study with other PLGA type scaffolds on its biological evaluation, *Fibers and Polymers* 13(6): 685–691.

Ghaedi, M., Soleimani, M and Shabani, I. 2012. Hepatic differentiation from human mesenchymal stem cells on a novel nanofiber scaffold, *Cellular & Molecular Biology Letters* 17: 89–106.

Ghosh, S., Laha, M., Mondal, S., Sengupta, S. and Kaplan, D.L. 2009. In vitro model of mesenchymal condensation during chondrogenic development, *Biomaterials* 30: 6530–6540.

Guo, H.F., Li, Z.S., Dong, S.W., Chen, W.J., Deng, L., Wang, Y.F. and Ying, D.J. 2012. Piezoelectric PU/PVDF electrospun scaffolds for wound healing applications, *Colloids and Surfaces B: Biointerfaces* 96: 29–36.

Hsu, F.Y., Hung, Y.S., Liou, H.M. and Shen, C.H. 2010. Electrospun hyaluronate collagen nanofibrous matrix and the effects of varying the concentration of hyaluronate on the characteristics of foreskin fibroblast cells, *Acta Biomaterialia* 6(6): 2140–2147

Jaiswal, A.K., Kadam, S.S., Soni, V.P and Bellare, J.R. 2013. Improved functionalization of electrospun PLLA/gelatin scaffold by alternate soaking method for bone tissue engineering, *Applied Surface Science. 268*: 477–488.

Jang, B.S., Jung, Y., Kwon, K., Mun, C.H. and Kim, S.H. 2012. Fibroblast culture on poly(L-lactide-co-ε-caprolactone) an electrospun nanofiber sheet, *Macromolecular Research* 20(12): 1234–1242.

Jao, W.C., Yang, M.C., Lin, C.H. and Hsu, C.C. 2012. Fabrication and characterization of electrospun silk fibroin/TiO_2 nanofibrous mats for wound dressings, *Polymers for Advanced Technologies* 23: 1066–1076.

Jayakumar, R., Prabaharan, M., Nair, S.V and Tamura, H. 2010. Novel chitin and chitosan nanofibers in biomedical applications, *Biotechnology Advances* 28: 142–150.

Kador, K.E., Montero, R.B., Venugopalan, P., Hertz, J., Zindell, A.N., Valenzuela, D.A., Uddin, M.S., Lavik, E.B., Muller, K.J., Andreopoulos, F.M. and Goldberg, J.L. 2013. Tissue engineering the retinal ganglion cell nerve fiber layer, *Biomaterials* 34(17): 4242–4250.

Kaina, L., Jiangnan, W., Xinqing, L., Xiaopeng, X. and Haiqing, L. 2012. Biomimetic growth of hydroxyapatite on phosphorylated electrospun cellulose Nanofibers. *Carbohydrate Polymers* 90: 1573–1581.

Khalil, K.A., Fouad, H., Elsarnagawy, T. and Almajhdi, F.N. 2013. Preparation and characterization of electrospun PLGA/silver composite nanofibers for biomedical applications, *International Journal of Electrochemical Science* 8: 3483–3493

Kim M. and Kim G.H. 2012. Electrospun PCL/phlorotannin nanofibres for tissue engineering: Physical properties and cellular activities, *Carbohydrate Polymers* 90: 592–601.

Kovacina, J.R., Wise, S.G., Li, Z., Maitz, P.K.M., Young, C.J., Wangb, Y. and Weiss, A.S. 2012. Electrospun synthetic human elastin: Collagen composite scaffolds for dermal tissue engineering, *Acta Biomaterialia* 8: 3714–3722.

Kretlow, J. D. and Mikos, A. G. 2008. From material to tissue: Biomaterial development, scaffold fabrication and tissue engineering. *AIChE Journal* 54: 3048–3067.

Krishnan, R., Rajeswari, R., Venugopal, J., Sundarrajan, S., Sridhar, R. Shayanti, M. and Ramakrishna, S. 2012. Polysaccharide nanofibrous scaffolds as a model for in vitro skin tissue regeneration, *Journal of Materials Science: Materials in Medicine* 23: 1511–1519.

Lee, J.H., Lee, Y.B., Rim, N.G., Jo, S.Y., Lim, Y.M. and Shin, H. 2011. Development and characterization of nanofibrous poly(lactic-*co*-glycolic acid)/biphasic calcium phosphate composite scaffolds for enhanced osteogenic differentiation, *Macromolecular Research* 19(2): 172–179.

Lee J., Yoo, J.J., Atala, A. and Lee, S.J. 2012a. The effect of controlled release of PDGF-BB from heparin-conjugated electrospun PCL/gelatin scaffolds on cellular bioactivity and infiltration, *Biomaterials* 33: 6709–6720.

Lee, J.S., Jin, G.H., Yeo, M.G., Jang, C.H., Lee, H. and Kim, G.H. 2012b. Fabrication of electrospun biocomposites comprising polycaprolactone/fucoidan for tissue regeneration, *Carbohydrate Polymers* 90: 181–188.

Lee, W.C., Oh, J.H., Park, J.C., Shin, H.I., Baek, J.H., Ryoo, H.M. and Woo, K.M. 2012c. Performance of electrospun poly(e-caprolactone) fiber meshes used with mineral trioxide aggregates in a pulp capping procedure, *Acta Biomaterialia* 8: 2986–2995.

Li, L., Li, G., Jiang, J., Liu, X., Luo, L. and Nan, K. 2012a. Electrospun fibrous scaffold of hydroxyapatite/poly (e-caprolactone) for bone regeneration, *Journal of Materials Science: Materials in Medicine* 23: 547–554.

Li, L., Qian, Y., Jiang, C., Lv, Y., Liu, W., Zhong, L., Cai, K., Li, S. and Yang, L. 2012b. The use of hyaluronan to regulate protein adsorption and cell infiltration in nanofibrous scaffolds, *Biomaterials* 33: 3428–3445.

Li, Y., Chen, F., Nie, J. and Yang, D. 2012c. Electrospun poly(lactic acid)/chitosan core–shell structure nanofibers from homogeneous solution, *Carbohydrate Polymers* 90: 1445–1451.

Ligia, M.M.C., Gabriel M.O., Bibin, M.C., Alcides, L.L., Sivoney, F.S. and Mariselma, F. 2013. Bionanocomposites from electrospun PVA/pineapple nanofibers/Stryphnodendron adstringens bark extract for medical applications. *Industrial Crops and Products* 41: 198–202.

Lin, Y.N., Chang, K.M., Jeng, S.C., Lin, P.Y. and Hsu, R.Q. 2011. Study of release speeds and bacteria inhibiting capabilities of drug delivery membranes fabricated via electrospinning by observing bacteria growth curves, *Journal of Materials Science: Materials in Medicine* 22: 571–577.

Lin, H.M., Lin, Y.H and Hsu, F.Y. 2012a. Preparation and characterization of mesoporous bioactive glass/ polycaprolactone nanofibrous matrix for bone tissues engineering, *Journal of Materials Science: Materials in Medicine* 23: 2619–2630.

Lin, T.C., Lin, F.H. and Lin, J.C. 2012b. In vitro feasibility study of the use of a magnetic electrospun chitosan nanofiber composite for hyperthermia treatment of tumor cells, *Acta Biomaterialia* 8: 2704–2711.

Ma, G., Fang, D., Liu, Y., Zhu, X. and Nie, J. 2012. Electrospun sodium alginate/poly(ethylene oxide) core–shell nanofibers scaffolds potential for tissue engineering applications, *Carbohydrate Polymers* 87: 737–743.

Macocinschia, D., Filip, D., Vlad, S., Butnaru, M. and Knieling, L. 2013. Evaluation of polyurethane based on cellulose derivative-ketoprofen biosystem for implant biomedical devices, *International Journal of Biological Macromolecules* 52: 32–37.

Mahoney, C., Mccullough, M.B., Sankar, J. and Bhattarai, N. 2012. Nanofibrous structure of chitosan for biomedical applications, *Journal of Nanomedicine & Biotherapeutic Discoverery* 2(1).

Matthews, J.A., Wnek, G.E., Simpson, D.G and Bowlin, G.L. 2002. Electrospinning of collagen nanofibers, *Biomacromolecules* 3: 232–238.

Maurice, N. C and Colin, B. 2013. Hyaluronic acid based scaffolds for tissue engineering—A review. *Carbohydrate Polymers* 92: 1262–1279.

McKenna, K.A., Hinds, M.T., Sarao, R.C., Wu, P.C., Maslen, C.L., Glanville, R.W., Babcock, D. and Gregory, K.W. 2012. Mechanical property characterization of electrospun recombinant human tropoelastin for vascular graft biomaterials, *Acta Biomaterialia* 8: 225–233.

Merrell, J.G., McLaughlin, S.W., Tie, L., Laurencin, C.T., Chen, A.F. and Nair, L.S. 2009. Curcumin loaded poly(ε-caprolactone) nanofibers: Diabetic wound dressing with antioxidant and anti-inflammatory properties, *Clinical and Experimental Pharmacology and Physiology* 36(12): 1149–1156.

Mobarakeh, L.G., Prabhakaran, P.M., Morshed, M., Esfahani, M.H.N and Ramakrishna, S. 2008. Electrospun poly(3-caprolactone)/gelatin nanofibrous scaffolds for nerve tissue engineering, *Biomaterials* 29: 4532–4539.

Nitya, G., Nair, G.T., Mony, U., Chennazhi, K.P. and Nair, S.V. 2012. In vitro evaluation of electrospun PCL/nanoclay composite scaffold for bone tissue engineering, *Journal of Materials Science: Materials in Medicine* 23: 1749–1761.

Norowski, P.A.,Babu, J., Adatrow, P.C., Godoy, F.G., Haggard, W.O. and Bumgardner, J.D. 2012. Antimicrobial activity of minocycline-loaded genipin-crosslinked nano-fibrous chitosan mats for guided tissue regeneration, *Journal of Biomaterials and Nanobiotechnology* 3: 528–532.

Park, J.M., Kim, M., Park, H.S., Jang, A., Min, J. and Kim, Y.H. 2013. Immobilization of lysozyme-CLEA onto electrospun chitosan nanofiber for effective antibacterial applications, *International Journal of Biological Macromolecules* 54: 37–43.

Phipps, M.C., Clem, W.C., Grunda, J.M., Clines, G.A. and Bellis, S.L. 2012. Increasing the pore sizes of bone-mimetic electrospun scaffolds comprised of polycaprolactone, collagen I and hydroxyapatite to enhance cell infiltration, *Biomaterials* 33: 524–534.

Prabhakaran, M.P., Venugopal, J.R. and Ramakrishna, S. 2009. Mesenchymal stem cell differentiation to neuronal cells on electrospun nanofibrous substrates for nerve tissue engineering, *Biomaterials* 30: 4996–5003.

Ruckh, T.T., Oldinski, R.A., Carroll, D.A., Mikhova, K., Bryers, J.D. and Popat, K.C. 2012. Antimicrobial effects of nanofiber poly(caprolactone) tissue scaffolds releasing rifampicin, *Journal of Materials Science: Materials in Medicine* 23: 1411–1420.

Sakai, S., Takagi, Y., Yamada, Y., Yamaguchi, T. and Kawakami, K. 2008. Reinforcement of porous alginate scaffolds by incorporating electrospun fibres, *Biomedical Materials* 3: 034102.

Shao, Z., Zhang, X., Pi, Y., Wang, X., Jia, Z., Zhu, J., Dai, L., Chen, W., Yin, L., Chen, H., Zhou, C. and Ao, Y. 2012. Polycaprolactone electrospun mesh conjugated with an MSC affinity peptide for MSC homing in vivo, *Biomaterials* 33: 3375–3387.

Sheikh, F.H., Barakat, N.A.M., Kanjwal, M.A., Park, S.J., Park, D.K. and Kim, H.Y. 2010. Synthesis of Poly(vinyl alcohol) (PVA) Nanofibers incorporating hydroxyapatite nanoparticles as future implant materials, *Macromolecular Research* 18(1): 59–66.

Shin, Y.M., Park, H. and Shin, H. 2011. Enhancement of cardiac myoblast responses onto electrospun PLCL fibrous matrices coated with polydopamine for gelatin immobilization, *Macromolecular Research* 19(8): 835–842.

Sundar, S.S. and Sangeetha, D. 2012. Fabrication and evaluation of electrospun collagen/poly (N-isopropyl acrylamide)/chitosan mat as blood-contacting biomaterials for drug delivery, *Journal of Materials Science: Materials in Medicine* 23: 1421–1430.

Suwantong, O., Opanasopit, P., Ruktanonchai, U. and Supaphol, P. 2007. Electrospun cellulose acetate fiber mats containing curcumin and release characteristic of the herbal substance, *Polymer* 48: 7546–7557

Szentivanyi, A., Chakradeo, T., Zernetsch, H. and Glasmacher, B. 2011. Electrospun cellular microenvironments: understanding controlled release and scaffold structured. *Advanced Drug Delivery Reviews* 63: 209–212.

Taepaiboon, P., Rungsardthong, U. and Supaphol, P. 2006. Drug-loaded electrospun mats of poly(vinyl alcohol) fibres and their release characteristics of four model drugs, *Nanotechnology* 17: 2317–2329.

Toncheva, A., Paneva, D., Manolova, N., Rashkov, I., Mita, L., Crispi, S. and Mita, D.G. 2012. Dual vs. single spinneret electrospinning for the preparation of dual drug containing non-woven fibrous materials, *Colloids and Surfaces A: Physicochemical and Engineering Aspects. 439*: 176–183.

Vimala, K., Mohan, Y.M., Varaprasad, K., Redd, N.N., Ravindra, S., Naidu, N.S. and Raju, K.M. 2011. Fabrication of curcumin encapsulated Chitosan-PVA silver nanocomposite films for improved antimicrobial activity, *Journal of Biomaterials and Nanobiotechnology* 2: 55–64.

Vrieze, S.D., Westbroeak, P., Camp, T.V and Langenhove, L.V. 2007. Electrospinning of chitosan nanofibrous structures: feasibility study, *Journal of Material Science* 42: 8029–8034.

Wang, H., Feng, Y., An, B., Zhang, W., Sun, M., Fang, Z., Yuan, W. and Khan, M. 2012a. Fabrication of PU/PEGMA crosslinked hybrid scaffolds by in situ UV photopolymerization favoring human endothelial cells growth for vascular tissue engineering, *Journal of Materials Science: Materials in Medicine* 23: 1499–1510.

Wang, H., Feng, Y., Behl, M., Lendlein, A., Zhao, H., Xiao, R., Lu, J., Zhang, L. and Guo, J. 2011a. Hemocompatible polyurethane/gelatin-heparin nanofibrous scaffolds formed by a bilayer electrospinning technique as potential artificial blood vessels, *Frontiers of Chemical Science and Engineering* 5(3): 392–400.

Wang, J., Ma, H., Jin, X., Hu, J., Liu, X., Ni, L. and Ma, P.X. 2011b. The effect of scaffold architecture on odontogenic differentiation of human dental pulp stem cells, *Biomaterials* 32: 7822–7830.

Wang, T.Y., Forsythe, J.S., Nisbet, D.R. and Parish, C.L. 2012b. Promoting engraftment of transplanted neural stem cells/progenitors using biofunctionalised electrospun scaffolds, *Biomaterials* 33: 9188–9197.

Wang, N., Burugapalli, K., Song, W., Halls, J., Moussy, F., Ray, A and Zheng, Y. 2013. Electrospun fibro-porous polyurethane coatings for implantable glucose biosensors, *Biomaterials* 34: 888–901.

Wepener, I., W. Richter., Papendorp, D.V. and Joubert, A.M. 2012. In vitro osteoclast-like and osteoblast cells' response to electrospun calcium phosphate biphasic candidate scaffolds for bone tissue engineering, *Journal of Materials Science: Materials in Medicine* 23: 3029–3040.

Wu, Y., Jia, W., An, Q., Liu, Y., Chen, J. and Li, G. 2009. Multiaction antibacterial nanofibrous membranes fabricated by electrospinning: An excellent system for antibacterial applications, *Nanotechnology* 20.

Wu, H., Fan, J., Chu, C.C. and Wu, J. 2010. Electrospinning of small diameter 3-D nanofibrous tubular scaffolds with controllable nanofiber orientations for vascular grafts, *Journal of Materials Science: Materials in Medicine* 21: 3207–3215.

Xiang, P., Li, M., Zhang, C.Y., Chen, D.L. and Zhou, Z.H. 2011. Cytocompatibility of electrospun nanofiber tubular scaffolds for small diameter tissue engineering blood vessels, *International Journal of Biological Macromolecules* 49: 281–288.

Xin, S., Li, X., Ma, Z., Lei, Z., Zhao, J., Pan, S., Zhou, X. and Deng, H. 2013. Cytotoxicity and antibacterial ability of scaffolds immobilized by polysaccharide/layered silicate composites, *Carbohydrate Polymers* 92: 1880–1886.

Xu, X., Zhong, W., Zhou, S., Trajtman, A. and Alfa, M. 2010. Electrospun PEG–PLA Nanofibrous membrane for sustained release of hydrophilic antibiotics, *Journal of Applied Polymer Science* 118: 588–595.

Xue, J., Feng, B., Zheng, R., Lu, Y., Zhou, G., Liu, W., Cao, Y., Zhang, Y. and Zhang, W.J. 2013. Engineering ear-shaped cartilage using electrospun fibrous membranes of gelatin/polycaprolactone, *Biomat erials* 34: 2624–2631.

Yuan, J., Xing, Z.C., Park, S.W., Geng, J. and Kang, I.K. 2009. Fabrication of PHBV/Keratin composite nanofibrous mats for biomedical applications, *Macromolecular Research* 17(11): 850–855.

Zahedi, P., Rezaeian, I., Siadat, S.O.R., Jafari, S.H and Supaphol, P. 2010. A review on wound dressings with an emphasis on electrospun nanofibrous polymeric bandages, *Polymers of Advanced Technologies* 21: 77–95.

Zhang, Y.Z., Venugopal, J., Huang, Z.M., Lim, C.T. and Ramakrishna, S. 2006. Crosslinking of the electrospun gelatin nanofibers, *Polymer* 47: 2911–2917.

Zhang, X.W., Kotaki, M., Okubayashi, S. and Sukigara, S. 2010. Effect of electron beam irradiation on the structure and properties of electrospun PLLA and PLLA/PDLA blend nanofibers, *Acta Biomaterialia* 6: 123–129.

Zhang, M., Wang, K., Wang, Z., Xing, B., Zhao, Q. and Kong, D. 2012. Small-diameter tissue engineered vascular graft made of electrospun PCL/lecithin blend, *Journal of Materials Science: Materials in Medicine* 23: 2639–2648.

Zhang, H., Jia, X., Han, F., Zhao, J., Zhao, Y., Fan, Y. and Yuan, X. 2013. Dual-delivery of VEGF and PDGF by double-layered electrospun membranes for blood vessel regeneration, *Biomaterials* 34: 2202–2212.

Zhao, J., Qiu, H., Chen, D.L., Zhang, W.X., Zhang, D.C. and Li, M. 2013. Development of nanofibrous scaffolds for vascular tissue engineering, *International Journal of Biological Macromolecules* 56: 106–113.

Zheng, F., Wang, S., Shen, M., Zhub, M. and Shi, X. 2013a. Antitumor efficacy of doxorubicin-loaded electrospun nano-hydroxyapatite–poly(lactic-co-glycolic acid) composite nanofibers, *Polymer Chemistry* 4: 933–941.

Zheng, F., Wang, S., Wen, S., Shen, M., Zhu, M. and Shi, X. 2013b. Characterization and antibacterial activity of amoxicillin-loaded electrospun nano-hydroxyapatite/poly(lactic-co-glycolic acid) composite nanofibers, *Biomaterials* 34: 1402–1412.

Zhou, Z.H., Yi, Q.F., Liu, L.H., Zhao, Y.M., Liu, X.P. and Zhou, J.N. 2010. Influence of bovine bone content on in vitro degradation of poly-lactic acid and bovine bone composite materials, *Journal of Macromolecular Science Part B-Physics* 49: 940–952.

Zhou, Y., Yang, H., Liu, X., Mao, J., Gu, S. and Xu, W. 2013. Electrospinning of carboxyethyl chitosan/poly(vinyl alcohol)/silk fibroin nanoparticles for wound dressings, *International Journal of Biological Macromolecules* 53: 88–92.

CHAPTER 12

FUNCTIONALIZATION OF SCAFFOLDS WITH BIOMOLECULES FOR VARIOUS TYPES OF TISSUE ENGINEERING APPLICATIONS

R. SELVAKUMAR, AMITAVA BHATTACHARYYA, J. GOPINATHAN, R. SOURNAVENI, and MAMATHA M. PILLAI

ABSTRACT

Tissue engineering is a multidisciplinary area which needs application of both life science as well as engineering research. Tissue engineering involves use of porous scaffold that can provide ambient conditions for the growth of target cells intended to grow inside or on the surface of the scaffold. Such growth inside the scaffold is mostly possible only when we follow a tissue engineering triad involving appropriate cells, relevant signaling molecules or biomolecules, and a proper porous scaffold. An effective cell adhesion, cell growth, and retention of differentiated cell's function in a scaffold depend on many factors such as biomimetic surface, oxygen tension, growth factor, immobilization or incorporation method of growth factor, controlled combinatorial activity of key signaling molecules or growth factors from scaffold or biomaterial, hydrophilicity of scaffold, etc. In this review, we have made an attempt to bring out the various types and strategies of biomolecular functionalization of scaffolds involved in various tissue engineering sectors like stem cell, vascular grafts, bone, skin, and nerve tissue engineering. We have highlighted the commonly used biomolecules for tissue engineering and biomedical applications and have described the mechanism of their action to enhance the cell attachment, growth, proliferation, and differentiation onto the scaffold. Emphasis has been given onto biochemistry of the biomolecules or growth factors that trigger the natural and regulated cascade of reactions leading to cell attachment and proliferation of cells into functional tissues.

12.1 INTRODUCTION

Tissue engineering is a highly multidisciplinary field and seeks experts from genetics, clinical medicine, materials science, mechanical engineering, and other related disciplines of both engineering and life sciences. This field mostly relies on the use of porous two- or three-dimensional (2-D or 3-D) artificial structures, known as scaffolds, to deliver the suitable environment for the rejuvenation of tissues and organs. These scaffolds basically act as a device for cell proliferation and tissue formation with the aid of chemical stimuli like growth factors and other biomolecules or physical stimuli in the form of bioreactor (Martin et al., 2004). These cell-seeded scaffolds are cultured in vitro for the formation of tissues which can be implanted into a damaged site by using the body's own systems, where integration of tissues is prompted in vivo. This blend of cells, scaffold, and signals is mentioned as tissue engineering triad. The word scaffold denotes the biomaterial structure before cells have been seeded (in vitro or in vivo). Scaffolds act as a template for tissue formation by allowing cells to migrate, adhere, and produce tissue. To renovate a tissue by tissue engineering, the following three components are required:

1) Cells – which are dissociated and harvested from the donor tissue
2) Biomaterials as scaffold – substrates in which cells are attached and proliferated for the formation of tissue
3) Growth factors – which promotes cell adhesion, proliferation, migration, and differentiation by up regulating or down regulating the synthesis of protein, growth factors, and receptors.

A prominent strategy to harvest tissues, implants, and prosthesis is to mimic the artificial extracellular matrix (ECM) by synthetically derived scaffold matrices which provide temporary mechanical support to facilitate the cell growth. This is considered as a better approach than the gene therapy and transplantation technology. The scaffold provides a temporary biomechanical profile until the cells produce their own matrix proteins and the full tissue is formed. All tissues arise from the biosynthetic efforts of specialized and differentiated cells. Hence, the in vitro response of cells on scaffold material is not always affirmative. On the other hand, cells do not exhibit the proper affinity to any 3-D surfaces randomly; several parameters need to be considered. The scaffolds should be optimized by physical and chemical cross-linking and suitable surface modifications with peptide molecules. To enhance the in vitro response of cells on 3-D matrices, oxygen, and nutrient diffusion into the interior of the scaffold is another important issue to be looked at. However, a smart peptide sequences with tunable properties could enhance the biomechanism of synthetically derived scaffold materials which exhibit strong resemblance to their counterpart (i.e., native tissues). This chapter will present the current state of the tissue engineering, with particular emphasis on the functionalization of bioactive scaffold material such as structure-function relationship, interaction between cells and artificial matrix, bioactive molecules in soliciting favorable cellular responses, and combinatorial responses of functional groups toward the in vitro cellular behavior.

12.2 IMPORTANCE OF SCAFFOLD MATRICES IN TISSUE ENGINEERING

Scaffold-based tissue engineering is considered to be one of the most promising strategies for tissue regeneration. So far, most of the tissue engineering works need the support of scaffold matrices. Scaffolds are extensively used to culture cells as they provide the structural basis for cells. Scaffold matrices can be used to achieve cell delivery with high loading and efficacy to specific sites. All scaffolds should maintain sufficient mechanical property for a period of at least 6 months so that the cells can attach, grow, proliferate, and secrete ECM to form the tissue. The concept is that the regenerated tissue should occupy the scaffold structure before the polymer degrades. Hence, the degradation rate of the scaffold should be slightly less than the tissue growth rate in the scaffold. A generic problem of scaffold usually comes from this point as the formation of tissue is very much patient or donor specific. The biocompatibility of scaffold materials actively participates in the signaling process for the requirement of safe degradation and also provides a substratum for cell migration into the defect sites of the tissue. The first application of such scaffolds for tissue reconstruction is an artificial skin for burn injury treatment which was introduced in 1990 (Miller and Peshwa, 1996). So far, scaffolds have been fabricated from a range of ceramics, natural and synthetic polymers, composite biomaterials, metals, and cytokine release materials. Decellularized native tissue is also used as a scaffold. Figure 12.1 shows a typical 3-D porous scaffold design suitable for cell culture.

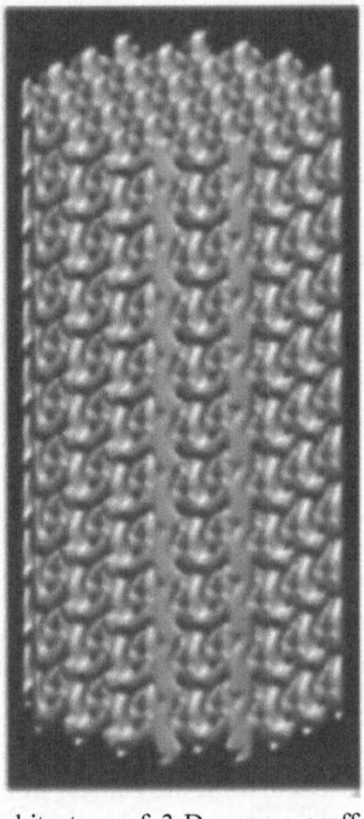

FIGURE 1 A designed architecture of 3-D porous scaffold created in the shape of cylindrical test specimens (Hollister and Lin, 2007)

12.3 PARAMETERS AFFECTING CELL GROWTH IN SCAFFOLD

Several factors affect the cell growth in scaffold. An ideal scaffold for tissue engineering should have the following criteria.

12.3.1 MECHANICAL STRENGTH

The structural integrity and mechanical strength of a scaffold is of utmost importance. Scaffolds are transplanted in the body along with the in vitro cultured tissue on it. When planted in the body, it should support the damaged part till the tissue is formed completely. The tensile properties of the tissues vary widely among bone, cartilage, blood vessels, tendons, and muscles. The property requirements also differ such as compressive, tensile, creep, stress-relaxation, and dynamic mechanical properties. The anisotropic mechanical properties of the natural tissue in different directions are also a matter of concern for fabricating appropriate scaffold. According to scientists tendons and ligaments have 200–500 times higher tensile property in parallel direction as compared to the perpendicular direction (*Liu et al., 2007*). Several other tissues also have anisotropic mechanical properties with respect to their structural functions.

12.3.2 POROSITY, PORE SIZE, SHAPE, AND DISTRIBUTION

Highly porous structure permits uniform cell distribution throughout the scaffold during cell seeding. The porous structure with an interconnected pore network supports suitable surface chemistry for cell attachment, migration, proliferation, and differentiation; minimizes diffusion constraints during in vitro cultivation of cells; regularizes the flow transport of nutrients, oxygen, metabolic waste, and controls degradation at a rate matching that in vitro tissue regeneration, followed by complete elimination of the foreign material to maximize in vivo biocompatibility. Pores are usually created by using some porogen or gas-foaming methods. The drawback of these techniques is the difficulty in controlling final pore size and structure. More precise control over pore structure and dimension can be achieved by rapid prototyping or 3D printing method. However, these methods are costly and all polymers cannot be used to form structures using these processes. Various chemicals have been used as porogen which depends upon the scaffold material as well as prostheses. Different types of cells show different growth behaviors depending on the material as well as overall and pore structures of scaffolds. Study on poly ε-caprolactone (PCL) scaffold with various pore sizes reveals that the chondrocytes and osteoblasts show better cell growth in the scaffold sections having larger pore sizes, while the fibroblasts show best cell growth in the scaffold section having pore size of 186–200 μm (Oh et al., 2007). Besides pore size requirements, generally scaffolds with high porosity (around 90%) facilitate better cell growth, infiltration, and ECM deposition. However, higher porosity results in lower mechanical strength; so, the level of porosity in the scaffold must lie within a critical range, enough to maintain its mechanical properties and provide optimal bioactivity.

12.3.3 DEGRADATION RATE OF SCAFFOLD MATERIALS

The degradation of scaffold is important along with the localization and controlled growth factor supplementation. Degradation rate depends on the property of the particular polymer material. It also depends on the molecular weight distribution of the polymer, process parameters and scaffold design. Degradation products may enhance the degradation rate of the polymer. Lactic acid released during poly-L-lactic acid (PLLA) degradation reduces the pH and hastens the degradation rate due to autocatalysis, resulting in a highly acidic atmosphere adjacent to the polymer which is not favorable for cells. PCL degrades slowly by hydrolysis of ester linkages, with removal of the resultant fragments by macrophages and giant cells. Polyglycolic acid (PGA) degrades through hydrolysis of its ester bonds to glycolic acid, which in turn is metabolized and eliminated as water and carbon dioxide. PGA loses its strength in vivo within 4 weeks and is completely absorbed within 6 months. Biodegradation rate of PGA can be controlled by copolymerization with other polymers like polylactic acid (PLA), PCL-co-PLA, polyhydroxyalkanoate (PHA), polyethylene glycol, etc. (Ravi and Chaikof, 2010). Cross-linking can improve the stability of some scaffolds like gelatin, polyvinyl alcohol, chitosan, etc. The pores and surface area also contribute on the degradation rate as they permit more diffusion of the lysozyme into the matrix and less area is exposed to enzyme degradation if the matrix contains fewer pores. The degradation of scaffold can be increased by introducing hydrophilicity in the scaffold using suitable techniques like plasma treatment. The surface modification of polymers by introducing polar groups containing esters like COOR, ether (-O-), Ketone (>CO), epoxy and carboxylic acid, acetyl, amide, or amine groups also influences the degradation rate of the scaffold. Introduction of such polar groups into the polymeric scaffold increases the crystallinity which in turn affects the mechanical property of the scaffold (Pramanik and Kar, 2012).

12.3.4 SURFACE PROPERTIES

The adhesion and proliferation of different types of cells on polymeric materials depend largely on surface characteristics such as wettability (hydrophilicity/hydrophobicity), chemistry, charge, roughness, and rigidity. These play a vital role in cell attachment and growth on the scaffold materials. Surface properties of the scaffold also depends on the type of modification carried out and the method involved (Figure 12.2). The surface wettability also influences the adsorption of protein onto the scaffold. The wettability of the scaffold is measured by its contact angle. The surface is more hydrophilic (super hydrophilic) when the contact angle is 0° and slowly moves toward hydrophobicity as the contact angle increase. If the contact angle is lesser than 90°, the scaffold surface is considered to be hydrophilic in nature and if it is greater than 90°, the surface will be hydrophobic. Increasing surface wettability influences hydrophilicity onto the surface of scaffold and induces attachment of proteins like fibronectin and cells like osteoblast and fibroblast. However, this attachment may also vary from one protein/cell to another. For instance albumin binding increases when the contact angle increases from 0° to 80° with increasing hydrophobic surface when compared to fibronectin which prefers hydrophilic surface (Yoshinari et al., 2009). The optimal contact angle showing good cell attachment onto various scaffolds and cells have been reported to be between 50° and 70° (Bernhardt et al., 2009; Kim et al., 2007; Xu and Siedlecki, 2007). The contact angle of various types of scaffold and cells studied are given in Table 12.1. Electron scanning chemical analysis and streaming potential are used for surface chemistry and charge analysis. The bioactivity of a scaffold can be enhanced by creating suitable functional groups and anchoring appropriate biomolecules on its surfaces. The gene expression, enzymatic activity, and mineral deposition study are usually carried out to estimate the impact of chemical composition of the scaffolds on cells (Khang, 2012).

TABLE 12.1 Various types of scaffolds and its contact angle used in tissue engineering

Scaffold	Contact angle (°)	Cells studied	Reference
Polycaprolactone (PCL) scaffold	129.97±0.4	Porcine chondrocytes	Kosorn et al. (2012)
Hydrolysed PCL scaffold	101.73±6.54	Porcine chondrocytes	Kosorn et al. (2012)
Plasma-treated PCL scaffold	48.03±12.89	Porcine chondrocytes	Kosorn et al. (2012)
Aligned PCL nanofibers	130	Mesenchymal stem cells	Jahani et al. (2012)
Plasma-treated aligned PCL nanofibers	<80	Mesenchymal stem cells	Jahani et al. (2012)
PCL scaffold	110.0±0.3	MG63 cells	So-Ra Son et al. (2013)
PCL/PMMA scaffold	95.0±0.4	MG63 cells	So-Ra Son et al. (2013)
Silk film-based scaffolds	87	Bone marrow mesenchymal stem cells and osteoblast	He et al. (2013)
Hydroxyapetite-coated silk-based scaffolds	42–76	Bone marrow mesenchymal stem cells and osteoblast	He et al. (2013)
Gelain-coated PCL nanofiber	0	Fibroblast	Safaeijavan et al. (2013)

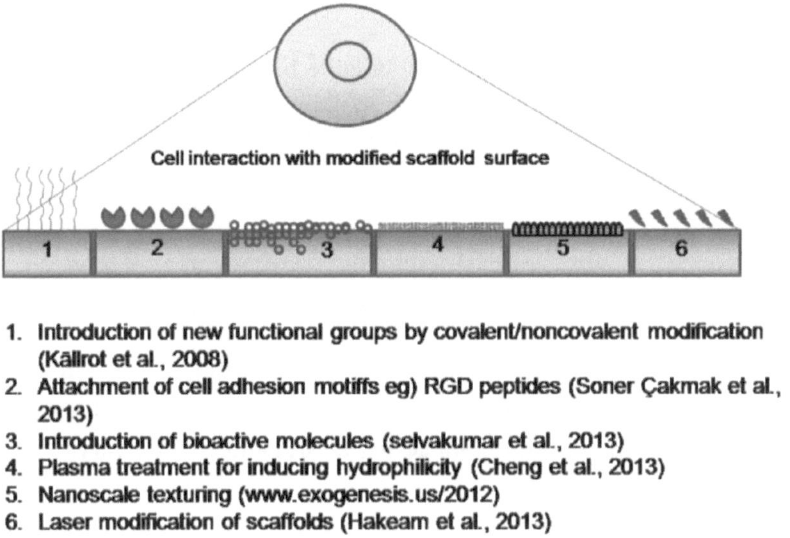

FIGURE 12.2 Schematic representation of surface modification of scaffold and cell interaction

12.3.5 PHYSICOCHEMICAL PROPERTIES

The physicochemical properties of ideal scaffolds can be summarized as given below (Kanani and Bahrami, 2010):

- ✓ Biocompatibility and biodegradability
- ✓ Easy processability and malleability into desired shapes
- ✓ Highly porous with a large surface to volume ratio
- ✓ Adequate mechanical strength and dimensional stability
- ✓ Ability to sterilize
- ✓ Reproducible fabrication of the complex structure
- ✓ Support cells to grow three dimensionally
- ✓ Stimulate cellular responses and induce, differentiate, and channel tissue growth.

These are not all, but they cover most of the areas.

12.3.6. ROLE OF FUNCTIONALIZED BIOMOLECULES

Functionalization of scaffolds with suitable biomolecules plays the most important role in tissue engineering. Without these biomolecules, tissue engineering research rarely succeeds. The role of such biomolecules, their mechanism to work, and the application areas have been discussed in detail in Section 12.5.

12.4 MATERIALS USED FOR SCAFFOLD CONSTRUCTION

Several materials have been used to prepare scaffolds. The scaffold materials can be classified into different categories as shown in Figure 12.3 depending on strategies for engineering tissues and cell delivery.

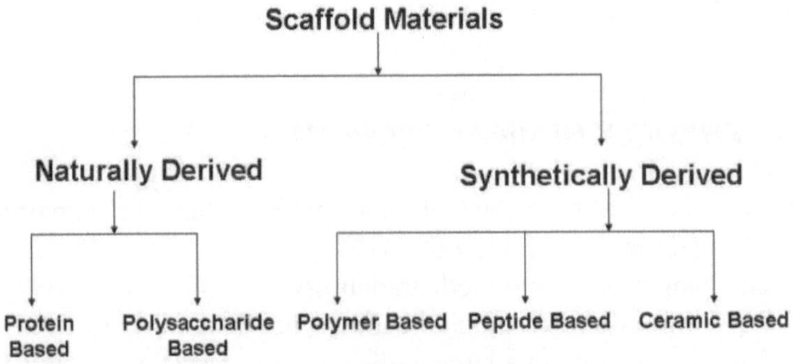

FIGURE 12.3 Classification of scaffold materials

All the scaffolds can be classified in three groups such as natural, synthetic, and ceramics. These materials interact with the biomolecules by forming various bonds like hydrogen, polar, Van der Waals and covalent bonds through functional groups like amide and carboxyl in collagen, gelatin, PCL, nitryl group in polyacrylonitrile (PAN), alcohol for polysaccharide, etc. Strong bonds like covalent bonds are not favorable as delivery of biomolecules will be affected. The major scaffold materials are briefly discussed below.

12.4.1 NATURAL POLYMERS

Although some natural polymers cause immunogenic responses during cell culture, most of them biocompatible and can be used extensively in tissue engineering. Natural polymers like alginate, collagens, gelatin, fibrin, keratin, albumin, gluten, elastin, fibroin, hyarulonic acid, cellulose, chitosan, pectin (pectinic acid), galactan, gellan, levan, emulsan, dextran, pullulan, heparin, silk, chondroitin 6-sulfate, PHAs, etc., have been used in tissue engineering as well as other implantable and nonimplantable medical applications (Garg et al., 2012).

Among the biocompatible natural polymers, collagen is the most abundant protein in mammals. It provides structural and mechanical support to tissues and organs, and fulfills biomechanical functions in bone, cartilage, skin, tendon, and ligament. Gelatin is another natural biopolymer. Gelatin contains Arg-Gly-Asp (RGD)-like sequence that promotes cell adhesion, migration, and form a polyelectrolyte complex. Gelatin blended with chitosan shows improved biological activity. Gelatin–chitosan scaffolds have been formed without or with cross-linkers such as glutaraldehyde and tested in regenerating various tissues including skin, cartilage, and bone (Liu et al., 2011). Cellulose is naturally occurring polysaccharide. Cellulose and its derivatives are widely used as tough versatile materials. They can be easily molded or drawn into fibers. Control over the crystallinity and hydrophilicity of cellulose-based materials improves their biodegradability. Several studies report the applicability of cellulose-based materials for culturing cells and for implantation in bone regeneration, hepatocyte culturing for an artificial liver, expansion of progenitor hematopoetic cells in culture and suppression of matrix metalloproteases (MMPs) action in wound healing. Cellulose-based materials are mostly biocompatible as they produce negligible inflammatory response reactions in vivo. Small amount of cationic groups can further improve their tissue compatibility (Entcheva et al., 2004).

12.4.2 SYNTHETIC POLYMERS FOR SCAFFOLDS

Synthetic polymers are divided into two categories namely biodegradable and nonbiodegradable. Some nonbiodegradable polymers are polyethylene terephthalate (PET), nylon, polypropylene, polyethylene, polymethyl methacrylate (PMMA), PAN, polyhydroxyethyl methacrylate (PHEMA), and poly (N-isopropylacrylamide) (PNIPAAm). The commonly used biodegradable synthetic polymers are PCL, polyglycolide or PGA, polylactide or PLA and its copolymer poly (lactide-co-glycolide) (PLGA), poly (trimethylene carbonate) (PTMC), polyphosphazene, polyanhydride, poly (propylene fumarate), polycyanoacrylate, polydioxanone (PDO), and biodegradable polyurethanes (Nair and Laurencin, 2007). For tissue engineering, biodegradable polymers are preferred as they degrade and excrete from body. The newly formed tissue replaces the polymer scaffold with time. Hence, a careful choice of polymer is required considering its degradation rate and degradation products as the tissue formation time and cell surrounding atmosphere are different for different cell lines.

Majority of biodegradable polymers belong to the polyester family. The most common and well known polymers in this class are PCL, PLA, and PGA. PCL is a semicrystalline rubbery polymer with a very low Tg of around ~60°C. Usually, PCL degrades by bulk hydrolysis like PLA. Also enzymatic degradation can occur under certain conditions. The degradation rate of PCL is significantly slower than PLA due to its high crystalline structure. The closely packed macromolecules allow limited fluid inflow which results in a very high in vivo degradation time (over 2 years). Thus, PCL is mostly suitable for long-term implants. Hutmacher (2000) listed the degradation time of almost all common biodegradable polymers. PLA degrades by bulk hydrolysis and leads to the production of lactic acid. PLA exists in different isomeric forms. In case of PLLA, degradation results in L (+) lactic acid which exists in human body, so this form is generally preferred. The degradation product L (+) lactic acid converts into pyruvic acid in liver and enters in tricarboxylic acid cycle. Finally, it is converted into water and carbon dioxide.

PTMC is easy for processing and cell culture in vitro as it degrades slowly in aqueous solutions. However, when tested in vivo, it shows high degradation rate (in the order of weeks) as it undergoes enzymatic degradation. One more advantage of PTMC over PLA is its degradation products do not decrease pH in the surrounding tissues of the scaffold (Zhang et al., 2006). PHAs are linear polyesters produced by bacterial fermentation of sugar or lipids. They can be modified for various degradation rates as well as mechanical properties. Polyhydroxyoctanoate (PHO) can provide the structural integrity and adequate mechanical support because of its slower degradation rate and high strength (Ravi and Chaikof, 2010).

12.4.3 BIOCERAMICS-BASED SCAFFOLDS

Bioceramics are produced by sintering or melting inorganic raw materials to create an amorphous or crystalline solid body which can be used as an implant. These ceramics mostly hard and resemble the natural inorganic component of bone. Alumina, zirconia, silicon nitride and carbons are inert bioceramics. Certain glass ceramics such as dense hydroxyapatites [$Ca_{10}(PO_4)_6(OH)_2$], are semi-inert type bioactive material. Examples of bioresorbable bioceramics are calcium phosphates, aluminum calcium phosphates, coralline, tricalcium phosphates ($Ca_3(PO_4)_2$), zinc calcium phosphorus oxides, zinc sulfate calcium phosphates, ferric calcium phosphorus oxides, and calcium aluminates. Synthetic apatite and calcium phosphate minerals, coral-derived apatite, bioactive glass, and demineralized bone particle (DBP) are mostly used in hard tissue engineering (*Oh et al., 2006*).

12.5 BIOMOLECULES USED FOR FUNCTIONALIZATION IN TISSUE ENGINEERING

Biomolecules (in tissue engineering aspect) are molecules of biological origin which have a defined role in regulating the proliferation and differentiation of cells (Depprich, 2009). These regulatory molecules generally include differentiation or growth factors, cytokines, proteins and peptides, carbohydrate-based biomolecules, receptors, lipids and its derivatives, etc. These biomolecules play major role in the assembly, structural integrity and functional parameters of the tissue engineering cells and construct. The controlled release of biomolecule from the scaffold depends on availability of biomolecule and the stability of biomolecule. The availability of biomolecule depends on various variables like cross-linking density, pH, molecular weight of the polymer, its degradation rate, and its affinity interaction with biomolecules. Similarly the stability of biomolecule depends on initiator/cross-linking content and the stability agent added to it. The stability of biomolecule is achieved by covalent bonding with polymer. Appropriate coupling methods should be selected based on the type of biomolecule to maintain the biological activity of the biomolecule. For example, binding of RGD peptide (Arg-Gly-Asp) to polymer nanofiber by covalent bonding increases cell adhesion (Kim et al., 2006). Another way of maintaining the stability of the biomolecule is to use core-shell nanofiber structure. The biomolecule is incorporated into the core polymer which maintains the functionality and stability of the biomolecule whereas the outer polymer takes care of the release profile (Jiang et al., 2006). In this section of this chapter, we have tried to bring out the basics of each biomolecules, its involvement and application in tissue engineering with reference to the current research work done in each area.

12.5.1 GROWTH FACTORS

Growth factors are biomolecules which are involved in cellular growth, differentiation, and proliferation. Growth factors are signaling molecules which enhances/triggers cell-to-cell interaction/communication by binding to the respective receptors available on the surface of the cell. In tissue engineering, there are numerous growth factors which have unique functions. The most common growth factors include vascular endothelial growth factor (VEGF), transforming growth factor (TGF), fibroblast growth factor (FGF), insulin-like growth factor (IGF), and epidermal growth factor (EGF). Each growth factor have specific role in cells and in tissue engineering constructs. The source, function, and receptors for major growth factors are given in Table 12.2.

TABLE 12.2 Details of major growth factors used in tissue engineering (Depprich, 2009; Kut et al., 2007)

Growth factor	Source	Receptor	Function
Vascular Endothelial Growth Factor (VEGF)	Platelets, leukocytes, and peripheral blood	VEGF receptor R1 and R2 (tyrosine kinase)	Induces neovascularization, stimulates cell survival and migration, involved in establishment of hematopoiesis
Transforming Growth Factor (TGF)	Platelet, bone, and extracellular matrix	Serine threonine sulfate	Stimulates proliferation of undifferentiated mesenchymal cells
Fibroblast Growth Factor (FGF)	Macrophages, mesenchymal cells, chondrocytes, and osteoblasts	Tyrosine kinase	Proliferation of mesenchymal cells, chondrocytes, and osteoblasts

TABLE 12.2 *(Continued)*

Growth factor	Source	Receptor	Function
Insulin-like Growth Factor (IGF)	Liver, osteoblasts, Bone matrix, chondrocytes, and myocytes	Tyrosine kinase	Proliferation and differentiation of osteoprogenitor cells
Epidermal Growth Factors (EGF)	Saliva, plasma, urine, and most of the body fluids	Tyrosine kinase	Mitogen for ectodermal, mesodermal, and endodermal cells, promotes proliferation and differentiation of epidermal and epithelial cells
Platelet-Derived Growth Factors (PDGFs)	Platelets, macrophages, endothelial cells, glial cells, astrocytes, myoblasts, and smooth muscle cells	Tyrosine kinase	Proliferation of mesenchymal cells, osteoblasts, and fibroblasts macrophage chemotaxis

12.5.1.1 VASCULAR ENDOTHELIAL GROWTH FACTOR (VEGF)

VEGF are secreted dimeric glycoproteins of 40 kDa molecular weight, involved in the process of angiogenesis (Olsson et al., 2006). Angiogenesis is the process of development of new blood vessels from preexisting vasculature and has a potent role in growth and development, wound healing and in tumorigenesis (Shibuya and Claesson-Welsh, 2006). VEGF helps in development of blood capillaries and are also called as vascular permeability factor (VPF). In mammals, five VEGF ligands of different splice variants have been reported so far which can bind with three types of receptor tyrosine kinases (VEGFR 1 to 3) and coreceptors like heparan sulfate proteoglycans (HSPGs) and neuropilins (Olsson et al., 2006). Such binding stimulates the production of biomolecules like endothelial nitric oxide synthase (eNOS), bFGF, and endothelial intercellular adhesion molecule (ICAM) which in turn promotes vessel permeability, proliferation/survival and migration respectively and ultimately leads to blood vessel formation.

Increased level of VEGF has been reported in many types of cancers and hence plays a vital role in formation of blood vessels in the newly formed cancerous cells (Hormbrey et al., 2002). The VEGF has been used in tissue engineering to induce establishment of vascular network between the developing tissues inside the scaffolds. VEGF can be administered locally or systemically. Since VEGF has the short circulation half life, high degradation rate and potential to bind to multiple sites in body tissue, localized delivery has resulted in failure of inducing angiogenesis (Huang et al., 2007). However, when VEGF is incorporated into polymers like PLGA in the form of scaffold or microspheres, localized delivery of VEGF has been achieved (Huang et al., 2007). VEGF is reported to preserve its activity when in it encapsulated or complexed with suitable polymers/immobilization matrix. Such complexation with suitable matrix provides high loading efficiency and controlled release and makes VEGF a suitable growth factor for applications in tissue engineering scaffolds (Des Rieux et al., 2011). Lin et al. (2012) developed intramyocardial injectable self-assembling peptide nanofibers in combination with VEGF to create an intramyocardial microenvironment with prolonged VEGF release and promote postinfarct neovascularization in rats. Such immobilization of VEGF into self-assembling peptide nanofibers showed sustained release of VEGF within myocardium for 14 days. This type of VEGF induced microenvironment attracts more cardiomyocytes to the spot and helps in cardiomyocyte regeneration. Seyednejad et al. (2012) fabricated VEGF-loaded nanofibrous scaffolds using hydroxyl-functionalized polyester (poly-hydroxymethylglycolide-co-ε-caprolactone) (pHMGCL)) and PCL. Such blending preserves the biological activity of VEGF and cause increased adherence of cells to the scaffold. These grafts showed increased polymer graft strength and prolonged biomolecule release, indicating the possible use of this technique in making functional vascular tissue engineering grafts.

12.5.1.2 TRANSFORMING GROWTH FACTOR (TGF)

TGF is a polypeptide present in human and bovine milk. These growth factors play a major role in epithelial cell growth and differentiation, apoptosis, embryo development, cellular homeostasis, carcinogenesis, and immune regulation (Donnet-Hughes et al., 2000). TGF β superfamily contains many ligands like bone morphogenetic proteins (BMPs), growth and differentiation factors (GDFs), antimullerian hormone (AMH), activin, nodal, and TGF-β's (Munir et al., 2004). BMPs are the main cause for osteoinductivity in bones. It comes under TGF-α family (Reddi et al., 1998). These ligands bind to type II receptors, which ultimately leads to activation of type I receptor followed by its phosphorylation. Such phosphorylation leads to further phosphorylation of R-SMADs (intracellular proteins that transduce extracellular signals from TGF-β ligands to the nucleus). These phosphorylated SMADs act as a transcription factor for regulation of targeted gene expression (Massagué et al., 2005). TGF-β2 with its serine/threonine kinase receptors can stimulate or inhibit endothelial cell proliferation and control synthesis and degradation of the ECM (Katsura et al., 2002). rh BMP2 belongs to member of TGF super family which is a promising factor for improvement of bone healing. It acts as hemotactic agents by initiating the recruitment of osteoprogenitor cells and mesenchymal stem cells (MSCs) toward bone defect sites (Sakou et al., 1999). TGF-βs also regulate cartilage fracture repair by ECM production. It is also considered as multifaceted cytokine that plays a key role in several downstream effects such as MSC differentiation, matrix production, and preventing dedifferentiation and controlled differentiation of stem cells (Qiao et al., 2005).

12.5.1.3 FIBROBLAST GROWTH FACTOR (FGF)

FGF are a type of heparin binding protein which interacts with cell-surface associated heparin sulfate proteoglycans. FGF plays a major role in cell proliferation and differentiation. They have a vital involvement in wound healing process by inducing fibrin formation and influences proliferation of endothelial cells by providing them growth and migration cues (Grazal-Bilska et al., 2002). There are around 22 members of FGF family in human beings which are structurally related to signaling molecules (Blaber et al., 1996; Ornitz and Itoh, 2001). Members of FGF1 to FGF10 bind normally with FGF receptors and activate various biological processes. However, FGF11 to FGF 14 do not bind to FGF receptors and are involved in other intracellular processes. Such FGFs are called as iFGFs (Itoh and Ornitz, 2008). FGF 18 is related to cell development, morphogenesis, and cartilage development (Moore et al., 2005). FGF 2-supplemented human MSCs showed longer life span with longer telomere size, higher proliferation rate, and greater chondrogenic potential (Handorf and Li, 2011).

12.5.1.4 PLASMA-DERIVED GROWTH FACTOR (PDGF)

PDGF are polypeptide growth factors produced by platelets during platelets activation due to the blood coagulation at the site of injury. Generally PGDF are dimeric glycoproteins with an approximate molecular weight of 30–35 kDa with two subunits (Westermark and Heldin, 1987). Antoniades (1981) isolated two types of PGDF glycoprotein namely PGDF I and II. Reduced PGDF I showed the presence of two inactive subunits with molecular weights of about 15 and 18 kDa. However, type II PDGF showed molecular weights of about 15 and 16 kDa. They are potent mitogens for cultured cells of mesenchymal origin and interact with smooth muscle cells, skin fibroblasts, 3T3 cells and glial cells. PGDF also acts as chemo attractants for inflammatory cells like monocytes, neutrophils, fibroblasts, and smooth muscles required for wound healing process (Kaplan et al., 1979a, b). PGDF is secreted by various normal cells like activated macrophages, endothelial cells and by smooth muscles. PGDF play a major role in various biological activities like wound healing, fetal development, cancer, atherosclerosis, and fibrosis (Ross et al., 1986).

12.5.1.5 INSULIN-LIKE GROWTH FACTOR (IGF)

IGF are proteins or growth hormones having sequence similarity with insulin molecule. IGFs are secreted by liver cells and are a part of cellular communication systems and secreted due to stimulation by growth hormones. The IGF family consists of IGF 1 and 2 ligands which can bind with its respective receptors. IGF 1 is actively involved in regulating neural development process especially neurogenesis, myelination, synaptogenesis, and dendritic branching and neuroprotection after neuronal damage. They also act as a major growth factor in adults (Gunnell et al., 2005). IGF 1 is also called "somatomedin C" and is involved in controlling the apoptotic process leading to delayed cardiomyocyte aging and cell death and in proper formation of cochlea (Höppener et al., 1985; Torella et al., 2004; Welch and Dawes, 2007). IGF 1 has a molecular weight of 7,649 daltons and is composed of 70 amino acids linked with three intramolecular disulfide bridges (Höppener et al., 1985). IGF 2 is another growth factor produced by liver, which is believed to play a major role in fetal development (especially during gestation). They are also known as Somatomedin-A. They act as a co-hormone along with follicle stimulating hormone (FSH) and luteinizing hormone (LH) by inducing granulosa cell proliferation during the follicular phase of the menstrual cycle and once ovulation is finished, they promote secretion of progesterone during the luteal phase. IGF 2 is also believed to play an active role in memory consolidation and enhancement (Chen et al., 2011).

12.5.1.6 EPIDERMAL GROWTH FACTORS (EGF)

EGF is a low molecular weight polypeptide (6,045 Da) containing 53 amino acids linked with three intramolecular disulfide bonds (Carpenter and Cohen, 1990). These factors are generally found in saliva, plasma, milk, urine, and in circulating macrophages and platelets. They play a major role in cell proliferation, differentiation, and healing of ulcers (oral and gastroesophageal) (Harris et al., 2003). Türkeri et al. (1998), reported expression of EGF in vessel walls of superficial bladder cancer and are correlated it with tumor reoccurrence.

12.5.2 OTHER PROTEINS AND PEPTIDES

12.5.2.1 BONE MORPHOGENIC PROTEINS (BMP)

BMP are cytokines or metabologens which has the ability to induce bone formation. In 1965, Prof. Marshall Urist of Bone Research Laboratory, University of California, coined this word as "bone morphogenic protein" which is also called as osteogenic proteins. BMPs are the first proteins which were approved for use in regenerative medicine and tissue engineering by regulatory bodies. There are 20 different types of BMP among which BMP 1 is a metalloprotease and BMP 2 to 7 belongs to TGF-β super family (Even et al., 2012). BMP react to the cell-surface BMP receptors and induce signal transduction by mobilizing the SMAD proteins which are important for the development of heart, cartilages, bone formation and in central nervous system. They also play a major role in early embryonic patterning and early skeletal formation (Reddi and Reddi, 2009). They are responsible for migration, proliferation, and differentiation of various types of cells. BMPs bind with the type I and type II serine/threonine kinase receptors leading to the formation of specific heterodimeric complexes.

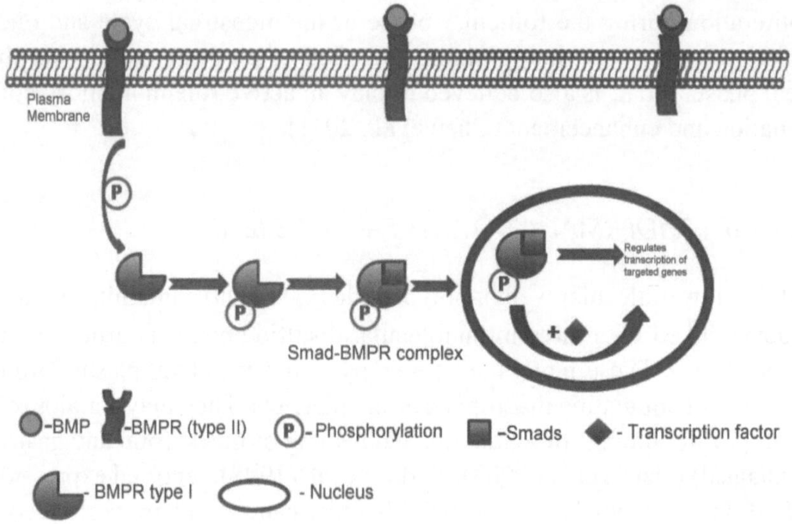

FIGURE 12.4 BMP receptor signaling system

Upon formation of complexes with type II kinase receptor, type I receptor is phosphorylated which in turn phosphorylates nuclear effective proteins called SMADs. Such phosphorylation induces intracellular signaling pathway. This R/C SMAD-BMPR complexes are then translocated to nucleus. Inside nucleus, SMADs directly or in combination with transcription factors regulate transcription of targeted genes (Figure 12.4) (Vukicevic and Sampath, 2008).

12.5.2.2 SYNTHETIC PEPTIDES

Recently, synthetic peptide-based materials have been proved to enhance cell adhesion onto tissue engineering scaffold and constructs and has been a promising approach in the areas of drug delivery, tissue repair and tissue engineering. These peptides can be used to meet out specific needs, owing to its biocompatibility and ability to form unique protein structures there by providing design flexibility (Lanza et al., 1997). Some of the recently reported synthetic peptides are RGD, RADA, EAK16, IKVAV, RGDS, YIGSR peptides, etc. RGD peptide is a tri-peptide made up of arginine, glycine, and aspartic acids. These peptides sequence (also called as Arginylglycylaspartic acid) acts a cell adhesion ligand for integrin cell adhesion receptors, there by enhancing cellular attachment (Jeschke et al., 2002). The RGD peptide attachment ultimately stimulates intracellular signals and gene expressions responsible for the cell growth, migration, and survival. This property along with its biocompatibility and lower toxicity makes it an ideal component for coating scaffold material to enhance cellular attachment. $RADA_n$ peptides are made up of arginine-alanine-aspartate-alanine amino acids and the number of repeats is denoted by n. These RADA peptides have been used to prepare scaffold which can influence the growth and differentiation of variety of cells from human, mammals, mouse, and chicken, and covering stem cells, progenitor cells and established cell lines (Zhao and Zhang, 2007). RADA peptides also act as a ligand for integrin receptor and can stimulate cellular attachment. RADA 16-I, a self-complementary peptide, contains 16 alternating amino acids and contains 50% charged residues. It efficiently responds to external stimuli under physiological conditions and maintains high water content which allows diffusion of wide range molecules. EAK16-II, a peptide molecule made up of 16 amino acids (AEAEAKAKAEAEAKAK) was originally found in a yeast protein called zuotin (Zhang et al., 1992). They had the property of self assembly and the scaffold made of this peptide was found to retain 99% of water (Zhang et al., 1993). They can form ß-sheets consisting of both polar and nonpolar surfaces. These oligopeptides having the property to form hydrogels have been used in preparation of tissue engineering scaffolds. IKVAV peptides are peptide made up of isoleucine-lysine-valine-alanine-valine amino acid sequence and are generally located in α-1 chain of laminarin, an extracellular membrane protein which promotes neuron attachment. It also promotes cell differentiation, collagen IV production, angiogenesis, and growth of axons and inhibits attachment of glial cells. These peptides have been shown to have self assembly property which can be used to make nanofiber gels for possible application in tissue engineering (Wu et al., 2006). These IKVAV peptides have been reported to act as a binding site for the β-amyloid precursor membrane protein from brain (Kibbey et al., 1993). RGDS peptide, an integrin interacting peptide, has serine amino acid in the chain of RGD peptide and is involved in stimulation of TGF β1 transcription and secretion through integrin activation. RGDS peptides activate caspases 8 and 9 directly and lead to pro apoptotic effect due to intracellular action (Aguzzi et al., 2004). Moon et al., (2009) during their study on pulmonary inflammation, reported that RGDS peptides inhibit lipopolysaccharides induced inflammatory cell migration into lungs along with inhibition of integrin signaling and MAP kinases activation. YIGSR peptide is a laminarin-based peptide (with following amino acid Tyr–Ile–Gly–Ser–Arg) found in the residues 929–933 of the β1 chain. This peptide has the property to inhibit tumor growth and metastasis (Fridman et al., 1990). When these peptides are polymerized or conjugated with polyethylene glycol, they significantly enhance the inhibition of tumor metastasis and are potent inhibitors of melanoma cell growth and metastasis by promoting apoptosis.

12.5.2.3 BONE SIALOPROTEIN (BSP)

BSP is a component of bone ECM and constitute approximately 8% of the noncollagenous protein in bone and cementum (Fisher et al., 1990). BSP is rich in sialic acid and has a molecular weight of 23 kDa. This protein is considered to be a breakdown product of sialoprotein which consists of sialic acid, glucosamine, galactosamine, and proteins (Herring, 1972). BSP has receptors in osteosarcoma cells and is found to enhance human fibroblast cell attachment (Oldberg et al., 1988). The BSP gene sequence is localized in chromosome 4 (Fisher et al., 1990).

12.5.2.4 OSTEOCALCIN

Osteocalcin is noncollagenous protein found in bone and dentins and is coded by bone gamma-carboxyglutamic acid-containing protein (BGLAP) gene (Puchacz et al., 1989). These proteins are secreted by osteoblast cells and play a vital role in bone building process and regulation of metabolic activities in human beings. Osteocalcin acts as a hormone and induces insulin secretion in beta cells in pancreas. Osteocalcin also induces synthesis of testosterone and has a direct role in male fertility. They are used as biomarker for bone formation.

12.5.2.5 ERYTHROPOIETIN (EPO)

EPO is a 30 kDa glycoprotein produced in kidney that controls erythropoiesis (Blood cell production). EPO is also called as erythropoetin or erthropoyetin. It acts as a protein signaling molecule for precursors involved in red cell production in bone marrow (Fisher, 2003). EPO is important for erythrocyte survival and differentiation. It has the ability to maintain vascular auto regulation and to attenuate primary (apoptotic) and secondary (inflammatory) causes of cell death (Calvillo et al., 2003). This hematopoietic growth factor has been found to mediate repair and regeneration after brain and spinal cord injury, including the recruitment of stem cells into the region of damage (Heeschen et al., 2003). EPO is also reported to be cardioprotective in nature.

12.5.2.6 EXTRACELLULAR MATRIX (ECM)-RELATED PROTEINS

ECM contains fiber forming proteins which play important role in cell-surface interaction such as migration, adhesion, phenotype differentiation and polarization, wound healing, and metastasis. The most reported ECM proteins are fibronectin, vitronectin, and laminin. Fibronectin is a 220 kDa multifunctional protein secreted by various cells like epithelial and mesenchymal cells, macrophages, hepatocytes, and intestinal epithelial cells (**Sitterley, 2008;** Underwood and Bennett, 1989). Fibronectin can bind to various components like cell surfaces, collagen, fibrinogen or fibrin, complement, glycosaminoglycans, proteoglycans, and heparin. Fibronectins play a major role in various cellular events like hemostasis, phagocytosis, cellular morphology control, cytoskeletal organization, embryonic differentiation and wound repair **(Sitterley, 2008).** Vitronectin is a 75 kDa glycoprotein with 459 amino acid residues which is present in blood serum and ECM. They inhibit damage caused by cytolytic complement pathway and also act as serine protease inhibitors. They play a vital role in cell spreading and adhesion process (Preissner and Seiffert, 1998). These proteins are called as VTN proteins which are coded by *VTN* gene and are believed to be involved in homeostasis and malignancy (Jenne and Stanley, 1987). The vitronectin activates plasminogen to regulate proteolysis process. Laminin is found in the basement membrane especially in the basal lamina and is involved in formation of protein networks in most of the cells. They are trimeric proteins that have α, ß and γ chain. There are five α chains, four ß chains and three γ chains reported so far. These chains can form network like structure which can bind to ECM and cell membranes and act as structural scaffolding in all the tissues of the organisms (Aumailley et al., 2005). The binding is mostly through integrin receptors and plasma membrane molecules like dystroglycan glycoprotein, Lutheran blood group glycoprotein, etc. Defective laminarins can lead to junctional epidermolysis bullosa (skin blistering disease) and nephritic syndrome (defects of the kidney filter) (Colognato and Yurchenco, 2000).

12.5.3 CARBOHYDRATE-BASED BIOMOLECULES

12.5.3.1 LECTIN

Lectins are protein molecules that can specifically bind to carbohydrates. Lectins are produced by plants (legumes, castor plant) and by animals. In animals, lectin regulates the adhesion of cells to glycoproteins and controls the protein content in blood. There are various types of lectins depending upon the carbohydrate molecule they attach (Table 12.3).

TABLE 12.3 Various types of lectins and its carbohydrate ligands (Source: *www.interchim.fr/ft/M/MS902z.pdf*)

Lectin symbol	Lectin name	Source	Ligand motif
Mannose binding lectins			
ConA	Concanavalin A	*Canavalia ensiformis*	α-D-mannosyl and α-D-glucosyl residues branched α-mannosidic structures (high α-mannose type, or hybrid type and biantennary complex type N-Glycans)
LCH	Lentil lectin	*Lens culinaris*	Fucosylated core region of bi- and triantennary complex type N-Glycans
GNA	Snowdrop lectin	*Galanthus nivalis*	α 1-3 and α 1-6 linked high mannose structures
ConA	Concanavalin A	*Canavalia ensiformis*	α-D-mannosyl and α-D-glucosyl residues branched α-mannosidic structures (high α-mannose type, or hybrid type and biantennary complex type N-Glycans)

TABLE 12.3 *(Continued)*

Lectin symbol	Lectin name	Source	Ligand motif
Galactose/N-acetylgalactosamine binding lectins			
RCA	Ricin, Ricinus communis Agglutinin, RCA120	*Ricinus communis*	Galβ1-4GalNAcβ1-R
PNA	Peanut agglutinin	*Arachis hypogaea*	Galβ1-3GalNAcα1-Ser/Thr (T-Antigen)
AIL	Jacalin	*Artocarpus integrifolia*	(Sia)Galβ1-3GalNAcα1-Ser/Thr (T-Antigen)
VVL	Hairy vetch lectin	*Vicia villosa*	GalNAcα-Ser/Thr (Tn-Antigen)
N-acetyl glucosamine binding lectins			
WGA	Wheat Germ Agglutinin, WGA	*Triticum vulgaris*	GlcNAcβ1-4GlcNAcβ1-4GlcNAc, Neu5Ac (sialic acid)
N-acetyl neuraminic acid binding lectins			
SNA	Elderberry lectin	*Sambucus nigra*	Neu5Acα2-6Gal(NAc)-R
MAL	Maackia amurensis leukoagglutinin	*Maackia amurensis*	Neu5Ac/ Gcα2,3Galβ1,4Glc(NAc)
MAH	Maackia amurensis hemoagglutinin	*Maackia amurensis*	Neu5Ac/ Gcα2,3Galβ1,3(Neu5Acα2,6) GalNac
Fucose binding lectins			
UEA	Ulex europaeus agglutinin	*Ulex europaeus*	Fucα1-2Gal-R
AAL	Aleuria aurantia lectin	*Aleuria aurantia*	Fucα1-2Galβ1-4(Fucα1-3/4) Galβ1-4GlcNAc, R2-GlcNAcβ1-4(Fucα1-6) GlcNAc-R1

Plant lectins have been studied for its ability to act as a mediator for bioadhesion of human chondrocyte cell lines C-28/I2 and T/C-28a2 (Toegel et al., 2007). Coating of these lectins have been proved to effective in enhancing the adhesion of cells like MSC, chondrocytes, and osteoblasts and induce resistance to protease and mechanical stimulus and hence find potential application in tissue engineering (Nishimura et al., 2004).

12.5.3.2 GALACTOSE

Galactose, a monosaccharide sugar is an epimer of glucose molecule. Galactose has been found immense application in liver tissue engineering. Cho et al. (2006) has written a complete review on how galactose-carrying polymers can be used for liver tissue engineering. Galactose conjugated substrates mediates adhesion through the galactose–asialoglycoprotein receptors (ASGPR) interaction and minimize the involvement of integrin mediated signaling pathway shown to induce the loss of hepatocyte phenotype (Ying et al., 2003). Various types of galactose molecules like lactose monohydrate (Kobayashi et al., 1985), lactobionic acid (Parkl et al., 2003) D-galactopyranoside (Ying et al., 2003), 1-amino-1-deoxy-aminohexyl-b-D-lactose (Donati et al., 2002) and1-amino-1-deoxy-b-D-galactose (Lopina et al., 1996) have been studied so far for introduction into ECM.

12.5.3.3 GLUCOSAMINE (GLCN)

GLCN is an amino sugar which acts as a precursor for the biosynthesis of glycosylated proteins and lipids. Being a part of chitin and chitosan, GLCN forms a part of exoskeleton in crustaceans and arthropods and is also present in cell wall of many fungi (Pigman et al., 1980). It enhances cartilage-specific proteoglycans biosynthesis and inhibits the activity of pro-inflammatory mediators to treat symptoms associated with cartilage degeneration and observed reduction in cell metabolism (Chan et al., 2005). Glucosamine sulfate (GS) has been routinely used as nutritional supplement for treatment of osteoarthritis (Richy et al., 2003).

12.5.3.4 GLYCOSAMINOGLYCANS (GAGS)

GAGs are long unbranched polysaccharides made of repeated disaccharide units. The repeated units contain uronic acids like glucuronic acid or iduronic acid along with N-acetyl glucosamine or N-acetyl galactosamine units (Esko et al., 2009). There are various types of GAGs like chondroitin sulfate (CS), dermatan sulfate, keratan sulfate, heparin/heparan sulfate, and hyaluronan. GAGs are synthesized in golgi bodies with monosaccharides as precursor followed by N-or O-sulfation, acetylation, deacetylation, or epimerization process (Muthana et al., 2012).

12.5.3.4.1 HYALURONIC ACID (HA)

HA is a nonsulfated linear anionic glycosaminoglycan molecules consisting of (1-β-4) D-glucuronic acid and (1-β-3) N-acetyl -D-glucosamine molecules. The molecular weight of the HA varies from one fluid to the other and generally ranges from 5 to 20,000 kDa (Saari et al., 1993). HA is synthesized by integral membrane protein. Hyaluronan are part of ECM and is present in various tissue sources like connective, epithelial, neural tissues, umbilical cord, rooster comb, synovial fluid, or vitreous humor. They are also produced by microbes through microbial fermentation process involving extracellular capsule material of group A *Streptococcal* sp (Wessels et al., 1991). In human beings, HA production is high in undifferentiated cells and during early embryogenesis and low during differentiation process (Liao et al., 2005). They are important for cell proliferation and migration. HA is reported to be present in synovial fluid and plays a major role in controlling the viscosity of the fluid. They interact with specific cell through CD44 receptor, receptor for HA-mediated motility (RHAMM), intracellular adhesion molecule-1 (ICAM-1) receptor and can promotes wound healing and induces chondrogenesis (Comper, 1996).

12.5.3.4.2 CHONDROITIN SULFATE (CS)

CS is GAGs having repeated dissacharide units of D-glucoronic acid N-acetyl galactosamine, sulfated at either 4 or 6 positions. CS have four different types based on the site of sulfation, namely chondroitin-4-sulfate, chondroitin-6-sulfate, chondroitin-2,6-sulfate and chondroitin-4,6-sulfate (called as CS A, C, D, and E, respectively). It is negatively charged molecule which interacts with positively charged molecules such as polymers or growth factors. It is one of the major components in cartilage ECM. CS is well known for its ability to increase matrix component production by human chondrocytes (Lee et al., 2005). CS attaches to the protein through their serine residues. CS are useful in tissue engineering due to its biological properties like tissue integration, anti-inflammatory activity, chondrogenesis, water and nutrient adsorption, restoration of joint functions, etc. (Wang et al., 2007).

12.5.4 RECEPTORS

12.5.4.1 SYNDECANS

Syndecans are transmembrane proteins which act as coreceptor for G protein coupled receptors. They have heparin sulfate and CS domains that can interact with various ligands of various growth factors like FGF, VEGF, and TGF-ß. There are four different types namely syndecans 1–4 (Carey, 1997). The expression of syndecans various from one cell to another and are cell specific. They contribute to the biological system by growth factor receptor activation, acting as a adhesion matrix by binding to ECM molecules such as collagens I, III, V, fibronectin, thrombospondin, and tenascin leading to cell-to-cell adhesion and finally in tumor suppression and progression (Elenius and Jalkanen, 1994). Syndecans are reported to be up regulated during osteoarthritis leading to damage of cartilage (Hass, 2009).

12.5.5 LIPIDS

12.5.5.1 CAVEOLAE

Caveolae are lipid raft which occur as a 50–100 nm invagination in plasma membrane in cells like adipocytes and endothelial cells. They appear as flask like structures rich in protein, lipids like cholesterol and sphingolipids (Anderson, 1998). They have an important role in endocytosis, oncogenisis, cell signaling, regulation of lipids and channels, calcium signaling and pathogenesis. They also behave as mechanosensors for flow sensation and stretch sensing (Parton and Simons, 2007).

12.6 FUNCTIONALIZATION OF SCAFFOLD FOR VARIOUS TISSUE ENGINEERING APPLICATIONS

Functionalization of scaffolds using suitable biomolecules is essential for tissue regeneration. Scaffold-based delivery of biomolecules such as growth factors, drugs, proteins, and peptides stimulate cell migration, growth, and differentiation. Growth factors are critical signaling molecules that control cell attachment, growth, and proliferation. The controlled delivery of growth factor enhances the regenerative process. Biomolecules can chemically immobilize or physically encapsulate on scaffold (Lee et al., 2011). As biomolecules are highly tissue specific, their immobilization, delivery, and activity in different tissue reconstruction are discussed below.

12.6.1 FUNCTIONALIZATION OF SCAFFOLD IN STEM CELL TISSUE ENGINEERING

Stem cells are undifferentiated cells having potential to differentiate into specialized cells under the influence of suitable biomolecules. The chemical composition of the scaffold is one of the key determinants for the effective differentiation of the stem cells. For example gelatin plays an important role for osteogenic differentiation of human mesencymal stem cells (hMSCs) (Rim et al., 2009). The cell-matrix interactions are vital for the regulation of lineage-specific differentiation of hMSCs. Cells in native tissue reside within an ECM environment where many biomacromolecules are aligned on a nanometer scale with complex structure. McBeath et al. (2004) reported that the different lineages of stem cells are regulated by the local tissue microenvironment. The hMSCs differentiate to adipocyte or osteoblast fate depending on its spread. If hMSCs are allowed to adhere, flatten, and spread, they undergo osteogenesis and if not allowed to spread, the round cells become adipocytes. The study reveals that the stem cell fate is controlled by the mechanical cues, embodied by cell shape, cytoskeletal tension and RhoA (Ras homolog gene family, member A) signaling. Dominant-negative RhoA committed hMSCs become adipocytes while osteogenesis is caused by constitutively active RhoA. The constitutive activation of the RhoA effector, ROCK (Rho-associated, coiled-coil containing protein kinase) induces osteogenesis independent of cell shape.

Actin-myosin generated tension influences this RhoA-ROCK signal. In addition to growth factors, the fate of stem cells is also moderated by tiny cell-permeable particle such as dexamethasone, vitamin C, sodium pyruvate, and retinoic acid (Ding et al., 2004). Furnémont et al. (2011) used a decellularized human myocardium as a biological composite scaffold for the tissue engineering of human mesenchymal progenitor cells (MPCs). Suspended cells in fibrin hydrogel were planted in it and researchers found that in vitro conditioning with a low concentration of TGF-β promotes arteriogenic profile of gene expression. The MPCs produced rapid vascular network in the infarct bed when seeded in the composite scaffold. The mechanisms are identified as the secretion of paracrine factors, such as SDF-1 and the migration of MPCs into ischemic myocardium.

12.6.2 FUNCTIONALIZATION OF SCAFFOLD IN VASCULARIZATION

Sufficient vascularization is a primary consideration for the engineering of large tissue constructs which maintains adequate perfusion. Rapid and high levels of vascularization of the cell-seeded scaffold are essential to meet the challenge. One approach to achieve vascularization is by incorporating a growth factor into the scaffold. Several angiogenic factors, such as VEGF, FGF, EGF, PDGF, and TGF promote the formation of new vascular beds from endothelial cells present within tissues. The localization and controlled release of these factors from a matrix might bring about enhanced vascularization of engineered tissues. An alternative approach to enhance the rate of vascularization is to transplant endothelial cells onto the scaffold. Type I collagen protein is major ECM component in the blood vessel. Collagen gels and fibers reconstructed from purified collagens show low inflammatory and antigenic responses during artificial blood vessel development (Nicolas and Gagnieu, 1997).

The main limitation in engineering in vitro tissues is the lack of vascularization systems while almost all tissues are vascularized in vivo. In prevascularization technique new vessels are generated by some biomolecules, such as growth factors, cytokines, peptides, and proteins. Naturally derived scaffolds, which contain vessels, are distinguished from synthetically manufactured matrices. The delivery of growth factors is usually involved to stimulate the formation of new blood vessels. Growth factors are powerful initiators for neo vascularization. The polysaccharide heparin, a blood anticoagulant binds to many of these growth factors and has been used as a delivery agent. Samuel I. Stupp's team at Northwestern University has studied self-assembling nanofiber scaffolds with heparin on their surfaces, which in turn binds to angiogenic growth factors (Rajangam et al., 2006). The nanostructures are able to display the heparin over a large surface area and activate proteins to promote angiogenesis. They activate the endothelial (progenitor) cells and stimulate them to migrate toward the factor gradient. Furthermore, they promote cell assembly, vessel formation and maturation. The main growth factors for up regulating angiogenic processes are VEGF, FGF, and hepatocyte growth factor (HGF). Successful immobilization of proteins or peptides on biomaterials for tissue engineering is a current method to enhance the vascularization process. Short peptide adhesion sequences derived from ECM proteins like fibronectin (e.g., RGD and REDV) or laminin (e.g., YIGSR) are well known to maintain cell adhesion.

12.6.3 FUNCTIONALIZATION OF SCAFFOLD IN BONE TISSUE ENGINEERING

Bones have the capability for self renewal as the bone marrow is rich in stem cells (mostly mesenchymal and endothelial) that are potent for multilineage differentiation. Immature cells are differentiated from osteoblastic lineage, which increases in the secretion of proteins like osteocalcin, bone sialoprotein and mineralization of the ECM. The control of morphogenesis in bone marrow is described in Figure 12.5. The circulating hormones and chondrocytes secrete growth factors that act on chondrocytes to regulate their proliferation and hypertrophy. Likewise the proliferated chondrocytes expressed VEGF followed by the formation of primary spongiosa which leads to bone auto induction. During the chondrogenic differentiation of MSCs, the coordinate expression of the core protein, multiple glycosyltransferases and modifying enzymes involve in the synthesis of CS. A periosteal bone collar has formed and formation of the primary center of the ossification has been initiated. The secondary center of ossification has formed at each end of the bone. Finally, skeletal maturity is observed with complete replacement of the growth plate cartilage by bone.

Osteoinductive growth factors are most promising in bone tissue engineering. BMPs are such growth factors which are capable of inducing new bone formation when a scaffold is functionalized with such molecules. The most important BMPs in bone research are BMP-2 and BMP-7 for specific orthopedic applications. BMP-2/7 heterodimer are 5–10 times more efficient for new bone formation and enhancing alkaline phosphatase activity than the homodimeric form of them. For this, BMP-2/7 heterodimer was mostly chosen as osteoinductive molecule (Madurantakam et al. 2011). Due to lacking of particular reactive sites in PLGA it cannot be efficiently used to deliver biomolecules on the cells. Fibrin and poly (lactic-co-glycolic) (PLGA) hybrid scaffold enhances the chondrogenesis of articular cartilage constructs in vitro (Shaban et al. 2008). The immobilized heparin with low dosage exhibits a stimulatory effect on osteoblasts proliferation where as high dosage of heparin does not (Jiang et al., 2010). Several studies have reported the increased mineralization of bone tissue in nanofibrous scaffolds compared to the control.A nanofibrous scaffold loaded with BMP-2 shows significant homogeneous ectopic bone formation (Hossein et al., 2007). Fujihara et al. (2005) verified that the incorporation of calcium carbonate ($CaCO_3$) in the electrospun PCL scaffold able to assist the bone cell regeneration.

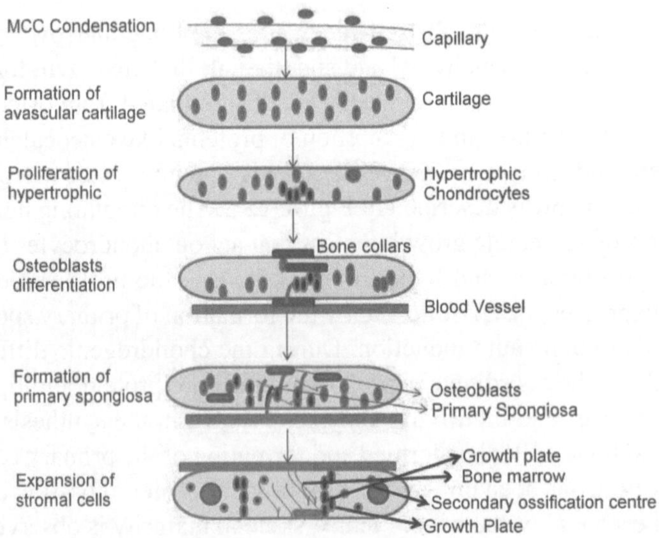

FIGURE 12.5 An endochondral bone formation

To achieve the desired mechanical stability, neat PCL and the mixture of PCL and $CaCO_3$ at different compositions, were employed. Kim et al. (2005) fabricated a nanocomposite nanofibrous scaffold by electrospinning of gelation and hydroxyapatite. They did electrospinning of the mixture under controlled conditions. The hydroxyapatite nanocrystals were dispersed in the gelatin fibers and showed significant improvements in cellular activity as compared to pure gelatin. The nanocomposite material mimics the human bone and it has good potential for bone tissue regeneration. Khoo et al. (2005) investigated the dose dependent effect of glucose on differentiation of human embryonic stem cells after formation of embryoid bodies (EBs). They found that the physiological glucose concentrations are suitable for the EB growth and FGF-2 enhanced the gene expression.

12.6.4 FUNCTIONALIZATION OF SCAFFOLD IN SKIN TISSUE ENGINEERING

The development of human skin equivalent (HSE) is a challenging task. The outermost layer of skin, epidermis mostly consists of keratinocytes. Several growth factors contribute the development of epidermis. Among them, keratinocyte growth factor (also known as FGF7) and TGFα are most important for proliferation and differentiation of cells. Epidermis is avascular in nature and gets nutrients by diffusion from dermis. All complex wounds affect the dermis. So, skin tissue regeneration research focuses mostly on the reconstruction of dermis. Fibroblasts are the major cellular component of dermis. Hence, 3-D scaffold-based fibroblast culture is dominating in this field. Mansbridge (2008) listed a number of commercially available scaffold materials for skin tissue regeneration. Self-assembling peptide (SAP) exhibits the biological activity of peptides by altered amino acid sequences. Bioactive wound dressing composed of SAP nanofiber scaffold with EGF shows superior performance than the scaffold without EGF (Schneider et al., 2008). Wound healing takes place faster in moist environment as provided by the hydrogel. The water vapor permeability of a wound dressing should prevent both excessive dehydration as well as building up of exudates. In wound healing, protein adsorption plays a vital role. It occurs rapidly after a material is implanted. These proteins can denature on hydrophobic surfaces, triggering the immune system and influencing wound healing. The bioactivation of the scaffolds with proper peptide sequence is essential for successful cell attachment and proliferation. Grafahrend et al. (2011) demonstrated this with a PLGA-based nanofibrous scaffold loaded with functional amphiphilic macromolecule based on star-shaped poly (ethylene oxide-stat-propylene oxide). Two peptides were adsorbed separately on such scaffolds and it was observed that the GRGDS peptide promotes fibroblast adhesion with excellent viability while GRGES completely resists the cell adhesion. The effect of electric field on cell growth has been examined by several researchers. Shi et al. (2008) prepared electrically conductive biodegradable polymer membranes by mixing conductive polypyrrole particles with PLLA solution. Human cutaneous fibroblasts were cultured on these membranes with and without electrical stimulation. It has been observed that the very small current has no significant effect on cell viability but cell viability has enhanced four folds (after 24 h of culturing) for the membranes having a higher electrical stimulation (100 mV/mm), than that of membranes without stimulation. They concluded that the electrical field stimulates the mitochondrial activity of human skin fibroblasts.

12.6.5 FUNCTIONALIZATION OF SCAFFOLD IN NERVE TISSUE ENGINEERING

Conventional nerve tissue repair approaches are based on a single component, either biomaterials or cells alone, due to the complexity of nervous anatomical system. The electrical properties of nerve cells and the effect of electrical stimulation on nerve cells are of utmost importance in nerve tissue engineering. The conductive polymers, polypyrrole (PPy) and polyaniline (PANI), along with their modifications may act as suitable scaffolds for nerve tissue engineering. The high degree of conjugation in the molecular backbone of polypyrrole makes it very rigid, insoluble. Hence, it is very difficult to process alone (Shi et al., 2004). Use of biomolecules enhances the migration and proliferation of neural cells. Neurotrophic factors including nerve growth factor (NGF), brain-derived neurotrophic factor (BDNF), neurotrophin-3 (NT-3), glial growth factor (GGF), and FGFs have been used for nerve tissue engineering (Landreth, 1999). RGD and other short peptide sequences such as IKVAV, YIGSR, RNAIAEIIKDI from laminin and HAV from N-cadherin were also used for neural tissue engineering (Schense et al., 2000). PANI is a good matrix to support the cell growth (Mattioli-Belmonte et al., 2003). Bidez III et al. (2006) investigated the adhesion and proliferation properties of H9c2 cardiac myoblasts on pure and HCl doped conducting PANI substrate. The electrical conductivity for doped PANI retains for 100 h in physiologic environment. Doping with CS or HA can cause an adverse effect on the physical properties of polypyrrole as its surface roughness increased results in poor cell adhesion, however, the cell proliferation enhanced (Gelmi et al., 2010; Gilmore et al., 2009). Guimard et al. (2007) described several other methods to modify polypyrrole such as physical adsorption, entrapping, surface binding and micro/nanopatterning.

12.7 CONCLUSION

Engineering or regenerating lost tissue requires huge research and engineering skill. In tissue engineering applications, synthetically derived matrices are used to replace the damaged tissue or organs. The artificial ECM supports to elevate the cellular behavior on scaffold matrices. Scaffolding with multifunctional properties is a complex process and tissue specific. As compared, native scaffolds mostly have poor physicochemical and biological properties which mediate cell morphogenesis. Hence, immobilization of cell-specific growth factors or biomolecules into the scaffold system can augment the cellular attachment, growth, proliferation, and differentiation of cells into appropriate tissues. The challenge is to correlate the biochemical properties of soft tissues and histological information with mechanical test results.

KEYWORDS

- **Biomolecules;**
- **ECM;**
- **Growth factors;**
- **Polymers;**
- **Scaffold;**
- **Tissue engineering**

REFERENCES

Aguzzi, M. S.; Giampietri, C.; De Marchis, F. RGDS peptide induces caspase 8 and caspase 9 activation in human endothelial cells. Blood. *2004,* 103, 4180–4187.

Anderson, R. G. The caveolae membrane system. Ann. Rev. Biochem. *1998*, 67, 199–225.

Antoniades, H. N. Human platelet-derived growth factor (PDGF): purification of PDGF-I and PDGF-II and separation of their reduced subunits. PNAS. *1981*, 78(12), 7314–7317.

Aumailley, M.; Bruckner-Tuderman, L.; Carter, W. G.; Deutzmann, R.; Edgar, D.; Ekblom, P.; Engel, J.; Engvall, E.; Hohenester, E.; Jones, J. C.; Kleinman, H. K.; Marinkovich, M. P.; Martin, G. R.; Mayer, U.; Meneguzzi, G.; Miner, J. H.; Miyazaki, K.; Patarroyo, M.; Paulsson, M.; Quaranta, V.; Sanes, J. R.; Sasaki, T.; Sekiguchi, K.; Sorokin, L. M.; Talts, J. F.; Tryggvason, K.; Uitto, J.; Virtanen, I.; von der Mark, K.; Wewer, U. M.; Yamada, Y.; Yurchenco, P. D. A simplified laminin nomenclature. Matrix Biol. *2005*, 4(5), 326–332.

Bernhardt, A.; Despang, F.; Lode, A.; Demmler, A.; Hanke, T.; Gelinsky, M.; Proliferation and osteogenic differentiation of human bone marrow stromal cells on alginate-getatine-hydroxyapatite scaffolds with anisotropic pore structure. J. Tissue Eng. Regen. Med. *2009*, (3), 54–62.

Bidez III, P. R.; Shuxi, Li.; Macdiarmid, A. G.; Venancio, E. C.; Yenwei; Lelkes, Peter I. Polyaniline, an electroactive polymer, supports adhesion and proliferation of cardiac myoblasts. J. Biomater. Sci. Polym. Edn. *2006*, 17(1–2), 199–212.

Blaber, M.; DiSalvo, J.; Thomas, K. A. X-ray crystal structure of human acidic fibroblast growth factor. Biochemistry. *1996*, 35(7), 2086–2094.

Çakmak, S.; Sera, A.; Umüşderelioğlu, M. RGD-bearing peptide-amphiphile-hydroxyapatite nanocomposite bone scaffold: an in vitro study. Biomed. Mater. *2013*, 8, 045014.

Calvillo, L.; Latini, R.; Kajstura, J.; Leri, A.; Anversa, P.; Ghezzi, P.; Salio, M.; Cerami, A.; Brines, M. Recombinant human erythropoietin protects the myocardium from ischemia-reperfusion injury and promotes beneficial remodeling. Proc. Natl. Acad. Sci. *2003*, 100, 4802–4806.

Carey, D. J. Syndecans: multifunctional cell-surface co-receptors. Biochem. J. *1997*, 327(1), 1–16.

Carpenter, G.; Cohen, S. Epidermal growth factor. J. Biol. Chem. *1990*, 265(14), 7709–7712.

Chan, P. S.; Caron, J. P.; Rosa, G. J.; Orth, M. W. Glucosamine and chondroitin sulfate regulate gene expression and synthesis of nitric oxide and prostaglandin E2 in articular cartilage explants. Osteoarthritis and Cartilage. *2005*, 13, 387–394.

Chen, D.; Sarah, S.; Ana, G. O.; Bernadette, S. R.; Gabriella, P.; Dhananjay, B. M.; Robert, D. B.; Cristina, M. A. A critical role for IGF-II in memory consolidation and enhancement. Nature. *2011*, 469, 491–497.

Cheng, Q.; Lee, B. L.; Komvopoulos, K.; Yan, Z.; Li, S. Plasma surface chemical treatment of electrospun poly(L-lactide) microfibrous scaffolds for enhanced cell adhesion, growth, and infiltration. Tissue Eng. Part A. *2013*, 19(9–10), 1188–1198.

Cho, C. S.; Seo, S. J.; Park, I. K.; Kim, S. H.; Kim, T. H.; Hoshiba, T.; Harada, I.; Akaike, T. Galactose-carrying polymers as extra cellular matrices for liver tissue engineering. Biomaterials. *2006*, 27, 576–585.

Colognato, H.; Yurchenco, P. Form and function: the laminin family of heterotrimers. Dev. Dyn. *2000*, 218(2), 213–234.

Comper, W. D. Extracellular Matrix, Molecular Components and Interactions, Harwood Academic Publishers, 1996, Volume 2.

Depprich, R. A. Biomolecule use in tissue engineering. Chapter 11. In Fundamentals of Tissue Engineering and Regenerative Medicine; Meyer, Ulrich.; Handschel, J.A.; Wiesmann, Hans Peter.; Meyer.; Thomas.; Springer Verlag Publication. 2009; p 121.

Des Rieux, A.; Ucakar, B.; Mupendwa, B. P.; Colau, D.; Feron, O.; Carmeliet, P.; Préat, V. 3D systems delivering VEGF to promote angiogenesis for tissue engineering. J. Control Release. *2011*, 150(3), 272–278.

Ding, S.; Schultz, P. G. A role for chemistry in stem cell biology. Nature Biotechnol. *2004*, 22, 833–840.

Donati, I.; Gamini, A.; Vetere, A.; Campa, C.; Paoletti, S. Synthesis, characterization, and preliminary biological study of glycoconjugates of poly (styrene-co-maleicacid). Biomacromolecules. *2002*, 3, 805–812.

Donnet-Hughes, A.; Duc, N.; Serrant, P.; Vidal, K.; Schiffrin, E. Bioactive molecules in milk and their role in health and disease: The role of transforming growth factor-β. Immunol. Cell Biol. *2000*, 78(1), 74–79.

Elenius, K.; Jalkanen, M. Function of the syndecans – A family of cell surface proteoglycans. J. Cell Sci. *1994*, 107, 2975–2982.

Entcheva, E.; Bien, H.; Yin, L.; Chung, C. Y.; Farrell, M.; Kostov, Y. Functional cardiac cell constructs on cellulose-based scaffolding. Biomaterials. *2004*, 25, 5753–5762.

Esko, J. D.; Kimata, K.; Lindahl, U. Proteoglycans and Sulfated Glycosaminoglycan. Essentials of Glycobiology. Cold Spring Harbor Laboratory Press. 2009, ISBN 0879695595.

Even, J.; Eskander, M.; Kang, J. Bone morphogenetic protein in spine surgery: current and future uses. J. Am. Acad. Orthop. Surg. *2012*, 20, 547–552.

Fisher, J. W. Erythropoietin: Physiology and pharmacology update. Exp. Biol. Med. (Maywood). *2003*, 228, 1–14.

Fisher, L. W.; McBride, O. W.; Termine, J. D.; Young, M. F. Human bone sialoprotein. Deduced protein sequence and chromosomal localization. J. Biol. Chem. *1990*, 265(4), 2347–2351.

Fridman, R.; Giaccone, G.; Kanemoto, T.; Martin, G. R.; Gazdar, A. F.; Mulshine, J. L. Reconstituted basement membrane (matrigel) and laminin can enhance the tumorigenicity and the drug resistance of small lung cancer cell lines. Proc. Natl. Acad. Sci. *1990*, 87, 6698–6702.

Fujihara, K.; Kotaki, M.; Ramakrishna, S. Guided bone regeneration membrane made of polycaprolactone/calcium carbonate composite nano-fibers. Biomaterials. *2005*, 26(19), 4139–4147.

Furnemont, A. F. G.; Martens, T. P.; Koeckert, M. S.; Wan, L.; Parks, J.; Arai, K.; Zhang, G.; Hudson, B.; Homma, S.; Vunjak-Novakovic, G. Composite scaffold provides a cell delivery platform for cardiovascular repair. Proc. Natl. Acad. Sci. *2011*, 108(19), 7974–7979.

Garg, T; Singh, O.; Arora, S.; Murthy, R. S. R. Scaffold: A novel carrier for cell and drug delivery. Crit. Rev. Therapeutic Drug Carrier Syst. *2012*, 29(1), 1–63.

Gelmi, A.; Higgins, M. J.; Wallace, G. G. Physical surface and electromechanical properties of doped polypyrrole biomaterials. Biomaterials. *2010*, 31(8), 1974–1983.

Gilmore, K. J.; Kita, M.; Han, Y.; Gelmi, A.; Higgins, M. J.; Moulton, S. E.; Clark, G. M.; Kapsa, R.; Wallace, G. G. Skeletal muscle cell proliferation and differentiation on polypyrrole substrates doped with extra cellular matrix components. Biomaterials. *2009*, 30 (29), 5292–5304.

Grafahrend, D.; Heffels, Karl-Heinz.; Beer, M. V.; Gasteier, P.; Moller, M.; Gabriele, B.; Dalton, P. D.; Jurgen, G. Degradable polyester scaffolds with controlled surface chemistry combining minimal protein adsorption with specific bioactivation. Nature Mater. *2011*, 10, 67–73.

Grazul-Bilska, A. T.; Luthra, G.; Reynolds, L. P.; Bilski, J. J.; Johnson, M. L.; Abdullah, S. A. Redmer, D. A.; Abdullah, K. M. Effects of basic fibroblast growth factor (FGF-2) on proliferation of human skin fibroblasts in diabetes mellitus. Exp. Clin. Endocrinol. Diab. *2002*, 110, 176–181.

Guimard, N. K.; Gomez, N.; Schmidt, C. E. Conducting polymers in biomedical engineering. Prog. Polym. Sci. *2007*, 32(8–9), 876–921.

Gunnell, D.; Miller, L. L.; Rogers, I.; Holly, J. M.; Alspac. Association of insulin-like growth factor I and insulin-like growth factor-binding protein-3 with intelligence quotient among 8- to 9-year-old children in the Avon longitudinal study of parents and children. Pediatrics. *2005*, 116(5), e681.

Haas, M. J. SDC4: OA joint effort. SciBX. *2009*, 1–2.

Handorf, A. M.; Wan-Ju, Li. Fibroblast growth factor-2 primes human mesenchymal stem cells for enhanced chondrogenesis. PLOS One. *2011*, 6(7), e22887.

Harris, R. C.; Chung, E.; Coffey, R. J. EGF receptor ligands. Exper. Cell Res. *2003*, 284(1), 2–13.

He, P.; Sahoo, S.; Ng, K. S.; Chen, K.; Toh, S. L.; Goh, J. C. H. Enhanced osteoinductivity and osteoconductivitythrough hydroxyapatite coating of silk-based tissue-engineered ligament scaffold. J. Biomed. Mater. Res. Part A. *2013,* 101A, 555–566.

Heeschen, C.; Aicher, A.; Lehmann, R.; Fichtlscherer, S.; Vasa, M.; Urbich, C.; Mildner-Rihm, C.; Martin, H.; Zeiher, A. M.; Dimmeler, S. Erythropoietin is a potent physiologic stimulus for endothelial progenitor cell mobilization. Blood. *2003*, 102(4), 1340–1346.

Herring, G. M. The biochemistry and physiology of bone (Bourne GH, ed). Academic Press, Orlando, FL. *1972*, Vol 1, pp 127–189.

Hollister, S. J.; Lin, C. Y. Computational design of tissue engineering scaffolds. Comput. Methods Appl. Mech. Eng. *2007*, 196, 2991–2998.

Hoppener, J. W.; de Pagter-Holthuizen, P.; Geurts van Kessel, A. H.; Jansen, M.; Kittur, S. D.; Antonarakis, S. E.; Lips, C. J.; Sussenbach, J. S. The human gene encoding insulin-like growth factor I is located on chromosome 12. Hum. Genet. *1985*, 69(2), 157–60.

Hormbrey, E.; Gillespie, P.; Turner, K.; Han, C.; Roberts, A.; McGrouther, D.; Harris, A. L. A critical review of vascular endothelial growth factor (VEGF) analysis in peripheral blood: is the current literature meaningful? Clin. Exp. Metastasis. *2002*, 19, 651–663.

Hossein, H.; Mohsen, H.; Ali, K.; Hisatoshi, K. Bone regeneration through controlled release of bone morphogenetic protein-2 from 3-D tissue engineered nano-scaffold. J. Contr. Release. *2007*, 117, 380–386.

Huang, M.; Vitharana, S. N.; Peek, L. J.; Coop, T.; Berkland, C. Polyelectrolyte complexes stabilize and controllably release vascular endothelial growth factor. Biomacromolecules. *2007*, 5, 1607–1614.

Hutmacher, D. W. Scaffolds in tissue engineering bone and cartilage. Biomaterials. *2000*, 21, 2529–2543.

Itoh, N.; Ornitz, D. M. Functional evolutionary history of the mouse Fgf gene family. Dev. Dyn. *2008*, 237(1), 18–27.

Jahani, H.; Kaviani, S.; Hassanpour-Ezatti, M.; Soleimani, M.; Kaviani, Z.; Zonoubi, Z. The effect of aligned and random electrospun fibrous scaffolds on rat mesenchymal stem cell proliferation. J. Cell (Yakhteh), *2012*, 14(1), 31–38

Jenne, D.; Stanley, K. K. Nucleotide sequence and organization of the human S-protein gene: repeating peptide motifs in the pexin family and a model for their evolution. Biochemistry. *1987*, 26(21), 6735–6742.

Jeschke, B.; Meyer, J.; Jonczyk, A.; Kessler, H.; Adamietz, P.; Meenen, N. M.; Kantlehner, M.; Goepfert, C.; Nies, B. RGD-peptides for tissue engineering of articular cartilage. Biomaterials. *2002*, 23(16), 3455–3463.

Jiang, H.; Hu, Y.; Zhao, P.; Li, Y.; Zhu, K. Modulation of protein release from biodegradable core-shell structured fibers prepared by coaxial electrospinning. J. Biomed. Mater. Res. B Appl. Biomater. *2006*, 79(1), 50–57.

Jiang, T.; Khan, Y.; Nair, L. S.; Abdel-Fattah, W. I.; Laurencin, C. T. Functionalization of chitosan/poly(lactic acid-glycolic acid) sintered microsphere scaffolds via surface heparinization for bone tissue engineering. J. Biomed. Mater. Res. *2010*, 93A(3), 1193–1208.

Källrot, M.; Edlund, U.; Albertsson, A.C. Surface functionalization of porous resorbable scaffolds by covalent grafting. Macromol. Biosci. *2008,* 8(7), 645–654.

Kanani, A. G.; Bahrami, S. H. Review on electrospun nanofibers scaffold and biomedical applications. Trends Biomater. Artif. Organs. *2010*, 24 (2), 93–115.

Kaplan, D. R.; Chap, F. C.; Stiles, C. D.; Antoniades, H. N.; Seher, C. D. Platelet alpha-granules contain a growth factor for fibroblasts. Blood. *1979a*, 53, 1043–1050.

Kaplan, K. L.; Broekman, M. J.; Cherooff, A.; Lesznik, G. R.; Drillings, M. Platelet alpha-granule proteins: studies on release and subcellular localization. Blood. *1979b*, 53, 604–618.

Katsura, N.; Ikai, I.; Mitaka, T.; Shiotani, T.; Yamanokuchi, S.; Sugimoto, S.; Kanazawa, A.; Terajima, H.; Mochizuki, Y.; Yamaoka, Y. Long-term culture of primary human hepatocytes with preservation of proliferative capacity and differentiated functions. J. Surg. Res. *2002*, 106, 115–123.

Khang, G. Handbook of Intelligent Scaffolds for Tissue Engineering and Regenerative Medicine. Biomaterials and manufacturing methods for scaffold in regenerative medicine; Pan Stanford Publishing, *2012*; pp 3–40.

Khoo, M. L. M.; McQuade, L. R.; Smith, M. S. R.; Lees, J. G.; Sidhu, K. S.; Tuch, B. E. Growth and differentiation of embryoid bodies derived from human embryonic stem cells: Effect of glucose and basic fibroblast growth factor. Biol. Reprod. *2005*, 73(6), 1147–1156.

Kibbey, M. C.; Jucker, M.; Weeks, B. S.; Neve, R. L.; Van Nostrand, W. E.; Kleinman, H.K. Beta-Amyloid precursor protein binds to the neurite-promoting IKVAV site of laminin. Proc. Natl. Acad. Sci. *1993*, 90, 10150–10153.

Kim, Hae-Won; Kim, Hyoun-Ee; Vehid, S. Stimulation of osteoblast responses to biomimetic nanocomposites of gelatin-hydroxyapatite for tissue engineering scaffolds. Biomaterials. *2005*, 26, 5221–5230.

Kim, T. G.; Park, T. G. Biomimicking extracellular matrix: cell adhesive RGD peptide modified electrospun poly(D,L-lactic-co-glycolic acid) nanofiber mesh. Tissue Eng. *2006*, 12, 221–233.

Kobayashi, K.; Sumitomo, H.; Ina, Y. Synthesis and functions of polystyrene derivatives having pendent oligosaccharides. Polym. J. *1985,* 17, 567–575.

Kosorn W.; Boonlom, T.; Paweena, U.; Pakkanun K.; Wanida J. Surface modification of polycaprolactone scaffolds by plasma treatment for chondrocyte culture. 4th International Conference on Chemical, Biological and Environmental Engineering (IPCBEE) *2012,* 43 IACSIT Press, Singapore. doi:10.7763/IPCBEE. 2012. V43. 10

Kut, C.; Gabhann, F. M.; Popel, A. S. Where is VEGF in the body? A meta-analysis of VEGF distribution in cancer. Brit. J. Cancer. *2007,* 97, 978–985.

Landreth, G. E. Classes of Growth Factors Acting in the Nervous System in: Siegel,G. J.; Agranoff, B. W.; Albers, R. W.; Fisher, S. K.; Uhler, M. D. Edited, Basic Neurochemistry: Molecular, Cellular and Medical Aspects. Philadelphia: Lippincott-Raven, *1999*; 6th edition.

Lanza, R. P.; Langer, R.; Chick, W. L.; eds. Textbook of Tissue Engineering. Austin, Texas: R.G. Landes Co. 1997; pp 89–100.

Lee, C. T.; Kung, P. H.; Lee, Y. D. Preparation of poly (vinyl alcohol)-chondroitin sulfate hydrogel as matrices in tissue engineering. Carbohydr. Polym. *2005*, 61(3), 348–354.

Lee, K.; Silva, E. A.; Mooney, D. J. Growth factor delivery-based tissue engineering: General approaches and a review of recent developments. J. R. Soc. Interface. *2011*, 8, 153–170.

Liao, Y. H.; Jones, S. A.; Forbes, B.; Martin, G. P.; Brown, M. B. Hyaluronan: pharmaceutical characterization and drug delivery. Drug Deliv. *2005*, 12, 327-342.

Lin C.C.; Anseth, K. S. PEG hydrogels for the controlled release of biomolecules in regenerative medicine. Pharmaceut. Res. *2009*, 3, 26.

Lin, Y. D.; Luo, C. Y.; Hu, Y. N.; Yeh, M. L.; Hsueh, Y. C.; Chang, M. Y.; Tsai, D. C.; Wang, J. N.; Tang, M. J.; Wei, E. I.; Springer, M. L.; Hsieh, P. C. Instructive nanofiber scaffolds with VEGF create a microenvironment for arteriogenesis and cardiac repair. Sci. Transl. Med. *2012*, 4(146), 1–11.

Liu, C.; Xia, Z.; Czernuszka, J. T. Design and development of three-dimensional scaffolds for tissue engineering. Chem. Eng. Res. Design. *2007*, 85, 1051–1064.

Liu, X.; Ma, L.; Mao, Z.; Gao, C. Chitosan-based biomaterials for tissue repair and regeneration. Adv. Polym. Sci. *2011*, 244, 81–128.

Lopina, S. T.; Wu, G.; Merrill, E. W.; Griffith-Cima, L. Hepatocyte culture on carbohydrate-modified star polyethylene oxide hydrogels. Biomaterials. *1996*, 17, 559–569.

Madurantakam, P. A.; Rodriguez, I. A.; Beckman, M. J.; Simpson, D. G.; Bowlin, G. L. Evaluation of biological activity of bone morphogenetic proteins on exposure to commonly used electrospinning solvents. J. Bioactive Compat. Polym. *2011*, 26(6), 578–589.

Mansbridge, J. Skin tissue engineering. J. Biomater. Sci. Polym. Edn. *2008*, 19(8), 955–968.

Martin, I.; Wendt, D.; Heberer, M. The role of bioreactors in tissue engineering. Trends Biotechnol. *2004*, 22, 80–86.

Massague, J.; Seoane, J.; Wotton, D. Smad transcription factors. Genes Dev. *2005*, 19(23), 2783–2810.

Mattioli-Belmonte, M.; Giavaresi, G.; Biagini, G.; Virgili, L.; Giacomini, M.; Fini, M.; Giantomassi, F.; Natali, D.; Torricelli, P.; Giardino, R. Tailoring biomaterial compatibility: in vivo tissue response versus in vitro cell behaviour. Int. J. Artif. Organs. *2003*, 26, 1077–1085.

McBeath, R.; Pirone, D. M.; Nelson, C. M.; Bhadriraju, K.; Chen, C. S. Cell shape, cytoskeletal tension, and RhoA regulate stem cell lineage commitment. Dev. Cell. *2004*, 6(4), 483–495.

Miller, W. M.; Peshwa, M. V. Tissue engineering, bioartificial organs, and cell therapies. J. Biotechnol. Bioeng. *1996*, 50, 347–348.

Moon, C.;Jeong, R. H.;Hyun-Jung, P.;Jong, S. H.;Jihee, L. K. Synthetic RGDS peptide attenuates lipopolysaccharide-induced pulmonary inflammation by inhibiting integrin signaled MAP kinase pathways. Respir. Res. *2009*,10, 18.

Moore, E.; Bendele, A.; Thompson, D.; Littau, A.; Waggie, K.; Reardon, B.; Ellsworth, J. Fibroblast growth factor-18 stimulates chondrogenesis and cartilage repair in a rat model of injury-induced osteoarthritis. Osteoarthritis and Cartilage. *2005*, 13(7), 623–631.

Munir, S.; Xu, G.; Wu, Y.; Yang, B.; Lala, P. K.; Peng, C. Nodal and ALK7 inhibit proliferation and induce apoptosis in human trophoblast cells. J. Biol. Chem. *2004*, 279(30), 31277–31286.

Muthana, S. M.; Campbell, C. T.; Gildersleeve, J. C. Modifications of glycans: Biological significance and therapeutic opportunities. ACS Chem. Biol. *2012*, 7(1), 31–43.

Nair, S.; Laurencin, C. T. Biodegradable polymers as biomaterials. Prog. Polym. Sci. *2007*, 32, 762–798.

Nicolas, F. L.; Gagnieu, C. H. Denatured thiolated collagen II. Cross-linking by oxidation. Biomaterials. *1997*, 18(11), 815–821.

Nishimura, H.; Nishimura, M.; Oda, R.; Yamanaka, K.; Matsubara, T.; Ozaki, Y.; Sekiya, K.; Hamada, T.; Kato, Y. Lectins induce resistance to proteases and/or mechanical stimulus in all examined cells–including bone marrow mesenchymal stem cells on various scaffolds. Exp. Cell Res. *2004*, 295(1), 119–127.

Oh, S.; Oh, N.; Appleford, M.; Ong, J. L. Bioceramics for tissue engineering applications – A review. Am. J. Biochem. Biotechnol. *2006*, 2(2), 49–56.

Oh, S. H.; Park, I. K.; Kim, J. M.; Lee, J. H. In vitro and in vivo characteristics of PCL scaffolds with pore size gradient fabricated by a centrifugation method. Biomaterials. *2007*, 28(9), 1664–1671.

Olsson, A. K.; Dimberg, A.; Kreuger, J.; Claesson-Welsh, L. VEGF receptor signalling – In control of vascular function. Mol. Cell Biol. Nat. Rev. *2006*, 7, 359–371.

Ornitz, D. M.; Itoh, N. Fibroblast growth factors. Genome Biol. *2001*, 2(3), 3005–3012.

Parkl, I. K.; Yang, J.; Jeong, H. J.; Bom, H. S.; Harada, I.; Akaike, T.; Kim, S. I.; Cho, C. S. Galactosylated chitosan as a synthetic extracellular matrix for hepatocytes attachment. Biomaterials. *2003,* 24, 2331–2337.

Parton, R.G.; Simons, K. The multiple faces of caveolae. Nat. Rev. Mol. Cell Biol. *2007*, 8(3), 185–194.

Pigman, W. W.; Horton, D.; Wander, J. D. The Carbohydrates. Vol IB. Academic Press. New York. 1980, pp 727–728.

Pramanik, S.; Kar, K.K. Functionalized poly(ether ether ketone): Improved mechanical property and acellular bioactivity. J. Appl. Polym. Sci. *2012*, 123, 1100–1111.

Preissner, K. T.; Seiffert, D. Role of vitronectin and its receptors in haemostasis and vascular remodeling. Thromb. Res.*1998*, 89(1), 1–21.

Puchacz, E.; Lian, J. B.; Stein, G. S.; Wozney, J.; Huebner, K.; Croce, C. Chromosomal localization of the human osteocalcin gene. Endocrinology. *1989*, 124(5), 2648–2650.

Qiao, B.; Padilla, S. R.; Benya, P. D. Transforming growth factor (TGF)-β-activated kinase 1 mimics and mediates TGF-β-induced stimulation of type II collagen synthesis in chondrocytes independent of *Col2a1* transcription and Smad3 signaling. J. Biol. Chem. *2005,* 280, 17562–17571.

Rajangam, K.; Behanna, H. A.; Hui, M. J.; Han Xiaoqiang.; Hulvat, J. F.; Lomasney, J. W.; Stupp, S. I. Heparin binding nanostructures to promote growth of blood vessels. Nano. Lett. *2006*, 6(9), 2086–2090.

Ravi, S.; Chaikof, E. L. Biomaterials for vascular tissue engineering. Regen. Med. *2010*, 5, 1, 107–120.

Reddi, A. H. Role of morphogenetic proteins in skeletal tissue engineering and regeneration. Nat. Biotechnol. *1998*, 16, 247–252.

Reddi, A. H.; Reddi, A. Bone morphogenetic proteins (BMPs): From morphogens to metabologens. Cytok. Growth Fact. Rev. *2009*, 20(5–6), 341–342.

Richy, F.; Bruyne, O.; Ethgen, O.; Cucherat, M.; Henrotin, Y.; Reginster, J. Y. Structural and symptomatic efficacyof glucosamine and chondroitin in knee osteoarthritis: A comprehensive meta-analysis. Arch. Int. Med. *2003*, 163(13), 1514–1522.

Rim, N. G.; Lee, J. H.; Jeong, S. I.; Lee, B. K.; Kim, C. H.; Shin, H. Modulation of osteogenic differentiation of human mesenchymal stem cells by poly[(L-lactide)-co-(epsilon-caprolactone)]/gelatin nanofibers. Macromol. Biosci. *2009*, 9(8), 795–804.

Ross, R.; Raines. E. W.; Bowen-Pope, D. F. The biology of platelet-derived growth factor. Cell. *1986*, 46, 155–169.

Oldberg, A.; Franzen, A.; Heinegaard, D. The primary structure of a cell-binding bone sialoprotein. J. Biol. Chem. *1988*, 263(36), 19430–19432.

Saari, H.; Konttinen, Y. T.; Friman, C.; Sorsa, T. Differential effects of reactive oxygen species on native synovial fluid and purified human umbilical cord hyaluronate. Inflammation. *1993*, 17, 403–415.

Sakou, T.; Onishi, T.; Yamamoto, T.; Nagamine, T.; Sampath, T.; ten Dijke, P. Localization of Smads, the TGF- β family intracellular signaling components during endochondral ossification. J. Bone Miner. Res. *1999*, 14, 1145–1111.

Safaeijavan, R., Soleimani, M., Divsalar, A., Eidi, A., & Ardeshirylajimi, A. Biological behavior study of gelatin coated PCL nanofiberous electrospun scaffolds using fibroblasts. J. Paramed. Sci. *2013*, 5(1), 67–73.

Schense, J. C.; Bloch, J.; Aebischer, P.; Hubbell, J. A. Enzymatic incorporation of bioactive peptides into fibrin matrices enhances neurite extension. Nat. Biotechnol. *2000*, 18(4), 415–419.

Schneider, A.; Garlick, J. A.; Egles, C. Self-assembling peptide nanofiber scaffolds accelerate wound healing. PLoS One. *2008*, 3(1), e1410.

Selvakumar, R.; Nazar Mohamed Mohaideen, S.; Aravindh. S.; Sabarinath, C.; Ananthasubramanian, M. Effect of biotin and galactose functionalized gelatin nanofiber membrane on Hep-2 cell attachment and cytotoxicity. J. Memb. Biol. *2013,* doi:10.1007/s00232-013-9608-x

Seyednejad, H.; Ji, W.; Yang, F.; van Nostrum, C. F.; Vermonden, T.; van den Beucken, J. J.; Dhert, W. J.; Hennink, W. E.; Jansen, J. A. Coaxially electrospun scaffolds based on hydroxyl-functionalized poly(ε-caprolactone) and loaded with VEGF for tissue engineering applications. Biomacromolecules. *2012*, 13(11), 3650–3660.

Shaban, M.; Kim, S. H.; Idrus, R. B.; Khang, G. Fibrin and poly(lactic-co-glycolic acid) hybrid scaffold promotes early chondrogenesis of articular chondrocytes: an in vitro study. J. Orthop. Surg. *2008*, 3, 17.

Shi, G.; Rouabhia, M.; Wang, Z.; Dao, L. H.; Zhang, Z. A novel electrically conductive and biodegradable composite made of polypyrrole nanoparticles and polylactide. Biomaterials. *2004*, 25, 2477–2488.

Shi, G.; Rouabhia, M.; Meng, S.; Zhang, Z. Electrical stimulation enhances viability of human cutaneous fibroblasts on conductive biodegradable substrates. J. Biomed. Mater. Res. A. *2008*, 15, 84(4), 1026–1037.

Shibuya, M.; Claesson-Welsh, L. Signal transduction by VEGF receptors in regulation of angiogenesis and lymphangiogenesis. Exp. Cell Res. *2006*, 312, 549–556.

Sitterley G., Attachment and matrix factors. BioFiles, 2008, 3(8), 7–8.

So-Ra Son; Nguyen-Thuy; Ba Linh; Hun-Mo Yang; Byong-Taek Lee. Invitro and invivo evaluation of electrospun PCL/PMMA fibrous scaffolds for bone regeneration. Sci. Technol. Adv. Mater. *2013*, 14, 015009 (10pp).

Toegel, S.; Harrer, N.; Plattner, V. E.; Unger, F. M.; Viernstein, H.; Goldring, M. B.; Gabor, F.; Wirth, M. Lectin binding studies on C-28/I2 and T/C-28a2 chondrocytes provide a basis for new tissue engineering and drug delivery perspectives in cartilage research. J. Contr. Rel. *2007*, 117(1), 121–129.

Torella, D.; Rota, M.; Nurzynska, D.; Musso, E.; Monsen, A.; Shiraishi, I. Cardiac stem cell and myocyte aging, heart failure, and insulin-like growth factor-1 over expression. Circ. Res. *2004*, 94, 514–524.

Turkeri, L. N.; Erton, M. L.; Cevik, I.; AkDas, A. Impact of the expression of epidermal growth factor, transforming growth factor alpha, and epidermal growth factor receptor on the prognosis of superficial bladder cancer. Urology. *1998*, 51(4), 645–649.

Underwood, P. A.; Bennett, F. A. A comparison of the biological activities of the cell-adhesive proteins vitronectin and fibronectin. Cull. Sri. *1989*, 93, 641–649.

Vukicevic, S.; Sampath, K. T. Bone morphogenetic proteins: From local to systemic therapeutics. Birkhauser Verlag AG, Basel-Boston-Berlin publication. *2008*.

Wang, D. A.; Varghese, S.; Sharma, B.; Strehin, I.; Fermanian, S.; Gorham, J.; Howard, F.; Brett, C.; Jennifer, H. E. Multifunctional chondroitin sulphate for cartilage tissue-biomaterial integration. Nat. Mater. *2007*, 6, 385–392.

Welch, D.; Dawes, P. J. Childhood hearing is associated with growth rates in infancy and adolescence. Pediatr. Res. *2007*, 62(4), 495–498.

Wessels, M. R.; Moses, A. E.; Goldberg, J. B.; DiCesare, T. J. Hyaluronic acid capsule is a virulence factor for mucoid group *A. streptococci*. PNAS. *1991*, 88(19), 8317–8321.

Westermark, B.; Heldin, C. H. Structure and function of platelet-derived growth factor. Acta Med. Scand. (Suppl.). *1987*, 715, 19–23.

Wu, Y.; Qixin, Z.; Jingyuan, Du.; Yulin, S.; Bin, Wu.; Xiaodong, G. Self-assembled IKVAV peptide nanofibers promote adherence of PC12 cells. J. Huazhong Univ. Sci. Technol. *2006*, 26(5), 594–596.

www.exogenesis.us (Exogenesis Corporation 20 Fortune Drive Billerica, MA USA 01821 (978) 439–0120).

Xu, L. C.; Siedlecki, C. A. Effects of surface wettability and contacttime on protein adhesion to biomaterial surfaces. Biomaterials *2007*, 28, 3273–3283.

Ying, L.; Yin C.; Zhuo, R. X.; Leong, K. W.; Mao, H. Q.; Kang, E. T.; Neoh, K. G. Immobilization of galactose ligands on acrylic acid graft-copolymerized poly(ethyleneterephthalate) film and its application to hepatocyte culture. Biomacromolecules. *2003*, 4, 157–165.

Yoshinari, M.; Wei, J.; Matsuzaka, K.; Inoue, T. Effect of cold plasma-surface modification on surface wettability and initial cell attachment. World Acad. Sci. Engg. Tech. *2009*, 34, 10–24.

Zhang, S.; Lockshin, C.; Herbert, A.; Winter, E.; Rich, A. Zuotin, a putative Z-DNA binding protein in Saccharomyces cerevisiae. EMBO. *1992*, 11, 3787–3796.

Zhang, S.; Holmes, T.; Lockshin, C.; Rich, A. Spontaneous assembly of a self-complementary oligopeptide to form a stable macroscopic membrane. Proc. Natl. Acad. Sci. *1993*, 90, 3334–3338.

Zhang, Z.; Kuijer, R.; Bulstra, S. K.; Grijpma, D. W.; Feijen, J. The in vivo and in vitro degradation behavior of poly(trimethylene carbonate). Biomaterials. *2006*, 27(9), 1741–1748.

Zhao, X. J.; Zhang, S. G. Designer self-assembling peptide materials. Macromol. Biosci. *2007*, 7(1), 13–22.

CHAPTER 13

ANTIMICROBIAL NANOMATERIALS FOR WOUND DRESSINGS

ANANTHA S. SELVARAJ and MALA RAJENDRAN

ABSTRACT

Sharp injury to the skin may occur due to myriad of reasons. Wound healing is a complex and dynamic process. Wound dressings accelerate wound healing and restore the structural and functional integrity of skin. Pathogens have developed resistance against conventional antimicrobial agents. Wound colonization by multidrug-resistant pathogens complicate and delays wound healing. In this era of antibiotic resistance, nanomaterials emerge as a technological tool to improve antimicrobial chemotherapy. Wound dressings target any one phase of wound healing and the choice of dressing varies based on the type and severity of wound. Wound dressings incorporating different nanomaterials are available in the market with varying degrees of clinical outcome. Among the battery of antimicrobial nanomaterials available for wound dressings, silver finds a unique position due to its broad-spectrum antimicrobial activity. It has comparatively low MIC and MBC than other antimicrobial nanomaterials. Application of silver nanomaterial impregnated surgical mesh block the attachment of microbes and break the nosocomial infection chain. Acticoat® is the first commercial wound dressing with silver nanoparticles. Sustained release of silver nanomaterial is facilitated by moisture absorbed by the polyester core. It is highly effective against drug-resistant pathogens such as vancomycin-resistant enterococci, methicillin-resistant *Staphylococcus aureus*, and fungi. Research is also focused on the use of zinc nanomaterial which functions as an antimicrobial and anti-inflammatory agent. Nanostructured cellulosescaffold with antiseptic property endowed by the incorporation of titanium dioxide nanomaterials are commercially available recently. Nanofibrous mats of chitosan fabricated by electrospinningcan mimic natural extracellular matrix and facilitates tissue regeneration. Cotton dressings are also laminated with a thin film of chitosan to provide structural support and antimicrobial property. This review highlights the types of wounds, wound dressings, nanomaterials, their formulation, and their impact on wound healing.

13.1 INTRODUCTION

13.1.1 STRUCTURE AND FUNCTION OF SKIN

Skin is the largest organ of the body, making up 16% of body weight, with a surface area of between 1.5 and 2.0 square meters (1,2). Skin functions as a mechanical barrier between internal organs and the external environment. It selectively permits exchange of water, electrolytes, and other substances. It functions as a physical as well as chemical protective barrier against ultraviolet (UV) radiations, toxic chemicals, and microorganisms (3). Skin is made up of three layers namely epidermis, dermis and hypodermis. Epidermis is the outer layer composed of a tough, supportive cell matrix and is comprised of a thin papillary layer and a thicker reticular layer. The main cells of the epidermis are the keratinocytes, which synthesize the protein keratin. Protein bridges called desmosomes connect the keratinocytes, which are in a constant state of transition from the deeper layers to the superficial. The dermis is the second layer providing the structural support of the skin, below which is a loose connective tissue layer. Hypodermis is a fat depot that maintains the moisture of the skin. Skin contains capillaries, lymphatic vessels, nerve endings, and glands which endow it with unique functions like a sensor that monitors and respond to changes in external environment, as a thermo regulator, an excretory organ of salt and water, and a store house of lipids. Thickness of the skin varies from 0.5 to 4.0 mm depending upon the function it serves. For example, the thickness of eye lids is thin and flexible compared to the thickness of palms of hands and soles of feet where it resists friction (4–8).

The epithelial surface of the skin acts as the first line of defense against invading organisms. Microbes adhered to the epithelium is removed by desquamation. When the skin is injured, inflammation is stimulated by inflammatory cytokines released by injured cells. This arrests the spread of infection and promotes wound healing (9). Innate immune cells recognize pathogen-associated molecular patterns (PAMPs) like peptidoglycan of bacteria and stimulate immune response and clearance of pathogens. In addition to this secretion, skin contains acidic constituents like lactic acid, propionic acid, and salts which inhibit microbial growth. It also contains lysozyme, a catalytic antimicrobial agent which breaks peptidoglycan layer of bacteria (10). Skin thus functions as a physical and chemical barrier against pathogens and the permeability of skin to nanomaterial is discussed in the last section of this review.

13.2 WOUND AND ITS CLASSIFICATION

A wound is an injury to the structural integrity of the skin impairing its function. Wound can be classified on many criteria's including, the origin, time taken to heal, depth, and appearance as illustrated in Figure 13.1. Etiology of wound can be internal or external. Wounds of internal origin are mainly due to impaired circulation, neuropathy, or medical illness. Wounds of external origin are due to an outside force or trauma that causes open or closed wound (11).

In case of closed wounds, the skin is intact and the underlying tissue is affected but not directly exposed to the outside environment. The most common types of closed wounds are contusions, hematomas and crush injuries. Contusions are type of sports injury. Hematomas are any injury that damages the small blood vessels and capillaries resulting in blood collection and pooling in a limited space. Crush injuries are caused by an external high-pressure force that squeezes part of the body between two surfaces. The degree of injury can range from a minor bruise to a complete destruction of the crushed area of the body (12). Concussion is the common type of traumatic brain injury (13). It is derived from the Latin word "*concutere*" meaning "to shake violently" (14). Laxation is a state of being luxated, as in the dislocation or displacement (15).

In case of open wounds, the skin is cracked open, leaving the underlying tissue exposed to the outside environment. Abrasions, lacerations, incisions, puncture, penetrating and gunshot wounds are the major types of open wounds. Abrasions are shallow irregular wounds of the upper skin layers due to brushing against either a rough surface or a smooth surface at high speed. Lacerations are tear-like wounds with irregularly torn edges that are usually deeper than abrasions and cause more pain and bleeding.

Incisions are the result of a surgical procedure or skin cut with a sharp object such as a scalpel or knife. Punctures are small rounded wounds that result from objects with thin pointed tips, such as needles, nails or teeth. Penetrating type of wound can be caused by any object or force that breaks through the skin to the underlying organs or tissue. Gunshot wounds are considered to be penetrating wounds that are exclusively caused by bullets from fire arms (16).

Injury to the epidermis is called superficial wound. Superficial wound heals within 5–7 days and the epidermis is replaced easily. Healing occurs without scarring. Injury to dermis is termed as partial thickness wound. The seriousness of a partial thickness wound depends on the extent of damage of the dermis. A deep and large partial thickness needs skin grafting. Injury to hypodermis is defined as full-thickness wound. A full-thickness wound destroys all three layers of skin, resulting in the loss of not only the skin but also the hair follicles, sweat glands, and the region where new skin cells are formed. For these reasons, full-thickness wound require skin grafts (17).

Acute wounds are caused by trauma but the wounds are usually healable within 8–12 weeks (18). These wounds can be caused by mechanical damage induced by sheer, blunting or stabbing action of hard objects, burn, irradiation, electrical shock or irritated with corrosive chemicals. Surgical wounds are also classified under the umbrella of acute wounds. Surgical wounds are further classified into four classes based on their cleanliness and infection by microbes as Clean, clean contaminated by microbes, contaminated wounds and dirty infected wounds (19). Chronic wounds are caused as a consequence of diseases such as diabetes, tumors, and severe physiological contaminations (20). Healing of these wounds could take more than 12 weeks (21,22).

In addition to the above wound types, wound can also be classified according to their appearance as epithelializing (clean, medium-to-high exudates), granulating (clean, exudating), slough-covered and necrotic (dry) wounds, respectively (23).

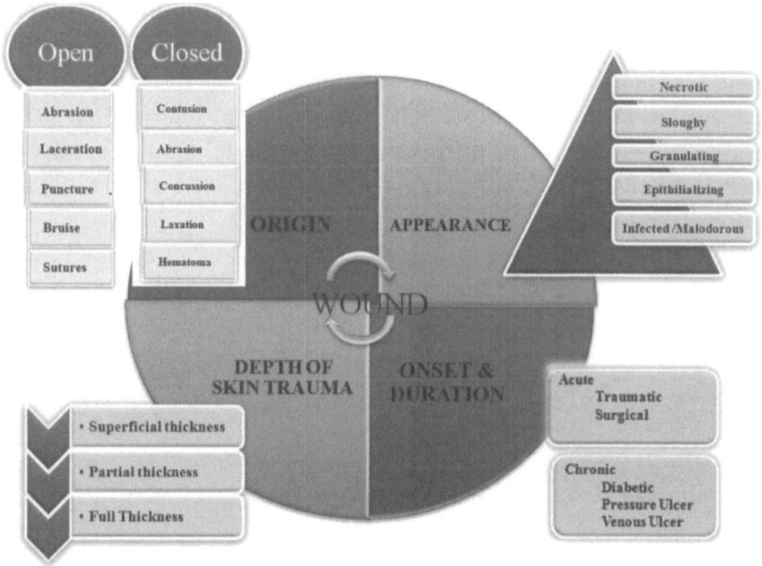

FIGURE 13.1 Classification of wound

13.2.1 PRINCIPLE OF WOUND HEALING

Wound healing is a biological process which occurs in four distinct but overlapping processes encompassing haemostatis, inflammation, proliferation and remodeling as represented in Figure 13.2 (24,25). Hemostasis, the first phase of wound healingbegins at the moment the tissue is injured. Platelets are chemotactically attracted to the sites of injury as the blood components spill into the site of injury. Contact of platelets with collagen and other components of extracellular matrix (ECM) trigger the release of clotting factors, growth factors and cytokines like platelet-derived growth factor and transforming growth factor beta (TGF-ß).This phase exists for 1–3 days. Hemostasis is followed by inflammation which lasts for 4–7 days. This phase is characterized by the infiltration of injured site by neutrophils. Neutrophils are cells of innate immunity which defend against microbes by phagocytosis of pathogens and remove foreign materials and damaged tissue. This antimicrobial activity by innate immune system provides nonspecific immunity against pathogens. As part of this inflammatoryphase, macrophages appear and continue the process of phagocytosis as well as releasing more PDGF and TGFß. Cardinal symptoms of inflammation are rubor, calor, dolar, and tumor. Proliferation follows inflammation. Fibroblasts migrate to begin the proliferativephase and deposit new ECM. Remodeling is the final phase whose duration varies from 12 to 30 weeks based on the nature of wound. The new collagen matrix then becomes cross-linked and organized during the final remodeling phase. Highly co-ordinated process of wound healing requires the release of versatile signaling molecules (26,27).

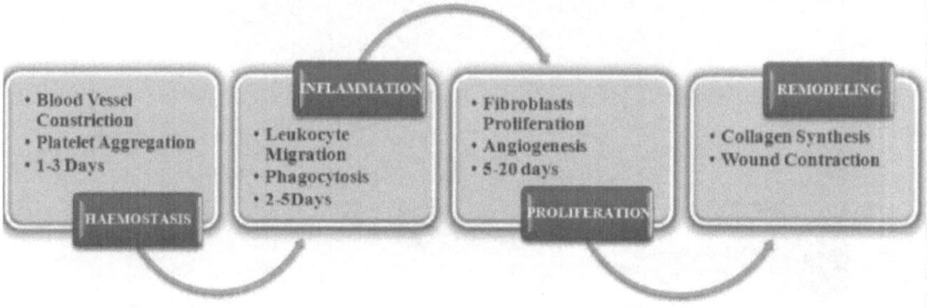

FIGURE 13.2 Phases of wound healing

13.2.2 WOUND DRESSINGS

Wound dressing is defined as an artificial skin used to cover the wound to protect it from external environment and infection. The main purpose of dressing is to achieve the highest rate of healing and aesthetic repair of the wound (28). Crude treatment for wound healing dates back to 1500 BC. Documents by Papyrus Ebers indicate wound treatment by covering wound with oiled frog skin, honey, lint and animal grease. Susrutu Samhita medical compendium authored by Susruta reported skin grafts being used as early 700 BC 700 (29–31). George (32) reported that the Sumerians were the first to fashion occlusive dressings, which are capable of maintaining a moist environment using clay. Gamgee wadding and tulle gras were the first manufactured dressings (33). Gamgee (1880) discovered that degreased cotton wrapped in bleached lint would absorb fluids and he introduced his first dressing in the 19th century. During the 1914–1918 World War, Lumiere in France developed cotton gauze impregnated with paraffin to prevent the dressing sticking to the wound. Only in the 1960s the difference between dry and moist healing were identified

(34,35). Continuous developments have led to extensive use of new bandages with improved performance. Today's wound dressing materials are usually based on synthetic polymers. Characteristics of an ideal wound dressing are outlined by Kumar et al. (36). It must be sterile, provide thermal insulation, mechanical and chemical protection against external environment and pathogens, absorb excess exudates in a wet wound, be nonadherent to the wound, and easily removed without trauma and pain. It must permit gaseous and fluid exchange, absorb wound odor and provide some debridement action. Above all, it must be available at low cost (36).

13.2.2.1 CLASSIFICATION OF WOUND DRESSINGS

Wound dressings can be classified from many dimensions (37,38). The main classification is based on its role in healing as passive, interactive and bioactive (39) wound dressings as represented in Table 13.1. Passive products just cover the wound and functions as a mechanical barrier to protect wound from the external environment. Interactive dressings are permeable to water vapor and atmospheric oxygen. Bioactive wound dressing provide desirable properties like enhanced antimicrobial activity, improved healing potential, etc. (40,41). Choice of the correct wound dressing depends on the type and severity of wound. No single dressing is suitable for all types of wounds.

TABLE 13.1 Classes of wound dressings

Dressing class	Dressing type		Wound type	Characteristics
Passive	Gauze	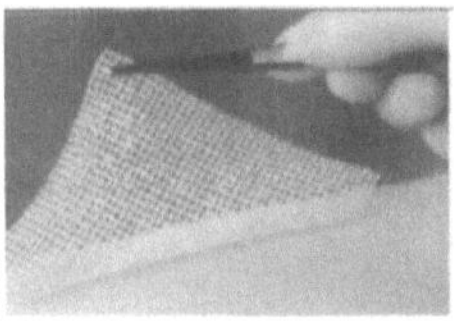	Clean, medium-to- high exudate (epithelializing)	Applicable to minor wounds or as secondary dressings. Stick to wounds.

Passive	Tulle		Burns, accident injuries, diabetic ulcers, skin grafting, sores, operative wounds, colostomies, etc.	Suitable for sensitive skin. Dressing does not stick to wound surface. Appropriate for flat wound. Example: Jelonet®, Paranet® and Komal
Interactive	Hydrogel	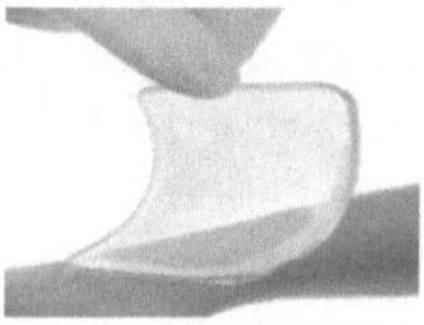 	Slough-covered, necrotic, burn and dry wounds and Slough-covered	Composed mainly of water in a complex network or fibers. Example: Tegagel®, Intrasite®
Interactive	Polyurehane or silicone foam	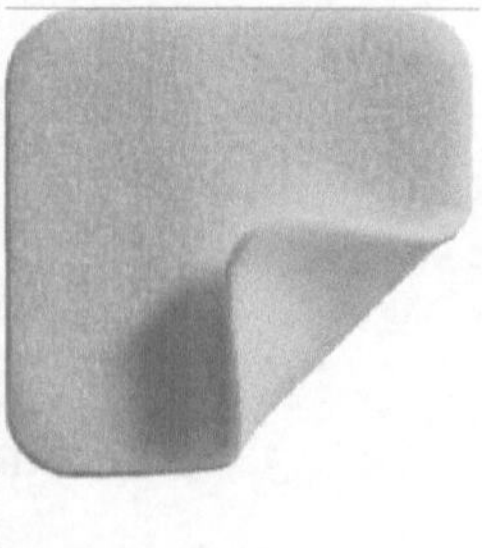	Clean, exudating (granulating)	Designed to absorb large amounts of exudates. Maintain a moist wound environment. Example: Allevyn® and Lyofoam®

TABLE 13.1 *(Continued)*

Bioactive	Semipermeable film	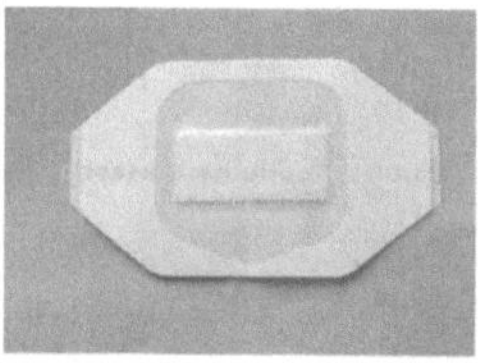	Clean, dry, low exudating (epithelialising)	Sterile transparent sheet of polyurethane coated with acrylic adhesive. Suitable for shallow wound with low exudates. Example: OpSite and Tegaderm.
Bioactive	Hydrocolloid	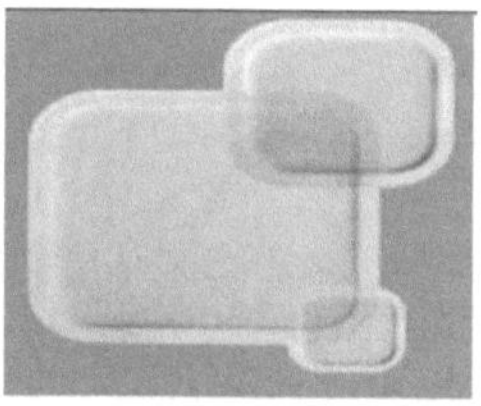	Clean, exudating (granulating) Slough-covered	Applicable to wounds with light to heavy exudates. Example: DuoDERM® and Tegasorb®
Bioactive	Alginate	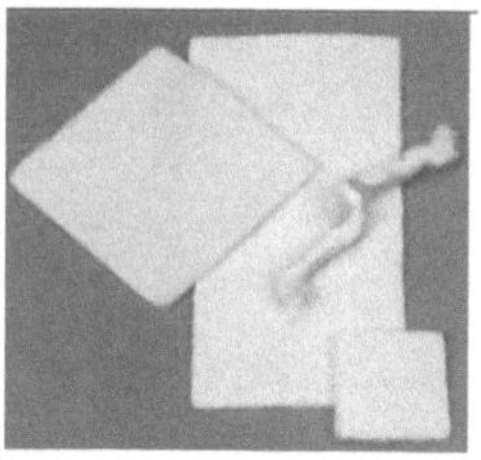	Clean, exudating (granulating)	Composed of calcium alginate. Helps in debridement of sloughing wounds. Example: Kaltostat® and Sorbsan®
Bioactive	Hydrofiber		Moderately exudating	Soft nonwoven pad or ribbon dressing made from sodium carboxymethylcellulose fibers. Applicable to deep wounds. Example:Aquacel

TABLE 13.1 *(Continued)*

Bioactive	Collagen	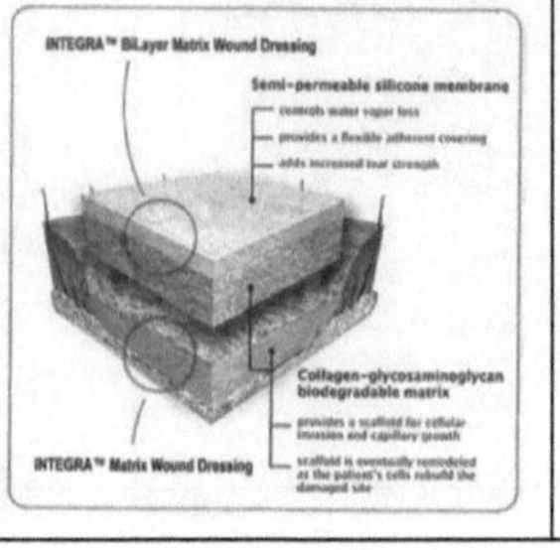	Moderately exudating and skin loss conditions	Promote deposition of newly formed collagen in the wound bed. Absorb exudates. Example: Promogran and Plurocol

13.2.2.1.1 GAUZE

Traditionally, gauze wound dressings are made from woven or nonwoven gauze. Gauze is highly permeable and relatively nonocclusive. Therefore, gauze dressings may promote desiccation in wounds with minimal exudate unless used in combination with another dressing or topical agent. They are inexpensive for one-time or short-term use. Gauze dressings come in many forms like squares, sheets, rolls, and packing strip. It can adhere to the wound surface and cause pain on removal.

13.2.2.1.2 TULLE

Tulle gras dressings have been used for many years for the management of wounds of all types. Structurally they comprise a gauze cloth impregnated with paraffin for nontraumatic removal or antiseptics such as chlorhexidine for prevention or treatment of infection. Tulle gras dressings are not absorbent and therefore require a secondary dressing.

13.2.2.1.3 HYDROGELS

Wichterle and Lim (42) developed the first synthetic hydrogel polymer by the copolymerization of 2-hydroxyethyl methcrylate and ethylenedimethacrylate. Since then, hydrogels have been used in many biological applications and in wound dressings (43). They contain hydrophilic sites that enable interaction with aqueous solutions and absorb and retain significant volumes of wound exudate. Hydrogels facilitate autolysis and rehydrate the wound. Hydrogels may be appropriate in diabetic ulcers that require debridement and are inappropriate for ischemic and gangrenous ulcers. Hydrogels can be fabricated as an amorphous gel or as a sheet.

Amorphous hydrogels are of the most benefit in the treatment of sloughy or necrotic wounds (44). Depending on the composition of the gel and the level of hydration of the target wound, hydrogel enhances autolytic debridement (45). Autolysis is most effective in a moist environment (46). Hydrogels have the ability to donate water molecules to dehydrated tissue while allowing the passage of water vapor and oxygen to the wound surface (47). This helps to increase the phagocytotic activity of leucocytes and dead tissues will be removed by autolysis. The use of an amorphous hydrogel requires application of a secondary dressing. They are formulations of water, polymers and other ingredients with no shape, designed to donate moisture to a dry wound and to maintain a moist healing environment. The high moisture content serves to rehydrate wound tissue.It is used for partial- and full-thickness wounds, wounds with necrosis and deep wounds with tunneling or sinus tracts ans are available in a wide variety of sizes. Impregnated hydrogel wound dressings are gauzes and nonwoven sponges, ropes, and strips saturated with an amorphous hydrogel. Hydrogel sheets are transparent, nonadherent dressing for the management of acute and chronic wounds with moderate to light exudate. It absorbs up to five times its own weight.

13.2.2.1.4 SEMIPERMEABLE FILM

Trudigan (48) advocated the use of a semipermeable film for dry wounds. The film reduces the escape of water vapor from the primary dressing, preventing it from drying out, thus helping to maintain a moist wound-healing environment (49).

13.2.2.1.5 FOAM

Foam wound dressings are sheets and other shapes of foamed polymer solutions with small, open cells capable of holding fluids. They may be impregnated or layered in combination with other materials. Absorption capability depends on thickness and composition of the foam. The area in contact with the wound surface is nonadhesive for easy removal. They are available with an adhesive border and/or a transparent film coating that acts as a bacterial barrier.

13.2.2.1.6 HYDROCOLLOID

Hydrocolloids are occlusive dressings designed to create and maintain a moist wound environment. They are capable of absorbing a moderate amount of wound exudates, resulting in a moist gel formation on the wound surface. However, oversaturation of the dressing may lead to leakage of the gelatinous substance, causing maceration of the surrounding skin. Therefore, hydrocolloids should be avoided on plantar ulcers of the foot, as the peri wound skin is susceptible to maceration. Additionally, hydrocolloids have been shown to retain growth factors under the dressing as well as promote granulation and epithelialization.

13.2.2.1.7 ALGINATE

Alginates are naturally occurring polysaccharides composed of mannuronic and guluronic acids. Alginate dressings are composed entirely of calcium alginate or a combination of calcium and sodium alginate. Alginates can absorb between 15 and 20 times their own weight in fluid. The absorptive capacity of alginates makes them eminently suited for the treatment of heavily draining wounds. However, caution should be used as the pooling exudate from alginate dressings may cause maceration of the surrounding tissues. When placed within the wound bed, alginate dressings react with serum and wound exudate to form a gel. This gel provides a moist wound environment and may trap bacteria, which can then be washed away during dressing changes. Alginates are highly permeable and nonocclusive. Therefore, they require a secondary dressing, most commonly gauze. Alginates are available in three forms. Alginate sheets may be placed on wound beds to absorb drainage. Alginate ropes are used to tightly fill wound tunnels or areas of undermining. And alginate-tipped applicators can be used to probe wounds, fill wound cavities and tunnels perform swab cultures, and measure wound depth.

13.2.2.1.8 HYDROFIBERS

Hydrofibers are dressings made from sodium carboxymethylcellulose. These dressings are similar to alginates in appearance, use, and precautions, and are often used interchangeably with alginates.

13.2.2.1.9 COLLAGEN

Collagen for collagen dressings are derived from bovine, porcine, equine, or avian sources, which is purified in order to render it nonantigenic. Collagen dressing may contain ingredients, such as alginates and cellulose derivatives that can enhance absorbency, flexibility, and comfort, and help maintain a moist wound environment. Collagen dressings have a variety of pore sizes and surface areas. Many collagen dressings are fabricated along with antimicrobial agents. Collagen dressing generally requires a secondary dressing. Collagen dressings are employed with different carriers and formulated as gels, pastes, polymers, oxidized regenerated cellulose (ORC), and ethylene diamine tetraacetic acid.Purocol collagen dressing contains ionic silver for providing antimicrobial activity.

13.2.3 INFECTION OF WOUND

Wound exposes the sterile subcutaneous tissue to external environment and provides a favorable route to the entry of microbes. Microorganisms can get access into a wound either by direct contact of air dispersal or by contamination (50). Infection is one of the major factors that interfere with wound healing. Microbes indirectly affects tissue repair by the production of endotoxins (51,52). Local infection if untreated can infect adjacent healthy tissues (53). Peculiarly bacteria exist as biofilms in wound surface which is thousand times resistant to antibiotics and it is a predominant challenge to wound healing (54,55). A biofilm is a community of adherent microorganisms embedded within a self-produced matrix of extracellular polysacharide. Microbes inside the biofilm are phenotypically different from free floating microbes and their gene expression and regulation are different from planktonic cells.Biofilm formation occurs in five different stages encompassing reversible attachment, irreversible attachment, multiplication, secretion of exopolysaccharides, and dispersal (56–58). Thousand-fold increase in resistance to antibiotic is due to the inactivation of antimicrobial agents by exopolysaccharides, overexpression of stress-responsive genes, oxygen gradients within the biofilm matrix, and differentiation of a subpopulation of biofilm cells into resistant dormant cells (59,60).

13.3 ANTIMICROBIAL AGENTS

Antimicrobial dressings are used only if the wound is burdened with infection. Antimicrobial agents are either microbiostatic or microbicidal depending on the dose of the agent. Antimicrobial agents can be disinfectants, antiseptics and antibiotics. Metals like silver, copper, titanium, magnesium, and biopolymer like chitosan also posses antimicrobial activity.

13.3.1 DISINFECTANTS

Disinfectants are used to remove microbes from inanimate substances. Disinfectants work by destroying the cell wall of microbes or interfering with the metabolism. Therefore they are not used in wound dressings.

13.3.2 ANTISEPTICS

Antiseptic in Greek means agents against putrefaction or sepsis. Antiseptics are generally distinguished from antibiotics by the fact that antibiotics can be transported through the lymphatic system to destroy bacteria within the body. Iodine is the widely used antiseptic. Cadexomer iodine and povidone–iodine are the form of iodine commonly used in dressings. It should not be used in patients with thyroid disorders. Tincture of iodine is iodine in alcohol and is used as a pre- and postoperative antiseptic. Sometimes it may induce scar formation and increase healing time. Novel iodine antiseptics containing povidone–iodine (an iodophor, complex of povidone, a water-soluble polymer, with triiodide anions I_3^-, containing about 10% of active iodine) are also used clinically. Iodine is a broad-spectrum antiseptic and it kills bacterial spores also. Alcohols, quarternary ammonium compounds, boric acid, brilliant green, chlorhexidine gluconate, hydrogen peroxide, mercurochrome, manuka honeyoctenidine dihydrochloride, phenol, polyhexanide, sodium chloride, sodium hypochlorite, calcium hypochlorite, and sodium bicarbonate are other antiseptics. Boric acid is used in suppositories to treat yeast infections of the vagina and in eyewashes as an antiviral agent to shorten the duration of cold sore attacks. It is also used in creams for burns and in trace amounts in eye contact solutions.

13.3.3 ANTIBIOTICS

The term "antibiosis," meaning "against life," was introduced by the French bacteriologist Jean Paul Vuillemin as a descriptive name of the phenomenon exhibited by these early antibacterial drugs (61,62). John Tyndall first described antagonistic activities by fungi against bacteria in England in 1875 (63). Antibiosis was first described in 1877 in bacteria when Louis Pasteur and Robert Koch observed that an airborne bacillus could inhibit the growth of *Bacillus anthracis* (64). In 1895, Vincenzo Tiberio, physician of the University of Naples discovered that a mold (*Penicillium*) in a water well had an antibacterial action (65,66). In 1928 Alexander Fleming observed antibiosis against bacteria by a fungus of the genus *Penicillium.* Fleming postulated the effect was mediated by antibacterial compound named penicillin, and that its antibacterial properties could be exploited for chemotherapy (67).

The term antibiotic was first used in 1942 by Waksman to denote chemical substances secreted by microorganisms, anatagonists to other organisms in high dilution (68). This definition excluded substances that kill bacteria but are not produced by microorganisms (such as gastric juices and hydrogen peroxide). It also excluded synthetic antibacterial compounds such as the sulfonamides. Many antibacterial compounds are relatively small molecules with a molecular weight of less than 2000 atomic mass units. Most of today's antibacterial agents are semisynthetic modifications of various natural compounds (69). These include, the beta-lactam antibacterials, which include the penicillins (produced by fungi in the genus *Penicillium*), the cephalosporins, and the carbapenems. Aminoglycosides are still isolated from living organisms. Sulfonamides, quinolones, and oxazolidinones are produced only by chemical synthesis.

Antibiotics target any one biological process of microbes including cell wall biosynthesis, replication, or protein biosynthesis as illustrated in Figure 13.4 and Table 13.2 (70)**. Since the discovery of penicillin in 1940, many antibiotics were** discovered and are available for treating infections. A vast range of antibiotics were incorporated in different wound dressings to be released slowly. Incorporation of gentamicin in collagen foams (71), ofloxacin in silicone gel sheets (72), minocycline in Chitosan (73), and tetracycline in fibrin were reported (74). Chlorhexidine Gauze Dressing, Fradiomycin Tulle Dressing, framycetin Gauze Dressing, and Nitrofurazone Gauze Dressing are also commercially available.

But w**idespread and inappropriate use of antibiotics has led to the development of resistance to antibiotics by microbes** (75)**.** Selective pressure exerted by antibiotics has led to the development of multidrug resistance among microbes. Hence, numerous classes of antimicrobial agents have become less effective (76,77). Multidrug resistance is now common in common pathogens such as *Escherichia coli, Staphylococcus aureus, Streptococcus pneumoniae, Klebsiella pneumonia,* and *Pseudomonas aeruginosa* (78). Antimicrobial resistance all over the world is very high with gradual loss of first line antimicrobials (79). Consequences of antibiotic resistance increase the health care costs, mortality, and morbidity (80–83). Broad-spectrum β-lactamase producing organisms were first observed in 1983 in *Klebsiella pneumoniae* (84). β-lactamase producing pathogens showed variable susceptibility profile to fluoroquinolones,ami noglycosides, and fourth-generation cephalosporins (85,86). Clofazimine and Dapsone are used against mycobacteria. The mechanism of action of antibiotics and their mode of resistance are represented in Figure 13.3 and summarized in Table 13.2.

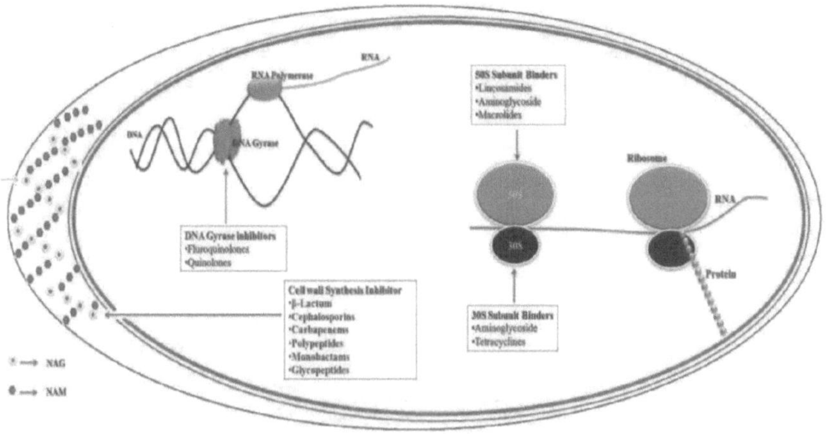

FIGURE 13.3 Mechanism of action of antibiotics

TABLE 13.2 Antibiotics and mechanism of resistance development

S.No	Antibiotic Class	Mechanism of action against microbes	Mode of resistance.
1.	β-Lactum	Disrupt the synthesis of the peptidoglycan layer of bacterial cell walls	Production of β lactamase mutations in genes encoding penicillin-binding proteins (87).
2.	Aminoglycoside	Binds to the bacterial 30S ribosomal subunit (some to the 50S subunit) and inhibit protein synthesis.	Modification by N-Acetyltransferases O-Adenyltransferases O-Phosphotransferases, etc. (88).
3.	Macrolides	Binds to the bacterial 50S subunit and inhibit protein synthesis.	Methylation or mutation that prevents the binding of the antibiotic to its ribosomal target, efflux of the antibiotic, and drug inactivation (89).

TABLE 13.2 *(Continued)*

4.	Sulphonamides	Competitive inhibitors of dihydropteroate synthetase, which catalyzes the conversion of para-aminobenzoate to dihydropteroate, a key step in folate synthesis. Folate is necessary for the cell to synthesize nucleic acids	Mutation leading to expression of dihydropteroate synthases (90).
5.	Cephalosporins	Disrupt the synthesis of the peptidoglycan layer of bacterial cell walls	Extended-spectrum β-lactamases and cephalosporinasesGupta (91).
6.	Carbapenems	Inhibition of cell wall synthesis	Porin mutation, efflux pump, zinc metallo enzymes, and β-lactamase carbapenamase (92).
7.	Fluroquinolones	Inhibit bacterial DNA gyrase or the topoisomerase IV enzyme, thereby inhibiting DNA replication.	Efflux mechanism, mutations in DNA gyrase, and topoisomerases (93).
8.	Tetracyclines	Inhibit the binding of aminoacyl-tRNA to the mRNA-ribosome complex and inhibit protein synthesis.	Tetracycline efflux, ribosome protection, and tetracycline modification (94).
9.	Quinolones	Inhibit the bacterial DNA gyrase or the topoisomerase IV enzyme, thereby inhibiting DNA replication.	Efflux pumps, mutations in DNA gyrase, and topoisomerase (95,96).
10.	Polypeptides	Inhibits isoprenyl pyrophosphate, a molecule that carries the building blocks of the peptidoglycan bacterial cell wall outside of the inner membrane [5]	Cell surface modifications (97).
12.	Monobactams	Disrupt the synthesis of the peptidoglycan layer of bacterial cell walls.	Extended-spectrum β-lactamases (98).

TABLE 13.2 *(Continued)*

13.	Lipopeptide	Bind to the membrane and cause rapid depolarization, and inhibit replication, transcription, and translation.	Synthesis of hydrolytic enzymes (99).
14.	Lincosamides	Bind to 50S subunit of bacterial ribosomes and inhibiting protein synthesis.	Ribosomal modification, efflux of the antibiotic(89).
15.	Glycopeptides	Inhibit peptidoglycan synthesis and cell wall synthesis.	Cell wall biosynthesis reprogramming, altered peptidoglycan precursor (100).

Infectious Disease Society of America (IDSA) on April 2013 has released a report on the weakness of present antibiotics to combat the growing ability of bacteria. Since 2009, only two new antibiotics were approved in United States, and the worsening fact is the decline in the number of new antibiotic discovery and approval. It reported that only seven antibiotics are currently in Phase 2 or Phase 3 clinical trials to treat the Gram-negative bacteria and the details are represented in Table 13.3. But it did not address the entire spectrum of the resistance developed by *E. coli*, *Salmonella*, *Shigella*, and the *Enterobacteriaceae* bacteria (101–103).

TABLE 13.3 Antibiotics in clinical trial

S.No	Antibiotic	Mechanism of action against microbes	Clinical trial
1.	Ceftolozane/tazobactam (CXA-201; CX101/ tazobactam)	Antipseudomonal cephalosporin/β-lactamase inhibitor combination	Phase 3
2.	Ceftazidime/avibactam (ceftazidime/NXL104)	Antipseudomonal cephalosporin/β-lactamase inhibitor combination	Phase 3
3.	Ceftaroline/avibactam (CPT-avibactam; ceftaroline/NXL104)	Anti-MRSA cephalosporin/β-lactamase inhibitor combination	Phase 3
4.	Imipenem/MK-7655	Carbapenem/β-lactamase inhibitor combination	Phase2

TABLE 13.3 *(Continued)*

5.	Plazomicin (ACHN-490)	Protein synthesis inhibitor	Phase 2
6.	Eravacycline (TP-434)	Synthetic tetracycline derivative/protein synthesis inhibitor	Phase 2 completed
7.	Brilacidin (PMX-30063)	Peptide defense protein mimetic (cell membrane disruption).	Phase 2

Bacterial biofilms and antibiotic resistance genes represent a tremendous hurdle in human health care. First, the NIH estimates that 3/4 bacterial infections are biofilm based. Bacteria within a biofilm are 1,000-fold more resistant to antibiotics and are inherently insensitive to the host immune response (104). Second, the dissemination of antibiotic resistance genes among diverse pathogenic bacteria, coupled with the dearth of new antibiotics has led to a situation in which many microbial infections are multidrug resistant and extremely difficult or impossible to treat. Examples of this include the outbreaks of methicillin-resistant *Staphylococcusaureus* (MRSA) infections among healthy individuals (105) and the multidrug-resistant strains of *Acinetobacter baumannii* (MDRAB) (106). Further it narrow downs the therapeutic options available for multidrug-resistant pathogens. Available antibiotics should be saved and spared for combating other infectious diseases.

13.3.4 HONEY

Medihoney™ is an antibacterial Honey Tulle Dressing. It inhibits antibiotic-resistant strains like methicillin-resistant *Staphylococci*, vancomycin-resistant *E.coli,* and *Acinetobacter*. It inhibits biofilm formation (107). It helps to reduce edema, pain and exudate associate with a protracted inflammatory response and inhibiting the bacteria that can cause malodor. It creates the surface wound pH to 6 and this acidic environment helps reduce protease activity, increase fibroblast activity, and increase oxygen release, consequently aiding wound healing (108).

13.4 NANOMATERIALS IN WOUND DRESSING

Nanotechnology spreads it root in all branches of science and engineering by tuning properties at nanoscale. Manipulating materials at nanoscale alters the properties of materials remarkably in all dimensions because of greater surface area to volume ratio. With the discovery of antibiotics, the fashion of using silver as an antimicrobial agent faded slowly. But with the emergence of antimicrobial resistance to existing antibiotics and with the advent of nanotechnology, many antimicrobial agents were reexplored at nanorange. Silver now regains its momentum in a new dimension as silver nanomaterial (109,110) and in wound dressings.

13.4.1 SYNTHESIS OF ANTIMICROBIAL METAL NANOPARTICLES

Metal nanoparticles have been proved to be potent broad antimicrobial agent against both Gram-positive and Gram-negative organisms (98–102, 111–115). Effectiveness of metal nanoparticles against HIV-1 (116,117), hepatitis B virus (118), respiratory syncytial virus (106, 119), herpes simplex virus type 1 (120,121), monkey pox virus (109,122), and influenza virus (123) were also documented.

An array of physical and chemical methods like microwave, UV rays, laser ablation, ultrasonic fields, aerosol technologies, lithography, and photochemical reduction techniques have been used to produce nanomaterials of desired morphology. But they are expensive and involve the use of hazardous chemicals. Therefore, there is an urgent need to produce nanomaterials using green methods. Reducing agents are used to reduce metals into nanoscale. Reduction of metallic ions leads to the formation of atoms. It is later agglomerated into oligomeric clusters. Strong reducing agents produce small monodisperse particles where as weak reducing agents produce large polydisperse particles (124). Metal nanoparticles can also be reduced by biomolecules like glucose, fructose and sucrose which are reducing sugars. An advantage of using biomolecules is that nanoparticles can be preserved without the use of stabilizing or capping agents. Nature has endowed us with rich biodiversity of plants and microorganisms. They can also be used as reducing agents. Synthesis of metal nanoparticles using chemical and biological reducing agents are summarized in Table 13.4.

TABLE 13.4 Synthesis of antimicrobial metal noparticles by reduction

S.No	Nanometal	Reducing agent	Reference
1	Ag	Sodium citrate	125
2	Ag	Sodium borohydride	126
3	Ag, Au, Pd and Cu	Dry ethanol	127
4	Au, Ag and Pd	poly(ethylene glycol)	128
5	Au, Ag, Pt, Pd	Glucose, fructose and sucrose	129
6	Ag	Enzymes	130,131
7	Ag	Plant extracts	132–152
8	Au	Plant extracts	153–156
9	Copper oxide	Streptomyces	157
10	Copper oxide	*E. coli*	158
11	Copper	Ginger extract	159

13.4.2 SILVER NANOMATERIAL IN WOUND DRESSING

Medical significance of silver was well known from ancient time onwards. People of middle age have recognized the antimicrobial significance of silver and they have used silver coins to purify water. Romans and Greeks have used silver for wound healing. Before the discovery of antibiotics silver was used to prevent wound infection during World War I. American traveling pioneers used silver coins to preserve water from microbial contamination (160). Silver is used in clinical practice like eye treatment and skin ulcers from 19th century (164). It was approved as an antibacterial agent by US FDA from 1920 (165). In 1954, silver was registered in the US as a pesticide for use in disinfectants, sanitizers and fungicides. Silver was used in 2007 to make the first antibacterial glass used in hospitals to fight infections (166). Now silver functionalized catheters are in use for short-term catheterization (167). Silver is toxic to bacteria at low concentrations from 10^{-5} to 10^{-7} Ag ions per cell (168). Silver can exist in neutral atom referred to as "elemental silver" or "metallic silver" and as a positively charged atom referred to as "ionic silver" or "silver cation" (Ag+). Silver cations (Ag+/ionic silver) are potent antimicrobials (169,170). More or less all wound product manufacturers produce silver dressings. Silver is used in combination with sulfadiazine as silver sulfadiazin cream for wound healing (171). Availability of silver sulfadiazin cream in 1968 after antibiotic discovery (172) revolutionized the management of infectious burn wounds. Silver nitrate has been used for the management of burn wounds (173–175). Exact mechanism of toxicity of silver to microbes has not yet been elucidated (176). Silver cations are thought to interact with multiple sites within the target cell. Many possible theories have been proposed for the antimicrobial activity of silver. One possible mechanism is that silver ion being positively charged binds with the negatively charged components of cell. For example, it can bind with negatively charged cell wall of bacteria and induce structural abnormalities and cause cell lysis. Likewise, it can also bind to anionic proteins, enzymes, DNA and RNA. It binds with the electron transport chain complex and interferes with energy transduction (177). In contrast to antibiotics that target any one biological process of microbes, silver is multidynamic, hitting multiple targets at a time. An advantageous feature of silver is that it hampers the irreversible adhesion of microbes to a substratum during biofilm formation by compromising intermolecular forces and further destabilizes biofilmmatrix (178,179). Therefore microbes which exhibit resistance to antibiotics will be generally susceptible to silver.

Among the battery of antimicrobial nanomaterials in research for wound dressings, silver finds a unique position due to its potent broad-spectrum antimicrobial activity. It has comparatively low MIC and MBC than its counterparts. Summary of wound dressings with silver nanomaterial is represented in Table 13.5. MIC and MBC values of silver nanomaterials were reported (180–181). Sibbald et al. (182) reported the bactericidal activity of silver nanoparticle embedded wound dressing. Atiyeh et al. (183) reviewed the wound-healing effect of silver nanomaterial in excisional wound model. The review reportedthat silver nanoparticles promoted the proliferation and migration of keratinocyte and they modulate the secretion of various cytokines. In addition to the direct effect on healing, silver nanoparticles decreased the activity of metalloproteinases and accelerated apoptosis. Facilitaion of angiogenesis by silver nanoparticles were also documented (184).

Silver nanomaterial was effective against broad-spectrum microbes and antibiotic-resistant bacteria in vitro (185,186). Sütterlin reported there might be a chance for bacteria to develop resistance to silver. Antimicrobial activity of silver nanomaterials differ in different dressings due difference in the release pattern and the size of the particle (187). Bleeker et al. (188) reported that efficiency of a nanomaterial wound dressing differes markedly due to the difference in the release pattern and kinetics of silver nanomaterial. This is one of the challenge in adopting and implementing regulations also. Liu et al. (189) suggested a different view that ionic silver might be the form that is used to transport nanosilver within the body and and nanosilver is reformed in body compartments again. Yang (190) reported the role of size and its solubility in water. Solubility of the nanomaterials vary from thr bulk as nanomaterials have greater surface energy. In nanomaterials higher proportion of the atoms are at the surface and there is a higher propensity for dissolution and reaction. Ma et al. (191) reported nil lattice strain in silver nanoparticles as small as 6 nm. Then solubility of smaller nanoparticles can be explained solely by the modified Kelvin equation. Xiu et al. (192) ruled out direct particle-specific antibacterial activity of nanosilver, and stated that silver ions are the definitive molecular toxicant. Kermanizadeh et al. (193) and Unrine et al. (194) evaluated the agglomeration behavior of silver nanomaterial and its scattering propery. Different types of stabilizing or capping agents are used to minimize or prevent the formation of aggregates. Their influence on the biological effects should also be evaluated (195). Dawson et al. (196) reported the separation of nanomaterials by sedimentation/centrifugation methods, such as DCS which separates particles in a gradient based on centrifugal force. Review by Bogumiła Reidy et al. (197) provided a critical assessment of the current understanding of silver nanoparticle toxicity.

Many theories were proposed to explain the antimicrobial activity of silver nanomaterial. In contrast to antibiotics which target any one biological process of microbes, silver nanomaterials attck multiple targets and are lethal to the growth of organisms. Electrostatic attraction between the positive charge on the Ag ion and negative charge on the cell membrane of microorganism is crucial for the antimicrobial activity of silver nanomaterial (198–200). In contrast, Sondi and Salopek-Sondi (113) document that the concentration of silver nanoparticle plays a major role in making the cell membrane leaky and in causing death of Gram-negative bacteria. A slightlymodified mechanism was proposed by Amro et al. They state that metal nanoparticle may damage the structural integrity of bacterial cell membrane by the formation of irregularly shaped pits in the outer membrane leading to the progressive release of lipopolysaccharide molecules and membrane proteins (201). Danilczuk et al.

(202) and Kim et al. (203) reported membrane damage through the formation of free radicals induced by silver nanoparticles. Another possible mechanism is that ilver nanoparticles bind to the electron donor groups in biological molecules containing sulfur, oxygen or nitrogen (e.g., enzymes) which, in turn, results in the loss of their function (204).

Large wounds are bridged with surgical mesh and they act as reinforcements to tissue repair. Nocosomial infections are very high in patients with implanted prosthetic materials (205). Application of silver nanomaterial impregnated surgical mesh warranty block and a break in the chain of nosocomial infection. Cohen (206) studied the impact of silver impregnated polypropylene mesh on *Staphylococci* and reported the beneficial effect on control of contamination and infection. Commercially available silver nanomaterial containing wound dressings (207–221) are given in Table 13.5.

TABLE 13.5 Silver nanomaterial wound dressings

S.No	Dressing type		Manufacture	Nano-antimicrobial agent
1	Acticoat*	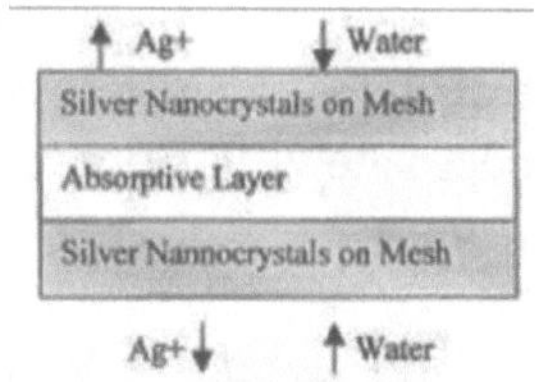	Smith & Nephew, Inc.	Elemental silver. ≤10 nm
2	Acticoat* 7		Smith & Nephew, Inc.	0.25±0.4 mg silver/mg high-density polyethylene
3	Acticoat* Flex 3	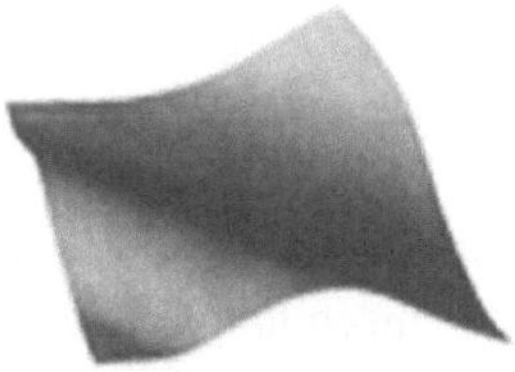	Smith & Nephew, Inc.	0.69 mg Nanocrystalline silver./cm^2

TABLE 13.5 *(Continued)*

S.No	Dressing type		Manufacture	Nano-antimicrobial agent
4	Acticoat* Flex 7		Smith & Nephew, Inc.	Nanocrystalline Silver 0.84 mg/ cm^2
5	Acticoat Surgical Absorbent Pad		Smith & Nephew, Inc.	Nanocrystalline Silver

Antifungal activity of silver nanoparticles, against *Penicillium citrinum* (222), *Aspergillus niger* (223), *Trichophyton mentagrophytes* (223), and *Candida albicans* (224) were also documented. Speshock et al. (225) proved that silver nanoparticles inactivate virus prior to entry into the cell.

13.4.2.1 COATING OF NANOCRYSTALLINE SILVER ONTO POLYETHYLENE MESH OF ACTICOAT

Acticoat®, is the first commercial dressing with silver nanoparticles. It is made up of two layers of polyamide ester membranes covered with nanocrystalline silver ions. Physical vapor deposition is used to coat nanocrystalline silver on to the membrane. Argon gas is introduced into a vacuum chamber that contains a silver cathode, the chamber acting as the anode. When an electric current is passed through the gas, positive argon ions are created which accelerate toward the negatively charged silver cathode. On impact, the argon ions knock out silver atoms that travel toward the substrate to be coated where they deposit and develop into nanocrystals when energy inputs are limited. These are only about 15 nanometers across, which is between 30 and 50 atoms. Thus a nanocrystalline structure is created with a significant grain-boundary component and increased surface area (226). These changes to the physical properties of the crystal lattice result in a metastable, high-energy form of elemental silver. Normal silver placed in water will not dissolve, but nanocrystalline silver dissolves to provide a continuous concentration in solution of around 70 ppm.

13.4.2.2 DELIVERY SYSTEM FOR NANOCRYSTALLINE SILVER

Dr. Burrell formulated the delivery system for nanocrystalline silver. The delivery system of acticoat wound dressing consists of three layers as represented in Figure 13.4.

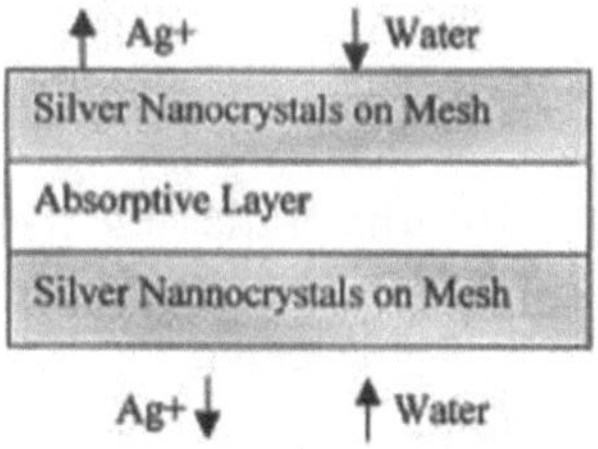

FIGURE 13.4 Silver delivery system of acticoat

An absorbent rayon/polyester core was laminated between an upper and lower layer of silver-coated, high-density polyethylene mesh by ultrasound welds. Concentration of silver was 0.25±0.4 mg silver/mg of polyethylene mesh. It has a unique physical structure, in combination with the oxygen atoms/molecules that are trapped in the crystal lattice, contribute to the enhanced solubility of the films which continue to release silver. Polyester core provides the necessary moisture for sustained release of nanocrystalline silver. The silver coating when wet with sterile water, produces a continual release of Ag+ and likely other silver radicals for days. Nanocrystalline silver system was highly effective against drug-resistant pathogens such as vancomycin-resistant enterococcus (VRE) and MRSA and fungi.

Acticoat™ 7 with silcryst nanocrystals: Acticoat™ 7 with silcryst nanocrystals is same as acticoat but it is made of five layers with two layers of an absorbent inner core sandwiched between three layers of silver-coated, low-adherent polyethylene net. It can release nanocrystalline silver in a sustained manner for 7 days.

13.4.3 COPPER NANOPARTICLES IN WOUND DRESSING

Copper and its alloys like brass, bronz, cupronickel and copper-nickel-zinc are natural antimicrobial materials. Antimicrobial properties of copper was exploited before 19th century. Copper is known to have activity against bacteria and fungi. The activity of Cu against Gram-positive cocci such as MRSA and Gram-negative bacilli causing food-associated disease, such as *Escherichia coli* O157, *Campylobacter jejuni* and *Salmonella* spp. Copper and its alloys are effective against extremely drug-resistant tubercle bacilli, nosocomial pathogens and yeast (227–229). Antibacterial activity of copper is a multifaceted attack. It drills hole in the outer membrane and makes the cell leaky causing death. Being charged species, it alters the transmembrane potential of bacteria as it binds to the cell membrane. This weakens the membrane and creates holes. Oxidative damage of copper in aerobic condition also drills holes. After membrane perforation, copper inhibit many enzymes and disturb cellular events like digestion, transport, signal transduction, cell division and energetic of a cell.

Copper nanomaterial is also a potent antimicrobial agent at a cheap cost. Green synthesis of copper nanomaterial by reduction with bacteria was reported (158). Textile fabrics can be functionalized with copper nanomaterials to prevent or to minimize infection with pathogenic bacteria in hospital environment (230). Incorporation of Copper nanoparticles into a polymeric nanofibrous structure via electrospinning technique was also reporetd (231). Mechanism of antimicrobial activity of copper nanomaterial is similar to copper but the nanosize remarkably increases the efficiency. Copper not only has an antimicrobial activity but also plays an important role in wound healing. Copper stimulates angiogenesis. It induces vascular endothelial growth factor expression (232). At a concentration of 1 μM it caused the stabilization of fibronectin which functions as a matrix on which new cells can be built in.Many enzymes working in wound healing are copper dependent including amine oxidases (233). ECM is composed of proteins like collagen and elastin. Collagen contains hydroxyl proline and hydroxyl lysine. Enzymes used in cross linking of lysine are copper dependent. Thus presence of copper facilitates remodeling phase of wound healing (234). Expression and distribution of alpha2beta1, alpha3beta1, alpha6beta4, and alphaVbeta5 integrins by basal layer keratinocytes were modified to enhance re-epithelialization phase.

13.4.4 ZINC NANOMATERIAL IN WOUND DRESSING

Zinc plays an important role in wound healing not only as an antibacterial agent but also as an and anti-inflammatory agent (235,236). It can be used in treatment of various dermatitis, diaper rashes, diaper wipes, blisters, and open-skin sores (237–239). Bactericidal effects of zinc against *Streptococci* and *Staphylococci* and in dentistry were demonstrated (240,241). In Södeberg et al. (242–244), it was found that Gram-positive bacteria were the most susceptible bacterial group to zinc ion and Gram-negative aerobic bacteria were usually not inhibited showed that nanoformulation of zinc is most potent than bulk zinc. The minimum inhibitory concentration for ZnO nanoparticles with smaller size was 0.08 mg mL-1 for *Staphylococcus aureus*, whereas that for larger particles of ZnO was 1.2 mg mL^{-1}. Vlad et al. (245) reported antifungal activity of a membrane ZnO-PU (polyurethane) at concentration of 5% and 10% ZnO up to 60 days. Such modified ZnO membranes present important antifungal properties and can be successfully used in food or biomedical applications, in order to block the formation of biofilms. Antimicrobial activity of zinc oxide nanomaterial coated with gentamycine as a wound dressing was evaluated against *Escherichia coli, Pseudomonas aeruginosa, Staphylococcus aureus, Bacillus cereus* and *Listeria monocytogenes*.

13.4.5 TITANIUM NANOMATERIAL IN WOUND DRESSING

Titanium dioxide is a photocatalyst under UV light. The photocatalytic properties of titanium dioxide were discovered by Akira Fujishima in 1967 and published in 1972. The process on the surface of the titanium dioxide was called the Honda-Fujishima effect (246,247). Titanium oxide absorbs light having an energy level higher than that of the band gap, and causes electrons to jump to the conduction band to create positive holes in the valence band. Positive holes have a strong oxidative decomposing power. When the water is oxidized by positive holes, hydroxy radicals (•OH), which have strong oxidative decomposing power, are formed. Then, the hydroxy radicals destroy microbes by oxidative decomposition.ORNL researchers have developed an improved bacterial cellulose wound dressings already marketed in the trade name Biofill. "Nanostructured" surface of cellulose act as a scaffold for tissue regeneration. It was provided with an antiseptic property by the addition of titanium dioxide nanoparticles into the cellulose wound dressing. When the dressing is illuminated with long wavelength UV light, the excitation of the nanoparticles causes the release of antimicrobial oxygen radicals, enhancing the therapeutic effect of the dressing.

13.4.6 IRON

The investigations of Berry et al. (248) showed iron suppressed the growth of oral bacteria. Donde et al. (249) established enhanced antibacterial and antifungal activity of Fe (II) and Hg (II) complexes of hydrazone as compared to parent ligand. Antibacterial effect of iron and different iron compounds were well documented (250–254).

13.4.7 CHITOSAN IN WOUND DRESSING

Chitosan is a naturally available abundant biopolymer. It is widely used as a vehicle for drug delivery. It's a potent antimicrobial agent. It forms a tough, water-absorbent, biocompatible film and this film can be also formed directly on a burn by application of an aqueous solution of chitosan acetate. Attention has long been paid to evaluate the application of chitosan for treatment of burns (255) open wounds, dermatitis and in ophthalmology (256). Electrostatic interaction between cationic amino-group of chitosan with anionic components of bacterial cell wall component N-acetylmuramic acid, sialic acid and neuraminic acid destroys the cell wall and inhibit all cellular events leading to the death of organisms. Complete inhibiton of the growth of *Candida tropicalis and Candida albicans* was demonstrated by Allan and Hadwiger (257).

Chitosan in nanofibrous form offers a grater surface area due to their extreme small size. This suits very well for medical textile products such as surgical facemasks, wound dressings, and drug delivery systems (258). They exhibit functional versatility, including desirable wound adherence, absorption, oxygen permeability, resorbability, and occlusivity (259–261). Electrospinning of chitosan into nanofibrous structure was well studied (Hasegawa, geng).It can also be casted as films (262,263). Electrospinning of chitosan by core/sheath technique was evaluated as chitosan is a polyelectrolye. Electrospinning of chitosan with polyethylene oxide shaeath and chitosan core was studied by Gorga and Ojha (264). Triton X-100TM (265) and trifluro acetic acid were also used as solvent for electrospinning of chitosan (266,267). 1,1,1,3,3,3-hexafluoro-2-propanol (HFIP) (268), trifluoroacetic acid (TFA) (269) and acetic acid were also used as solvents. But the toxicity of the solvents was not completely eliminated and there is a possibility of their remnance in the nanofiber also. So to reduce the toxicity of solvents, water-soluble Chiotosan were used to prepare nanofibers such as carboxyethyl chitosan/polyvinyl alcohol (PVA) (270) and quaternary chitosan (271) for wound dressing applications. Chitosan can be processed to form nanofibers without the use of organic solvents or toxic acids such as CS–hydroxybenzotriazole (hobt)/PVA (272) and CS–ethylenediaminetetraacetic acid (EDTA)/PVA (273).

Nanofibrous mats can be fabricated by electrospinning. This nanomesh mimics a natural ECM, which acts as scaffold that permits cell adhesion, proliferation and differentiation (274–276). As mentioned earlier, chitosan alters the bacterial cell membrane permeability, causing the leakage of intracellular constituents such as electrolytes, UV-absorbing material, proteins, amino acids, glucose, and lactate dehydrogenase leading to the death of cells. It is well known that chitosan films can be used for treating burns and wound healing (277–279). Chitosan will gradually depolymerize into N-acetyl-d-glucosamine, which initiates fibroblast proliferation, assists in ordered collagen deposition and stimulates increased levels of natural hyaluronic acid synthesis at the wound site (280,281).

Electrospinning was also used to fabricate nanofibrous PVA scaffolds for wound healing **(282–284).** One of the significant disadvantages of electrospinning method is the lower productivity (285). Electrospun polyurethane–dextran nanofiber mats containing ciprofloxacin was also studied by Afeesh et al. (286). Prabaharan and Jayakumar (287) studied the status of electrospun materials for drug delivery. Abdelgawad et al. (288) has developed a novel chitosan/PVA nanofiber mats loaded with silver nanoparticles by electrospinning for antimicrobial applications. Li1 et al. (289) investigated the outcome of preclinical study on Silver nanoparticle/chitosan oligosaccharide/PVA nanofiber wound dressings. Maleki et al. (290) fabricated a drug-carrying nanofibrous web wound dressing exploiting the unique properties of honey. Nanofiber meshes were produced from aqueous mixtures of PVA and honey via electrospinning. Leung et al. (291) focused their attention to overcome the inherent disadvantage of alginate scaffold by employing glutaraldehyde double cross-linking to an electrospun alginate nanofiber nonwoven mat. Bidgoli et al. (292) evaluated the release of silver nanomaterial from wound dressings and its systemic toxicity. Their study reported elevated levels of serum alanine transaminases.

13.4.7.1 CHITOSAN NANOGAUZE

Chitosan is a most suitable biopolymer for the fabrication of wound dressing materials, due to its inherent biocompatibility, biodegradability, antimicrobial activity, wound healing property, and antitumor effect (294). Chitosan is widely used for drug delivery and bone healing (295–297). Wound dressing films made of a mixture of chitosan and minocycline hydrochloride was formulated for the treatment of severe burns by Aoyagi et al. (298). Chitosan membrane possesses an ideal characteristic of a transparent film suitable for monitoring the progress of wound healing (299). Nanofiber webs form highly effective filters for the contaminants, particulates, and microorganisms without sacrificing air and moisture permeability. In addition to electrospinning a wide variety of techniques like drawing, template synthesis, phase separation and self-assembly are available to fabricate nanofibrous architecture of chitosan (300). But electrospinning offers many advantages over other methods for wound dressing material by providing very large surface area to volume ratio and flexibility in surface. Nanomembranes of chitosan conform well to the ideal characteristics of a wound dressing like exudates absorption, and oxygen permeability. They mimic the function of a skin by functioning as a chemical and mechanical barrier to infectious organisms (301) and other environmental pollutants. They conform to the wound bed, provide appropriate adherence/nonadherence to the healing tissue (301), Only few polymers were used to prepare electrospun nanofiber for wound healing such as PVA, poly(l-lactic acid) (PLA), polycaprolactone (PCL), gelatin and chitosan (CS) (302).

Nanofiber coatings can be done onto conventional textile bandages for structural support. Till now no study has completely characterized pure chitosan nanofibers in wound dressings. Deposition of thin film or lamination of secondary cotton wound dressing with chitosan nanofibers as a primary dressing layer by plasma treatment might influence the adhesion property (303,304), mechanical property of wound dressing (305–308), hydrophobicity (309) and crystallinity (310,311). Vitchuli et al. (310) employed atmospheric plasma treatment to improve the adhesion between nylon nanofibers and nylon/cotton fabric substrates. Lysozyme-loaded chitosan–Ethylene diamine tetra acetic acid/PVA nanofibers enhanced the healing of wounds (311,312), solubility in water and synergistic antibacterial effects with EDTA (313). Lysozymes break the cell wall architecture of bacteria and destabilize their structural integrity causing death.

13.5 CHALLENGES AND NEW APPROACHES IN NANOMATERIAL WOUND DRESSINGS

Fabrication of wound dressings evolved from basic cotton dressings to advanced nanomaterial and skin substitute dressings at a faster pace. Modern technology makes wound dressings to provide an optimum environment for enhanced epithilialization in an unhampered way. Effectiveness of a wound dressing is determined not only by the dressing but also by the nature and type of the wound being treated, pathophysiological conditions of the patient and the stage of healing. Greatest breakthrough in the history of wound care is the incorporation of silver nanomaterial incorporated wound dressings. An added advantage of silver nanomaterial is its dual function as an antimicrobial and an anti-inflammatory agent. Different dressings employ different strategies for the delivery of silver (314,315). Silver nanomaterial should be released in sufficient concentrations to destroy bacterial and fungal cells without damage to the host cell. Generally dressings are fabricated with high concentration of silver nanomaterial but designed to release it in a sustained manner. In wound management, silver quantities should be sufficient to provide sustained bactericidal action (316). Challenges encountered in the use of silver nanomaterial dressing are the number of highly reactive soluble silver ions released and not the concentration of silver in the wound dressing. Release pattern from different dressings differ. Silver is released from silver carbomethylcellulose, silver calcium phosphate and silver chloride at very low concentration (317). Ag^0 is the metallic or uncharged form of silver found in crystalline, silver structures. Ag^0 is not readily deactivated by biological systems like ionic forms (316). Once it is depleted from the wound bed, it is released from the source again and this repeats the continued release till the source is depleted completely or the dressing is removed.

From clinical point of view it is meaningless to release silver or silver nanomaterial at low concentrations in a controlled fashion. This will again facilitate the development of resistance among microbes against silver similar to antibiotics. Moreover, their penetration rate into the wound bed, their inactivation and clearance by the host cells also decide the efficacy of the dressing (318). Clinically resistance to silver was reported in contrast to the literature available. Similar to the operation of efflux pump responsible for antibiotic resistance, silver can also be bound as a intracellular complex and excreted by efflux pump (316). Use of sublethal dose of silver could induce resistance and bactericidal concentrations don't induce resistance as it will completely destroy all pathogens and hence resistance development is dose dependent (319). Therefore release profile of silver must be at bactericidal concentration.

Another major concern of silver nanomaterials is their cytotoxicity to fibroblasts. Reports have documented decreased function of mitochondria exposed to silver nanoparticles at 5–50 mg/ml. Cytotoxicity of silver nanomaterial to liver is mediated by significant depletion of glutathione (GSH), reduced mitochondrial membrane potential, and increased reactive oxygen species production (320). Nanocrystalline silver released from Acticoat was reported to inhibit keratinocyte growth and it was suggested not to use Acticoat as a topical dressing on cultured skin grafts (321). It was also found to delay re-epithelialization (322). Transient discoloration of the skin was observed by the use of Acticoat. Ingestion of colloidal silver was reported to cause brain toxicity (323).

13.5.1 CYTOTOXICITY OF NANOMATERIALS

Skin is one of the potentialroutes of exposure to nonmaterial for both producers and consumers. Earlier it was assumed that skin is a stringent mechanical barrier and injury to the skin is essential for the penetration of nanomaterials (324,325). Now study by Ryman-Rasmussen et al. (326) demonstrated that quantum dots can penetrate intact skin and become localized within the skin layers by 8 hours. Recent investigations report the permeability of skin to nanomaterials and their harmful impact on cell viability (327,328). Direct entry of nanomaterials through skin by the application of sun screens containing nanomaterial is harmful to the cells.

Use of silver nanomaterial containing wound dressings directly expose the damaged skin to nanomaterials and has risk implications. Toxicity of nanomaterial is a constantly debatable issue because of the complexity in assessing the toxicity and the difference in the nature of nanoparticle, their detection and penetration analysis. In addition, toxicity assessment is further complicated by the method of sample preparation; its mode of formulation as a powder, liquid, colloid, suspension, aerosol, etc.; and their stability, ability to aggregate, and solubility in biological fluids. The size of nanomaterial influences the magnitude of toxicity. Study by Pan et al. focused on the influence of size of nanomaterial on toxicity of connective tissue fibroblasts and melanoma cells by using 0.8–15 nm gold nanoparticles (329). The results concluded that smaller the size of the nanomaterial, greater will be the toxicity as the size of the material dectermined the mechanism of toxicity. For example, it was reported that 1.2 nm gold particles caused cell death by apoptosis, whilst the 1.4 nm particle caused cell death by necrosis (329). Thus it was proved that even a smallest change in the size of the nanomaterial will have remarkable difference in the extent of toxicity by altering the particle behavior. Skin is amphoteric in nature and its isoelectric point ranges from 3 to 4. At physiological pH (7.4), skin has therefore negatively charge. (330). Skin as a selectively permeable barrier permits cations at this pH, Marro et al. (331). Results of Monteiro-Riviere et al. (328) demonstrated the permeability of epidermal keratinocytes to multiwalled carbon nanotubes and its ability to induce the release of IL-8 release.

In vitro methods to study the penetration of compounds into dermis have been published by the OECD in the form of OECD TG 428, Skin Absorption: In vitro Method (332). Advances in tissue engineering have developed the many reconstituted three-dimensional human skin models such as EpiDerm® and EpiSkin® (333). These can be used to assess the toxicity of nanomaterials to human system. Explant skin tissues grown in cell culture flasks can also be used to study dermal absorption of nanomaterials. Cohen et al. (334) investigated the toxicity of copper nanomaterials using cultured human skin tissue. Isolated perfused porcine skin flap (IPPSF) can also be used as an in vitro model for toxicity assessment (335). Use of single cell culture monolayer of keratinocytes is also entertained in the study for evaluationg the penetration and absorption of nanomaterials (336–338). Volatilability, dose and , time of exposure to nanomaterial, integrity and surface area of the experimental skin models influence the outcome of the study and they should be controlled properly (339).

In vivo models for dermal absorption study are published by OECD as OECD TG 427 (340). The guidelines suggests that dry material or liquid preparation can be tested on the skin of animals at a concentration of 1–5 mg/cm^2 for a solid or up to 10 μL/cm^2 for liquids. Toxicity can be assessed by using a finite dose or an infinite dose of a nanomaterial. Infinite dose is preferred if the investigation is aimed to estimate the maximal flux which is essential to determine the permeability coefficients. According to Labouta et al. 47% of the dermal penetration studies were performed on rat, mouse and pig (341). Even though studies were conducted in animals, entire results could not be predicted to apply to human system due to difference in dermal layer thickness, lipid composition and hence the difference in penetration and absorption of compounds which differ markedly (342). It is generally considered that the dermal permeability between species can be ranked as follows in descending order from the more permeable to the less permeable: Rabbit > Rat > Pig > Monkey > Human (343). Depending on the compound studied, these differences in terms of skin permeability could range from more than 4 times for the pig and up to 9 times more permeable for the rat compared to human skin (343). Other animal models like human skin grafts onto immune-deficient mice (SCID mice) were also used to study the influence of nanomaterial (344). Several nanomaterials have already been tested on humans such as TiO_2 (344) and silver nanoparticles (345).

13.6 CONCLUSION

Nanomaterials find their wide application in daily life of technologically advanced era. Nanomaterials offer a great potential because of their size effect and greater surface area. Reasearch has been focused to fabricate wound dressings with nanomaterials not only to battle against wound colonizers but also to enhance tissue regeneration and remodeling. At present only few nanomaterial wound dressings are commercially available including acticoat and silverlon. Synthesis of nanomaterials by green methods are economical and avoid the use of toxic solvents and downstream processing to remove them. Enormous studies were undertaken to simplify the synthesis of silver nanomaterials using easily available edible and medicinal herbs. Among all nanomaterials, silver ranks first in the race for the application of wound dressings because of its wide spectrum activity against pathogens at a smaller concentration than its counterparts. Chitosan nanofibrous mats are also in the final stage to step into common use. Exixting cotton gauze dressings can also be coated with silver nanomaterials or can be laminated with a chitosan film. But the challenge with chitosan lamination is its limited adherence. Nanomaterial wound dressings other than silver are in pipeline and they will find their own place after cytotoxity evaluation in animal models. Application of silver or any other nanomaterials in wound dressing is unique in many ways that it directly exposes the wounded skin to nanomaterials which is otherwise a selective barrier against foreign bodies. In spite of enormous research on the use of silver nanomaterial in wound dressing, several key questions need to be answered more elaborately with technical data.

The most important one is the quantification of nanomaterials. Generally the present techniques estimate the concentration of silver only and not exclusively silver nanomaterial alone. It includes all molecular forms of silver. In vitro kinetics of nanomaterial release might be different from in vivo systems due to the interaction with biological fluids. The fate of silver ions inside a living system is different from nanomaterial which is a particulate form. Extent of inactivation of silver nanomaterials by complex formation with biomolecules inside the living system should also be taken into account while calculating the overall effectiveness of the nanomaterial. This will be reflected in the toxicological assessment which is based on concentration of the experimental analyte. During evaluation of experimental results the fate of nanomaterial still completely eliminated from the body by the excretory system shold be considered. Mostly studies in in vivo systems ignore the blood profile during the exposure to nanomaterials. Only a few studies reported elevations in liver function tests. Kidney function tests should also be monitored during the entire experimental duration for assessing the potency of nanomaterials. Above all the possibility of bacteria developing resistance against silver should be studied elaborately as silver nanomaterial is widely used in daily walk of life. Even though nanomaterial wound dressings are available, they are not at an affordable price for people of all economic classes. Technology should be developed to fabricate advanced wound dressings taking into consideration the cost effectiveness also.

KEYWORDS

- **Antimicrobial agent**
- **Chitosan**
- **Nanocrystalline silver**
- **Nanomaterials**
- **Wound dressings**

REFERENCES

1. Ro, B.I.; Dawson, T.L. The role of sebaceous gland activity and scalp micro-oral metabolism in the etiology of seborrheic dermatitis and dandru€. J. Investigat. Dermatol. SP. *2005*, 10(3), 194±7.
2. Gawkrodger, D.J. Dermatology, An Illustrated Colour Text. 3rd ed. Edinburgh: Churchill Livingstone. 2002.
3. Proksch, E.; Brandner, J.M.; Jensen, J.M. The skin: an indispensable barrier. Exper. Dermatol. *2008*, 17(12), 1063–1072.
4. Stücker, M.A.; Struk, P.; Altmeyer, M.; Herde, H.; Baumgärtl; Lübbers, D.W. The cutaneous uptake of atmospheric oxygen contributes significantly to the oxygen supply of human dermis and epidermis. J. Physiol. *2002,* 538(3), 985–994.
5. Alibardi, L. Adaptation to the land: The skin of reptiles in comparison to that of amphibians and endotherm amniotes. J. Exp. Zoolog. B. Mol. Dev. Evol. *2003*, 298(1), 12–41.
6. Madison, K.C. Barrier function of the skin: "la raison d'être" of the epidermis. J. Invest. Dermatol. *2003*, 121(2), 231–241.
7. Proksch, E.; Brandner, J.M.; Jensen, J.M. The skin: An indispensable barrier. Exp. Dermatol. *2008,* 17(12),1063–1072.
8. Tortora,G.J.; Derrickson,B. Principles of Anatomy and Physiology. Published by John Wiley and Son Inc. 11th edition, 2006.
9. Stvrtinová, V.; Ján, J.; Ivan, H. Inflammation and Fever *from Pathophysiology: Principles of Disease*. Computing Centre, Slovak Academy of Sciences: Academic Electronic Press. 1995.
10. Janeway, C.; Paul, T.; Mark, W.; Mark, S. *Immunobiology; Fifth Edition.* New York and London: Garland Science. 2001. ISBN 0-8153-4101-6.
11. http://www.woundcarecenters.org/wound-basics/different-types-of-wounds.html
12. http://www.clinimed.co.uk/Wound-Care/Education/Wound-Essentials/Wound Classification.aspx
13. Brooks, D.; Hunt, B. Current concepts in concussion diagnosis and management in sports: A clinical review. BC Med. J. *2006*, 48(9), 453–459.
14. Pearce, J.M. Observations on concussion. A review. Europ. Neurol. 2007, 59(3–4), 113–119.

15. http://medical-dictionary.thefreedictionary.com/luxation
16. http://en.wikipedia.org/wiki/Wound
17. Flanagan, M. Assessment criteria. Nursing Times. *1994*, 90(35), 76–86.
18. http://nursing.uchc.edu/unit_manuals/perioperative/or/docs/Surgical%20Wound%20 Classification.pdf
19. Percival, N.J. Classification of wounds and their management. Surgery. *2002*, 20(5), 114–117.
20. Moore, K.; McCallion, R.; Searle, R.J.; Stacey, M.C.; Harding, K.G. Prediction and monitoring the therapeutic response of chronic dermal wounds. Int. Wound. J. *2006*, 3(2), 89–96.
21. Harding, K.G.;Morris, H.L.;Patel, G.K. Science, medicine and the future: healing chronic wounds. BMJ. *2002,* 19,324(7330), 160-163.
22. Ferreira, M.C.; Tuma; Carvalho, V.F.; Kamamoto. Review complex wounds. Clinics. *2006,* 61(6), 571–578.
23. Musset, J.H.; Winfield, A.J. Wound Management, Stoma, and Incontinence Products, (Eds. A. J. Winfield R. M. E Richards, Churchill Livingstone, London, 1998, pp. 176–187.
24. Strodtbeck, F. Physiology of wound healing. Newborn Infant Nursing Rev. *2001*, 1, 43–52.(Initial)
25. Boateng, J.S.; Matthews, K.H.; Stevens, H.N.; Eccleston, G.M. Wound healing dressings and drug delivery systems: A review. J. Pharm. Sci. *2008*, 97(8), 2892–2923.
26. Martin, P. Wound healing: aiming for perfect skin regeneration. Science. *1997*, 276, 75–81.
27. Ratner, B.D.; Hoffman, A.S.; Schoen, F.J.; Lemons, J.E. Biomaterials Science: An Introduction to Materials in Medicine. Academic Press, California, 1996.
28. Thomas, S. Wound Management and Dressing. Pharmaceutical Press, London, 1990.
29. Szycher, M.; Lee, S.J. Modern wound dressings: a systematic approach to wound healing. J. Biomater. Appl. *1992*, 7, 142–213.
30. Inngjerdingen, K.; Nergård, C.S, Diallo, D.; Mounkoro, P.P.; Paulsen, B.S. An ethnopharmacological survey of plants used for wound healing in Dogonland, Mali, West Africa. J. Ethnopharmacol. *2004*, 92(2–3), 233–244.
31. Mensah, A.Y.; Houghton, P.J.; Dickson, R.A.; Fleischer, T.C.; Heinrich, M.; Bremner, P. *In vitro* evaluation of effects of two ghanaian plants relevant to wound healing. *Phytothe. Res. 2006*, 20(11), 941–944.
32. George G. Wound management. Richmond: PJB Publications, 1996.
33. Gamgee, J.S. Absorbent and medicated surgical dressings. Lancet. *1880*, 1, 127–128.
34. Winter, G.D. Formation of the scab and the rate of epithelialisation of superficial wounds in the skin of the young domestic pig. Nature. *1962,* 193(4812), 293–294.
35. Hinman, C.D.; Maibach, H. Effect of air exposure and occlusion on experimental human skin wounds. Nature. *1963,* 200(4904), 377–378.
36. Kumar, P.T.; Lakshmanan, V.K.; Biswas, R.; Nair, S.V.; Jayakumar, R. Synthesis and biological evaluation of chitin hydrogel/nano ZnO composite bandage as antibacterial wound dressing. J. Biomed. Nanotechnol. *2012*, 8(6), 891–900.
37. van Rijswijk, L. Ingredient-based wound dressing classification: A paradigm that is passe and in need of replacement. J. Wound Care. *2006*, 15(1), 11–14.

38. Thomas, S. Wound and Wound healing in: Wound Management and Dressings. Pharmaceutical Press, London, 2004.
39. Queen, D.; Orsted, H.; Sanada, H.; Sussman, G. A dressing history. *Int. Wound J. 2004*, 1, 59–77.
40. Falabella, A.F. Debridement and wound bed preparation. Dermatol. Ther. *2006*, 19(6), 317–325.
41. Ovington, L.G. Advances in wound dressings. Clin. Dermatol. *2007*, 25(1), 33–38.
42. Wichterle, O.; Lim, D. Hydrophilic gels for biological use. Nature. *1960*, 185, 117–118.
43. Wheeler, J.C, Woods, J.A.; Cox, M.J., Cantrell, R.W.; Watkins, F.H.; Edlich, R.F. Evolution of hydrogel polymers as contact lenses, surface coatings, dressings anddrug delivery systems. J. Long Term Eff. Med. Implants. *1996*, 6, 207–217.
44. **Pudner, R.** *Amorphous hydrogel dressings in wound management. J. Comm. Nursing. 2001, 15(6), 43–46.*
45. **Thomas, S.; Leigh, I.M.** *Wound dressings. In: Leaper, D.J., Harding, K.G. (eds) Wounds: Biology and management. Oxford: Oxford University Press.1998.*
46. **Tong, A.** *Theidentification and treatment of slough. J. Wound Care. 1999, 8(7), 338–339.*
47. **Jones, V.; Milton, T.** *When and how to use hydrogels. Nurs. Times. 2000, 96(23), 2–4.*
48. **Trudigan, J.** *Investigating the use of aquaform hydrogel in wound management. B. J. Nurs. 2000, 9(14), 943–948.*
49. **Thomas, S.; Banks, V.; Fear, M. et al.** *A study to compare two film dressings used as secondary dressings. J. Wound Care. 1997, 6(7), 333–336.*
50. Rahman, M.H.; Anson, J. Peri-operative anti-bacterial prophylaxis. Pharma. J. *2004*, 272, 743–745.
51. Konturek, P.C.; Bzozowski, T.; Konturek, S.J.; Kwiecien, S.; Dembinski, A.; Hahn, E.G. Influence of bacterial lipopolysaccharide on healing of chronic experimental ulcer in rat. Scand. J. Gastroenterol. *2001*, 36, 1239–1247.
52. Power, C.; Wang, J.H.; Sookai, S.; Street, J.T.; Redmond, H.P. Bacterial wall products induce down regulation of vascular endothelial growth factor receptors on endothelial cells via a CD 14-dependent mechanism: implications for surgical wound healing. J. Surg. Res. *2001*, 101, 138–145.
53. Dow, G.; Browne, A.; Sibbald, R.G. Infection in chronic wounds: controversies in diagnosis and treatment. Ostomy. Wound Manage. *1999*, 45, 23–40.
54. Ceri, H.; Olson, M.E.; Stremick, C.; Read, R.R.; Morck, D.; Buret, A. The Calgary biofilm device: New technology for rapid determination of antibiotic susceptibilities of bacterial biofilms. J. Clin. Microbiol. *1996*, 37, 1771–1776.
55. Rhoads, D.D.; Wolcott, R.D.; Percival. S.L. Biofilms in wounds: Management strategies. J. Wound Care. *2008*, 17, 502–508.
56. Rogers,S.A.; Huigens, R.W.; Cavanagh, J.; Melander, C. Synergistic effects between conventional antibiotics and 2-aminoimidazole-derived antibiofilm agents. J. Antimicrob. Chemother. *2010,* 2112–2118
57. Perez, F.; Hujer, A.M.; Hujer, K.M.; Decker, B.K.; Rather, P.N.; Bonomo, R.A. Global challenge of multidrug-resistant *Acinetobacter baumannii*. Antimicrob. Agents. Chemother. *2007*, 51, 3471–3484.

58. Fux, C.A.; Costerton, J.W.; Stewart, P.S.; Stoodley, P. Survival strategies of infectious biofilms. Trends Microbiol. *2005*, 13, 34–40.
59. Musk, D.J.; Hergenrother, P.J. Chemical countermeasures for the control of bacterial biofilms: Effective compounds and promising targets. Curr. Med. Chem. *2006*, 13, 2163–2177.
60. Keren, I.; Kaldalu, N.; Spoering, A.; Wang, Y.; Lewis. K. Persister cells and tolerance to antimicrobials. FEMS Microbiol. Lett. *2004*, 230, 13–18.
61. Calderon, C.B.; Sabundayo, B.P. Antimicrobial Classifications: Drugs for Bugs. In Schwalbe R, Steele-Moore L, Goodwin AC. Antimicrobial Susceptibility Testing Protocols. CRC Press. Taylor & Frances group. ISBN 978-0-8247-4100-6, 2007.
62. Foster, W.; Raoult, A. Early descriptions of antibiosis. J. R. Coll. Gen. Pract. *1974*, 24 (149), 889–894.
63. Kingston, W. Irish contributions to the origins of antibiotics. Int. J. Med. Sci. *2008*, 177(2), 87–92.
64. Landsberg, H. Prelude to the discovery of penicillin. Isis. *1949*, 40(3), 225–227.
65. http://www.almanacco.rm.cnr.it/reader/cw_usr_view_recensione?id_articolo=1704&giornale=1679
66. Salvatore De Rosa, Introduttore: Fabio Pagan. "Vincenzo Tiberio, vero scopritore degli antibiotici – Festival della Scienza" (in (Italian)). Festival2011.festivalscienza.it. Retrieved October 19, 2012.
67. Fleming, A. Classics in infectious diseases: on the antibacterial action of cultures of a penicillium, with special reference to their use in the isolation of *B. influenzae* by Alexander Fleming, Reprinted from the B. J. Experiment. Pathol. Rev. Infect. Dis. *1980*, 2(1), 129–139.
68. Waksman, S.A. What is an antibiotic or an antibiotic substance? Mycologia. *1947*, 39(5), 565–569.
69. von Nussbaum, F.; Brands, M.; Hinzen, B.; Weigand, S.; Häbich, D.. Medicinal chemistry of antibacterial natural products – Exodus or revival? Angew. Chem. Int. Ed. *2006*, 45(31), 5072–5129.
70. Pelczar, M.J.; Chan, E.C.S.; Krieg, N.R. Host-Parasite Interaction; Nonspecific Host Resistance, In: Microbiology Concepts and Applications, 6th ed., McGraw-Hill Inc., New York, U.S.A. pp. 478–479, 1999.
71. Rutten, J.T.; Nijhuis, P.H.A. Prevention of wound infection in colorectal surgery by local application of gentamycine collagen sponge. Eur. J. Surg. *1997*, 163, 31–35.
72. Sawada, Y.; Tadashi, O.; Masazumi, K.; Kazunobu, S.; Koichi, O.; Sasaki, J. An evaluation of a new lactic acid polymer drug delivery system: A preliminary report. Br. J. Plast. Surg. *1994*, 47, 158–161.
73. Ganlanduik, S.; Wrigtson, W.R.; Young, S.; Myers, S.; Polk, H.C. Absorbable delayed antibiotic beads reduce surgical wounds. Am. J. Surg. *1997*, 63, 831–835.
74. Kumar, T.R.S.; Bai, M.V.; Krishnan, L.K. Freeze dried fibrin discs as biodegradable drug release matrix. Biologicals. *2004*, 32, 49–55.
75. McGowan, J.E. Resistance in non-fermenting gram-negative bacteria: multidrug resistance to the maximum. Am. J. Infect. Control. *2006*, 34, 29–37.
76. Oskay, M.; Oskay, D.; Kalyoneu, F. Activity of some plant extracts against multidrug resistant human pathogens, Iran. J. Pharm. Res.*2009*, 8(4), 293–300.

77. Pearson, C. Antibiotic resistance fast-growing problem worldwide. Voice of America. Archived from the original on 2 December 2008. Retrieved 2008,12-29.
78. Nabeela, N.; Munazza, A.; Shekh.; A.R.; Zaid, A.P. Urinary tract infections associated with multidrug resistant enteric bacilli: Characterization and genetical studies. J. Pak. Pharma. Sci. *2004,* 17(2), 115–123.
79. World Health Organization. Antimicrobial resistance in the European Union and the World. Lecture delivered by Dr. Margaret Chan, Director-General of WHO at the conference on Combating antimicrobial resistance: Time for action. Copenhagen, Denmark, March 14th, 2012.
80. Webb, G.F.; D'Agata, E.M.; Magal, P.; Ruan, S. A model of antibiotic-resistant bacterial epidemics in hospitals. Proc. Natl. Acad. Sci. *2005*, 102, 13343–13348.
81. Maragakis, L.L.; Perencevich, E.N.; Cosgrove, S.E. Clinical and economic burden of antimicrobial resistance. Expert. Rev. Anti. Infect. Ther. *2008,* 6, 751–763.
82. Martinez, J.L.; Fajardo, A.; Garmendia, L.; Hernandez, A.; Linares, J.F. Martinez-Solano, L.; Sanchez. M.B. A global view of antibiotic resistance. FEMS Microbiol. Rev. *2009,* 33, 44–65.
83. van der Horst, M.A.; Schuurmans, J.M.; Smid, M.C.; Belinda, B.; Koenders, B.B.; Kuilel, B.H. De Novo acquisition of resistance to three antibiotics by *Escherichia coli.* Microb. Drug Resist. *2011,* 17(2), 141–147.
84. Knothe, H.; Shah, P.; Kremery, V.; Anatal, M.; Mitsuhasi, S. Transferable resistance to cefotaxime, cefoxitin, cefamandole, and cefuroxime in clinical isolates of *Klebsiella pneumoniae* and *Serratia marcescens*. Infection. *1983,* 11, 315–317.
85. Lautenbach, E.; Patel, J.B.; Bilker, W.B.; Edelstein, P.H.; Fishman, N.O. Extended-spectrum ß-lactamases– producing *Escherichia coli* and *Klebsiella pneumoniae*: Risk factors for infection and impact on resistance of outcomes. Clin. Infect. Dis. *2001*, 32, 1162–1171.
86. Kariuki, S.; Revathi, G.; Corkill, J.; Kiiru, J.; Mwituria, J.; Mirza, N.; Hart, C.A.. *Escherichia coli* from commonly-acquired urinary tract infections resistant to flouroquinolones and extended spectrum beta-lactams. J. Infect. Dev. Count. *2007*, (1), 257–262.
87. Albrich, W.; Monnet, D.L; Harbarth, S. Antibiotic selection pressure and resistance in *Streptococcus pneumoniae* and *Streptococcus pyogenes*. Emerg. Infect. Dis. *2004*, 10 (3), 514–517.
88. Davies, J.; Wright, G. Bacterial resistance to aminoglycoside antibiotics. Trends Microbiol. *1997*, 5, 234–239.
89. Leclercq, R. Mechanisms of resistance to macrolides and lincosamides: Nature of the resistance elements and their clinical implications. Clin. Infect Dis. *2002*, 15, 34(4), 482–492.
90. Sköld, O. Resistance to trimethoprim and sulfonamides. Vet. Res. *2001*, 32(3–4), 261–273.
91. Gupta, N.; Limbago, B.M.; Patel, J.B.; Kallen, A.J. Carbapenem-resistant Enterobacteriaceae epidemiology and prevention. Clin. Infect. Dis. *2011*, 53(1), 60–67.
92. Arnold, R.S.; Thom, K.A., Sharma, S.; Phillips, M.; Kristie Johnson, J.; Morgan, D.J. Emergence of *Klebsiella pneumoniae* carbapenemase-producing bacteria. South. Med. J. *2011,* 104(1):40–45.

93. Hooper, D.C. Mechanisms of fluoroquinolone resistance. Drug Resist. Updat. *1999*, 2(1), 38–55.
94. Roberts, M.C. Epidemiology of tetracycline-resistance determinants. Trends Microbiol. *1994*, 2, 353–357.
95. Robicsek, A.; Jacoby, G.A.; Hooper, D.C. The worldwide emergence of plasmid-mediated quinolone resistance. Lancet. Infect. Dis. *2006*, 6(10), 629–640.
96. Morita, Y.; Kodama, K.; Shiota, S.; Mine, T.; Kataoka, A.; Mizushima, T.; Tsuchiya. T. NorM, a putative multidrug efflux protein, of Vibrio parahaemolyticus and its homolog in *Escherichia coli*. Antimicrob. Agents. Chemother. *2000*, 42(7), 1778–1782.
97. Nize, V. Antimicrobial peptide resistance mechanisms of human bacterial pathogens. Curr. Issues Mol. Biol. *2006*, 8, 11–26.
98. Luzzaro, F.; Pagani, L.; Porta, F.; Romero, E. Extended-spectrum beta-lactamases conferring resistance to monobactams and oxyimino-cephalosporins in clinical isolates of Serratia marcescens. J. Chemother. *1995,* 7(3), 175–178.
99. Hoefler, B.C.; Gorzelnik, K.V.; Yang, J.Y.; Hendricks, N.; Dorrestein, P.C.; Straigh, P.D. Enzymatic resistance to the lipopeptide surfactin as identified through imaging mass spectrometry of bacterial competition. Proc. Natl. Acad. Sci. *2012,* 109(32), 13082–13087.
100. Pootoolal, J.; Neu, J.; Wright, G.D. Glycopeptide antibiotic resistance. Annu Rev. Pharmacol. Toxicol. *2002*, 42, 381–408.
101. Ramadoss, N.S.; Alumasa, J.N.; Cheng, L.; Wang, Y.; Li, S.; Chambers, B.S.; Chang, H.; Chatterjee, A.K.; Brinker, A.; Engels,I.H.; Keiler, K.C. Small molecule inhibitors of trans-translation have broad-spectrum antibiotic activity. Proc. Natl. Acad. Sci. U S A. *2013*, 110(25), 10282–10287.
102. Steenhuysen, J. Drug pipeline for worst superbugs 'on life support': Report. Reuters. Retrieved 23 June 2013. n.reuters.com/article/2013/04/18/us-antibiotics-superbugs
103. Boucher, H.; Talbot, G.; Benjamin, D.; Bradley, J.; Guidos, R.; Jones, R.; Murray, B.; Bonomo, R.; Gilbert, D.; Infectious Diseases Society of America. 10 X'20 Progress – Development of new drugs active against gram-negative bacilli: An update from the Infectious Diseases Society of America. Clin. Inf. Dis. (Oxford University Press), *2013*, 56(12), 1685–1694.
104. Musk, D.J.; Hergenrother, P.J. Chemical countermeasures for the control of bacterial biofilms: Effective compounds and promising targets. Curr. Med. Chem. *2006,* 13, 2163–2177.
105. Kuehn, B.M. MRSA infections rise. J. Am. Med. Assoc. *2007*, 298, 1389.
106. Perez, F.; Hujer, A.M.; Hujer, K.M.; Decker, B.K.; Rather, P.N.; Bonomo, R.A. Global challenge of multidrug-resistant *Acinetobacter baumannii*. Antimicrob. Agents Chemother. *2007*, 51, 3471–3484.
107. Bateman, S.; Graham, T. The use of Medihoney™ Antibacterial Wound Ge™l on surgicalwounds post-CABG. Wounds UK. *2007*, 3(3), 76–83.
108. Gethin. G. The signifi cance of surface pH in chronic wounds. Wounds UK. *2007,* 3(3), 52–54
109. Klasen, H.J. A historical review of the use of silver in the treatment of burns. II. Renewed interest for silver. Burns. *2000*, 26, 131–138.

110. Lara, H.H.; Ayala-Nuñez, N.V.; Ixtepan-Turrent, L.; Rodriguez-Padilla, C. Mode of antiviral action of silver nanoparticles against HIV-1. J. Nanobiotechnol. *2010*, 8, 1–10.
111. Morones, J.R.; Elechiguerra, J.L.; Camacho, A.; Holt, K.; Kouri, J.B.; Ramírez, J.T.; Yacaman, M.J. The bactericidal effect of silver nanoparticles. Nanotechnology. *2005*, 16, 2346–2353.
112. Kim, J.S.; Kuk, E.; Yu, K.N.; Kim, J.H.; Park, S.J.; Lee, H.J.; Kim, S.H.; Park, Y.K.; Park, Y.H.; Hwang, C.Y.; Kim, Y.K.; Lee, Y.S.; Jeong, D.H.; Cho, M.H. Antimicrobial effects of silver nanoparticles. Nanomedicine. *2007*, 3, 95–101.
113. Sondi, I.; Salopek-Sondi, B. Silver nanoparticles as antimicrobial agent: A case study on *E. coli* as a model for Gram-negative bacteria. J. Colloid Interf. Sci. *2004*, 275, 177–182.
114. Shahverdi, A.R.; Fakhimi, A.; Shahverdi, H.R.; Minaian, S. Synthesis and effect of silver nanoparticles on the antibacterial activity of different antibiotics against *Staphylococcus aureus* and *Escherichia coli*. Nanomedicine. *2007*, 3, 168–171.
115. Rai, M.; Yadav, A.; Gade, A. Silver nanoparticles as a new generation of antimicrobials. Biotechnol. Adv. *2009*, 27, 76–83.
116. Elechiguerra, J.L.; Burt, J.L.; Morones, J.R.; Camacho-Bragado, A.; Gao, X.; Lara, H.H.; Yacaman, M.J. Interaction of silver nanoparticles with HIV-1. J. Nanobiotechnol. *2005*, 29, 3–6.
117. Sun, R.W.; Chen, R.; Chung, N.P.; Ho, C.M.; Lin, C.L.; Che, C.M. Silver nanoparticles fabricated in Hepes buffer exhibit cytoprotective activities toward HIV-1 infected cells. Chem. Commun. (Camb*) 2005*, 40, 5059–5061.
118. Lu, L.; Sun, R.W.; Chen, R.; Hui, C.K.; Ho, C.M.; Luk, J.M.; Lau, G.K.; Che, C.M. Silver nanoparticles inhibit hepatitis B virus replication. Antivir. Ther. *2008*, 13, 253–262.
119. Sun, L.; Singh, A.K.; Vig, K.; Pillai, S.; Shreekumar, R.; Singh, S.R. Silver nanoparticles inhibit replication of respiratory sincitial virus. J. Biomed. Biotechnol. 2008, 4, 149–158.
120. Baram-Pinto, D.; Shukla, S.; Perkas, N.; Gedanken, A.; Sarid, R. Inhibition of herpes simplex virus type 1 infection by silver nanoparticles capped with mercaptoethane sulfonate. Bioconjug. Chem. *2009*, 20, 1497–1502.
121. Baram-Pinto, D.; Shukla, S.; Gedanken, A.; Sarid, R. Inhibition of HSV-1 attachment, entry, and cell-to-cell spread by functionalized multivalent gold nanoparticles. Small. *2010,* 6, 1044–1050.
122. Rogers, J.V.; Parkinson, C.V.; Choi, Y.W.; Speshock, J.L.; Hussain, S.M. A preliminary assessment of silver nanoparticles inhibition of monkeypox virus plaque formation. Nanoscale Res. Lett. *2008*, 3, 129–133.
123. Papp, I.; Sieben, C.; Ludwig, K.; Roskamp, M.; Böttcher, C.; Schlecht, S.; Herrmann, A.; Haag, R. Inhibition of influenza virus infection by multivalent sialic-acid-functionalized gold nanoparticles. Small. *2010*, 6, 2900–2906.
124. Shirtcliffe, N.; Nickel, U.; Schneider, S. Reproducible preparation of silver sols with small particle size using borohydride reduction: For use as nuclei for preparation of larger particles. J. Colloid Interf. Sci. *1999,* 211, 122–129.
125. Lee, P.C.; Meisel, D. Adsorption and surface-enhanced Raman of dyes on silver and gold sols. J. Phys. Chem. *1982*, 86, 3391–3395.

126. Creighton, J.A.; Blatchford, C.G.; Albrecht, M.G. Plasma resonance enhancement of Raman scattering by pyridine adsorbed on silver or gold sol particles of size comparable to the excitation wavelength. J. Chem. Soc. Faraday Trans. 2 Mol. Chem. Phys. *1979*, 75, 790–798.
127. Ayyappan, S.; Srinivasa, G.R.; Subbanna, G.N.; Rao, C.N.R. Nanoparticles of Ag, Au, Pd, and Cu produced by alcohol reduction of the salts. J. Mater. Res. *1997*, 12, 398–401.
128. Longenberger, L.; Mills, G. Formation of metal particles in aqueous solutions by reactions of metal complexes with polymers. J. Phys. Chem. *1995*, 99, 475–480.
129. Panigrahi, S.; Kundu, S.; Kumar Ghosh, S.; Nath, S.; Pal, T. General method of synthesis for metal nanoparticles. J. Nanopart. Res. *2004*, 6, 411–414.
130. Kalimuthu, K.; Suresh Babu, R.; Venkataraman, D.; Bilal, M.; Gurunathan, S. Biosynthesis of silver nanocrystals by *Bacillus licheniformis*. Colloids Surf. B Biointerf. *2008*, 65, 150–153.
131. Kalishwaralal, K.; Deepak, V.; Ram Kumar Pandian, S.; Kottaisamy, M.; Barathmani Kanth, S.; Kartikeyan, B.; Gurunathan, S. Biosynthesis of silver and gold nanoparticles using *Brevibacterium casei*. Colloids Surf. B Biointerf. *2010*, 77, 257–262.
132. Jegadeeswaran, P.; Shivaraj, R.; Venckatesh, R. Green synthesis of silver nanoparticles from extract of *Padinatetrastromatica* leaf. Dig. J. Nanomat. Bios. *2012*, 7, 991–998.
133. White, II G.V.; Kerscher, P.; Brown, M.R.; Morella, D.J.; McAllister, W.; Dean, D.; Kitchens, L.C. Green synthesis of robust, biocompatible silver nanoparticles using *Garlic* Extract. J. Nanomater. *2012*, 12, 730–746.
134. Sulochana S.; Krishnamoorthy, P.; Sivaranjani, K. Synthesis of silver nanoparticles using leaf extract of *Andrographis paniculata*. J. Pharmacol. Toxicol. *2012*, 7, 251–258.
135. Kora, A.J.; Arunachala, J. Green fabrication of silver nanoparticles by gum Tragacanth (*Astragalus gummifer*): A dual functional reductant and stabilizer. J. Nanomater. *2012*, 86–97.
136. Jain, D.; Daima, H.K.; Kachhwaha, S.; Kothari, S.L. Synthesis of plant-mediated silver nanoparticles using papaya fruit extract and evaluation of their antimicrobial activities. Dig. J. Nanomat. Bios. *2009*, 4, 557–563.
137. Palaniselvam K.; Velanganni, A.A.J.; Govindan, S.N.; Karthi. Leaf assisted bioreduction of silver ions using leaves of *Centella asiatica* L. and its bioactivity. E-J. Life Sci. *2012*, 1, 46–49.
138. Dwivedi, A.D.; Gopal, K. Plant- mediated biosynthesis of silver and gold nanoparticles. J. Biomed. Nanotechnol. *2011*, 7, 163–164.
139. Vanaja, M.; Annadurai, G. *Coleus aromaticus* leaf extract mediated synthesis of silver nanoparticles and its bactericidal activity. Appl. Nanosci., *2012*, 121–129.
140. Ghosh, S.; Patil, S.; Ahire, M.; Kitture, R.; Kale, S.; Pardesi, K.; Cameotra, S.S.; Bellare, J.; Dhavale, D.D.; Jabgunde, A.; Chopade, A.B. Synthesis of silver nanoparticles using *Dioscorea bulbifera* tuber extract and evaluation of its synergistic potential in combination with antimicrobial agents. Int. J. Nanomed. *2012*, 7, 483–496.
141. Maheswari, R.U.; Prabha, A.L.; Nandagopalan, V.; Anburaja V. Green synthesis of silver nanoparticles by using rhizome extract of *Dioscorea oppositifolia* L. and their

anti microbial activity against human pathogens. IOSRJ. Pharma. Biol. Sci. *2012,* 1, 38–42.
142. Gnanajobitha, G.; Annadurai, G.; Kannan, C. Green synthesis of silver nanoparticles using *Elettariacardamomom* and assessment of its antimicrobial activity. Int. J. Pharma. Sci. Res. *2012*, 3, 323–330.
143. Dinesh, S.; Karthikeyan, S.; Arumugam, P. Biosynthesis of silver nanoparticles from *Glycyrrhiza glabra* root extract. Arch. Appl. Sci. Res. *2012,* 4, 178–187.
144. Bindhu, M.R.; Umadevi, M. Synthesis of monodispersed silver nanoparticles using *Hibiscus cannabinus* leaf extract and its antimicrobial activity. Spectrochim. Acta A Mol. Biomol. Spectrosc. *2012,* 101, 184–190.
145. Sable, N.; Gaikwad, S.; Bonde, S.; Gade, A.; Rai M.M. Phytofabrication of silver nanoparticles by using aquatic plant *Hydrilla verticilata*. Nus. Biosci. *2012*, 4, 45–49.
146. Sivakumar, P.; Nethradevi, C.; Renganathan, S. Synthesis of silver nanoparticles using *Lantana camara* fruit extract and its effect on pathogens. Asian J. Pharm. Clin. Res. *2012,* 5, 97–101.
147. Ramteke, C.; Chakrabarti, T.; Sarangi, B.K.; Pandey, R-A. Synthesis of silver nanoparticles from the aqueous extract of leaves of *Ocimum sanctum* for enhanced antibacterial activity. J. Chem. *2013*, Article ID 278925, 7 pages, doi:10.1155/2013/278925
148. Mary, E.J.; Inbathamizh, L. Green synthesis and characterization of nano silver using leaf extract of *Morinda pubescens*. Asian J. Pharm. Clin. Res. *2012*, 5, 159–162.
149. Ashok kumar, D. Rapid and green synthesis of silver nanoparticles using the leaf extracts of *Parthenium hysterophorus*: A novel biological approach. Int. Res. J. Pharma. *2012*, 3, 169–173.
150. Mallikarjuna, K.; Dillip, G.R.; Narashima, G.; Sushma, N.J. and Raju, B.D.P. Phytofabrication and characterization of silver nanoparticles from *Piper betel* broth. Res. J. Nanosci. Nanotech. *2012*, 2, 17–23.
151. Garg, S. Rapid biogenic synthesis of silver nanoparticles using black pepper (*Piper nigrum*) corn extract. Int. J. Innov. Biol. Chem. Sci. *2012*, 3, 5–10.
152. Patil, D.C.; Patil, V.S.; Borase, P.H.; Salunke, K.B.; Salunkhe, B.R.; Larvicidal activity of silver nanoparticles synthesized using *Plumeria rubra* plant latex against *Aedesaegypti* and *Anopheles stephensi*. Parasitol. Res. *2012*, 110, 1815–1822.
153. Aromal, S.A.; Vidhu, V.K.; Philip, D. Green synthesis of well-dispersed gold nanoparticles using *Macrotylomauniflorum*. Spectrochim. Acta A Mol. Biomol. Spectrosc. *2012*, 85, 99–104.
154. Thirumurugan, A.; Jiflin, G.J.; Rajagomathi, G.; Tomy N.A.; Ramachandran S.; Jaiganesh, R. Biotechnological synthesis of gold nanoparticles of *Azadirachta indica* leaf extract. Int. J. Biol. Tech. *2010*, 1, 75–77.
155. Boruah, S.K.; Boruah, P.K.; Sarma, P.; Medhi, C.; Medhi, O.K. Green synthesis of gold nanoparticles using *Camelliasinensis* and kinetics of the reaction. Adv. Mat. Lett. *2012*, 3, 481–486.
156. Fazaludeena, M.F.; Manickamb, C.; Ashankytyc, M.A.I.; Ahmedd, M.Q.; Quaser Zafar Beg. Synthesis and characterizations of gold nanoparticles by *Justicia gendarussa* Burm F leaf extract. J. Microbiol. Biotech. Res. *2012*, 2, 23–34.
157. Usha, R.; Prabu, E.; Palaniswamy, M.; Venil, C.K.; Rajendran, R. Synthesis of metal oxide nano particles by Streptomyces Sp for development of atimicrobial textiles. GJBB. *2010*, 5, 153–160.

158. Singh,V.; Ajay, R.; Patil, A.; Anand, P.; Milani, W.N. Biological synthesis of copper oxide nano particles using *Escherichia coli* Gade. Curr. Nanosci. *2010*, 6(4), 365–369.
159. Subhankari, I.; Nayak, P.L. Antimicrobial activity of copper nanoparticles synthesised by ginger (*Zingiber officinale*) extract. World J. Nano Sci. Technol. *2013*, 2(1), 10–13.
160. Information and history of silver. 2010. http://www.cctechnologies.com.au/PDF/Information%20and%20history%20of%20 silver.pdf. Date accessed Feb-2010
161. Russell, A.D.; Russell, N.J. Biocides: Activity, Action and Resistance. Fifty Years of Antimicrobials: Past Perspectives and Future Trends, P.A. Hunter, G.K. Darby, and N.J. Russell, eds., Cambridge University Press, Cambridge. 1995.
162. History of silver. 2010. http://www.painreliefwellness.com/clients/2842/documents/History_of_Silver_Usa ge.pdf. Date accessed Feb-2010.
163. Wijnhoven, S.W.P.; Peijnenburg, W.J.G.M.; Herberts, C.A.; Hagens, W.I.; Oomen, A.G; Zijverden, M.; Sips, A.J.A.M.; Geertsma, R.E. Nanosilver – A review of available data and knowledge gaps in human and environmental risk assessment. Nanotoxicology. *2009*, 3(2), 109–138.
164. Foot Defense. 2010. Silver: A study of this precious metal and its use in foot care. http://www.acor.com/Downloads/ebSilverFoot.pdf. Date accessed Feb-2010.
165. Wikipedia. 2010. The free encyclopedia. http://en.wikipedia.org/wiki/Silver. Date accessed Feb-2010.
166. AGCGlass. 2007. http://www.agcflatglass.eu/AGC+Flat+Glass+Europe/English/Homepage/News/Press+room/PressDetail-Page/page.aspx/979?pressitemid=1031. Date accessed Feb-2010.
167. Sanjay, S.; Meddings, J.A.; Calfee, D.; Kowalski, C.P.; Krein, S.L. Catheter-associated urinary tract infection and the medicare rule changes. Ann. Int. Med. *2009,* 150, 877–884.
168. Los Alamos National Labs. Chemistry Division. 2010. "Silver": http://www.periodic.lanl.gov/elements/47.html. Last accessed: May, 2010.
169. Ovington, L.G. The truth about silver. Ostomy Wound Mang. *2004*, 50, 1–10.
170. Lansdown, A. B.; Williams,A.; Chandler, S. Benfield. S. Silver absorption and antimicrobial efficacy of silver dressings. J. Wound Care. *2005,* 14, 155–160.
171. Lansdown, A.B.G. A review of the use of silver in wound care: Facts and fallacies. Br. J. Nurs. *2004,* 13, 6–19.
172. Fox, C.L. Silver sulphadiazine: A new topical therapy for *Pseudomonas aeruginosa* in burns. Archiv. Surg. *1968*, 96, 184–188.
173. Moyer, C.A.; Brentano, L.; Gravens, D.L.; Margraf, H.W.; Monafo, W.W. Jr. Treatment of large human burns with 0.5 per cent silver nitrate solution. Arch Surg. *1965*, 90, 812–867.
174. Price, W.R.; Wood, M. Silver nitrate burn dressing. Treatment of seventy burned persons. Am. J. Surg. *1966*, 112, 674–680.
175. Sawhney, C.P.; Sharma, R.K.; Rao, K.R.; Kaushish, R. Long-term experience with 1 percent topical silver sulphadiazine cream in the management of burn wounds. Burns. *1989*, 15, 403–406.
176. Drug and Therapeutics Bulletin. Silver dressings – Do they work? 201, 48, 38–42.

177. **Lansdown, A.;** *Silver,* **B.** *2. Toxicity in mammals and how its products aid wound repair.* J. Wound Care. *2002,* **11,** 173–*177.*
178. Chaw, K. C.; Manimaran, M.; Tay, F. Role of silver ions in destabilization of intermolecular adhesion forces measured by Atomic Force Microscopy in *Staphylococcus epidermidis b*iofilms. Antimicrob. Agents. Chemother. *2005,* 49(12), 4853–4859.
179. Sutherland, I.W. The biofilm matrix—An immobilized but dynamic microbial environment. Trends Microbiol. *2001*, 9, 222–227.
180. Wright, J.B.; Lam, K.; Burrell, R.E. The comparative efficacy of two antimicrobial barrier dressings: in vitro examination of two controlled release of silver dressings. Wounds. *1998*, 10, 179–188.
181. Yin, H.Q.; Langford, R.; Burrell, R.E. Comparative evaluation of the antimicrobial activity of Acticoat TM antimicrobial barrier dressing. J. Burn. Care Rehabil. *1999*, 20(3), 195–200.
182. Sibbald, J.; Contreras-Ruiz, P.; Coutts, M.; Fierheller, A.; Rothman, Woo, K. Advances skin. Wound Care. *2007*, 20, 549–558.
183. Atiyeh, B.S.; Costagliola, S.; Hayek, S.N.; Dibo, S.A. Effect of silver on burn wound infection control and healing: Review of the literature. Burns. *2007,* 33, 139–148.
184. Wright, J.B.; Lam, K.; Buret, A.G.; Olson, M.E.; Burrell, R.E. Early healing events in a porcine model of contaminated wounds: effects of nanocrystalline silver on matrix metalloproteinases, cell apoptosis, and healing. Wound Repair Regeneration *2002*, 10, 141–151.
185. Yu, D.G.; Zhou,J.; Chatterton, N.P.; Li, Y.; Huang, J.; Wang, X. Polyacrylonitrile nanofibers coated with silver nanoparticles using a modified coaxial electrospinning process. Int. J. Nanomed. *2012*, 7, 5725–5732.
186. Eid, K.A.; Azzazy, H.M. Controlledsynthesis and characterization of hollow flower-like silver nanostructures. Int. J. Nanomed. *2012*, 7, 1543–1550.
187. Tijing, L.D.; Ruelo, M.T.G.; Amarjargal, A.; Pant, H.R.; Park, C.H.; Kim, C.S. One-step fabrication of antibacterial (silver nanoparticles/poly (ethylene oxide))-polyurethane bicomponent hybrid nanofibrous mat by dual-spinneret electrospinning. Mater. Chem. Phys. *2012*, 134(2–3), 453–458.
188. Bleeker, A.J.; Cassee, F.R.; Geertsma, R.E.; de Jong, W.H.; Heugens, E.H.W.; Koers-Jacquemijns, M.; van de Meent, D.; Oomen, A.G.; Popma, J.; Rietveld, A.G. et al. Interpretation and Implications of the European Commission Recommendation on the Definition of Nanomaterial, RIVM Letter Report; RIVM: Bilthoven, The Netherlands, 2012.
189. Liu, J.; Wang, Z.; Liu, F.D.; Kane, A.B.; Hurt, R.H. Chemical transformations of nanosilver in biological environments. ACS Nano. *2012*, 6, 9887–9899.
190. Yang, X.; Gondikas, A.P.; Marinakos, S.M.; Auffan, M.; Liu, J.; Hsu-Kim, H.; Meyer, J.N. Mechanism of silver nanoparticle toxicity is dependent on dissolved silver and surface coating in caenorhabditis elegans. Environ. Sci. Technol. *2012,* 46, 1119–1127.
191. Ma, R.; Levard, C.; Marinakos, S.M.; Cheng, Y.; Liu, J.; Michel, F.M.; Brown, G.E., Jr.; Lowry, G.V. Size-controlled dissolution of organic-coated silver nanoparticles. Environ. Sci. Technol. *2012*, 46, 752–759.

192. Xiu, Z.; Zhang, Q.; Puppala, H.L.; Colvin, V.L.; Alvarez, P.J.J. Negligible particle-specific antibacterial activity of silver nanoparticles. Nano Lett. *2012*, 12, 4271–4275.
193. Kermanizadeh, A.; Pojana, G.; Gaiser, B.K.; Birkedal, R.; Bilaničová, D.; Wallin, H.; Jensen, K.A.; Sellergren, B.; Hutchison, G.R.; Marcomini, A.; Stone, V. In vitro assessment of engineered nanomaterials using a hepatocyte cell line: Cytotoxicity, pro-inflammatory cytokines and functional markers. Nanotoxicology. *2012*, 7, 301–313.
194. Unrine, J.M.; Colman, B.P.; Bone, A.J.; Gondikas, A.P.; Matson, C.W. Biotic and abiotic interactions in aquatic microcosms determine fate and toxicity of ag nanoparticles. Part 1. Aggregation and dissolution. Environ. Sci. Technol. *2012*, 46, 6915–6924.
195. Suresh, A.K.; Pelletier, D.; Wang, W.; Morrell-Falvey, J.L.; Gu, B.; Doktycz, M.J. Cytotoxicity induced by engineered silver nanocrystallites is dependent on surface coatings and cell types. Langmuir. *2012*, 28, 2727–2735.
196. Dawson, K.A.; Anguissola, S.; Lynch, I. The need for in situ characterisation in nanosafety assessment: Funded transnational access via the qnano research infrastructure. Nanotoxicology. *2012*, 7, 1–4.
197. Reidy, B.; Haase, A.; Luch, A.; Dawson, K.; Lynch, I. Mechanisms of silver nanoparticle release. Trans. Toxic. A Crit. Rev. Curr. Knowledge Recommend. Futur. Stud. Appl. Mater. *2013,* 6, 2295–2350.
198. Hamouda, T.; Myc, A.; Donovan, B.; Shih, A.; Reuter, J.D.; Baker, Jr J.R. A novel surfactant nanoemulsion with a unique non-irritant topical antimicrobial activity against bacteria, enveloped viruses and fungi. Microbiol. Res. *2000*, 156, 1–7.
199. Dibrov, P.; Dzioba, J.; Gosink, K.K.; Hase, C.C. Chemiosmotic mechanism of antimicrobial activity of Ag(+) in *Vibrio cholerae*. Antimicrob. Agents. Chemother. *2002*, 46, 2668–2670.
200. Dragieva, I.; Stoeva, S.; Stoimenov, P.; Pavlikianov, E.; Klabunde, K. Complex formation in solutions for chemical synthesis of nanoscaled particles prepared by borohydride reduction process. Nanostruct. Mater. *1999*, 12, 267–270.
201. Amro, N.A.; Kotra, L.P.; Wadu-Mesthrige, K.; Bulychev, A.; Mobashery, S.; Liu, G. High-resolution atomic force microscopy studies of the *Escherichia coli* outer membrane: structural basis for permeability. Langmuir. *2000*, 16, 2789–2796.
202. Danilczuk, M.; Lund, A.; Saldo, J.; Yamada, H.; Michalik, J. Conduction electron spin resonance of small silver particles. Spectrochim. Acta Part A *2006*, 63, 189–191.
203. Kim, J.S.; Kuk, E.; Yu, K.N.; Kim, J.H.; Park, S.J.; Lee, H.J.; Kim, S.H.; Park, Y.K.; Park, Y.H.; Hwang, C.Y.; Kim, Y.K.; Lee, Y.S.; Jeong, D.H.; Cho, M.H. Antimicrobial effects of silver nanoparticles. Nanomedicine. *2007*, 3, 95–101.
204. Uchida, M.; Yamamoto, T.; Taniguchi, A. Reaction of silver ions and some amino acids. J. Antibact. Antif. Agents. *2003*, 31, 695–704.
205. Darouiche, R.O. Treatment of infections associated with surgical implants. N. Engl. J. Med. *2004*, 350, 1422–1429.
206. Cohen, M.S.; Stern, J.M.; Vanni, A.J.; Kelley, R.S.; Baumgart, E.; Field, D.; Libertino, J.A.; Summerhayes, I.C. Surgical Infections. *2007*, 8, 397–403.
207. Roberts, C.; Ivins, N.; Widgerow, A. ACTICOATTM and ALLEVYNTM Ag made easy. Wounds Int. *2011*, 2(2): Available from http://www.woundsinternational.com

208. Maple, P.A.C.; Hamilton-Miller, J.M.T.; Brumfitt, W. Comparison of the in-vitro activities of the topical antimicrobials azelaic acid, nitrofurazone, silver sulphadiazine and mupirocin against methicillin-resistant *Staphylococcus aureus*. J. Antimicrob. Chemother. *1992*, 29, 661–668.
209. Smith and Nephew, Data On File : Report No: IV288/31/06. G R Micro, London, *2007*.
210. Wright, J.B.; Lam, K.; Hansen, D.; Burrell, R.E. Efficacy of topical silver against fungal burn wound pathogens. Am. J. Inf. Control. *1999*, 27(4), 344–350.
211. Livermore, D.M.J. Has the era of untreatable infections arrived? J. Antimicrob. Chemother. *2009*, 64 Suppl 1:i, 29–36.
212. Kumarasamy, K.K.; Toleman, M.A.; Walsh, T.R.; Bagaria, J.; Butt, F.; Balakrishnan, R.; Chaudhary, U.; Doumith, M.; Giske, C.G.; Irfan, S.; Krishnan P.; Kumar, A.V.; Maharjan, S.; Mushtaq, S.; Noorie, T.L.; Paterson, D.L.; Pearson, A.; Perry, C.; Pike, R.; Rao, B.; Ray, U.; Sarma, J.B.; Sharma, M.; Sheridan, E.; Thirunarayan, M.A.; Turton, J.; Upadhyay, S.; Warner, M.; Welfare, W.M.; Livermore, D.; Woodford. N. Emergence of a new antibiotic resistancemechanism in India, Pakistan, and the UK: A molecular, biological, and epidemiological study. Lancet. Infect. Dis. *2010*, 10, 597–602.
213. European Centre for Disease Prevention and Control. Risk assessment on the spread of Carbapenemase producing Enterobacteriaceae (CPE) through patient transfer between healthcare facilities , with special emphasis on cross border transfer Stockholm: ECDC; 2011. sockholm, September 2011 ISBN 978-929193-317-4. doi:10.2900/59034
214. Smith and Nephew Research Centre Work Report # WRP-TW141-89. Efficacy of Silver-Containing Wound. Barriers vs. NDM-1 producing *Enterobacteriaceae* and *Acinetobacter* spp. Russell Hope, Shazad Mushtaq, Rachael Adkin, Aiysha Chaudhry, Neil Woodford and David Livermore. 2011.
215. Wright, J.B.; Lam, K.; Burrell, R.E. Wound management in an era of increasing bacterial antibiotic resistance: a role for topical silver treatment. Am. J. Inf. Contr. *1998*, 26(6), 572–577.
216. Wright, J.B.; Lam, K.; Olson, M.E.; Burrell, R.E. Is antimicrobial efficacy sufficient? A question concerning the benefits of new dressings. Wounds. *2003*, 15, 133–142.
217. Tredget, E.E.; Shankowsky, H.A.; Groeneveld, A.; Burrell, R. A matched-pair, randomized study evaluating the efficacy and safety of Acticoat silver-coated dressing for the treatment of burn wounds. J. Burn Care Rehabil. *1998*, 19, 531–537.
218. Yin, H.Q.; Langford, R.; Burrell, R.E.;Comparative evaluation of antimicrobial activity of acticoat antimicrobial barrier dressing. J. Burn Care Rehabil. *1999*,20(3), 195-200.
219. Guerrero, R.; Menezes, J.; Reilly, D.A.; Garner, W.L.Wound care with actocoat in patients with non burn skin loss. J. Inves. Med. *2000*, 48(1S), 319–322.
220. Voigt, D.W.; Paul, C.N. The use of acticoat silver impregnated telfa dressing in a regional burn and wound care center.The clinicians view.In: Wounds-A compendium of clinical Research and Practice. Wayne, PA: Health Management Publication Inc: 2001. pp. 11–20.

221. Kirsner, R.; Orsted,Wright, B. Matrix metallo proteinases in normal and impaired wound healing: A potential role of nanocrystalline silver. Wounds. *2001*, 13(supplC), 5–15.
222. Zhang, Y.; Peng, H.; Huang, W.; Zhou, Y.; Yan, D. Facile preparation and characterization of highly antimicrobial colloid Ag or Au nanoparticles. J. Colloid Interface. Sci. *2008*, 325, 371–376.
223. Kim, K.J.; Sung, W.S.; Moon, S.K.; Choi, J.S.; Kim, J.G.; Lee, D.G. Antifungal effect of silver nanoparticles on dermatophytes. J. Microbiol. Biotechnol. *2008*, 18, 1482–1484.
224. Kim, K.J.; Sung, W.S.; Suh, B.K.; Moon, S.K.; Choi, J.S.; Kim, J.G. Antifungal activity and mode of action of silver nano-particles on *Candida albicans.* Biometals. *2009*, 22, 235–242.
225. Speshock, J.L.; Murdock, R.C.; Braydich-Stolle, L.K.; Schrand, A.M.; Hussain, S.M. Interaction of silver nanoparticles with Tacaribe virus. J. Nanobiotechnol. *2010*, 8, 19–27.
226. Pulsed Laser Deposition of Thin Films, edited by Douglas B. Chrisey and Graham K. Hubler, John Wiley & Sons, 1994 ISBN 0-471-59218-8.
227. Noyce, J.O.; Michels, H.; Keevil, C.W. Potential use of copper surfaces to reduce survival of epidemic methicillin-resistant *Staphylococcus aureus* in the healthcare environment. J. Hosp. Infect. *2006*, 63, 289–297.
228. Fau´ndez, G.; Troncoso, M.; Navarrete, P.; Figueroa, G. Antimicrobial activity of copper surfaces against suspensions of *Salmonella enterica* and *Campylobacter jejuni*. BMC Microbiol. *2004*, 4(19), 685–690.
229. Singh, J.A.; Upshur, R.; Padayatchi, N. XDR-TB in South Africa: No time for denial or complacency. PLoS Med. *2006*, 4(1), e50. doi:10.1371/journal.pmed.0040050
230. Tenaud, I.; Sainte-Marie, I.; Jumbou, O.; Litoux, P.; Dreno, B. In vitro modulation of keratinocyte wound healing integrins by zinc, copper and manganese. Br. J. Dermatol. *1999,* 140, 26–34.
231. Adomaviciene, M., Stanys, S.; Demsar, A.; Godec, M. Insertion of Cu nanoparticles into polymeric nanofibrous structure via electrospinning technique. Fibres and Textiles in Eastern Europe. *2010*, 1(78), 17–20.
232. Sen,C.K.; Khanna, S.; Venojarvi, M.; Trikha, P. Ellison, E.C.; Hunt, T.K.; Roy, S. Copper-induced vascular endothelial growth factor expression and wound healing. Am. J. Physiol. Heart. Circ. Physiol. *2002,* 282, H1821–H1827.
233. Wozniak, J.; Bieganski, T.; Maslinski, C. Diamine oxidase activity duringwound healing in guinea pig skin [proceedings]. Agents Actions. *1979*, 9, 45–47.
234. Kobayashi, H.; Ishii, M.; Chanoki M.; Yashiro, N.; Fushida, H.; Fukai, K.; Kono, T.; Hamada, T.; Wakasaki, H.; Ooshima, A. Immunohistochemical localization of lysyl oxidase in normal human skin. Br. J. Dermatol. *1994*, 131, 325–330.
235. Padmavathy, N.; Vijayaraghavan, R. Enhanced bioactivity of ZnO nanoparticles—An antimicrobial study. Sci. Technol. Adv. Mat. *2008*, 9(3). doi:10.1088/1468-6996/9/3/035004
236. Atmaca, S.; Gul, K.; Cicek, R. The effect of zinc on microbial growth. Tr. J. Med. Sci. *1998*, 28, 595–597.

237. Shalumon, K.T.; Anulekha, K.H.; Nair, S.V.; Nair, K.P. Chennazhi, Jayakumar, R. Sodium alginate/poly (vinyl alcohol)/nano ZnO composite nanofibers for antibacterial wound dressings. Int. J. Biol. Macromol. *2011*, 49(3), 247.
238. Kumar, P.T.S.; Lakshmanan, V.K.; Anilkumar, T.V.; Ramya, C.; Reshmi, P.; Unnikrishnan, A.G.; Nair, S.V.; Jayakumar, R. Flexible and microporus chitosan hydrogel/ nanoZno composite bandages for wound dressing : *invitro* and *invivo* evaluationACS Appl. Mat. Interf. *2012*, 4(5), 2618.
239. Rajashri, S.K.; Amruta, B.; Vrinda, P.B.; Koyar, S.R. Bactericidal action of N doped ZnO in sunlight. Biointerf. Res. Appl. Chem. *2011*, 1(2), 57–63.
240. Paetzold, O.H.; Wiese, A. Experimentelle Untersuchungen tiber die antimikrobielle Wirkung *yon* Zinkoxid. Arch. Derm. Res. *1975*, 253(23),151–159.
241. Moore, L.J. Evaluation of the patient for temporomandibular joint surgery. Oral Maxillofac. Surg. Clin. North Am. *2006*, 18, 291–301.
242. Södeberg, T.A.; Sunze, B.; Holm, S.; Elmr, T.; Hallmans, G.; Sjöberg, S. Antibacterial effect of zinc oxide in vitro. Scand. J. Plast. Reconstr. Hand. Surg. *1990*, 24, 193–197.
243. Södeberg, T.A.; Holm, S.; Gref, R.; Hallmans, G. Scand. The effect of zinc on microbial growth. J. Plast. Reconstr. Hand. Surg. *1991*, 25, 19–24.
244. Liu, Y.; He, L.; Mustapha, A.; Li, H.; Hu, Z.Q.; Lin, M. Antimicrobial activities of zinc oxide nanoparticles against *Escherichia coli* O157:H7. J. Appl. Microbiol. *2009*, 107, 1193–1201.
245. Vlad, S.; Tanase, C.; Macocinschi, D.; Ciobanu, C.; Balaes, T.; Filip, D.; Gostin, I.N.; Gradinaru, L.M. Antifungal behaviour of polyurethane membranes with zinc oxide nanoparticles. Dig. J. Nanomater. Bios. *2012,* 7(1), 51–58.
246. Yu Tatsukawa. Discovery and applications of photocatalysis—Creating a comfortable future by making use of light energy. Jpn. Nanonet. Bullet. *2005*, 44 http://green.wikia.com/index.php?title=Titanium_dioxide
247. Fujishima, A.; Honda, K. Electrochemical photolysis of water at a semiconductor electrode. Nature. *1972,* 238(5358), 37–38.
248. Berry, C.W.; Moore, T.J.; Safar, J.A.; Henry, C.A.; Wagner, M.J. Antibacterial activity of dental implant metals. Impl. Dent. *1992*, 1(1), 59–65.
249. Donde, K.J.; Patil, V.R.; Malve, S.P. Antimicrobial studies of hydrazone complexes of Hg (II) and Fe (II) divalent metal ions. Acta Poloniae Pharm. *2003*, 60(3), 173–175.
250. Diarra, M.S.; Lavoie, M.C.; Jacques, M.; Darwish, I.; Dolence, E.K.; Dolence, J.A.; Ghosh, M.; Ghosh, M.; Miller, M.J.; Malouin, F. Species selectivity of new siderophore-drug conjugates that use specific iron uptake for entry into bacteria. Antimicrob. Agents. Chemother. *1996,* 40(11), 2610–2617.
251. Gvozdyak, R.I.; Shvets, T.M.; Kushchevskaya, N.F.; Denis, R.O. The antibacterial activity of preparations with highly dispersed iron. Mikrobiolohichnyi Zhurnal (Kiev), *1996,* 58(6), 45–49.
252. Bacchi, A.; Carcelli, M.; Pelagatti, P.; Pelizzi, C.; Pelizzi, G.; Zani, F. Antimicrobial and mutagenic activity of some carbono- and thiocarbonohydrazone ligands and their copper (II), iron (II) andzinc (II) complexes. J. Inorg. Biochem. *1999*, 75(2), 123–133.

253. Popova, T.P.; Alexandrova, R.I.; Tudose, R.; Costisor, O. Preliminary in vitro investigations on antimicrobial activity of two copper complexes. Comptes Rendus de l'Académie Bulgare des Sciences. *2004*, 57(6), 105–110.
254. Popova, T.P.; Alexandrova, R.I.; Tudose, R.; Mosoarca, E.-M. Costisor, O. Antibacterial activity in vitro of four cobalt (II) complexes with Mannich type ligands. Comptes Rendus de l'Académie Bulgare des Sciences. *2006*, 59, 551–556.
255. Burke J.F.; Bondoc, C.C. Dermatology in general medicine, eds. T.B. Fitzpatrik et al., McGraw–Hill, Inc., New York, 931, 1979.
256. Fatt I. Soft contact lenses. In: Clinical and applied technology, ed. M. Reben, Willey Medical Publications, New York, 83, 1978.
257. Allan, C.R.; Hadwigar, L.A. Studies on the fungistatic activity of chitosan. Exp. Mycology. *1974*, 3, 258.
258. Jayaraman, K.; Kotaki, M.; Zhang, Y.; Mo; Xiu, M.; Ramakrishna, S. Recent advances in polymer nanofibers. J. Nanosci. Nanotechnol. *2004*, 4(1–2), 52–65.
259. Duan, Y.Y.; Jia, J.; Wang, S.H.; Yan, W.; Jin, L.; Wang, Z.Y. Preparation of antimicrobial poly(e-caprolactone) electrospun nanofibers containing silver-loaded zirconium phosphate nanoparticles. *J. Appl. Polym. Sci. 2007*, 106(2), 1208–1214.
260. Venugopal, J; Ramakrishna, S. Applications of polymer nanofibers in biomedicine and biotechnology. Appl. Biochem. Biotechnol. *2005*, 147–157.
261. Jia, J.; Duan, Y.Y.; Wang, S.H.; Zhang, S.F.; Wang, Z.Y. Preparation and characterization of antibacterial silver-containing nanofibers for wound dressing applications. J. US-China Med. Sci. *2007*, 4, 52–54.
262. Hasegawa, M.; Isogai, A.; Kuga, S.; Onabe, F. Preparation of cellulose-chitosan blend film using chloral/dimethylformamide. Polymer. *1994*, 35(5), 983–987.
263. Geng, X.; Kwon, O.; Jang, J. Electrospinning of chitosan dissolved in concentrated acetic acid solution. Biomat. *2005*, 26, 5427–5432.
264. Ojha, S.; Gorga, R. Fabrication and characterization of electrospun chitosan nanofibers formed via templating with polyethylene oxide. Biomacromol. *2008*, 9, 2523–2529.
265. Bhattarai, N.; Edmondson, D.; Veishe, O.; Matsen, F.A.; Zhang, M. Electrospun chitosan-based nanofibers and their cellular compatibility. Biomat. *2005*, 26(31), 6176–6184.
266. Ohkawa, K.; Cha, D.; Kim, H.; Nishida, A.; Yamamoto, H. Electrospinning of chitosan. Macromol. Rapid Commun. *2004*, 25, 1600–1605.
267. Ohkawa, K.; Minato, K.I.; Kumagai, G.; Hayashi, S.; Yamamoto, H. Chitosan nanofiber. Biomacromol. *2006*, 7, 3291–3294.
268. Min, B.M.; Lee, S.W.; Lim, J.N.; You, Y.; Lee, T.S.; Kang, P.H.; Park, W.H. Chitin and chitosan nanofibers: Electrospinning of chitin and deacetylation of chitin nanofibers. Polymer. *2004*, 45, 7137–7142.
269. Sangsanoh, P.; Supaphol, P. Stability improvement of electrospun chitosan nanofibrous membranes in neutral or weak basic aqueous solutions. Biomacromol. *2006*, 7, 2710–2714.
270. Zhou, Y.; Yang, D.; Chen, X.; Xu, Q.; Lu, F.; Nie, J. Electrospun water-soluble carboxyethyl chitosan/poly(vinyl alcohol) nanofibrous membrane as potential wound dressing for skin regeneration. Biomacromol. *2008*, 9, 349–354.

271. Ignatova, M.; Manolova, N.; Rashkov, I. Novel antibacterial fibers of quaternized chitosan and poly(vinyl pyrrolidone) prepared by electrospinning. Eur. Polym. J. *2007*, 43, 1112–1122.
272. Charernsriwilaiwat, N.; Opanasopit, P.; Rojanarata, T.; Ngawhirunpat, T. Fabrication and characterization of chitosan–ethylenediaminetetraacetic acid/polyvinyl alcohol blend electrospun nanofibers. Adv. Mater. Res. *2011*, 195, 648–651.
273. Charernsriwilaiwat, N.; Opanasopit, P.; Rojanarata, T.; Ngawhirunpat, T.; Supaphol, P. Preparation and characterization of chitosan–hydroxybenzotriazole/polyvinyl alcohol blend nanofibers by the electrospinning technique. Carbohydr. Polym. *2010*, 81, 675–680.
274. Schindler, M.; Ahmed, I.; Kamal, J.; Nur-EKamal, A.; Grafe, T. H.; Chung, H. Y.; Meiners S. A synthetic nanofibrillar matrix promotes in vivo-like organization and morphogenesis for cells in culture. Biomaterials. *2005*, 26(28), 5624–5631.
275. Ma, Z.W.; Kotaki, M.; Inai, R.; Ramakrishna, S. Potential of nanofiber matrix as tissue-engineering scaffolds. Tissue Eng. *2005*, 11(1–2), 101–109.
276. Dvir, T.; Gang, T.; Cohen, S. Designer scaffolds for tissue engineering and regeneration. Is. J. Chem. *2005*, 45, 487–494.
277. Qin, Y. Chitin and chitosan as wound dressing materials. Text. Horiz. *1994*, 14(6), 19–21.
278. Ueno; Mori, H.; Fujinaga, T. Topical formulations and wound healing applications of chitosan. Adv. Drug Del. Rev. *2001*, 105–115.
279. Raymond, L.; Morin, F.G.; Marchessault, R.H. Degree of deacetylation of chitosan using conductometric titration and solid-state NMR. Carb. Res. *1993*, 246, 331–336.
280. Jayakumar, R.; Prabaharan, M.; Kumar, P.T.S.; Nair, S.V.; Tamura, H. Biomaterials based on chitin and chitosan in wound dressing applications. Biotechnol. Adv. *2011*, 29, 322–337.
281. Paul, W.; Sharma, C.P. Chitosan and alginate wound dressings: a short review. Trends Biomater. Artif. Organs. *2004*, 18, 18–23.
282. Charernsriwilaiwat, N.; Opanasopit, P.; Rojanarata, T.; Ngawhirunpat, T. Lysozyme-loaded, electrospun chitosan-based nanofiber mats for wound healing. Int. J. Pharm. *2012*, 427(2), 379–384.
283. Cencetti, C.; Bellini, D.; Pavesio, A., Senigaglia, D.; Passariello, C.; Virga, A.; Matricardi, P. Preparation and characterization of antimicrobial wound dressings based on silver, gellan, PVA and borax. Carbohydr. Polym. *2012*, 90(3), 1362–1370.
284. Nitanan, T.; Akkaramongkolporn, P.; Rojanarata, T.; Ngawhirunpat, T.; Opanasopit, P. Neomycin-loaded poly (styrene sulfonic acid-co-maleic acid)(PSSA-MA)/polyvinyl alcohol (PVA) ion exchange nanofibers for wound dressing materials. Int J Pharm. *2013*, 448(1),71–78.
285. Suet wai Leung. 2013. Electro-spinning of nano structure materials for wound dressing BA (Hons) Scheme in Fashion and Textiles (fashion technology specialism). Institute of textiles & clothing, The Hong Kong Polytechnic University 2013.
286. Afeesh, R.; Unnithana, Nasser, A.M.; Barakat, P.B.; Pichiahd T.; Gnanasekarane, G.; Nirmalab, R.; Chad, Y-S.; Junge, C-H.; El-Newehyfg, M., Kimb, H.Y. Wound-dressing materials with antibacterial activity from electrospun polyurethane–dextran nanofiber mats containing ciprofloxacin HCl Carbohydrate Polymers. *2012*, 90(4), 1786–1793.

287. Prabaharan, M.; Jayakumar, R.; Nair, S.V. Electrospun nanofibrous scaffolds-current status and prospectus in drug delivery. Advan. Polym. Sci. *2012*, 246, 241–262.
288. Abdelgawada, A.M.; Hudsona, S.M; Rojasb, O.J. Antimicrobial wound dressing nanofiber mats from multicomponent (chitosan/silver-NPs/polyvinyl alcohol) systems . Carbohydr. Polym. *2013* (In Press).
289. Li1, C.; Fu, R.; Yu, C.; Li, Z.; Guan, H.; Hu, D.; Zhao, D.; Lu, L. Silver nanoparticle/chitosan oligosaccharide/poly(vinyl alcohol) nanofibers as wound dressings: a preclinical study. Int. J. Nanomed. *2013*, 8, 4131–4145.
290. Maleki, H.; Gharehaghaji, A.A.; Dijkstra, P.J. A novel honey-based nanofibrous scaffold for wound dressing application. J. Appl. Polym. Sci. *2013*, 127(5), 4086–4092.
291. Leung, V.; Hartwell, R.; Elizei, S.S.; Yang, H.; Ghahary, A.; Ko, F. Postelectrospinning modifications for alginate nanofiber-based wound dressings. J. Biomed. Mater. Res. Part B. *2013, (Press).*
292. Bidgoli, S.; Mahdavi, M.; Rezayat, S.; Korani, M.; Amani, A.; Ziarati, P. Toxicity assessment of nanosilver wound dressing in wistar rat. Acta Med. Iran. *2013*, 51(4), 203–208.
293. Rinaudo, M.C. Chitosan: Properties and applications. Prog. Polym. Sci. *2006*, 31, 603–632.
294. Biagini, G.; Bertani, A.; Muzzarelli, R.; Damadei, A.; DiBenedetto, G.; Belligolli, A.; Riccotti, G.; Zucchini, C.; Rizzoli, C. Wound management with N-carboxybutyl chitosan. Biomaterials. *1991*, 12, 28–286.
295. Abhay, S. P. US Pat. *1998*, 5(836), 970.
296. Su, C.H.; Sun, C.S.; Juan, S.W.; Hu, C.H.; Ke, W.T.; Sheu, M.T. Fungal mycelia as the source of chitin and polysaccharides and their applications as skin substitutes. Biomat. *1997*, 18(17), 1169–1174.
297. Aoyagi, S.; Onishi, H.; Machida, Y. Novel chitosan wound dressing loaded with minocycline for the treatment of severe burn wounds. Int. J. Pharm. *2007*, 330(1–2), 138–145.
298. Mi, F.L.;, Shyu, S.S.; Wu, Y.B.; Lee, S.T.; Shyong, J.Y.; Huang, R.N.; Fabrication and characterization of a sponge-like asymmetric chitosan membrane as a wound dressing. Biomat. *2001,* 22(2), 165–173.
299. Huang, Z.M.; Zhang, Y.Z.; Kotaki, M.; Ramakrishna, S. A review on polymer nanofibers by electrospinning and their applications in nanocomposites. Compos. Sci. Technol. *2003*, 63, 2223–2253.
300. Kokabi, M.; Sirousazar, M.; Hassan, Z.M. Macromolecular nanotechnology: PVA–clay nanocomposite hydrogels for wound dressing. Euro. Polymer. J. *2007,* 43,773–781.
301. Zhong, S.P.; Zhang, Y.Z.; Lim, C.T. Tissue scaffolds for skin wound healing and dermal reconstruction. WIREsNanomed. Nanotechnol. *2010*, 2, 510–525.
302. Zahedi, P.; Rezaeian, I.; Ranaei-Siadat, S.-O., Jafari, S.-H., Supaphol, P.A review on wound dressings with an emphasis on electrospun nanofibrous polymeric bandages. Polym. Adv. Technol. *2010*, 21, 77–95.
303. Hwang, Y.J.; An, J.S.; McCord, M.G.; Park, S.W.; Kang, B.C. The effects of helium atmospheric pressure plasma treatment on low-stress mechanical properties of polypropylene Nonwoven fabrics. Text. Res. J. *2005*, 75, 771–778.

304. McCord, M.G.; Qiu, Y.; Zhang, C.; Hwang, Y.J.; Bures, B. The effect of atmospheric pressure helium plasma treatment on surface and mechanical properties of ultrahigh modulus polyethylene fibers. Adhes. Sci. Technol. *2002*, 16, 99–107.
305. McCord, M.G.; Hwang, Y.J; Kang, B.C. Helium/oxygen atmospheric pressure plasma treatment on poly(ethylene terephthalate) and poly(Trimethylene terephthalate) fabrics: Comparison of low-stress mechanical properties. Fiber Polym. *2005*, 6, 113–120.
306. Cai, Z.; Qiu, Y.; Hwang, Y.; Zhang, C.; McCord, M. G. The use of atmospheric pressure plasma treatment in sesizing PVA on viscose fabrics. J. Indus. Text. *2003*, 32, 223–232.
307. Hwang, Y.J.; McCord, M.G.; Kang, B.C. Helium/oxygen atmospheric pressure plasma treatment on poly(ethylene terephthalate) and poly(trimethylene terephthalate) knitted fabrics: Comparison of low-stress mechanical/surface chemical properties. Fiber Polym. *2005*, 6, 113–120.
308. McCord, M.G.; Hwang, Y.J.; An, J.S.; Park, S.W.; Kang, B.C. The effects of helium atmospheric pressure plasma treatment on low-stress mechanical properties of polypropylene nonwoven fabrics. Textile Res. J. *2005*, 75, 771–778.
309. McCord, M.G.; Hwang, Y.J.; Qiu, Y.; Canup, L.K.; Bourham, M.A. Surface analysis of cotton fabrics fluorinated in radio-frequency plasma. J. Appl. Polym. Sci. *2003*, 88, 2038–2047.
310. Vitchuli, N.; Shi, Q.; Nowak, J.; Nawalakhe, R.; Sieber, M.; Bourham, M.; McCord, M.; Zhang, X. Plasma-electrospinning hybrid process and plasma pretreatment to improve adhesive properties of nanobers on fabric surface. Plasma Chem. Plasma Process. *2011*, 32(2), 275–291.
311. Hughey, V.L.; Wilger, P.A.; Johnson, E.A. Antibacterial activity of hen egg white lysozyme against Listeria monocytogenes Scott A in foods. Am. Soc. Microbiol. *1989*, 55, 631–638.
312. Mecitoflu, C.; Yemenicioflu, A.; Arslanoflu, A.; ElmacÂ, Z.S.; Korel, F.; Cetin, A.E. Incorporation of partially purified hen egg white lysozyme into zein films for antimicrobial food packaging. Food Res. Int. *2006*, 39, 12–21.
313. Branen, J.K.; Davidson, P.M. Enhancement of nisin, lysozyme, and monolaurin antimicrobial activities by ethylenediaminetetra acetic acid and lactoferrin. Int. J. Food Microbiol. *2004*, 90, 63–74.
314. Warriner, R.; Burrell, R. Infection and the chronic wound: A focus on silver. Adv. Skin Wound Care. *2005*, 18(8), 2–12.
315. Kirsner, R.S.; Orstead, H.; Wright, J.B. Matrix metalloproteinases in normal and impaired wound healing: A potential role for nanocrystalline silver. Wounds. *2001,* 13(3 Suppl C), 5–12.
316. Dunn, K.; Edwards-Jones, V. The role of Acticoat TM with nanocrystalline silver in the management of burns. Burns. *2004*, 30(Suppl 1), S1–S9.
317. Burrell, R.E. A scientific perspective on the use of topical silver preparations. Ostomy Wound Manag. *2003*, 49(5A Suppl), 19–24.
318. Taylor, P.L.; Ussher, A.L.; Burrell, R.E. Impact of heat on nanocrystalline silver dressings. Part I. Chemical and biological properties. Biomat. *2005*, 26(35), 7221–7229.

319. Li, X.Z.; Nikaido, H.; Williams, K.E. Silver-resistant mutants of *Escherichia coli* display active efflux of Ag^+ and are deficient in porins. J. Bacterial. *1997*, 179, 6127–6132.
320. Hussain, S.M.; Hess, K.L.; Gearhart, J.M.; Geiss, K.T.; Schlager, J.J. In vitro toxicity of nanoparticles in BRL 3A rat liver cells. Toxicol. In Vitro. *2005*, 19(7), 975–983.
321. Lam, P.K.; Chan, E.S.; Ho, W.S.; Liew, C.T. In vitro cytotoxicity testing of a nanocrystalline silver dressing (Acticoat) on cultured keratinocytes. Br. J. Biomed. Sci. *2004*, 61(3), 125–7.
322. Innes, M.E.; Umraw, N.; Fish, J.S.; et al. The use of silver coated dressings on donor site wounds: a prospective, controlled matched pair study. Burns. *2001*, 27, 621–627.
323. Mirsattari, S.M.; Hammond, R.R.; Sharpe, M.D.; Leung, F.; Young, G.B. Myoclonic status epilepticus following repeated oral ingestion of colloidal silver. Neurology. *2004*, 62(8), 1408–1410.
324. Oberdorster, G.; Oberdorster, E.; Oberdorster, J. Nanotoxicology: An emerging discipline evolving from studies of ultrafine particles. Environ. Health Perspect. *2005*, 113, 823–839.
325. Tinkle, S.S.; Antonini, J.M.; Rich, B.A.; Roberts, J.R.; Salmen, R.; DePree, K.; Adkins, E.J. Skin as a route of exposure and sensitization in chronic beryllium disease. Environ. Health Perspect. *2003,* 111, 1202–1208.
326. Ryman-Rasmussen, J.P.; Riviere, J.E.; Nancy, A. Monteiro-riviere. penetration of nano material into skin. Toxicol. Sci. *2006*, 91(1), 159–165.
327. Shvedova, A.A.; Castranova, V.; Kisin, E.R.; Schwegler-Berry, D.; Murray, A.R.; Gandelsman, V.Z.; Aynard, A.; Baron, P. Exposure to carbon nanotube material: Assessment of nanotube cytotoxicity using humankeratinocyte cells. J. Toxicol. Environ. Health. *2003*, A66, 1909–1926.
328. Monteiro-Riviere, N.A.; Nemanich, R.J.; Inman, A.O.; Wang, Y.Y.; Riviere, J.E. Multi-walled carbon nanotube interactions with human epidermal keratinocytes. Toxicol. Lett. *2005*, 155, 377–384.
329. Pan, Y.; Neuss, S.; Leifert, A.; Fischler, M.; Wen, F.; Simon, U.Size-dependent cytotoxicity of gold nanoparticles. Small. *2007*, 3, 1941–1949.
330. Banga, A.K. Therapeutic peptides and proteins: Formulation, processing and delivery systems. Second Edition edn, (Taylor & Francis Group). 2006.
331. Marro, D.; Guy, R.H.; Delgado-Charro, M.B. Characterization of the iontophoretic permselectivity properties of human and pig skin. J. Control Release. *2001*, 70, 213–217.
332. OECD. *Test No. 428: Skin Absorption: In Vitro Method.* (OECD Publishing).
333. Schäfer-Korting, M.; Bock, U.; Gamer, A.; Haberland, A.; Haltner-Ukomadu, E.; Kaca, M.; Kamp, H.; Kietzmann, M., Korting, H.C., Krächter, H.U., Lehr, C.M., Liebsch, M., Mehling, A., Netzlaff, F., Niedorf, F.; Rübbelke, M.K.; Schäfer, U.; Schmidt, E.; Schreiber, S.; Schröder, K.R.; Spielmann, H.; Vuia, A. Reconstructed human epidermis for skin absorption testing: results of the German prevalidation study. Alternatives to Laboratory Animals, ATLA. *2006,* 34, 283–294.
334. Cohen, D.; Soroka, Y.; Ma'or, Z.; Oron, M.; Portugal-Cohen, M.; Brégégère, F.M.; Berhanu, D.; Valsami-Jones, E.; Hai, N.; Milner, Y. Evaluation of topically applied copper(II) oxide nanoparticle cytotoxicity in human skin organ culture. Toxicol. In vitro Int. J. in association with BIBRA. *2013,* 27, 292–298.

335. Riviere, J.E.; Bowman, K.F.; Monteiro-Riviere, N.A.; Dix, L.P.; Carver, M.P. The isolated perfused porcine skin flap (IPPSF). I. A novel in vitro model for percutaneous absorption and cutaneous toxicology studies. Fundamental Appl. Toxicol. Off. J. Soc. Toxicol. *1986*, 7, 444–453.
336. Mortensen, L.J.; Ravichandran, S.; Delouise, L.A. The impact of UVB exposure and differentiation state of primary keratinocytes on their interaction with quantum dots. Nanotoxicology. *2013,* 7, 1244–1254.
337. Saathoff, J.G.; Inman, A.O.; Xia, X.R.; Riviere, J.E.; Monteiro-Riviere, N.A. In vitro toxicity assessment of three hydroxylated fullerenes in human skin cells. Toxicol. In vitro : Int.J. published in association with BIBRA. 2011, **25**, 2105-2112.
338. Shukla RK, Sharma V, Pandey AK, Singh S, Sultana S, Dhawan A. ROS-mediated genotoxicity induced by titanium dioxide nanoparticles in human epidermal cells. Toxicology in vitro: Int. J. published in association with BIBRA, *2011,* 25, 231–241.
339. OECD. *Guidance Document for the Conduct of Skin Absorption Studies*. (OECD Publishing).
340. OECD. *Test No. 427: Skin Absorption: In Vivo Method.* (OECD Publishing).
341. Labouta, H.I.; Schneider, M. Interaction of inorganic nanoparticles with the skin barrier: current status and critical review. Nanomed. Nanotechnol. Biol. Med. *2013*, 9, 39–54.
342. Chilcott, R.P.; Price, S.; Interscience, W. Principles and practice of skin toxicology. 2008, Wiley Online Library.
343. Magnusson, B.M.; Walters, K.A.; Roberts, M.S. Veterinary drug delivery: potential for skin penetration enhancement. Adv. Drug Deliv. Rev. *2001*, 50, 205–227.
344. Gontier, E.; Ynsa, M.D.; Bíró, T.; Hunyadi, J.; Kiss, B.; Gáspár, K. Is there penetration of titania nanoparticles in sunscreens through skin? A comparative electron and ion microscopy study. Nanotoxicol. *2008,* 2, 218–231.
345. Vlachou, E.; Chipp, E.; Shale, E.; Wilson, Y.T.; Papini, R.; Moiemen, N.S. The safety of nanocrystalline silver dressings on burns: a study of systemic silver absorption. Burns: J. Int. Soc. Burn Injur. *2007,* 33, 979–985.

CHAPTER 14

CUTANEOUS WOUND CARE: GRAFTS TO TISSUE-ENGINEERED SKIN SUBSTITUTES

ROBIN AUGUSTINE, BHAVANA VENUGOPAL, NANDAKUMAR KALARIKKAL, and SABU THOMAS

ABSTRACT

The skin is the outer most covering of the human body. It is the largest organ of the integumentary system. Trauma or injury to skin is common and it is necessary to take immediate action as early as possible to restore the functions of skin after an injury. This book chapter gives an overview about skin, its structure, and essential functions. Recent advancements in the wound care with a focus on skin substitutes and skin grafts have been presented. Advancement of polymeric substances as skin substitutes aiding the healing, acting as physical barriers, and giving other functionalities of the natural skin is also discussed.

14.1. INTRODUCTION

Skin is the largest and most dynamic organ of the body, with a constant state of change, as cells of the outer layers are continuously shed and replaced by inner cells. It has a surface area of 1.8 m^2. Amongst the several vital functions, most important are to constitute a physical barrier to the environment, allowing and limiting the inward and outward passage of water, electrolytes, and various substances while providing protection from microorganisms, ultraviolet irradiation, toxic factors, and mechanical injuries. Skin, being always in direct contact with the external environment and other stresses makes them highly susceptible to damage and/or injury. Damages to this barrier will cause loss of water and protein, and bacterial invasion in the underlying tissue. Extensive damages in the skin therefore require immediate action taken as early as possible (Prasanna et al., 2004; Xiao et al., 2002). The current trend of burn wound care has been shifted from merely achieving satisfactory survival rate to improvement in the long-term form and function of the healed burn wounds and quality of life. The change in the trend has demanded the emergence of various skin substitutes in the management of burn injury.

Therefore numerous efforts are being made to enhance the regeneration of the skin. Initial studies were based on the use of grafts of allo or auto origin to promote healing. As a recent development, polymeric skin substitutes have been developed to act as smart skin substitutes by performing many of the functions of natural skin. As a future scope, nanotechnology is also taking the pleasure of advancing the skin substitute's efficiency. Nanocomposites of metals such as silver and zinc with inherent antimicrobial activity are now incorporated with skin substitutes. In the time ahead, it is expected that the researchers would find a better substitute that would render the patient a scar free skin.

14.2 STRUCTURE OF SKIN

Skin is the largest organ of the human body. The skin of an average adult body covers a surface area of about 20 square feet and weighs more than 10% of the total body mass (Moore and Chien, 1988). It has several functions, the most important being to form a physical barrier to the environment, allowing and limiting the inward and outward passage of water, electrolytes, and various substances while providing protection against microorganisms, ultraviolet radiation, toxic agents, and mechanical damages. Skin has the following three layers:

- The epidermis, the outermost layer of skin, provides a waterproof barrier and creates our skin tone.
- The dermis, beneath the epidermis, contains tough connective tissue, hair follicles, and sweat glands.
- The deeper subcutaneous tissue (hypodermis) is made of fat and connective tissue.

The epidermis, which consists of several layers, is mostly made of dead epithelial skin cells, which include the following:

- Stratum basale
- Stratum spinosum
- Stratum granulosum
- Stratum licidum
- Stratum corneum

The bottom layer, the stratum basale, has cells that are shaped like columns. In this layer, the cells divide and push already formed cells into higher layers. As the cells move into the higher layers, they flatten and eventually die. The top layer of the epidermis, the stratum corneum, is made of dead, flat skin cells that shed about every 2 weeks and has been demonstrated to constitute the principal barrier to percutaneous penetration (Blank, 1969; Scheuplein et al., 1971). The exquisite barrier properties of the stratum corneum can be ascribed to its unique structure and composition.

Directly beneath the epidermis, is the other primary skin layer, which is called the dermis. This layer has small blood vessels, nerve endings, oil and sweat glands, and hair follicles (Schaefer et al., 1996). The dermis also contains collagen and elastic tissue, which function to keep the skin firm and strong. There is an extra layer underlying the dermis called the subcutaneous layer, which is made up of fatty tissue that acts as a foundation for the dermis. The hypodermis or subcutaneous fat tissue supports the dermis and epidermis and provides thermal isolation and mechanical protection to the body. As the first line of defense against the external environment, the epidermis is continuously replenishing and shedding tens of thousands of dead cells every minute to protect the body from: mechanical impact (Sherwood et al., 2004) fluids, radiation, and infection. It plays a crucial role in the regulation of the body temperature and serves as a sensory organ transmitting external environmental information, such as pain and heat (Barry, 1983; Williams and Barry, 1991).

14.3 CUTANEOUS WOUNDS AND WOUND CARE

A cutaneous wound is defined as a defect or break in the skin, resulting from physical or thermal damage or as a result of the presence of an underlying medical or physical condition. Cutaneous wounds may be classified by several methods viz; their etiology, location, type of injury or presenting symptoms, wound depth, and tissue loss or clinical appearance of the wound. In general, cutaneous wounds are classified as follows:

- Superficial (loss of epidermis only)
- Partial thickness (involve the epidermis and dermis)
- Full thickness (involve the dermis, subcutaneous fat, and sometimes bone) [Flanagan, M. (1994) Assessment Criteria. Nursing Times: 90; 35, 76–86]

Based on the nature and repair process of wounds, they can also be classified as chronic wounds and acute wounds (Boateng et al., 2008). Acute wounds are tissue injuries that heal within 8–12 weeks. The primary causes of acute wounds are mechanical injuries (friction contact between skin and hard surfaces), burns, and chemical injuries. Chronic wounds heal slowly and leave serious scars. There can be different reasons that chronic wound do not heal as fast as acute wounds. Among most common are diabetes, infections, and poor primary treatment (Boateng et al., 2008).

The following are the common chronic wounds:

- Venous ulcers
- Arterial ulcers
- Diabetic foot ulcers
- Pressure ulcers
- Vasculitis (Inflammation of a blood vessel)
- Pyoderma gangrenosum (Fonder et al., 2008).

A wound is colonized when growth and death of bacterial in the wound is balanced by the host. If the host is not able to keep the bacterial growth in balance, the wound will enter the infection phase (bacterial load in excess of 1015). Symptoms of an infected wound are erythema, edema, warmth, pain, and exudate. Infections of chronic wounds are often polybacterial with *Staphylococcus aureus* and anaerobes being the most common (Fonder et al., 2008).

A successful wound treatment requires wound care, which involve the assessment of the patient, since systemic problems can withhold the wound healing process. Characterization of wound for its size, depth, and exudates is next in line. Adequate nutrient intake is the most overlooked requirement in wound care. Inadequate calorie supply can impair the normal wound healing process. Prevention of infections of any sort demands a great priority as well.

14.4 CUTANEOUS WOUND HEALING

Cutaneous wound healing has evolved significantly in the past century from magical spells, incantations, and potions to the use of sophisticated wound dressings, hemostats, hyperbaric oxygen chambers, and the very recent regenerative dermal substitutes. Despite the emergence of new therapies and the evolution of wound care management, there is still a pressing need for more enhanced and efficient wound healing treatments. Currently, several therapies involving the administration of growth factors and stem cells to wound sites are being investigated to accelerate the wound healing process.

A wound, on a scientific prospective, is a severance of the normal anatomical structure and function of the skin (Lazarus et al., 1994) whereas, wound healing is a biological process that the body finds to overcome the situation, and is always related to the physiological parameters. The wound healing process comprehends several physiological and biological stages (Boateng et.al. 2008), which include; coagulation (hemostasis), inflammation (early and late), proliferation, and remodeling (maturation). Nevertheless, wound healing is not linear and often wounds can progress both forwards and back through the phases depending on intrinsic and extrinsic forces at work within the patient. The phases of wound healing are as follows:

- Coagulation phase
- Inflammatory phase
- Proliferation phase
- Maturation phase

Key events involved in each step of wound healing are given in Figure 14.1.

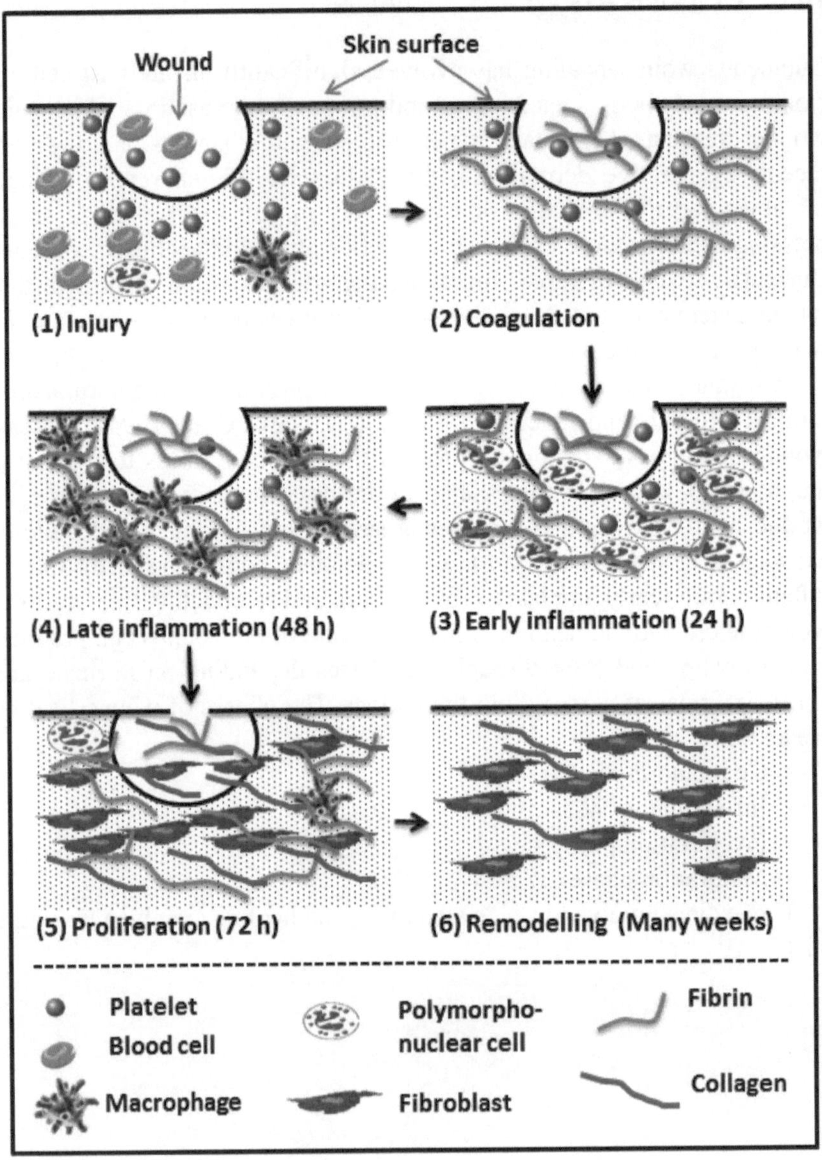

FIGURE 14.1 Stages of wound healing

Upon an injury or trauma to the skin, the **coagulation** cascade, inflammatory pathways, and immune system are activated immediately. They have a crucial role in preventing further blood and fluid loss, removing dead tissue, as well as avoiding infection (Gurtner et al., 2008). Homeostasis is rapidly established directly after injury with vascular constrictions and formation of a platelet plug (consisting of platelet, red- and white blood cells), followed by a fibrin-fibronectin matrix (Guo and DiPietro, 2010; Rodero and Khosrotehrani, 2010; Valenick et al., 2005). The fibrin matrix will act as a scaffold for cells which are infiltrating into the wound site. The platelet plug and surrounding tissue to the wound release pro-inflammatory cytokines and growth factors, including transforming growth factor (TGF), platelet-derived growth factor (PDGF), fibroblast growth factor (FGF), as well as epidermal growth factor (EGF) (Guo and DiPietro, 2010).

The **inflammatory** phase involves the body's natural response to injury. After the initial wounding, the blood vessels in the wound bed contract and a clot is formed. Once haemostasis has been achieved, blood vessels dilate to allow essential cells; antibodies, white blood cells, growth factors, enzymes, and nutrients to reach the wounded area. This leads to a rise in exudate levels. The characteristic signs of inflammation can be seen; erythema, heat, edema, pain, and functional disturbance. The predominant cells at work are the phagocytic cells; "neutrophils and macrophages"; mounting a host response and autolyzing any devitalized "necrotic/sloughy" tissue (Grose and Werner, 2004; Guo and DiPietro, 2010).

As the inflammation is reduced, the body starts to repair the injury. Therefore **proliferation phase** characterized by cellular proliferation and re-epithelialization of the wound (Guo and DiPietro, 2010, Gurtner et al., 2008). Keratinocytes migrate from the wound edges to the injured dermis by the production of specific proteases, such as collagenase, to degrade the extracellular matrix (ECM) (Gurtner et al., 2008; Pilcher et al., 1997). Fibroblasts and endothelial cells (ECs) support the capillary growth which supply oxygen and nutrients to the new/healing tissue, a process known as angiogenesis (Guo and DiPietro, 2010; Gurtner et al., 2008). Granulation tissue will be visible near the wound by the end of the first week and this tissue will continue to grow until the wound heals. The tissue contains new blood vessels and other components to fill in the damaged tissue. Granulation tissue is normally bright red, moist, soft to the touch, and has a bumpy appearance.

Remodelling (maturation phase) is the concluding phase of wound healing process which happens within 2–3 weeks after injury and can last for years (Guo and DiPietro, 2010; Gurtner et al., 2008). During the remodeling phase, healing process halts or scales down and most ECs, macrophages, and myofibroblasts will either undergo apoptosis or leave the wound while type I collagen replaces type III collagen (Gurtner et al., 2008). The vascular density returns to normal and the ECM regenerates to a scaffold similar to that of normal tissue (Guo and DiPietro, 2010; Gurtner et al., 2008).Cellular activity reduces and the number of blood vessels in the wounded area regress and decrease.

14.5 FACTORS THAT AFFECT WOUND HEALING

Factors that can slow the wound healing process include:

- **Dead skin (necrosis)** – dead skin and foreign materials interfere with the healing process.
- **Dryness** – wounds (such as leg ulcers) that are exposed to the air are less likely to heal. The various cells involved in healing, such as skin cells and immune cells, need a moist environment.
- **Infection** – an open wound may develop a bacterial infection. The body fights the infection rather than healing the wound.
- **Age** – wounds tend to take longer to heal in elderly people.
- **Smoking – cigarette smoking** impairs healing and increases the risk of complications.
- **Haemorrhage** – persistent bleeding will keep the wound margins apart.
- **Mechanical damage** – for example, a person who is immobile is at risk of bedsores because of constant pressure and friction.
- **Diet** – poor food choices may deprive the body of the nutrients it needs to heal the wound, such as vitamin C, zinc, and protein.
- **Medical conditions** – such as diabetes, anemia, and some vascular diseases that restrict blood flow to the area, or any disorder that hinders the immune system.
- **Medicines** – certain drugs or treatments used in the management of some medical conditions may interfere with the body's healing process.
- **Varicose veins** – restricted blood flow and swelling can lead to skin break down and persistent ulceration.

14.6 SKIN SUBSTITUTES

Skin grafting has been considered as the most conventional treatment that can restore the functions of the skin immediately after an injury (Boucard et al., 2007). Grafts are the primitive forms of the modern skin substitutes. The technique skin grafting was so deep rooted that it was also prevalent in the times of *Koomas,* a caste recognized for pottery and tile making in India. The ancient Indians followed a simple technique rather a crude way by pounding the skin slices obtained from a donor using a wooden slipper until it is swollen and inflamed. Skin grafting has evolved from the initial autograft and allograft preparations to biosynthetic and tissue-engineered living skin replacements. This has been fostered by the dramatically improved survival rates of major burns where the availability of autologous normal skin for grafting has become one of the limiting factors.

Since the time immemorial, relentless researches were promoted to find a new resolution in the form of skin substitutes in skin trauma care. The biosynthetic materials and tissue-engineered living skin replacements are the new strategies in skin repair.

Skin substitutes can be classified into many types. Mainly they are classified into epidermal equivalents, dermal components, and composite skins (those possessing distinct dermal and epidermal components) (Balasubramani et al., 2001). Use of both natural and synthetic polymers is considered as the recent advancement in field of tissue-engineered skin substitutes (Lanza et al., 2011). Biodegradable polymers are endorsed with a great value in tissue engineering since they disintegrates when the functional and original tissue regenerates in the process of healing. In medical industry, the most recent trend is the application of polymeric substitutes incorporated with drug and antimicrobial substances which prevent any retardation on the healing or regeneration processes. The growth factors and other ECM substances immobilized biopolymers, also serve as an active drug delivery system. Not a long ago, Nanotechnology is extending its interests to bring out the smartest skin substitutes ever. Each day is a new hope for the future wherein scientists are relentlessly working on new ways to draw out a better way of relief.

Skin substitutes are heterogeneous group of wound coverage materials that aid in wound closure and replace the functions of the skin, either temporarily or permanently, depending on the product characteristics. These substances serve as alternatives to the standard wound coverage in circumstances when standard therapies are not desirable (Shores et al., 2007). Skin substitutes are employed to assist in wound closure, control associated pain and replace the skin function to encourage healing of the injury. Skin substitutes have important functions in the treatment of deep dermal and full-thickness wounds of various etiologies (Halim et al., 2010). Treating wounds with "skin substitutes" dates back to 1880 when Joseph Gamgee described an absorbent dressing made of cotton wool sandwiched between layers of gauze (Ho, 2002).

A variety of biosynthetic and tissue-engineered human skin equivalents (HSE) are manufactured under an array of trade names and marketed for various purposes. All of these products are procured, produced, manufactured or processed in sufficiently different manners that they cannot be addressed and evaluated as equivalent products. However, an optimal skin substitute will provide immediate replacement of both the lost dermis and epidermis, with permanent wound coverage (Sheridan and Moreno, 2001). Any skin substitute should maintain a moist environment at the wound interface, allow gaseous exchange, act as a barrier to microorganisms, and remove excess exudates. It should also be nontoxic, nonallergenic, nonadherent and easily removed without trauma, it should be made from a readily available biomaterial that requires minimal processing, possesses antimicrobial properties, and promotes wound healing.

Other key characteristics of an ideal skin substitute are as follows (Shores et al., 2007):

- Quality guaranteed with long shelf life
- Can be used off the shelf
- Nonanitgenic
- Durable
- Flexible
- Prevents water loss
- Bacterial barrier
- Drapes easily
- Easy to use and totally secure
- Growths with a child
- Can be applied in one operation

- Does not become hypertonic
- Efficient and cost effective

14.7 CLASSIFICATION OF SKIN SUBSTITUTES

Skin substitutes are often categorized as either temporary or permanent and are likewise thought as products that provide temporary wound coverage or wound closure. In actuality, few (if any) of the products used to treat burns today are permanent skin substitutes. Deciding whether a product is a temporary or permanent covering and whether or not it produces formal wound closure as opposed to simple coverage is somewhat arbitrary and can be confusing (Hansbrough and Franco, 1998).

Skin substitutes can be obtained from human tissue (allografts) or animal tissue (xenografts), or using membranes fabricated from natural or synthetic polymers. To date, there is no ideal skin substitute available that fulfills all the above-mentioned features. Currently, tissue engineering and biotechnology are gearing toward the direction of creating an optimal skin substitute.

14.7.1 TEMPORARY SKIN SUBSTITUTES

Temporary skin substitutes is a collection of varied and topically applied agents that are thought to offer more protective covering to a healing burn. Such skin substitutes include products that have inherent healing properties of their own or have added biological active substances, which presumably are able to promote wound healing. Temporary skin substitutes provide transient physiologic wound closure, including protection from mechanical trauma, physical barrier to bacteria, and creation of a moist wound environment (Sheridan and Moreno, 2001). One of the most widely used temporary skin substitutes is frozen cadaver skin (Burke et al., 1981).

The most common uses for temporary skin substitutes include the following:

1. For dressing on donor sites to facilitate epithelialization and pain control
2. For dressing on clean superficial wounds until epithelialization

3. To provide temporary physiological closure of deep dermal and full-thickness wounds after excision while awaiting autografting
4. As sandwich graft technique over the widely meshed autografts
5. As a "test" graft in questionable wound beds (Halim et al., 2010).

14.7.2 PERMANENT SKIN SUBSTITUTES

When the skin is irreversibly damaged, we nevertheless have no materials that will replace it perfectly, and an inexpensive, dependable, and durable permanent skin substitute will revolutionize burn care (Sheridan and Tompkins, 1999). Permanent skin substitutes are used to achieve a permanent wound closure, replace the skin components, and provide a higher quality skin replacement than the thin autologous skin graft. The split-thickness autograft is still the best current solution. There are, nevertheless, a number of membranes that offer more or less level of permanent skin replacement. These membranes can be classified as epidermal, dermal, and composite substitutes.

The technology for epidermal replacement was built up in the 1970s. Epithelial cells are secured from a full-thickness skin biopsy and cultured in a medium containing growth factors over a layer of mouse fibroblasts. Colonies of epithelial cells expand into full sheets which are recaptured until confluent thin layers of undifferentiated cells are obtained. These sheets are then attached to a petroleum gauze carrier for easier handling. Many of the imperfections associated with epithelial wound closure may be attributable to the absence of a dermal component. Outcomes were mixed, with the first effort which involved leaving behind from a split-thickness allograft the vascularized allergenic dermis after mechanical excision of the epidermis and upper dermis. The first synthetic dermal analogue was probably a biodegradable polygalactin mesh seeded with allergenic fibroblasts, but it was not successful in clinical tests. Commonly avaiable skin substatutes in the market are given in Table 14.1.

TABLE 14.1 Classes of commercially available skin substitutes

Substitute Type	Commercial Forms	Description	Uses
Autograft	Epicel®	Cultured epidermal autograft	Severe deep dermal, full-thickness burns
	MySkin™	Cultured epidermal autograft	For burns, ulcers, and other nonhealing wounds
	Cultured Skin Substitutes	Cultures composite autograft	For large burns and other congenital skin disorders.
	Bioseed®-S	Autologus keratinocyte fibrin glue suspension	Treatment of chronic leg ulcers
	CellSpray®	Cultured epithelilal autograft suspension	To treat superficial burns
	Stratagraft®	Cultured composite autograft	Burns and severe skin wounds.
	Recell®	Autologus cell therapy device	To treat burns, scalds, traumatic wounds, scars
Allograft	Lyphoderm™	Lysate of cultured human keratinocyte	For chronic leg ulcers
	ICX-SKN	Cultured dermal allograft	To cover surgically excised partial thickness burns.
	Alloderm®	Cadaver skin with acellular dermal matrix and intact basement membrane	For ENT/head and neck plastic reconstruction
Acellular Allogra	OASIS®	Processed dermal xenograft	For partial and full-thickness wounds and trauma wounds

TABLE 14.1 *(Continued)*

Xenograft and Biosynthetic	Permacol™	Processed dermal xenograft	For temporary coverage of partial thickness burns
	Matriderm®	Bovine dermal collagen and elastin	For burns and reconstruction
	Biobrane®	Porcine dermal collagen bonded to semipermeable silicone membrane	To cover partial thickness burns and skin graft donor sites
	Integra®	Two layered skin substitute comprsisng bovine collagen and an outer silicone layer.	For surgically excised deep and full-thickness burns
	EZ Derm™	Porcine derived xenograft with collagen crosslinked to an aldehyde	For partial thickness wounds, donor sites, and sandwich autografts and full-thickness wounds

14.7.3 BIOLOGICAL SKIN SUBSTITUTES

These skin substitutes, which act temporarily like skin, have the advantages of being relatively abundant in supply and not expensive. The biological skin substitutes have a more intact and native ECM structure which may allow the construction of a more natural new dermis. They also show excellent re-epithelialization characteristics due to the presence of a basement membrane. The most widely used biological substitute worldwide is a cadaveric skin allograft, porcine skin xenograft, and amnion.

14.7.4. XENOGRAFT

Xenografts are skin substitutes harvested from the animals for use as temporary graft in human. Porcine skin allograft, which is the highly processed pig skin consisting only the dermal layer, is the commonly used xenograft in the modern practice of burn care. It is primarily used in the coverage of partial thickness burns and excised wounds prior to skin grafting.

14.7.5 SKIN ALLOGRAFT

The cadaveric skin allograft application is one of the most commonly applied skin substitutes in trauma management in many major burn centers all over the globe. There are even reports on skin allograft from living donors. Depending on the methods of processing and storage, there are two main types of cadaveric skin allografts, cryo-preserved allograft, and glycerol-preserved allograft (GPA). The GPA is the most popular in clinical practices (Khoo et al., 2010).

14.7.6 AMNION

The amnion is a thin semitransparent tissue found in the innermost layer of the fetal membrane. It has been utilized as biological dressings for burns since 1910. Since it is obtained from human placenta, amnion is one of the most effective substitutes to be used in healing or covering partial thickness burns, as well as other superficial wounds.

14.7.7 CULTURED EPITHELIAL AUTOGRAFTS (CEA)

Keratinocytes can be grown in culture to produce thin epithelial sheet grafts (Liu et al., 2010). The autologous keratinocytes are isolated, cultured, and expanded into sheets over periods of 3–5 weeks. The use of suspension keratinocytes in fibrin glue has reduced the time for clinical use to 2 weeks (Wood et al., 2006).

14.7.8 SYNTHETIC SKIN SUBSTITUTES

The hunt for a true skin substitute is an ongoing process because of the inherent difficulties with allograft and xenograft. The term "skin substitute" entered the burn vernacular in the 1980s. A giant leap ahead in the evolution of skin substitutes came from Burke et al. and the development of Integra® (Integra Life Sciences, Plainsboro, NJ). Their historic and groundbreaking work was exhibited in 1981. According to Ravage, the development of Integra was the outcome of a collaborative research by these two pioneers that dates back almost 30 years (Ravage et al., 2004).

Synthetic skin substitutes are made out of nonbiological materials and polymers that are not present in a normal skin (van der Veen et al., 2010). These constructs are designed to be stable, biodegradable, and provide an adequate environment for the regeneration of tissue. It should maintain its three-dimensional (3-D) structure for at least 3 weeks to allow ingrowths of blood vessels, fibroblast, and coverage by epithelial cells. Biodegradation should preferably take place over this point. The process should not bring up any massive foreign body reaction as the process is prone to increase the inflammatory response, which may be associated with profound scarring. It should likewise be composed of incompatible materials to avoid immunoreactive processes.

The contrived nature of these skin substitutes has some distinct advantages and disadvantages when compared to natural biological structures. The composition and properties of the product can be controlled more precisely. Various additives such as growth factors and matrix components can be added to heighten the effect. These products could also avoid complications due to potential disease transmission. However, to add to the side of disadvantage, these synthetic skin substitutes generally lack basement membrane and their architecture do not resemble the native skin. The role of nonbiological components can be problematic when intended to produce a biologically compatible material (van der Veen et al., 2010). However, there is also a substantial number of synthetic substitutes undergoing in vitro or animal testing (Keong and Halim, 2009; Powell and Boyce, 2009). Amongst the synthetic skin substitutes available in the market are Biobrane®, Dermagraft®, Integra®, Apligraft®. The use of synthetic polymers has never been a solid answer to the problem of a skin substitute. A high incidence of infection and a relatively low capacity for inducing vascularization and epithelialization are frequently reported. Nevertheless, useful insights into the necessities for a satisfactory skin replacement have been revealed through the usage of synthetic polymers.

Despite being clinically useful, skin grafts have many restrictions including the availability of the donor, especially in conditions of extensive skin loss, immune rejection in allogeneic skin grafts, pain, scarring, slow healing, and infection (Horch et al., 2005; Lee, 2000). To overcome the bottleneck, scientists are in a constant stride to find an efficient skin substitute that replaces the need for a natural skin graft.

14.8 TISSUE-ENGINEERED SKIN

Tissue engineering has emerged as a new and promising field for the treatment of skin lesions, combining scaffolds, cells, and biomolecular signals, such as growth factors. It is an interdisciplinary field of Nanomedicine in which bilateral and medical science understands of pathological tissue and the principles employed to reach this understanding are applied to the improving or sustaining of tissue function through the growth of biological reserves.

Most tissue-engineered skin is created by expanding normal skin cells in the laboratory. Such engineered skin can be used for long time healing against the synthetic materials that can solely be used for short time healing, because the materials must eventually be removed to be replaced by natural skin cells (MacNeil, 2007). An ideal tissue scaffold should have certain attributes such as being biocompatible, mechanical properties to match the tissue, an appropriate surface chemistry, and be highly porous with a network of interconnected pores that will allow cells to attach and be able to transport nutrients and waste (Hutmacher et al., 2001). A scaffold for tissue engineering with a large surface to volume ratio will have higher opportunities for the cells to attach and electrospun nanofibers have this property and are found to promote cell adhesion and migration (Min et al., 2004).

14.9 ELECTROSPINNING OF POLYMERS

Electrospinning is a simple and effective method for producing fibers from tens of nanometers to micrometers (Doshi et al., 1995). In this process, a polymer solution from a reservoir is ejected to a small opening of a capillary by means of Coulombic repulsion of charges that are accumulated at the tip of a pendant droplet as soon as an electrical potential applied between the capillary and a collecting device increases beyond a critical value (Formhals, 1934). As the charged jet travels to the collector, it readily dries out, forming nonwoven fibrous mats depositing on the collector.

A schematic representation of electrospinning process is given in Figure 14.2.

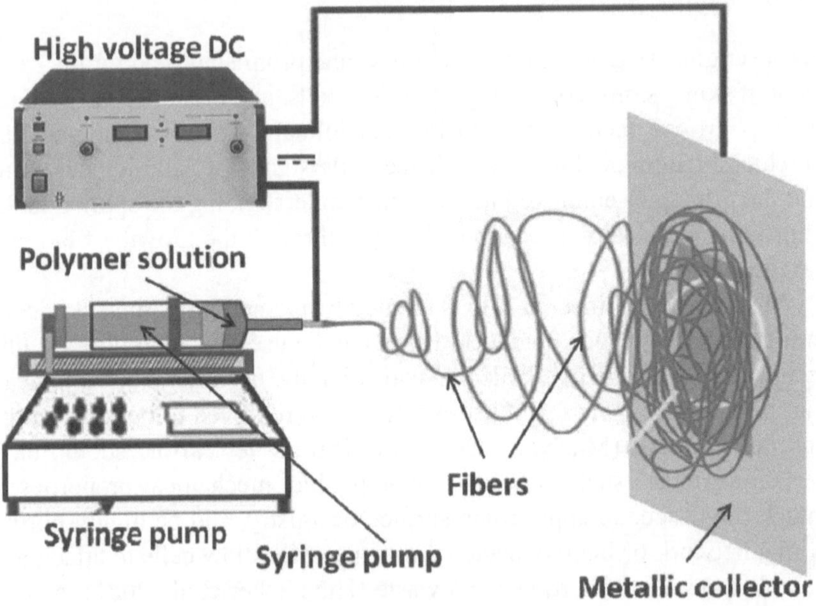

FIGURE 14.2 Schematic representation of electrospinning process.

Electrospinning technology, which can easily mass-produce thin nanofibrous membranes with good conformability, could provide a solution to the manufacture of skin substitutes. Electrospun nanofibers resemble the native topographical features of the natural ECM and may thus promote the cell's natural functions in a biomimetic fashion. These materials possess various properties that make them suitable as skin substitute application such as high oxygen permeability, variable pore size, a high surface area to volume ratio, and morphological similarity to the ECM (Smith and Ma, 2004; Zahedi et al., 2010; Zhou et al., 2007). Various natural and synthetic polymer/polymer blends have been electrospun into nanofibers to generate potential wound dressing materials. The ability to incorporate a variety of bioactive molecules (such as antimicrobials, anti-inflammatories) into the nanofibers can increase the desirable wound healing (and wound dressing) properties.

14.9.1. ELECTROSPUN MEMBRANES AS SKIN SUBSTITUTES

Many researchers shown that electrospun membranes endowed with excellent capability of supporting fibroblast and/or keratinocytes attachment and proliferation with characteristic phenotypic shape (Fang et al., 2011). Poly(ε-caprolactone) (PCL) and collagen core-sheath nanofibrous membranes have demonstrated to have good biocompatability on fibroblasts (Zhang et al., 2005a; Zhao et al., 2007). Strategies like improving hydrophilicity by blending of hydrophilic polymers like polyethylene glycol (PEG) with poly-L-lactic acid (PLLA) nanofibers and the biological reactivity of fibroblast cells also tried (Bhattarai et al., 2006). The presence of hydrophilic nanofibers (chitosan/PVA) in PLGA increased the absorption of nutrient fluid during cell culture and thus promoted fibroblast attachment, proliferation, migration, and infiltration in the fiber matrix (Duan et al., 2006). A previous report suggests that the proliferation of epidermal skin cells was enhanced when aligned PLLA nanofibers were used (Kurpinski et al., 2010). Fiber density and skin tissue regeneration has a direct relation and has been investigated on electrospun gelatin skin substitutes. Cell migration was limited to the upper regions of the skin substitute if the interfiber distance was less than 5.5 μm, and the distances between 5 and 10 μm have favored the proliferation of the cells deep into the scaffold (Powell and Boyce, 2008).

Recently, electrospun PCL membranes incorporated with zinc oxide (ZnO) nanoparticles have been demonstrated to use as skin substitutes with enhanced cell proliferation and wound healing capacity (Augustine et al., 2014a).

14.9.2 ADVANTAGES OF ELECTROSPUN MEMBRANES AS SKIN SUBSTITUTES

The skin substitute materials fabricated by electrospinning process have special properties as compared to the dressings produced by conventional methods (Zhang et al., 2005b). Using electrospun nanofibers as scaffold for skin substitutes has many benefits such as:

1. Absorptibility

High surface area to volume ratio of electrospun fibers, make them to absorb huge amount of water than a typical film dressing. This facilitates effective exudate management. Fluid absorption of electrospun membranes can be between 17.9% and 213%, where a standard film absorbs only 2.3% water. Good absorptive properties of a scaffold will help maintain a moist environment (Zhang et al., 2005b; Williams, 2007).

2. Functionability

An interesting feature of the electrospun membranes is that they can be made bioactive. They can be used in delivery system for a variety of biological agents such as drugs or other active components to the wound, which could improve the healing in a controlled way (Liang et al., 2007; Zhang et al., 2005a).

3. Hemostasis

Due to the small pores and high effective surface area, electrospun membranes are capable to promote hemostasis without the usage of external hemostatic agents (Wnek et al., 2003; Zhang et al., 2005b).

4. Comformability

Fine fiber fabrics are easier to fit to a complicated 3-D structure of a wound compared with thicker fibers, therefore electrospun fibermats will provide excellent conformability.

5. Semipermeability

The porous architecture of electrospun membranes will facilitate good respiration for the cells and does not lead to wound dehydration. This will give some control of the moist environment. At the same time, the pores are so small that the fibers will protect the wound from bacteria invasion (Huang et al., 2003; Zhang et al., 2005b).

6. Scar-free healing

It is believed to be possible to heal wound without leaving scars when using electrospun fiber mats. Even though this would be hard to achieve a research group have been working on this, but this is still only a work in process (Zhang et al, 2005b).

10. Revascularization through skin substitutes

An active blood vessel network is necessary for the integration of any implanted scaffold with existing host tissue. Rapid vascularization is essential to ensure the success of skin substitutes. Novel strategies for vascularizing skin substitutes by incorporating ECs have recently demonstrated encouraging preclinical and clinical results (Black et al., 1998; Drosou et al., 2005; Greenberg et al., 2005; Kearney, 2001; Marston, 2004; Supp et al., 2002). A model have been reported of augmented vascularization of human skin substitutes formed from decellularized dermis and human neonatal foreskin keratinocytes induced by seeding the underside of the grafts with human umbilical vein endothelial cells (HUVEC) prior to orthotopic transplantation to the backs of immunodeficient (C.B-17 SCID/Bg) mice (Schechner, 2003). Both formation of human EC-lined vessels within the engrafted construct and inosculation with mouse microvessels were demonstrated. The concentration of vascularization and the frequency of successful engraftment were increased further when the seeded HUVEC had been retrovirally transduced to constitutively express the antiapoptotic protein Bcl-2, a manipulation that increases the capacity of HUVEC to form mature vessels (Enis et al., 2005; Schechner et al., 2000). Our group demonstrated that ZnO nanoparticles can enhance angiogenesis in tissue engineering scaffolds (Augustine et al., 2014b). ZnO nanoparticles are known for their capability to generate reactive oxygen species (ROS) which can induce angiogenesis through growth factor mediated mechanisms. This study showed that electrospun PCL scaffold incorporated with ZnO nanoparticles are able to make extensive vasculature through the scaffold.

11. Nanoparticles as antimicrobial agents in skin substitutes

The nanoscience and nanotechnology are rapidly evolving with a wide variety of nanomaterial. The nanomaterials are nothing but insoluble particles that are 100 nm in size (Weir et al., 2008). The range of products extends from the mere household necessities to the lifesaving biomedical products (Jin et al., 2010). Surprisingly in the past twenty decades, an alarming trend was witnessed in the use of nanoparticles for the purpose of therapeutics alone. Nanoparticles are used in bioapplications such as therapeutics (Kreuter and Gelperina, 2008), antimicrobial agents (Raghupathi et al., 2011), transfection vectors (Tan et al., 2007) and fluorescent labels (Su et al., 2008). The greatest advantage in the use of nanoparticles is by the virtue of their very small size which elevates their promptness to incorporate in various topical treatments practiced in clinics (Przyborowski and Berthiaume, 2013). The nanosize results in specific physiochemical properties that make them extremely different from their counterparts of large size.

The distressing increase in resistance of microorganisms toward the potent antibiotics has spurred a tremendous activity toward investigating bactericidal properties of nanoparticles (Singh et al., 2012). Most importantly, the nanoparticles tackle multiple biological pathways found in a broad spectrum of microorganisms which may require many concurrent mutations to achieve resistance against the nanoparticle's antimicrobial activity. A plethora of studies that took part in the arena has plowed up the impact of any nanoparticle is leveraged by its size and shape (Pal et al., 2007).

Of all the products of nanosize in commerce, silver containing products are most abundant, primarily influenced by the broad-spectrum antimicrobial activity of silver along with low toxicity toward mammalian cell (Jones et al., 2004). Silver has been used for the treatment of wound and bacterial infection since olden days in the form metallic silver, silver nitrate, and silver sulfadiazine (Dibrov et al., 2002). Zinc oxide stands next in line for their advantage silver nanoparticles such as low production cost, white appearance, and UV-blocking properties (Dastjerdi and Montazer, 2010). The nano-zinc oxide multilayer deposited on cotton fabrics showed excellent antibacterial activity against *S. aureus* (Zhang et al., 2013).

12. The future skin substitutes: Genetic modification of tissue-engineered skin

Tissue-engineered skin is one of most advanced tissue constructs, yet it has several bottlenecks. Gene therapy, if applied in wound healing might promote tissue regeneration or prevent healing abnormalities such as formation of scars and keloids. Moreover, gene-enhanced skin substitutes are anticipated to have great potential as cell-based devices to deliver therapeutics locally or systemically. Although significant progress has been made in the development of gene transfer technologies, several challenges have to be dealt before clinical application of genetically modified skin tissue. The prominent engineering challenges include methods for improved efficiency and targeted gene delivery; efficient gene transfer to the stem cells that constantly regenerate the dynamic epidermal tissue; and development of novel biomaterials for controlled gene delivery. The potential of gene therapy to generate the next generation of skin substitutes with enhanced capacity for treatment of burns, chronic wounds, and even systemic diseases is the next novel strategy that is being looked upon (Andreadis et al., 2007).

13. Conclusion

Frequent damages to the skin necessitate the use of skin grafts or skin substitutes to immediately restore the functions of the skin. For the past few decades, introduction of allografts and autografts played important role in treating wounds and traumas. Cultured cells and tissue-engineered grafts were also developed as a part of the growing interest in the field of skin regeneration upon healing. In the long run, natural polymers also came up with their contribution which proved satisfactory and enhanced the efficiency of the skin substitutes. They provided reduced healing time with ease of drug delivery. Recently, nanocomposites of metals such as silver and zinc were introduced with an aim to provide antimicrobial aid against the potential infection during the wound healing. It is anticipated that, in future, the developments in skin substitutes will bring a massive change in the classic wound and skin care.

KEYWORDS

- **Coagulation**
- **Nanofibers**
- **Skin grafting**
- **Skin substitutes**
- **Tissue engineering**
- **Wound healing**

REFERENCES

Andreadis, Stelios T. "Gene-modified tissue-engineered skin: the next generation of skin substitutes." *Tissue Engineering II*. Springer, Berlin Heidelberg, 2007. 241–274.

Augustine, Robin, Edwin Anto Dominic, Indu Reju, Balarama Kaimal, Nandakumar Kalarikkal, and Sabu Thomas. "Electrospun polycaprolactone membranes incorporated with ZnO nanoparticles as skin substitutes with enhanced fibroblast proliferation and wound healing." *RSC Advances* 4, (2014a), 24777–24785.

Augustine, Robin, Edwin Anto Dominic, Indu Reju, Balarama Kaimal, Nandakumar Kalarikkal, and Sabu Thomas. "Investigation on angiogenesis and its mechanism using zinc oxide nanoparticles-loaded electrospun tissue engineering scaffolds." *RSC Advances* (2014b), DOI:10.1039/C4RA07361D

Balasubramani, Manimalha, T. Ravi Kumar, and Mary Babu. "Skin substitutes: a review." *Burns* 27, no. 5 (2001): 534–544.

Barry, B. W. "Structure, function, diseases, and topical treatment of human skin." *Dermatological formulations. Percutaneous absorption.* New York: Marcel Dekker, Inc. p 9 (1983).

Bhattarai, Shanta Raj, Narayan Bhattarai, Periasamy Viswanathamurthi, Ho Keun Yi, Pyoung Han Hwang, and Hak Yong Kim. "Hydrophilic nanofibrous structure of polylactide; fabrication and cell affinity." *Journal of Biomedical Materials Research Part A* 78, no. 2 (2006): 247–257.

Black, Annie F., François Berthod, Nicolas L'heureux, Lucie Germain, and François A. Auger. "In vitro reconstruction of a human capillary-like network in a tissue-engineered skin equivalent." *The FASEB Journal* 12, no. 13 (1998): 1331–1340.

Blank I. H. "Transport across the stratum corneum". *Toxicology and Applied Pharmacology Supplement* 3, (1969): 23–29.

Boateng, Joshua S., Kerr H. Matthews, Howard N. E. Stevens, and Gillian M. Eccleston. "Wound healing dressings and drug delivery systems: a review." *Journal of Pharmaceutical Sciences* 97, no. 8 (2008): 2892–2923.

Boucard, Nadège, Christophe Viton, Diane Agay, Eliane Mari, Thierry Roger, Yves Chancerelle, and Alain Domard. "The use of physical hydrogels of chitosan for skin regeneration following third-degree burns." *Biomaterials* 28, no. 24 (2007): 3478–3488.

Dastjerdi, Roya, and Majid Montazer. "A review on the application of inorganic nano-structured materials in the modification of textiles: focus on anti-microbial properties." *Colloids and Surfaces B: Biointerfaces* 79, no. 1 (2010): 5–18.

Dibrov, Pavel, Judith Dzioba, Khoosheh K. Gosink, and Claudia C. Häse. "Chemiosmotic mechanism of antimicrobial activity of Ag+ in *Vibrio cholerae*." *Antimicrobial Agents and Chemotherapy* 46, no. 8 (2002): 2668–2670.

Doshi, Jayesh, and Darrell H. Reneker. "Electrospinning process and applications of electrospun fibers." *Journal of Electrostatics* 35, no. 2 (1995): 151–160.

Drosou, Anna, Robert S. Kirsner, Tomoaki Kato, Naveen Mittal, Ahmed Al-Niami, Barbara Miller, and Andreas G. Tzakis. "Use of a bioengineered skin equivalent for the management of difficult skin defects after pediatric multivisceral transplantation." *Journal of the American Academy of Dermatology* 52, no. 5 (2005): 854–858.

Duan, Bin, Xiaoyan Yuan, Yi Zhu, Yuanyuan Zhang, Xiulan Li, Yang Zhang, and Kangde Yao. "A nanofibrous composite membrane of PLGA–chitosan/PVA prepared by electrospinning." *European Polymer Journal* 42, no. 9 (2006): 2013–2022.

Enis, David R., Benjamin R. Shepherd, Yinong Wang, Asif Qasim, Catherine M. Shanahan, Peter L. Weissberg, Michael Kashgarian, Jordan S. Pober, and Jeffrey S. Schechner. "Induction, differentiation, and remodeling of blood vessels after transplantation of Bcl-2-transduced endothelial cells." *Proceedings of the National Academy of Sciences of the United States of America* 102, no. 2 (2005): 425–430.

Fang, Jian, Wang, Xungai and Lin, Tong. Functional applications of electrospun nanofibers, in Nanofibers – production, properties and functional applications, InTech–Open Access Publisher, Rijeka, Croatia, 2011, pp. 287–326.

Fonder, Margaret A., Gerald S. Lazarus, David A. Cowan, Barbara Aronson-Cook, Angela R. Kohli, and Adam J. Mamelak. "Treating the chronic wound: a practical approach to the care of nonhealing wounds and wound care dressings." *Journal of the American Academy of Dermatology* 58, no. 2 (2008): 185–206.

Formhals, Anton. "Process and apparatus fob pbepabing." U.S. Patent 1,975,504, issued October 2, 1934.

Greenberg, Shari, Alexander Margulis, and Jonathan A. Garlick. "In vivo transplantation of engineered human skin." In *Epidermal Cells*. Humana Press, 2005, pp. 425–429.

Grose, Richard, and Sabine Werner. "Wound-healing studies in transgenic and knockout mice." *Molecular Biotechnology* 28, no. 2 (2004): 147–166.

Guo, Shujuan, and Luisa A. DiPietro. "Factors affecting wound healing." *Journal of Dental Research* 89, no. 3 (2010): 219–229.

Gurtner, Geoffrey C., Sabine Werner, Yann Barrandon, and Michael T. Longaker. "Wound repair and regeneration." *Nature* 453, no. 7193 (2008): 314–321.

Halim, Ahmad Sukari, Teng Lye Khoo, and Shah Jumaat Mohd Yussof. "Biologic and synthetic skin substitutes: an overview. *Indian Journal of Plastic Surgery* 43, no. Suppl (2010): S23.

Hansbrough, J. F., and E. S. Franco. "Skin replacements. Wound healing: state of the art." *Clinical Plastic Surgery* 10 (1998): 407–422.

Ho, Wai-Sun. "Skin substitutes: an overview." *Annals of the College of Surgeons of Hong Kong* 6, no. 4 (2002): 102–108.

Horch, Raymund E., Marc G. Jeschke, Gerald Spilker, David N. Herndon, and Jürgen Kopp. "Treatment of second degree facial burns with allografts—preliminary results." *Burns* 31, no. 5 (2005): 597–602.

Huang, Zheng-Ming, Y-Z. Zhang, M. Kotaki, and S. Ramakrishna. "A review on polymer nanofibers by electrospinning and their applications in nanocomposites." *Composites Science and Technology* 63, no. 15 (2003): 2223–2253.

Hutmacher, Dietmar W., Thorsten Schantz, Iwan Zein, Kee Woei Ng, Swee Hin Teoh, and Kim Cheng Tan. "Mechanical properties and cell cultural response of polycaprolactone scaffolds designed and fabricated via fused deposition modeling." *Journal of Biomedical Materials Research* 55, no. 2 (2001): 203–216.

Jin, Xue, Minghua Li, Jinwen Wang, Catalina Marambio-Jones, Fubing Peng, Xiaofei Huang, Robert Damoiseaux, and Eric MV Hoek. "High-throughput screening of silver nanoparticle stability and bacterial inactivation in aquatic media: influence of specific ions." *Environmental Science & Technology* 44, no. 19 (2010): 7321–7328.

Jones, Samantha A., Philip G. Bowler, Michael Walker, and David Parsons. "Controlling wound bioburden with a novel silver-containing Hydrofiber® dressing." *Wound Repair and Regeneration* 12, no. 3 (2004): 288–294.

Kearney, J. N. "Clinical evaluation of skin substitutes." *Burns* 27, no. 5 (2001): 545–551.

Keong, Lim Chin, and Ahmad Sukari Halim. "In vitro models in biocompatibility assessment for biomedical-grade Chitosan derivatives in wound management." *International Journal of Molecular Sciences* 10, no. 3 (2009): 1300–1313.

Khoo, T. L., A. S. Halim, A. Z. Saad, and A. A. Dorai. "The application of glycerol-preserved skin allograft in the treatment of burn injuries: an analysis based on indications." *Burns* 36, no. 6 (2010): 897–904.

Kreuter, Jorg, and Svetlana Gelperina. "Use of nanoparticles for cerebral cancer." *Tumori* 94, no. 2 (2008): 271.

Kurpinski, Kyle T., Jacob T. Stephenson, Randall Raphael R. Janairo, Hanmin Lee, and Song Li. "The effect of fiber alignment and heparin coating on cell infiltration into nanofibrous PLLA scaffolds." *Biomaterials* 31, no. 13 (2010): 3536–3542.

Lanza, Robert, Robert Langer, and Joseph P. Vacanti, eds. *Principles of tissue engineering*. Academic Press, 2011.

Lazarus, GS, DM Cooper, DR Knighton, DJ Margolis, RE Pecoraro, G. Rodeheaver, and MC Robson. "Definitions and guidelines for assessment of wounds and evaluation of healing." *Archives of Dermatology* 130, no. 4 (1994): 489–493.

Lee, Kwang Hoon. "Tissue-engineered human living skin substitutes: development and clinical application." *Yonsei Medical Journal* 41, no. 6 (2000): 774–779.

Liang, Dehai, Benjamin S. Hsiao, and Benjamin Chu. "Functional electrospun nanofibrous scaffolds for biomedical applications." *Advanced Drug Delivery Reviews* 59, no. 14 (2007): 1392–1412.

Liu, J., Z. Bian, A. M. Kuijpers-Jagtman, and J. W. Von den Hoff. "Skin and oral mucosa equivalents: construction and performance." *Orthodontics & Craniofacial Research* 13, no. 1 (2010): 11–20.

MacNeil, Sheila. "Progress and opportunities for tissue-engineered skin." *Nature* 445, no. 7130 (2007): 874–880.

Marston, William A. "Dermagraft®, a bioengineered human dermal equivalent for the treatment of chronic nonhealing diabetic foot ulcer." *Expert Review of Medical Devices* 1, no. 1 (2004): 21–31.

Min, Byung-Moo, Gene Lee, So Hyun Kim, Young Sik Nam, Taek Seung Lee, and Won Ho Park. "Electrospinning of silk fibroin nanofibers and its effect on the adhesion and spreading of normal human keratinocytes and fibroblasts in vitro." *Biomaterials* 25, no. 7 (2004): 1289–1297.

Moore, L., and Y. W. Chien. "Transdermal drug delivery: a review of pharmaceutics, pharmacokinetics, and pharmacodynamics." *Critical Reviews in Therapeutic Drug Carrier Systems* 4, no. 4 (1988): 285.

Pal, Sukdeb, Yu Kyung Tak, and Joon Myong Song. "Does the antibacterial activity of silver nanoparticles depend on the shape of the nanoparticle? A study of the gram-negative bacterium *Escherichia coli*." *Applied and Environmental Microbiology* 73, no. 6 (2007): 1712–1720.

Pilcher, Brian K., Jo Ann Dumin, Barry D. Sudbeck, Stephen M. Krane, Howard G. Welgus, and William C. Parks. "The activity of collagenase-1 is required for keratinocyte migration on a type I collagen matrix." *The Journal of Cell Biology* 137, no. 6 (1997): 1445–1457.

Pitt, Colin G., A. Robert Jeffcoat, Ruth A. Zweidinger, and Anton Schindler. "Sustained drug delivery systems. I. The permeability of poly (ϵ-caprolactone), poly (DL-lactic acid), and their copolymers." *Journal of Biomedical Materials Research* 13, no. 3 (1979): 497–507.

Powell, H. M., and S. T. Boyce. "Fiber density of electrospun gelatin scaffolds regulates morphogenesis of dermal–epidermal skin substitutes." *Journal of Biomedical Materials Research Part A* 84, no. 4 (2008): 1078–1086.

Powell, Heather M., and Steven T. Boyce. "Engineered human skin fabricated using electrospun collagen–PCL blends: morphogenesis and mechanical properties." *Tissue Engineering Part A* 15, no. 8 (2009): 2177–2187.

Prasanna, Mita, Prabodh Mishra, and C. Thomas. "Delayed primary closure of the burn wounds." *Burns* 30, no. 2 (2004): 169–175.

Przyborowski, Melissa, and Francois Berthiaume. "Nanoparticles for skin wound healing." *Nano LIFE* 3, no. 03 (2013).

Raghupathi, Krishna R., Ranjit T. Koodali, and Adhar C. Manna. "Size-dependent bacterial growth inhibition and mechanism of antibacterial activity of zinc oxide nanoparticles." *Langmuir* 27, no. 7 (2011): 4020–4028.

Ravage B. *Burn Unit: Saving Lives After the Flames*. Cambridge, MA; Da Capo Press: 2004.

Rodero, Mathieu P., and Kiarash Khosrotehrani. "Skin wound healing modulation by macrophages." *International Journal of Clinical and Experimental Pathology* 3, no. 7 (2010): 643.

Schaefer, Hans, and Thomas E. Redelmeier. *Skin barrier: principles of percutaneous absorption*. Vol. 19. Basel: Karger, 1996.

Schechner, Jeffrey S., Anjali K. Nath, Lian Zheng, Martin S. Kluger, Christopher CW Hughes, M. Rocio Sierra-Honigmann, Marc I. Lorber et al. "In vivo formation of complex microvessels lined by human endothelial cells in an immunodeficient mouse." *Proceedings of the National Academy of Sciences* 97, no. 16 (2000): 9191–9196.

Schechner, Jeffrey S., Saara K. Crane, Feiya Wang, Anya M. Szeglin, George Tellides, Marc I. Lorber, Alfred LM Bothwell, and Jordan S. Pober. "Engraftment of a vascularized human skin equivalent." *The FASEB Journal* 17, no. 15 (2003): 2250–2256.

Scheuplein R.J., Blank I.H. "Permeability of the skin". *Physiological Review* 51 (1971): 702–747.

Sheridan, R. L., and Carlos Moreno. "Skin substitutes in burns." *Burns* 27, no. 1 (2001): 92.

Sheridan RL, Tompkins RG. "Skin substitutes in burns." Burns 25 (1999): 97–103

Sherwood, Lauralee. "Human physiology: from cells to systems." (2004): 1. 6th Edition, Thomson Brooks, Stamford.

Shores, Jaimie T., Allen Gabriel, and Subhas Gupta. "Skin substitutes and alternatives: a review." *Advances in Skin & Wound Care* 20, no. 9 (2007): 493–508.

Singh, Gagandeep, Eadaoin M. Joyce, James Beddow, and Timothy J. Mason. "Evaluation of antibacterial activity of ZnO nanoparticles coated sonochemically onto textile fabrics." *Journal of Microbiology, Biotechnology and Food Sciences* 2 (2012).

Smith, L. A., and P. X. Ma. "Nano-fibrous scaffolds for tissue engineering." *Colloids and Surfaces B: Biointerfaces* 39, no. 3 (2004): 125–131.

Su, Junfeng, Jun Zhang, Li Liu, Yalou Huang, and Ralph P. Mason. "Exploring feasibility of multicolored CdTe quantum dots for in vitro and in vivo fluorescent imaging." *Journal of Nanoscience and Nanotechnology* 8, no. 3 (2008): 1174–1177.

Supp, Dorothy M., K. A. I. L. A. Wilson-Landy, and Steven T. Boyce. "Human dermal microvascular endothelial cells form vascular analogs in cultured skin substitutes after grafting to athymic mice." *The FASEB Journal* 16, no. 8 (2002): 797–804.

Tan, Wee Beng, Shan Jiang, and Yong Zhang. "Quantum-dot based nanoparticles for targeted silencing of HER2/neu gene via RNA interference." *Biomaterials* 28, no. 8 (2007): 1565–1571.

Valenick, Leyla V., Henry C. Hsia, and Jean E. Schwarzbauer. "Fibronectin fragmentation promotes α4β1 integrin-mediated contraction of a fibrin–fibronectin provisional matrix." *Experimental Cell Research* 309, no. 1 (2005): 48–55.

van der Veen, Vincent C., Martijn van der Wal, Michiel CE van Leeuwen, Magda MW Ulrich, and Esther Middelkoop. "Biological background of dermal substitutes." *Burns* 36, no. 3 (2010): 305–321.

Weir, Emma, Antoin Lawlor, Aine Whelan, and Fiona Regan. "The use of nanoparticles in antimicrobial materials and their characterization." *Analyst* 133, no. 7 (2008): 835–845.

Williams, A. C., and B. W. Barry. "Skin absorption enhancers." *Critical Reviews in Therapeutic Drug Carrier Systems* 9, no. 3–4 (1991): 305–353.

Williams, Tasha R. "Fabrication and characterization of electrospun tecophilic scaffolds for gene delivery." PhD diss., University of Akron, 2007.

Wnek, Gary E., Marcus E. Carr, David G. Simpson, and Gary L. Bowlin. "Electrospinning of nanofiber fibrinogen structures." *Nano Letters* 3, no. 2 (2003): 213–216.

Wood, F. M., M. L. Kolybaba, and P. Allen. "The use of cultured epithelial autograft in the treatment of major burn injuries: a critical review of the literature." *Burns* 32, no. 4 (2006): 395–401.

Xiao-Wu, Wu, David N. Herndon, Marcus Spies, Arthur P. Sanford, and Steven E. Wolf. "Effects of delayed wound excision and grafting in severely burned children." *Archives of Surgery* 137, no. 9 (2002): 1049.

Zahedi, Payam, Iraj Rezaeian, Seyed-Omid Ranaei-Siadat, Seyed-Hassan Jafari, and Pitt Supaphol. "A review on wound dressings with an emphasis on electrospun nanofibrous polymeric bandages." *Polymers for Advanced Technologies* 21, no. 2 (2010): 77–95.

Zhang, Y. Z., J. Venugopal, Z-M. Huang, C. T. Lim, and S. Ramakrishna. "Characterization of the surface biocompatibility of the electrospun PCL-collagen nanofibers using fibroblasts." *Biomacromolecules* 6, no. 5 (2005a): 2583–2589.

Zhang, Yanzhong, Chwee Teck Lim, Seeram Ramakrishna, and Zheng-Ming Huang. "Recent development of polymer nanofibers for biomedical and biotechnological applications." *Journal of Materials Science: Materials in Medicine* 16, no. 10 (2005b): 933–946.

Zhang, Desuo, Ling Chen, Di Fang, Guoyang William Toh, Xinxia Yue, Yuyue Chen, and Hong Lin. "In situ generation and deposition of nano-ZnO on cotton fabric by hyperbranched polymer for its functional finishing." *Textile Research Journal* (2013).

Zhao, Pengcheng, Hongliang Jiang, Hui Pan, Kangjie Zhu, and Weiliam Chen. "Biodegradable fibrous scaffolds composed of gelatin coated poly (ϵ-caprolactone) prepared by coaxial electrospinning." *Journal of Biomedical Materials Research Part A* 83, no. 2 (2007): 372–382.

Zhou, Yingshan, Dongzhi Yang, Xiangmei Chen, Qiang Xu, Fengmin Lu, and Jun Nie. "Electrospun water-soluble carboxyethyl chitosan/poly (vinyl alcohol) nanofibrous membrane as potential wound dressing for skin regeneration." *Biomacromolecules* 9, no. 1 (2007): 349–354.

INDEX

D

E

O

T